INFORMATION TECHNOLOGY AND APPLICATIONS

PROCEEDINGS OF THE 2014 INTERNATIONAL CONFERENCE ON INFORMATION TECHNOLOGY AND APPLICATIONS (ITA2014), XI'AN, CHINA, 8–9 AUGUST 2014

Information Technology and Applications

Editor

Xiaolong Li
College of Technology, Indiana State University, Terre Haute, Indiana, USA

CRC Press
Taylor & Francis Group
Boca Raton London New York Leiden

CRC Press is an imprint of the
Taylor & Francis Group, an **informa** business

A BALKEMA BOOK

CRC Press/Balkema is an imprint of the Taylor & Francis Group, an informa business

© 2015 Taylor & Francis Group, London, UK

Typeset by V Publishing Solutions Pvt Ltd., Chennai, India

Published by: CRC Press/Balkema
      P.O. Box 11320, 2301 EH Leiden, The Netherlands
      e-mail: Pub.NL@taylorandfrancis.com
      www.crcpress.com – www.taylorandfrancis.com

ISBN: 978-1-138-02677-3 (Hbk)
ISBN: 978-1-315-68721-6 (eBook PDF)

Table of contents

II *Computer science*

Preface

The 2014 International Conference on Information Technology and Application (ITA2014) was held in Xi'an, China, from August 8 to August 9, 2014. ITA2014 is to offer scholars, professionals, academics and graduate students to present, share, and discuss their studies from various perspectives in the aspects of Information Technology and Applications. This proceedings tends to collect the up-to-date, comprehensive and worldwide state-of-art knowledge on information technology and application. All of accepted papers were subjected to strict peer-reviewing by 2–4 expert referees. The papers have been selected for this proceedings based on originality, significance, and clarity for the purpose of the conference. The selected papers and additional late-breaking contributions to be presented will make an exciting technical program on conference. The conference program is extremely rich, featuring high-impact presentation. We hope this conference will not only provide the participants a broad overview of the latest research results on information technology, but also provide the participants a significant platform to build academic connections.

The Technical Program Committee worked very hard to have all papers reviewed before the review deadline. The final technical program consists of 88 papers. There are one keynote speech and 2 invited sessions. The keynote speaker is internationally recognized leading expert in the research field of information technology, who have demonstrated outstanding proficiency and have achieved distinction in their profession. The proceedings were published by CRC Press/Balkema.

We would like to express our sincere gratitude to all the members of Technical Program Committee and organizers for their enthusiasm, time, and expertise. Our deep thanks also go to many volunteers and staffs for the long hours and hard work they have generously given to ITA2014. Finally, we would like to thank all the authors, speakers, and participants of this conference for their contributions to ITA2014, and also looks forward to welcoming you to ITA2015.

I *Information theory, information processing methods and information systems*

Information Technology and Applications – Li (Ed.)
© 2015 Taylor & Francis Group, London, ISBN 978-1-138-02677-3

Consumption intent analysis in Chinese microblog

Junkang Mao
*Key Laboratory of Shanghai Education Commission for Intelligent Interaction and Cognitive Engineering,
Shanghai Jiao Tong University, Shanghai, China*

Tianfang Yao
Department of Computer Science and Engineering, Shanghai Jiao Tong University, Shanghai, China

ABSTRACT: With the development of Chinese microblog and E-commerce, the messages with opinions and intents expressed by users would have a significant impact on many domains. In this paper, we study on how to identify the consumption intent in Chinese microblog. We use text classification based on NB, SVM and EM-based transfer learning algorithm to identify consumption intent. And we compare NB, SVM and EM algorithm methods. Due to the freedom and irregular of the microblog posts, we consider more about preprocessing microblog raw data, which bring new challenges to the traditional text classification methods.

Keywords: intent analysis; transfer learning; classification; microblog

1 INTRODUCTION

With the development of the Internet and the mobile devices, the microblogging services such as Sina Weibo have played an important role in our life. There have been a large number of researches in microblog, e.g., sentiment analysis [1–2] and social network analysis [3]. In this paper, we study on how to identify user consumption intents which have a big impact on the e-commerce and other business areas. Some users' microblog posts contain consumption intents. For example, "今天看了 Moto X的评测,有点兴趣,想买." (Today I saw the Moto x evaluation, somewhat interested in it, want to buy.). Such messages will help manufacturers collect the potential consumers or improve the accuracy of advertising.

As for the intent identification of social media data, research situation at home and abroad are both at an initial stage. The general text that can be classified contains traditional blog, forums and so on. Microblog is timeliness, theme-inclusive. Also it only contains at most 140 words [4]. Users express freely and the messages contain noisy data. People write posts every day and traditional texts are much different from microblog posts, which have emoticons composed by some characters, mixed writing in English and URLs. There has extensive and profound Chinese grammar and various combinations of varying significance, which also make it more difficult to deal with Chinese texts. Using machine learning methods that can train the normal texts to classify the irregular microblog posts cannot get an ideal result, especially for cross-domain messages. Thus, for processing of microblog, we take into account these characteristics to find appropriate method.

The intents can be explicit and implicit. The explicit intents would be stated obviously in posts. If one user wrote, "我想买苹果5." (I want to buy Iphone5). we have no need to deduce, because it can be indented easily. On the other hand, some intents like "苹果手机怎么样,好用么?" (How about the Iphone, is it easy to use) are so relatively implicit that different people will get different meanings. The implicit microblog posts are always used in computational advertising or recommender system.

2 RELATED WORK

The most related study is to get the intents in the domain of web search [4–5], where user intent is a hot issue. But the task is to classify the input text which includes 2 or 3 words when the user uses a search engine. It's different from our work.

Go and Bhayani [6] tried an approach for classifying sentiment in Twitter. They used distant supervision to achieve high accuracy in Twitter, which means extra information such as emoticons can be the features. Because Twitter posts may not provide sufficient word occurrences. Striam and Fuhry [7] proposed a new feature selection method to process the raw data in Twitter. Pennacchiotti and Popsecu [8] also tried to find the special feature to help Twitter user classification.

Microblogging interest mining and personalized recommendations are both related to our work. For microblogging interest mining, Banerjee et al. [9] use keywords to analyze user's interest in microblog that are treated as source data. Ma and Zeng et al. [10] tried to combine multiple social media platforms to do the interest mining. Chen and Nairn [11] thought that they can find the users' interests through the distribution of interests of followers. Personalized recommendation generally uses two recommended methods [12] which are content-based filtering and collaborative filtering. Online news recommendation system developed by GroupLens [13] and Google News personalized recommendation developed by Das and Datar [14] and so on are all based on the collaborative filtering. Hybrid model news recommendation system developed by Chiang and chen [15] and book recommendations developed by Roy [16] and other similar systems are based on content-based filtering.

Consumption intents in microblog are related to different domains, resulting in unbalanced text features. That's one of the difficulties faced by this task. In general, for one domain, we can use traditional statistical methods, such as Naive Bayes, Support Vector Machine, Maximum Entropy and so on. In the case of sufficient labeled corpus, we can get good results. But for different domains, different features would cause a decline in accuracy. Transfer learning provides the possibility to improve the recognition accuracy of consumption intent.

In this paper, we use brief transfer leaning method to classify the cross-domain micro-blog text. Transfer learning is a machine learning framework. It's much different from the traditional supervised learning, unsupervised learning and semi-supervised learning. This method can build a compact and efficient model through labeled data from source domains and unlabeled data or a small amount of labeled data from target domains, and the model will be applied to the target domains. Transfer learning doesn't need training data and test data which should obey the same distribution in traditional machine learning, so it can share and migrate the common information among the similar domains. Transfer learning can be divided into three categories: Transductive Transfer Learning, Unsupervised Transfer Learning and Inductive Transfer Learning.

Pan and Yang [17] described cross-domain transfer learning methods. Aue and Gamon [18] tried to train a model of several domains to classify the target data. There are many related studies on such transfer learning methods [19–21]. In addition, there are more complex transfer learning studies trying to dig the common features of the source

and target areas of the field, then corresponding to these two areas [22]. Some researchers have tried to use the topic model in transfer learning [23–24].

Chen and Liu et al [25] proposed that consumption intent recognition in forum posts would face two problems. First, only a few sentences in forum posts could express the intent, while others are noisy data. This leads to the common features of different domains is too limited to be dug out. Second, in different domains, there have the common features when users express positive intents. But expressing negative intents of different domains will result in almost no common features, thus the training data have imbalance features. They used Co-Class algorithm to resolve the problems.

3 THE TECHNIQUE

3.1 *SVM classifier*

The SVM model treats the data as the real vector x, we want to find the maximum-margin hyperplane that divides two classifiers. Therefore, we can use the follow formulas:

$$\arg\min_{(w,b)} \frac{1}{2} \| w \|^2 \tag{1}$$

$$y_i(w^* x_i - b) \geq 1 \tag{2}$$

where y belongs to {-1, 1}. According to the Karush–Kuhn–Tucker condition, the solution can be expressed as a linear combination of the training vectors.

$$w - \sum_{i=1}^{n} \alpha_i y_i x_i \tag{3}$$

A few α values will be greater than zero. The corresponding x values are exactly the support vectors.

3.2 *EM algorithom*

EM (Dempster, Laird, & Rubin, 1977) is a popular class of iterative algorithms for maximum likelihood estimation in problems with incomplete data. It works by pretending that we know what we're looking for the model parameters. The EM algorithm consists of two steps, the Expectation step (E-step) and the Maximization step (M-step). First, E-step basically fills in the missing data, we use our current model parameters to estimate some measures, the ones we would have used to compute the parameters, and M-step re-estimates the parameters. This process iteratively repeats these two steps until convergence. In this paper we use EM to train a Naive Bayes classifier.

Navie Bayes classifier gets the maximum probability of a sentences.

$$classifier(s) = \arg\max_c p(c|s) \qquad (4)$$

P(c|s) get be calculated by Bayes rules.

$$p(c|s) = \frac{p(s|c)p(c)}{p(s)} \qquad (5)$$

Assuming that every word x_i is mutually independent, then we get (6).

$$p(s|c) = \prod_{x_i \in X} p(x_i|c) \qquad (6)$$

We use Laplace smoothing to get $p(x_i|c)$.

3.3 *Feature selection*

As for the feature selection, Information Gain (IG) is a popular algorithm for text classification. IG is based on entropy reflecting importance of the feature by measuring presence or absence of each feature, which is defined as:

$$H(D) = -\sum_{k=1}^{k} \frac{|C_K|}{|D|} \log_2 \frac{|C_K|}{|D|} \qquad (7)$$

$$H(D|A) = -\sum_{i=1}^{n} \frac{|D_i|}{|D|} \sum_{k=1}^{k} \frac{|D_{ik}|}{|D_i|} \log_2 \frac{|D_{ik}|}{|D_i|} \qquad (8)$$

$$IG(D,A) = H(D) - H(D|A) \qquad (9)$$

3.4 *Data preprocessing*

We use Sina API to get the microblog posts. For the raw posts, we think that the URLs, emoticons and the tags are not related to the content, so we just remove them. Thus we build the template that every posts will be checked if it contains the invalid information.

In Sina Weibo, "zombie fans" are existing all the time. They are fake followers, usually with no avatar

and fans. Their posts have no value for our task. So processing the raw data is quite important. We collect about 2000 fake fans as training data, and use NB classifier to build the model to filter the raw data. This has reduced our workload, so we just check the processed data and remove the posts that cannot be found out. Figure 1 just shows the detail.

4 EXPERIMENT

We only identify explicit consumption intents, which are clearly expressed and not ambiguous in each post. We chose bigrams with 1000 features and 3000 features. The SVM and NB classifiers are used for one domain training, and the EM algorithm method is used for four domains training. We collect four domains from Sina Weibo. The tables below show the details.

Table 1. Dataset where the '+' means it's positive and the '−' means it's negative.

Domains	Training data +	Training data −	Total
Cellphone	700	700	1400
Automobile	700	700	1400
Laptop	700	700	1400
TV	700	700	1400

Table 2. Dataset where the '+' means it's positive and the '−' means it's negative.

Domains	Test data +	Test data −	Total
Cellphone	200	200	400
Automobile	200	200	400
Laptop	200	200	400
TV	200	200	400

Table 3. F scores for each domain in 1000 features.

Methods	Cellphone	Automobile	Laptop	TV
NB	0.733	0.697	0.732	0.657
SVM	0.690	0.654	0.711	0.650
EM	0.756	0.712	0.728	0.683

Table 4. F scores for each domain in 3000 features.

Methods	Cellphone	Automobile	Laptop	TV
NB	0.713	0.703	0.728	0.630
SVM	0.688	0.641	0.709	0.623
EM	0.725	0.692	0.716	0.650

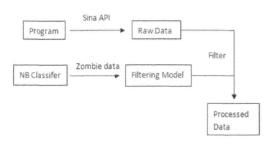

Figure 1. Preprocess the raw data.

We can find that NB classifier is simple but effective, it's better than SVM classifier in this experiment. EM method gets best result. And the experiment with 1000 features is a little better than the one with 3000 features.

5 CONCLUSION

This paper studied on how to identify consumption intent in Chinese microblog. We've tried popular supervised machine learning methods and the brief transfer leaning methods to get an acceptable result.

There are still some areas that can be optimized. We need try more methods to filter the fake fans. Moreover, some contents such as URLs, emoticons and so on can be the features to some degree. In this paper, we just use the simple transfer learning method, it can be developed in the future.

REFERENCES

[1] Pang B, Lee L. Opinion mining and sentiment analysis[J]. Foundations and trends in information retrieval, 2008, 2(1–2): 1–135.

[2] Pang B, Lee L. A sentimental education: Sentiment analysis using subjectivity summarization based on minimum cuts[C]//Proceedings of the 42nd annual meeting on Association for Computational Linguistics. Association for Computational Linguistics, 2004: 271.

[3] Wasserman S. Social network analysis: Methods and applications[M]. Cambridge university press, 1994.

[4] Brenes D J, Gayo-Avello D, Pérez-González K. Survey and evaluation of query intent detection methods[C]//Proceedings of the 2009 workshop on Web Search Click Data. ACM, 2009: 1–7.

[5] Hu J, Wang G, Lochovsky F, et al. Understanding user's query intent with wikipedia[C]//Proceedings of the 18th international conference on World wide web. ACM, 2009: 471–480.

[6] Go A, Bhayani R, Huang L. Twitter sentiment classification using distant supervision[J]. CS224 N Project Report, Stanford, 2009: 1–12.

[7] Sriram B, Fuhry D, Demir E, et al. Short text classification in twitter to improve information filtering[C]//Proceedings of the 33rd international ACM SIGIR conference on Research and development in information retrieval. ACM, 2010: 841–842.

[8] Pennacchiotti M, Popescu A M. A Machine Learning Approach to Twitter User Classification [C]//ICWSM. 2011.

[9] Banerjee N, Chakraborty D, Dasgupta K, et al. User interests in social media sites: an exploration with micro-blogs[C]//Proceedings of the 18th ACM conference on Information and knowledge management. ACM, 2009: 1823–1826.

[10] Ma Y, Zeng Y, Ren X, et al. User interests modeling based on multi-source personal information fusion and semantic reasoning[M]//Active Media Technology. Springer Berlin Heidelberg, 2011: 195–205.

[11] Chen J, Nairn R, Nelson L, et al. Short and tweet: experiments on recommending content from information streams[C]//Proceedings of the SIGCHI Conference on Human Factors in Computing Systems. ACM, 2010: 1185–1194.

[12] Schafer J B, Frankowski D, Herlocker J, et al. Collaborative filtering recommender systems[M]//The adaptive web. Springer Berlin Heidelberg, 2007: 291–324.

[13] Resnick P, Iacovou N, Suchak M, et al. GroupLens: an open architecture for collaborative filtering of netnews[C]//Proceedings of the 1994 ACM conference on Computer supported cooperative work. ACM, 1994: 175–186.

[14] Das A S, Datar M, Garg A, et al. Google news personalization: scalable online collaborative filtering[C]//Proceedings of the 16th international conference on World Wide Web. ACM, 2007: 271–280.

[15] Mooney R J, Roy L. Content-based book recommending using learning for text categorization[C]//Proceedings of the fifth ACM conference on Digital libraries. ACM, 2000: 195–204.

[16] Chiang J H, Chen Y C. An intelligent news recommender agent for filtering and categorizing large volumes of text corpus[J]. International Journal of Intelligent Systems, 2004, 19(3): 201–216.

[17] Pan S J, Yang Q. A survey on transfer learning[J]. Knowledge and Data Engineering, IEEE Transactions on, 2010, 22(10): 1345–1359.

[18] Aue A, Gamon M. Customizing sentiment classifiers to new domains: A case study[C]//Proceedings of recent advances in natural language processing (RANLP). 2005, 1(3.1): 2.1.

[19] Dai W, Xue G R, Yang Q, et al. Transferring naive bayes classifiers for text classification[C]//Proceedings of the national conference on artificial intelligence. Menlo Park, CA; Cambridge, MA; London; AAAI Press; MIT Press; 1999, 2007, 22(1): 540.

[20] Tan S, Wu G, Tang H, et al. A novel scheme for domain-transfer problem in the context of sentiment analysis[C]//Proceedings of the sixteenth ACM conference on Conference on information and knowledge management. ACM, 2007: 979–982.

[21] Yang H, Callan J, Si L. Knowledge Transfer and Opinion Detection in the TREC 2006 Blog Track[C]//TREC. 2006.

[22] Blitzer J, Dredze M, Pereira F. Biographies, bollywood, boom-boxes and blenders: Domain adaptation for sentiment classification[C]//ACL. 2007, 7: 440–447.

[23] Gao S, Li H. A cross-domain adaptation method for sentiment classification using probabilistic latent analysis[C]//Proceedings of the 20th ACM international conference on Information and knowledge management. ACM, 2011: 1047–1052.

[24] He Y, Lin C, Alani H. Automatically extracting polarity-bearing topics for cross-domain sentiment classification[C]//Proceedings of the 49th Annual Meeting of the Association for Computational Linguistics: Human Language Technologies-Volume 1. Association for Computational Linguistics, 2011: 123–131.

[25] Chen Z, Liu B, Hsu M, et al. Identifying Intention Posts in Discussion Forums[C]//Proceedings of NAACL-HLT. 2013: 1041–1050.

Information Technology and Applications – Li (Ed.)
© 2015 Taylor & Francis Group, London, ISBN 978-1-138-02677-3

Analysis of the micro-lecture construction of mechanical engineering network education

Lei Xu

School of Distance and Continuing Education, Dalian University of Technology, Dalian, China

ABSTRACT: With the increasing popularity of mobile terminals, vigorous development of network educational resources and the arrival of the era of MOOCs, micro-lecture has become a hot issue. Network education institutions have done a lot of attempts for micro-lecture. The concept and characteristics of micro-lecture is introduced, and design and development emphasis on micro-lecture of mechanical engineering which considering characteristics of mechanical professional and evaluation of micro-lecture of mechanical engineering courses is proposed, to explore how to promote development of micro-lecture of network education in mechanical engineering better.

Keywords: network education; mechanical engineering; micro-lecture

1 THE CONCEPT AND FEATURES OF MICRO-LECTURE

The concept of micro-lecture was raised by senior instructional designer, the College online services manager David Penrose, who worked in New Mexico San Juan College in 2008. David Penrose is known as the one minute professor. This term of micro-lecture is not a simple micro-lecture refers to the micro-teaching and the micro-content. It makes use of constructivist methods to form online learning or mobile learning for the purpose of practical teaching content.

In Foshan City, Guangdong Province, China, Mr. Tiesheng Hu who works in Education Information Network Center, thought of new information technology in education, research results of his "micro-lecture" got the recognition and praise of scholars and experts, he is the first person who proposed and defined the concept of micro-lecture and the first person who studied the region of micro-lecture. He is considered to be the domestic founder of micro-lecture. Compared to micro-lecture proposed by David Penrose, Tiesheng Hu deepened the concept of micro-lecture from the perspective of educational information resources. The concept of micro-lecture proposed by David Penrose substantially changed the form of online teaching courses, but he did not consider the "active" of resources, i.e., whether the micro courses are same with the traditional open educational resources, which facing the same problem of low utilization, and Tiesheng Hu believe that only those which meet the needs of teachers and students, and

has semi-structured, dynamically generated teaching resources that can improve the utilization of micro-lecture resources. The audience of micro-courses will be expanded from students to teachers, that students can choose targeted learning content autonomously; teachers can choose targeted micro-lecture as teaching resources and to promote their professional development. Micro-lecture could be positioned as the supplement and expand resources for traditional classroom learning and corresponds to mobile learning, ubiquitous learning concepts. The specific form of micro-courses should be classified also.

Micro-course is the whole process around a knowledge point or a wonderful teaching and learning activities. It takes dapper online video as the main carrier, and also includes teaching instructional design, material courseware, teaching reflection, exercise testing and students' feedback, teachers' reviews and other auxiliary teaching resources which related to teaching topics. Micro-course forms the subject of constructivism system thematic, depth, structured application resource unit of "micro-environment." Therefore, the micro-courses is not only different from single resource type of traditional teaching cases, courseware, instructional design, teaching reflection, teaching resources, but also a new teaching resources base on the inheritance and development of traditional teaching type.

Micro-course has the following main features: the "micro" characteristics: a learning content refining, short and efficient notable features, continuous accumulation of micro-learning; "micro" platform: more suitable for mobile terminals learning, including notebook computers, tablet PCs, smart mobile phones; "micro" environment: to provide effective conditions for "micro-learning", to build a comfortable environment for ubiquitous learning, which can increase the driving force of the depth of learning supported by micro-learning process; "micro" goal: a micro-lesson learning of thematic attributes, associated system of micro-course knowledge and learning goals reached echoes.

Core content of micro-course is multimedia teaching videos. It also includes topics related to the teaching of instructional design, material courseware, teaching modules, practice tests and feedback from students, review of teachers and other teaching support resources, which work together to build a semi-structured, a prominent theme of resource unit applications with a certain structure and presentation.

2 THE MECHANICAL CHARACTERISTICS AND MICRO-COURSE DESIGN AND DEVELOPMENT

Mechanical engineering is rich in professional knowledge, many knowledge points, involving a wide range of professional area, strong practice is also a distinctive feature. Professional content of mechanical engineering based on mechanical technology, and combine with modern science and technology of computers, automation, sensor testing, requires students to master the broad basics technical knowledge, including mechanics, mechanical, electrical and electronic technology, mechanical engineering materials, mechanical design, mechanical manufacturing, automation infrastructure, mechatronics and business management basics. Meanwhile mechanical engineering must pay attention to professional teaching theory with practice teaching and theory teaching.

As a complex system engineering, development of micro-course content, generally experiences topics designed, lesson shooting, post-production, published online implementation, evaluation feedback and other sectors, in order to ensure the teaching quality and the teaching effect.

Combine with subject characteristics of mechanical engineering, teaching design of micro-course includes the following aspects:

2.1 Topic design

The content of mechanical engineering micro-course should be divided into several micro-topics. The topic of micro-course shall be concise; each topic can be a knowledge point. Teaching content

of the topic should be clear, for importing before class, or for critical knowledge points teaching, or for breaking through emphasis and difficult, or for extending knowledge after-lass, or for specific mechanical practitioners. It could be the explanation of basics of mechanical, description of using textbook, exam questions succinctly, induction of important test sites, teaching of experimental methods, or demonstration of experimental procedure.

2.2 Courseware length design

Micro-course is a reproduction of classroom teaching process, comparing with the traditional classroom; the difference is that it's short. Micro-course is generally about 5 to 8 minutes, the longest one should not be more than 15 minutes. According to the characteristics of different mechanical courses taught, it is important to design the length of micro-course properly so as to ensure that learners can access the most essential knowledge in fragments time.

2.3 Course structure design

Teaching process of micro-courses should be brief and complete. It should include proposing the issue of teaching, arrangements of teaching activities; problem-solving of students and other areas. A combination of mechanical engineering theoretical knowledge and specific projects could make it easier for learners to accept.

2.3.1 Bring out the theme as soon as possible, to attract students
You can enter the knowledge points straightly, or set up a problem to introduce knowledge; or introduce knowledge from the engineering practice related to phenomena or problems; or introduce the basic knowledge from previous teaching content; You can also set a topic to introduce the knowledge point; But regardless which method and which means, the methods and way of cut topic require striking and novel; and also require that the topic should be compactly, cut topic should be quickly.

2.3.2 Lectures organized clarity
Though micro-course teaching process, it is necessary to expand knowledge, focus emphasis of the knowledge point, reveal the contents of the trunk of the knowledge, and delete dispensable illustration. In the process of teaching knowledge point, arguments should be refined and simplified. It is important that arguments should be strived fully and accurately, and it might not lead to new questions.

2.3.3 *Summary clear and concise*

Summary of micro-course is essential. Good summary can be taught to play the role of focusing on the vital of teaching content. It can enhance students' impression of what they learn, reduce students' burden of memory. Good summary can raise a lesson to a new grade, learners can feel clearly that the basic principles or basic concepts they have mastered through the lesson.

2.4 *Teaching resources design*

Teaching content of micro-course generally requires that the selected point is clear; theme is prominent and intact relatively. For example: design of an mechanical experiment of micro-course should integrate design of teaching, multimedia teaching materials and courseware, experimental procedures, experimental reports, student feedback and expert reviews, and other related teaching support resources, constitutes a clear theme, compact, diverse types of "theme unit resource package", to create a true "micro-teaching laboratory resources and the environment" through the main line of multimedia video clips.

2.5 *Teaching expression design*

In the micro-course, because of the limited time, accurate and concise language is seemed to be more important. In the process of the design of mechanical micro-courses, it is necessary to practice in advance, which combine with the content what to say, what content will be expressed, which style of expressions, gestures, and facial expressions should be used, and which keywords should paid attention. The language should be vivid, charm, and concise. Meanwhile, you can also combine charts, 3D animation, etc., break through the restrictions of language, to make the course clear at a glance.

3 CONSTRUCTION OF COMPREHENSIVE EVALUATION INDEX SYSTEM OF MECHANICAL MICRO-COURSES

Micro-course curriculum is not only static text, more importantly; it is a dynamic curricular activity. Curriculum evaluation is a process, which gives a value judgment on the constituent elements of each course curriculum activities and links of themselves.

The evaluation of micro-course development includes evaluation of the background of the preparatory phase, evaluation of design itself in the design phase, diagnostic evaluation of the implementation effect in use phase.

Combination of mechanical engineering features, the evaluation indicators of implementation of micro-course are as follows:

3.1 *Reasonable instructional design*

Topic should be "small and fine." Target is clear, aims mainly at specific knowledge of mechanical courses, exercises, experimental activities and other aspects of the teaching, such as analysis, reasoning, question answering test.

Instructional design should be reasonable. There are clear learning objectives for teaching or learning in common, typical representatives of principle or method designed to address the key, difficult and doubtful, test sites and other issues effectively in the teaching and learning process. The ideas of teaching organization should be clear. Choosing the appropriate multimedia expression for teaching content is also very important.

3.2 *Complete and accurate teaching content*

Scientifically correct. Teaching content must be correct, no scientific errors, the contents expressed are accurate.

Logic clear. Organization and arrangement of teaching content should be rich in logical, meet the cognitive characteristics of mechanical practitioners learners, content aims at specific issues directly, the main themes should be highlighted.

Resource should be complete. Micro lesson plans should be designed around the selected themes, focused, results-oriented; PPT courseware of micro-course should be designed to be visually distinct, simple, and have good teaching effects. It could focus on teaching objectives, reflect the main teaching content, and instruct videos reasonably. The design of micro-exercises should be targeted and hierarchy, design difficulty levels of subjective and objective exercise; micro-discussion should be observed and analyzed after completion of the micro-course, and it also should be strived to be objective, reasonable and enlightening.

3.3 *Clear and reasonable teaching process*

Cutting the theme should be quick. To cut the theme quickly for attracting students, the methods and approaches of cutting the theme should be striking and novel.

Course clues should be clear. In the teaching process of the micro-course, the clues should be clear. The argument should be listed fine and simple; the argument should be fully and accurately, in a limited time, complete the prescript task of teaching successfully.

Table 1. Evaluation system of mechanical micro-courses.

Content	Proportion
Reasonable instructional design	25%
Complete and accurate teaching content	20%
Clear and reasonable teaching process	20%
Specific courseware form	15%
Fine teaching effect	20%

Summary finishing should be fast. Play the contents of the lecture focused on the vital role that can enhance the impression of the students what they learn.

PPT courseware should be simple and reasonable. PPT courseware can display the main point of the teacher, to help people grasp the main points of the lecture content, the entire basic program and content of teaching could be looks like concise, complete and beautiful, highlight the point, clear clues, give learners a complete visual effect.

Language decent. Medium of instruction should be accurate and concise, lively, generous nature, contagious and logical.

3.4 *Specific courseware form*

Structural should be in integrity. The beginning of the video should display the title, which will indicate the lectures content of the subject, courses, audience and other information. It should have a certain independence and integrity, the main teaching sector should have subtitles tips.

Technical should have specifications. Length of the courseware should be generally not more than 15 minutes, the image must be clear and stable, and composition should be reasonable, sound clear, sound and picture synchronization.

3.5 *Fine teaching effect*

Forms should be novel. The design of micro-course have innovation or innovative ideas, teaching methods should be creative, the overall effect are impressive.

Strong interest. Teaching process should be easy to understand, vivid, exciting and fun, various means could be used to stimulate motivation, inspire attention, promote thinking, and bring up capacity.

Goal achievement. Completing the setting objectives of teaching, solving practical problems of teaching effectively, and promoting students to raise the level of knowledge and bring up innovation capabilities.

Application. Micro Courseware should be welcomed after release, it must have high click rate, good user feedback, and greater space of promotion. Evaluation system of mechanical micro-courses is shown in Table 1.

4 CONCLUSION

Currently, using mobile terminals to work, learn and play have become an indispensable part of people's daily activities. With the rapid development of information technology and mobile communications, micro-course have unlimited growth potential. Micro-course is in micro-micro-features, although it is also deficient in breadth, depth and complexity, it can meet the demand of distance learners, who need different disciplines to learn knowledge, personalized learning and mobile learning. It not only can fill gaps and also can strengthen and consolidate the knowledge. Beginning with the concept of micro-course, content and features are discussed, according to the characteristics of mechanical engineering, mechanical engineering design and development strategies and evaluation methods of micro-courses are proposed, how to develop high-quality micro-course of mechanical professional better is explored. With a view to promote a new situation of the development and construction of micro-course resources, it is hoped that the micro-course education could penetrate into network education deeply, create new ways of teaching and learning, which is more suitable for the information age, and meet the opportunities, which brought by application of micro-course for education and social development.

REFERENCES

Guo Shaoqing Jianjun Huang, Research on the micro-course design and development, Modern Educational Technology, 2013, 5:31–35.

Jianxin Fan, Distance learning curriculum resources building micro-course and design, Journal of Jiangsu Radio & Television University, April 2013:5–8.

Qiaofang Li, WenMei Yang, Analysis of micro-course of research status and development trend, China's Education Technology and Equipment, October 2013:12–14.

YueMing Liang, Jinming Liang, From resource development to applications: Micro-course status and trends, China Educational Technology, August 2013:71–76.

YueMing Liang, QiaoQiao Cao, Baohui Zhang, Micro-course design model—Based on the comparative analysis of domestic and micro-course, Open Education Research, February 2013:65–73.

Information Technology and Applications – Li (Ed.)
© 2015 Taylor & Francis Group, London, ISBN 978-1-138-02677-3

Construction and design of network database system of wild flowers of Beijing

Kai Huang, Qun Zhao, Zhihuan Chen & Zhaojie Liu
Landscape Department, Beijing University of Agriculture, Beijing, China

ABSTRACT: There are many kinds of wild flowers in Beijing. It is an important task to let these valuable resources to be known to students. So wild flower resources of Beijing are investigated in detail. And we summarize 100 kinds of wild flower data. Using c/s architecture, SQL Server as background database, through VB interface, network database system of wild flowers of Beijing is built. Through this database, students can query data of wild flowers of Beijing. Based on the database, readers can understand these flowers, and can develop and utilize these resources much better.

Keywords: wild flower; database; construction

1 INTRODUCTION

China is rich in natural resources. There is a wide range of plants and rich resources of wile flowers. These resources have unique ornamental value and great potential exploitation and attract more and more attention. The wild flowers are an important part of local natural landscape and vegetation. They are not only ancestors of existing cultivated flowers, but also important resources and raw data for cultivating new varieties. But at present, less than 1/3 wild flowers are utilized, 2/3 wild flowers are not to be exploited.

2 WILD FLOWERS RESOURCES IN BEIJING

2.1 *Current status of wild flowers in Beijing*

Utilization of wild flowers of Beijing are less both in variety and in quantity, such as Bittersweet, Philadelphus pekinensis, Mountain plum, Red clove, Berberis poiretii etc.. Multi wild flowers are planted in garden, not obtained in production and landscaping widely. Other wild ornamental woody flowers are not used such as Dongling Hydrangea. At the aspect of utilization of wile herbaceous flowers, in addition to a small portion of perennial flowers such as fern has a small amount of application.

2.2 *Effect of wild flowers*

Wild flowers are ornamental, and also are resources of breeding. Some of them can be used

as medicine, some can be extracted to be essence and industrial fiber, starch, oil and other raw material. At the same time, wild flower landscape is also a valuable tourism resource. And development of wild flowers resources in Beijing is still in its initial stage. In order to be more convenient to use Beijing wild flowers data, we build Beijing wild flowers database. Through the database, we hope that more and more people will know about Beijing wild flowers. And then wild flowers can be protected more reasonable, and will play an important role in city greening.

3 CONSTRUCTION OF NETWORK DATABASE OF BEIJING WILD FLOWERS

3.1 *Acquisition of wild flowers in Beijing*

According to incomplete statistics, there are 227 species, belonging to 71 families. Through investigation of wild flowers in Beijing, reading relevant literature, 124 kinds of ornamental wild flowers which have breeding value are collected. Each plant is introduced with Chinese name, Latin name, alias, a morphological description and garden use. And each plant is taken photo.

3.2 *Construction of network database of wild flowers*

Access 2003 is one of the important components of Microsoft Office System series office suite software. It is the most commonly used database software in practical work. Its function is strong enough to manage and process general data. And it is easy to study. Using Access 2003, users do not need to have professional computer technology and database knowledge, users can easily create, design and display database products.

Microsoft SQL Server is a comprehensive database platform. It is integrated Business Intelligence (BI) tools for enterprise data management. It provides more secure and reliable storage for relational data and structured data. Using it can make construction and management for high availability and high performance data applications.

Data in Access export to SQL Server. Then SQL Server database is connected to VB by ADO. ADO is a Microsoft database application programming interface. It is built in OLE DB on top of a high level database access technology. ADO interface is simpler, more flexible. For new project, ADO should be used as data access interface. ADO encapsulates interface provided by OLE DB. Compared to OLE DB provider, ADO interface allows programmers to interact data at a high level. ADO technology can be applied to not only relation database, but also non relational database.

So through connecting SQL Server, network database of wild flowers is built.

4 CONCLUSION

Discovery, collection, preservation, utilization of wild flower resources is an important strategic method to rich flower species and improve quality of flower varieties. Establishment of Beijing wild flower database is intended to facilitate reader to query data of wild flowers of Beijing. Based on understanding of these flowers, readers will have awareness of protection. Wild flower resources can be exploited regularly.

REFERENCES

[1] Beijing wild flower. Chen zhihuan, Chen hubiao, Duan bihua. China agricultural science and Technology Press. Beijing. 2005.
[2] Floriculture. Baomanzhu. Chinese Agriculture Press. Beijing. 1998.
[3] Economic management of garden. Huangkai, Zhang xiangping. Meteorological Press. Beijing. 2001.
[4] Flora China. Chinese Plant Science Research Insititute. Beijing: Science Press, 1995.
[5] Perennial flowers. Fei yanliang. Beijing: China Forestry Press, 1999.
[6] Ornamental flower cultivation. Liu lian. Beijing: Intellectual Property Press, 2001.
[7] China flowers. Chen junyu, Cheng xuke. Shanghai: Shanghai Culture Press, 2003.

Information Technology and Applications – Li (Ed.)
© *2015 Taylor & Francis Group, London, ISBN 978-1-138-02677-3*

A research on the construction of the E-learning platform in art and design teaching

Jiaofang Shi
Suzhou Institute of Art and Design Technology, Suzhou, Jiangsu, China

ABSTRACT: In this paper, a research on the construction of the E-learning platform in art and design teaching is presented. It is based on the need of students studying in this area, and aims to build a personalized E-learning environment, which is customer-oriented instead of information-oriented. This paper gives analysis on resource and technical support that facilitates the learning and designs the E-learning platform featured by "close interaction", supporting collaborative learning using embedded social networks, enabling graphic discussion on interactive whiteboards as well as providing management tools to motivate learning, etc. The construction of the E-learning platform where information is presented, gathered and shared is expected to further improve efficiency in art and design teaching.

Keywords: art and design; learning platform; close interaction

1 INTRODUCTION

Students of the art and design major are generally required to be actively engaged in open learning and to have a strong motivation for studying as innovation is often prompted by an open perspective and the passion to explore. In the practice of digital teaching nowadays, however, the efficiency and quality of E-learning in the Web 1.0 era are widely questioned by teachers and students. Firstly, web-based learning is rarely effective. The usefulness and availability of resources are questionable; digital learning services neither support open learning exploration, nor cater to the needs of different curriculums and students of diversified learning methods and habits. Secondly, the digital environment designed for educators is not "learning-oriented". Thirdly, although interaction is often considered essential, E-learning has still been limited to non-real-time BBS Q & A for decades. What students and teachers really need, however, is timely, convenient and unimpeded interaction, which is here defined as "close interaction".

The core idea of the booming new technique Web2.0 is "creative commons" [1] [2], information is created commonly, gathered online, and shared freely. The concept of open construction inspired the design of a new generation of digital learning platforms in art and design field. This platform aims to build a personalized E-learning environment, which is customer-oriented instead of information-oriented.

Art and design courses require strong resource and technical support, which could enable the presentation, gathering and sharing of the information. The emphasis on the design works consequently brings a need for intuitive and convenient "close interaction". Therefore, the design of the platform is supposed to focus on learning supporting services while actively responding to the expectations of teachers and students. Learning support may include resource and technical support (e.g. digital resources, digital libraries, technical support, etc.), academic support (e.g. course introductions, counseling, learning assessments, etc.), management support (e.g. credit transfer, financial support, career counseling, etc.) [3]. We believe that providing a Web2.0 environment where information resource construction is paid due attention and the method of "creative commons" is applied, would largely promote effective E-learning [4] [5].

2 LEARNING SERVICES

2.1 Learning resource supporting services

In the 21st century, the impact of education technology on learning has become so profound and astonishingly extensive. New forms of learning have emerged: flipped classroom, ubiquitous learning, mobile learning, micro-lessons, micro-video learning, etc. They all have their own features and demand technology support to facilitate learning. In these forms of learning, the mind-set of learning has also changed: emphasis is laid on big data, large networks and small classrooms. The significance of related thinking in studying professional courses is drawing more attention. Technology nowadays is no longer the supplement or means for learning, but the basis of learning.

The National Education Technology Plan 2010 (NETP) released by the U.S. Department of Education, Office of Educational Technology has built a "technically supported learning model in the 21st century" [6]. According to this learning model, learning can be based on learning communities, professional and authoritative resources or social networks. In this case, the concept for construction of resources should be altered from teaching-oriented to learning-oriented. It should be centred on students' needs and based on common learning experiences, i.e. the capacities that all possess, the concepts that all would understand, so as to provide diversified, selective, group-based or personalized learning resources for students with the help of a variety of information techniques.

First, in order to meet the demands from students selecting different courses, of different habits and in different stages of learning, standardized and modulated resources should be provided as modulated resources facilitate personal learning and standardized resources would be more easily extended and shared for its compatibility.

Second, the concept of learning supporting services demands the resources to be no longer static, but dynamic. Art and design learning requires collisions of thoughts and fresh ideas. Therefore, learners are not recipients, but contributors to the construction of the resources. Learning platform should also provide learners an environment where the method of "creative commons" is applied.

Third, learning supporting services also require the provision of extensive, fragmentary resources. The learning model of the 21st century demonstrated that learning communities and social networks are deeply intertwined in the Web2.0 era. People learn anywhere, anytime, online or offline. Meanwhile, information in this age is more frequently presented as fragmentary instead of systematical, urging the construction of "micro learning resources".

Fourth, different learning situations demand different categories of resources. Huang Ronghuai classified learning into the following five types: learning through lectures in class (the most typical type), self-studying, learning through group work, learning through practice and work-based learning [7]. More attention is paid to the first and the second type in construction of conventional learning platforms while there is a little research on the other types which is equally important. To provide resources for collaborative learning, it is essential to analyze the design and implementation of learning activities and stress on the potentials of techniques.

Take the design of theme activities for example. M. David Merrill, in the article "First Principles of Instruction", raised five first principles in learning: 1) Learning is promoted when learners are engaged in solving real-world problems. 2) Learning is promoted when existing knowledge is activated as a foundation for new knowledge. 3) Learning is promoted when new knowledge is demonstrated to the learner 4) Learning is promoted when new knowledge is applied by the learner. 5) Learning is promoted when new knowledge is integrated into the learner's world. And particularly, in the elaboration of the first principle "problem-centred", learning would be promoted when 1) learners are shown the task or the problem, 2) learners are engaged at the problem or task level instead of the operation or action level, 3) learners solve a progression of problems that are explicitly compared to one another [8]. Teachers should classify and summarize the knowledge included in a specific course, according to the expected learning outcomes of the whole course and of each unit. Teachers should then design a series of specific tasks that is to be completed by the students, according to the difficulty of the content and the time sequence. It is also to be expected that teacher specify whether each task is more suitable for independent learning or group learning and offer a detailed description of the preparation that is required (to acquire the basic knowledge, to learn to manage necessary learning tools and technology, etc.) for completing each task as well as the expected learning outcomes after completing each theme activity. In independent learning and cooperative learning, the design of the theme activity should both motivate and activate the students. Students should be encouraged to make full use of a variety of learning resources (library resources, e-library resources, learning platform resources, network resources, etc.) and learning environments (campus E-learning platforms, virtual learning communities, blog sites, QQ groups, libraries, classrooms, computer rooms, practical training centres, etc.) to learn in a relaxed and happy way, and improve their problem-solving ability and creativity through practices.

2.2 Supporting learners with introductions and help based on "close interaction"

Traditional course introductions generally focus on the summary of the content, the schedule, and the demands for students. Its potentials to attract students' attention in online and offline study, to stimulate students' curiosity, interest, and potentials for exploring, and to vividly present latest research and applications in relation to the course are mostly neglected. On the other hand, traditional introductions are often limited to monotonous, inefficient non-real-time messages and BBS posts in terms of interaction. With more high-quality resources such as online opening class videos in our reach, course introductions are expected to provide more multimedia information on the background, latest research progress, extensive reading materials, real-life applications as well as remaining problems relevant to the subject. The combination of the "close interaction" online, possibly in the form of video conferences, and the face-to-face counselling offline would enable instant, efficient communication, impelling students to be more actively engaged in their study.

Helping service is featured by its timeliness and practicality. There are two kinds of need for helping services that deserve special attention: firstly, that of the students having difficulties doing procedural homework or applying their acquired knowledge and skills to solving new problems, and secondly, that of the students participating in interactive learning activities, who would want responses and evaluations from teachers and other students for their opinions so as to get answers as well as develop a sense of accomplishment. "Close interaction" aims to overcome the time and space limitations of traditional teaching and promote the development of individuality and collaboration.

A new term "Digital Native", coined in Marc Prensky's 2001 work Digital Natives, Digital Immigrants has gained immense popularity these days, referring to a person who was born in the digital age under the domination of information technologies. College students nowadays are "Digital Natives", they were born in a world with Internet and have interacted with digital technology from an early age. Therefore, they are easily distracted by the overflowing information on the Internet when they are learning online. And information overload reduces the efficiency of learning. Thus, it is crucial that we provide management tools to motivate learning. Developing self-management functions on the learning platform, for example, would largely promote productivity and help students form good habits.

3 THE DESIGN OF A DIGITAL LEARNING PLATFORM BASED ON ART AND DESIGN LEARNING SUPPORTING SERVICES

Based on the foregoing analysis on the student-centred learning supporting services, we aim to build a digital learning platform, embedded with social network collaborative learning systems, and characteristically enabling the presentation,

gathering and sharing of the information. The platform is designed for students in the art and design major, and great emphasis is placed on constructing supporting resources of visual arts. It is constructed on the ground of basic resources, and uses intelligent learning tools, such as push notifications for resource recommendation or update, intelligent search, and intelligent presentation of other links, etc.; the learning platform has also strengthened technical support for socialization, such as micro-blogging interactions, topics on Wikipedia, interactive graphics, video sessions, collaboration systems, etc. Moreover, considering the wide use of intelligent terminals, the learning platform supports smart phones, tablet PCs, PDAs, etc. as access points.

3.1 Learning resource supporting services of the platform

Supporting resources are standardized and modulated. Students in the art and design major are diversified in terms of their knowledge, learning environment, ability, motivation, and especially creativity and talent. Therefore, learning resources of different levels are strongly required. To meet this demand, we provide on the platform relatively independent modules of different themes arranged with a thread running through the whole session. Functional designs also include allowing useful links to Q & A, teacher demonstrations, homework evaluations, etc. to pop up as hot words; integrating practice resources such as procedural assignments, topic extended materials, etc. with original resources; etc.

Resources are also categorized in terms of the learning situations they serve. For group discussion, resource support includes task description, study guide, topic discussion, presentation and evaluation, while technical support focuses on interactive conversation, review, commentary, presentation and feedback. For learning through practice, resource support consists of teachers' demonstration, operating instruction, guidance, presentation and commentary, while technical support places more emphasis on Video-On-Demand (VOD), evaluation, Q & A, help and feedback. For work-based learning, resource support pays more attention to job descriptions, industry standards, procedures, guidance, presentation and commentary, with technical support providing services for VOD, evaluation, Q & A, help, interactive conversation, and project presentation.

3.2 Social learning platform

Furthermore, "close interaction" fully takes participants' socialization need into consideration.

Collaborative learning system embedded with social networks would not only facilitate the construction and sharing of resources, but also provide "close interaction"-based learning introductions and help that students demand.

First, the platform is equipped with a "topics on Wikipedia" module, propelling students to broaden their knowledge as well as collaborate to construct the information resource of a particular field and sort the information with "labels" and "tags". In art and design learning, the discussions of the topics, mutual assessments, case studies, and creative expressions are all fresh resources, displaying the process of learning and creating, and thereby motivating and inspiring students in their study. The resources that are commonly created are at the same time commonly shared, intelligently, as illustrated by the functions of intelligent recommendation and intelligent search. Learning resources are constructed with the help of the semantic web and ontology technology, ensuring that knowledge and theme are no longer separated, but closely related. On this platform, when the student is seeking for answer to a specific question, the system would automatically recommend relevant information. Such divergent, extensive resource links would largely benefit the students, especially those in the field of art and design.

Second, the platform is embedded with microblogging services. Students may connect to each other using real names or establish links to the outside. This design aims to encourage students to gather resources, share ideas, and play an active role in learning. It also inspires their creativity and provides them opportunities to display their works so that the boring learning environment becomes more friendly and appealing.

Third, "close interaction" is further facilitated with the modules of interactive graphics, video sessions, etc. Given the need of exchanging graphical information in art and design learning, we designed the "interactive whiteboard", on which students and teachers can all mark and review on the graphic and engage in convenient and efficient "graphic discussions". Visualized interaction gives timely feedbacks, and thus emotionally stimulates learners by spurring a sense of accomplishment and the feeling of belonging.

3.3 Learning cooperation and learning assessment

Collaborative learning modules help group work as well. Students share ideas with the help of social network, brainstorm, and exchange sketches and materials during the process of creating. Online and offline communication streamline teamwork.

Learning self-management module is another functional feature of the platform. It is comprised of functions such as reminder, statistical data, personal records, etc. Teachers could use functions and tools, for example, to remind students to finish procedural assignments in time or to actively participate in learning activities of various situations, according to the teaching schedule. TAs, on the other hand, could conduct individualized monitoring and supervision on students' learning process, online and offline, referring to the statistical data on the platform (such as student logins, homework submission, online learning behavior, participation in discussions, presentations, etc.). Students could also check their online learning behavior records and score records for self-management and improvement.

The purpose of curriculum assessment is to examine and evaluate the outcomes of students' learning for a particular course. Considering the various forms of learning, course assessment should not be limited to students' test scores, for they reflect only the amount of knowledge they have acquired. "Better teach a man to fish than give a man a fish." Similarly, the learning of a course should not be limited to memorizing the knowledge on the textbooks. More emphasis should be laid on the development of learning methods, the improvement of the learning ability and comprehensive abilities of individuals. Therefore, curriculum assessment methods should not only be diversified, but also motivate well-rounded development of students.

Learning process assessment includes online learning assessment, cooperative learning assessment, class performance, attendance. Online learning assessment mainly refers to user login frequency, recorded study time, study time allocation, course forum participation, including the quantity and quality of the posts, etc. Cooperative learning assessment concerns face-to-face panel discussion, group activity contribution, recorded online communication, etc. And class performance considers in-class discussion participation, interaction with other students, and the performance of giving presentations. Class attendance records whether students have arrived late, left early or skipped classes.

Assessment of learning performance takes individual learning performance, cooperative learning performance and classwork performance into consideration. Individual learning performance measures the quality of homework and assignments students have completed independently on the learning platform, such as question sets, essays and theses. Participation and awards in related competitions will also be noted. Cooperative learning performance considers the quality of work done by groups in theme activities, such as lab reports, survey reports, team project websites, etc.

And classwork generally refers to assignments that are to be finished in class or after class in order to help consolidate students' knowledge obtained in lectures.

During the operation of this "student-centred" digital learning platform based on learning supporting services in art and design major, we have received positive feedbacks. It is illustrated that resources designed and constructed for learning significantly stimulate students' motivation, interest, and promote better performance, and introductions and help based on "close interaction" as well as management support provide an open environment for learning.

REFERENCES

[1] Tim O'Reilly (09/30/2005). What is Web2.0 Design Patterns and Business Models for the Next Generation of Software, http://oreilly.com/web2/archive/what-is-web-20.html.

[2] Tim O'Reilly (09/30/2005). What is Web2.0, http://www.donews.com/Content/200511/a9f1b26851114-b4cabb16d75cd129c17.shtm.

[3] Qi Wenxin, Du Ruo, Li Zhe, Wang Guowei. The Practice and Reflection of Web2.0-based Course Learning Supporting Services Model[J]. Chinese Distance Education. 2010(11):49–53.

[4] Zhou Chunhong. The Examination and Practice of Blended Learning Mode based on Blackboard Learning Platform[J]. E-education Research. 2011,(2):87–98.

[5] http://www.etc.edu.cn/academist/hkk/blending.htm.

[6] Liang Linmei. Changing American education: Technology Facilitates Learning[J]. Open Education Research. 2010(8):35–41.

[7] Huang Ronghuai, Chen Geng, Zhang Jinbao, Chen Peng, Li Song. Five Laws on Technology Promoting Learning[J]. Open Education Research. 2010(2):11–19.

[8] M. David Merrill. First Principles of Instruction [J]. Educational Technology Research and Development, Vol. 50, No. 3(2002), 43–59.

Information Technology and Applications – Li (Ed.)
© 2015 Taylor & Francis Group, London, ISBN 978-1-138-02677-3

Query processing with knowledge association context

Yuangang Yao
China Information Technology Security Evaluation, Beijing, China

Xiaoyu Ma
Patent Examination Cooperation Center of the Patent Office, SIPO, Beijing, China

Hui Liu, Jin Yi, Haiqiang Chen & Xianghui Zhao
China Information Technology Security Evaluation, Beijing, China

ABSTRACT: During query processing, context is the knowledge graph behind user query inputs. But current context based methods cannot satisfy the demand of semantics and flexibility in query processing. That is because the context models are either fixed or informalized. In this paper, we introduce a new context based query processing approach, in which the context is explicitly defined. It can infer implicit or potential information behind query inputs. For given query inputs, the approach dynamically builds context models and refines queries based on semantic inference. After the query processing, queries are translated into executable semantic search instructions for query engine. We value the proposed approach by experiments in industry search. The result shows that it improves precision and recall in retrieval by analyzing the user query context and the processing is flexible.

Keywords: query context; semantic association; context inference; query processing

1 INTRODUCTION

Query processing is the entrance to retrieve information. It analyzes query inputs to understand user intents, and translates them into executable instructions for search engines. Context based query processing approaches try to capture the meanings behind keywords by the support of related background knowledge. In current context-based approaches, flexibility and explicitness of query processing are not considered comprehensively (Bai 2007, Cao 2009, Ingwersen 2005, Tablan 2008). Furthermore, other knowledge related works and consequent user activities, such as user feedback, navigation, and knowledge sharing, also need supports of context to make them more effective and efficient (Xu 2000). This paper proposes a systematic context-based query processing approach, which considers both and explicitness of query processing. In our approach, context inference explores implied or potential information behind query inputs to generate query candidates. It is also used to rerank generated semantic queries for precision and recall in retrieval.

2 QUERY PROCESSING APPROACH WITH CONTEXT

2.1 *The framework*

Our query processing approach is based on knowledge association (Yao 2013) and context. The knowledge association is the global knowledge network about some domain area, and the context is the relevant local knowledge about a specific knowledge based application. We build semantic context model and take advantage of semantic analysis and inference operations for query processing. Figure 1 shows the context-based query processing framework.

Query analysis firstly analyzes user query inputs to recognize terms in queries, such as the named entities like concepts. Query keywords are expressed as multiple word spaces containing the possible understandings in lexicon. Query structuring identifies the entities and structures queries based on knowledge association. It constructs initial semantic structured queries that are consistent with knowledge descriptions in knowledge association. Context construction dynamically build context for specific query. The context represents query

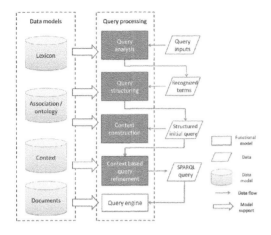

Figure 1. Context-based query processing framework.

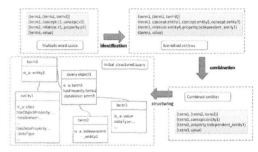

Figure 2. Query structuring processing.

entities as core entities in it, and specifies knowledge scope for context-based query processing. Query refinement constructs sub-queries, ranks them and generates standard queries. Query objects and conditions are specified by context inference considering both precision and recall in retrieval. Then query engine execute the generated query instructions to retrieve and return back the results.

2.2 Query analysis and structuring

For common users, their typical query inputs are key words, natural language sentences or fragments. Query analysis use basic syntactic analysis technologies to parse the inputs, and then extracts the structure and keywords in the inputs. Query structuring then identifies, combines, and structures the terms in query. Semantic-based structuring finally converts user inputs into structured descriptions with semantic annotations (in practice, we choose OWL (McGuinness 2004)), and relates them with knowledge association. Figure 2 shows the query structuring processing.

2.3 Context construction and inference

Context construction builds the basic context model for some specific query task. It models the structured query elements with formalized descriptions and supplements metadata about the elements. After that, the initial structured query is transformed into a local contextual knowledge about the query elements. Then we enrich the context for more query intent information. We define three context enrichment operations including context supplement, expansion and contraction to enrich context at different levels (Yao 2011). Besides context enrichment, we also define three context shifting operations, including context lifting and lowering, core entity changing,

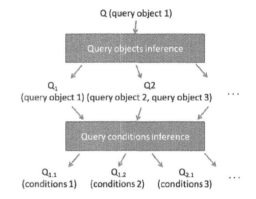

Initial structured query: Q
Constructed query candidates: $Q_{1.1}$, $Q_{2.1}$, $Q_{1.2}$ ...

Figure 3. Generating query candidates in query construction.

and merging for context modification according to user feedbacks. The context construction and inference processing build rich query background knowledge using semantic associated knowledge entities. It covers the local knowledge graph about user query, which implies the intents behind query inputs.

2.4 Context based query refinement

Considering both precision and recall in retrieval, we deploy three processes for query refinement with context-based inference. They are query construction, query ranking and query translation. Query construction generates a set of potential query candidates according to user query and context model. Query is refined in several ways. For example, the enhancement refines query by adding constrains, descriptions, relations, etc., and the identification refines query by choosing query objects and conditions that are not mutually exclusive. These derived query candidates are potential query intents at different levels. Figure 3 shows the general steps to generate query candidates in query construction. Query ranking lists candidates in order through computing their probability weights. According to context, the constraints precision and relevance to query intents are considered. Query translation finally converts the ranked query candidates into executable SPARQL (Prez 2006) queries for semantic retrieval.

3 CASE STUDY

We apply query processing with context inference to product design in enterprise environment. Document resources are heterogeneous including technical reports, standards, management documents, design documents, drawings, etc. Lexicon for query processing is extracted based on WordNet (Fellbaum 1998), domain dictionary and ontologies. These ontologies include domain ontology, product ontology, design ontology, service ontology and enterprise ontology. Multidimensional association model represents knowledge network in enterprise environment, which enriches concepts, relations and properties in ontologies. Context models are dynamically constructed according to specific queries.

An example query Q1 is provided to explain the processing: "specification of loudspeaker in motorcar by designer who designs YD1367", where "YD1367" is a loudspeaker ID number.

User inputs are parsed into keywords and then recognized according to lexicon. Then terms are organized in multiple spaces. The first line is structure of initial query inputs, and the rest is multiple word space to represent words and understandings in lexicon. Figure 4 illustrates an example of query analysis and structuring. The first line is "((((specification, loudspeaker), motorcar), designer), design YD1367)" and the second line is "(specification, concept: specification, concept: spec, concept: description, concept: introduction)". These recognized terms are further mapped to association for more explicit semantic understandings, so the keyword "specification" is interpreted as "(specification, Class IRI = "#specification", Class IRI = "#introduction")". In query entity combination, we select entity candidates considering their relevance in knowledge association to construct initial query. Then query elements semantic structured associated with association model.

To construct context for query inference, we should identified its core entities and contextual association. Core entities are contextual entities in initial structured query, and contextual association is inferred by context operations we introduce in section 2.3. The sample of the contextual association and core entities is depicted in Figure 5.

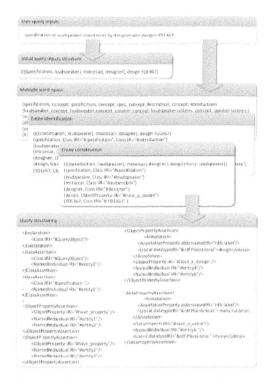

Figure 4. Query analysis and structuring for Q1.

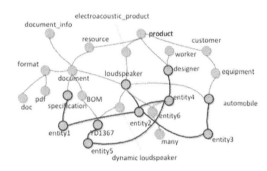

Figure 5. Query context for Q1 and Q2.

Entity2 and entity4 have a relation in query, we try to find contextual knowledge about it in context construction. According to property inheritance and axiom-based inference, the relation between class product and class designer may be basis for this property. So the two classes and corresponding property are involved in the context, and the unidentified property can be further inferred.

With context model, query is refined. We infer query objects and list candidates in order according to probability computation. We choose query object "entity1" for further processing. For query

21

conditions, query entities are further confirmed based on context inference. For example, to detail relation between entity1 and entity2, we consider some related descriptions in context, which contain "#specification have individual #entity1", "#loudspeaker have individual #entity2", "#product have resources #resource", and "#loudspeaker have documents #document". Then according to property inheritance, the relation is identified as "#entity2 have document #entity1". During inference, the relevance weights are calculated to estimate the probability.

At last the query translation converts candidates into SPARQL queries. In the example, one top ranked query is shown in Figure 6.

In the use case above, query intent is explicit after inference. If we replace "YD1367" with "many loudspeaker" (Q2), the inference will be ambiguous. In the modified query, there are two new identified entities, entity6 for "loudspeaker" and data property "have a value" for "many". Context is also modified as shown in Figure 5, but the contextual knowledge does not represent real user intents. "many" should be an constraints not a direct data value of entity6. During context-based query reference, the inconsistency is detected, so besides directly translation of "#entity6 have a value "many"", this query condition will also be translated as entity6 with description or annotation "many". The best way for this kind of query is to calculate the numbers of entities that satisfies the constrains, which we will consider in the future.

We select 150 typical queries in product design, and compute the Mean Reciprocal Rank (Voorhees 1999) (MRR) of query processing with context. Contexts are built to connect query items; furthermore, contexts are expanded for more details and abstract information. In our experiment, most of the queries are 3 to 5 words (62%), and the average length of queries is 3.9. In Table 1 we group the queries into three query sets, and compute the MRR respectively. The results show that the MRR falls down as the query length increases. That is because in our current context-based method, more entities

Table 1. Query processing results with context inference.

Query set	Query length	Total	MRR
QS1	1–2	49	0.886
QS2	3–5	140	0.774
QS3	6 or above	36	0.625
Average	3.9	–	0.775

in contexts mean more different understandings of query inputs.

4 CONCLUSION

In this paper, we propose a context inference method for query processing to improve the understanding of user query inputs. With the help of semantic context inference, the processes include query analysis, query structuring, context construction, and query refinement.

Our main contributions in the paper are concluded as follows. 1. We use knowledge association to support context. It is a knowledge enhancement focusing on both metadata and specific concepts in resources. 2. The context model is constructed using background knowledge within and around query contents in knowledge association, which is dynamic and flexible. 3. The proposed context inference improves current query graph matching methods in query processing, and the inference can also used in other semantic processing scenarios. 4. User query intents can be detected and retain in semantic query language SPARQL according to proposed query refinement processing, which improve the retrieval accuracy.

In the next step of the work, we will apply the context model to initial query analysis to further improve the user intent capture precision.

REFERENCES

Bai, J., Nie, J.Y., Cao, G., Bouchard, H., 2007. Using Query Contexts in Information Retrieval. Proceedings of the 30th *International* ACM SIGIR Conference on Research and Development in Information Retrieval. Amsterdam, The Netherlands. 15–22.

Cao, H., Hu, D.H., Shen, D., Jiang, D., Sun, J.T., Chen, E., *Yang, Q.,* 2009. Context-aware Query Classification. Proceedings of the 32nd International ACM SIGIR Conference on Research and Development in Information Retrieval, Boston, MA. 3–10.

Fellbaum, C., 1998. WordNet: An Electronic Lexical Database. The MIT Press, Cambridge, MA.

Ingwersen, P., 2005. Selected Variables for IR Interaction in Context: Introduction to IRiX SIGIR 2005 Workshop. *Proceedings* of the ACM SIGIR 2005 Workshop on Information Retrieval in Context, *Salvador*, Brazil. 6–9.

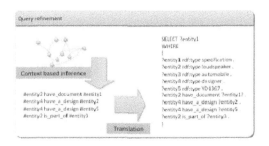

Figure 6. Query refinement for Q1.

McGuinness, D.L., Van Harmelen, F., 2004. OWL web ontology language overview. W3C recommendation. Chicago.

Prez, J., Arenas, M., Gutierrez, C., 2006. Semantics and Complexity of SPARQL. In The Semantic Web-ISWC 2006 (pp. 30–43). Springer Berlin Heidelberg.

Tablan, V., Damljanovic, D., Bontcheva, K., 2008. A natural language query interface to structured information. The Semantic Web: Research and Applications, European Semantic Web Conference 2008, 5021:361–375.

Voorhees, E.M., 1999. TREC-8 Question Answering Track Report. Proceedings of the 8th Text Retrieval Conference, Gaithersburg, Maryland. 77–82.

Xu, J., Croft, W.B., 2000. Improving the effectiveness of information retrieval with local context analysis. ACM Transactions on Information Systems, 18(1):79112.

Yao, Y., Lin, L., 2011. Semantic Context Enrichment and Shifting Based on Multi-dimensional Association. International Conference on Computers, Networks, Systems and Industrial Engineering, Jeju, Korea. 26–30.

Yao, Y.G., Lin, L.F., Wang, F., Zhang., W.Y., 2013. Multi-perspective modeling: managing heterogeneous manufacturing knowledge based on ontologies and topic maps. International Journal of Production Research, 51(11):3252–3269.

Information Technology and Applications – Li (Ed.)
© *2015 Taylor & Francis Group, London, ISBN 978-1-138-02677-3*

A new type of optimization method based on benchmark learning

Anshi Xie

School of Public Policy and Management, Tsinghua University, Beijing, China

ABSTRACT: According to the benchmark learning theory in the business management, the Benchmarking Learning Algorithm (BLA), was proposed in this paper. First, by right of imitation and learning, all the individuals within population were able to approach to the high yielding regions in the solution space, and seek out the optimal solutions quickly. What is more, the premature convergence problem got completely overcame through new optimal solution policy. Finally, BLA is able to accurately detect the slight changes of the environments and track the trajectory of the extreme points in the search space. And thus, it is naturally adaptable for the dynamic optimization problems. The comparative experiments for dynamic optimization problem showed that BLA is robust and able to perform friendly interactive learning with the environments, whose search speed, optimization ability and dynamic tracking ability were far superior to other similar methods.

Keywords: benchmark learning; search pattern; evolutionary algorithm; swarm intelligence; dynamic environments

1 INTRODUCTION

Intelligent computation, also known as natural computation, is kind of optimization model, which was inspired by the principles of the natural world, especially the biological world. Many optimization algorithms are included in the field of intelligent computation, which mainly consists of Evolutionary Algorithms (EAs) and Swarm Intelligences (SIs) et al. EAs include four main branches such as Genetic Algorithm (GA), Evolutionary Programming (EP), Evolutionary Strategy (ES), Genetic Programming (GP). SIs include Ant Colony Optimization (AC-O) particle Swarm Optimization (PSO), Artificial Fish Swarm Algorithm (AFSA), Shuffled Frog Leaping Algorithm (SFLA). Besides EAs and SIs, Simulated Annealing (SA), Taboo Search (TS) and Predatory Search (PS) should be included within intelligent computation as well.

These intelligent optimization algorithms mentioned above have special distinguishing features from each other, but they all have some common deficiencies. First of all, they all try to search the optimal solution by the individual's random drift in the solution space, yet the search direction and search purpose of the random drift are indeterminate and uncertain. Furthermore, they are all population convergence-oriented. That is to say, at the end of the search process, all the individuals are apt to converge to a certain point within the solution space. This point should have been the global optimal solution; however, all the individuals are likely to converge to a local optimal solution, because

the population diversity cannot be maintained due to the convergence strategy. Finally, they are all designed and studied for static optimization problems and their search process are always passively adaptable, yet most of practical applications are dynamic and changeable. Because unable to maintain the population diversity, and unable to detect and make quick responses to the slight changes in the environments, they always loss the adaptabilities for the environments and cannot track the trajectories of the extreme value point in the solution space, and cannot adapt to solving dynamic optimization problems.

Benchmark [1] originally means that a surveyor's mark on a permanent object of predetermined position and elevation used as a reference point. As a kind of management idea and management method, benchmarking learning originated from enterprise management domain, and it means that some outstanding enterprise can be set as a standard, by which other companies can be measured or judged, and improved consequently.

Based on benchmarking concept, the Benchmarking Learning Algorithm (BLA), was proposed in this paper.

2 THE BENCHMARKING LEARNING ALGORITHM (BLA)

2.1 *External benchmarking learning*

Let X_E^{best} be the best individual, whose evaluation function value is the maximal or the minimal according to the optimization purpose, in the whole ecological system, that is, the global optimal individual and the external benchmarking, let G_E^{best} be its corresponding gene expression, let G_K^i be the gene expression of X_K^i, which is the ist individual within niche population P_K, then, the external learning rate of X_K^i can be given as rule (1):

$$\begin{cases} \max f(x): & Grate_K^i = Grate' + f_K^i / \tilde{f}_K - 1 \\ \min f(x): & Grate_K^i = Grate' + \tilde{f}_K / f_K^i - 1 \end{cases} \quad (1)$$

whereas, $Grate'$ stands for the initial value of the external learning rate, f_K^i stands for the value of the evaluation function of X_K^i, $\tilde{f}_K$ stands for the average value of P_K.

If binary encoding method was put to use, external benchmarking learning was carried out by X_K^i means that the gene-bits in G_K^i, which are different from that in G_E^{best}, would be replaced by the gene-bits in G_E^{best} with a probability of $Grate_K^i$. That is to say, X_K^i took the initiative to narrow the Hamming distance with X_E^{best}.

If float-point encoding method was involved, external benchmarking learning was conducted by X_K^i means that with a probability of $Grate_K^i$, G_K^i would be updated according to rule (2) as below. That is to say, X_K^i took the initiative to reduce the Euclidean distance with X_E^{best}.

$$G_K^i = G_K^i + \lambda(G_E^{best} - G_K^i) \quad (2)$$

whereas, $\lambda \in [0,1]$, stands for the shift step length of individual X_K^i. Experiments show that the optimization effect will be better if λ is proportional to the search space, or fixed dynamically according to the evaluation function value in the process of learning. But this is not the focal point of this paper, so it will not be took into further discussion.

2.2 Internal benchmarking learning

Let X_K^{best} be the best individual, whose evaluation function value is the maximal or the minimal according to the optimization purpose, in niche population P_K, namely, the local optimal individual and the internal benchmarking; let G_K^{best} be its corresponding gene expression; let G_K^i be the gene expression of X_K^i, which is the ist individual in niche population P_K. Then, the internal learning rate of X_K^i can be given as rule (3):

$$\begin{cases} Binary: & Brate_K^i = Brate' - HD_{k,h} / Length + 1 \\ Float: & Brate_K^i = Brate' - ED_{k,h} / Radius + 1 \end{cases} \quad (3)$$

whereas, $Brate'$ stands for the initial value of the internal learning rate; $HD_{K,h}$ stands for the

Hamming distance between X_K^i and X_K^{best}; $Length$ stands for the length of the gene expression encoding; $ED_{K,h}$ stands for the Euclidean distance between X_K^i and X_K^{best}, namely, $ED_{k,h} = \sqrt{\sum_1^n (x_i^{best} - x_i)^2}$. $Radius$ stands for the diameter of the search space, that is, $Radius = \sqrt{\sum_1^n (b_i - a_i)^2}$. Here, x_i is the ist dimension of the gene expression, and $x_i \in [a_i, b_i]$.

Similar to external benchmarking learning, if binary encoding method was put to use, internal benchmarking learning was carried out by X_K^i means that the gene-bits in G_K^i, which are different from that in G_K^{best}, would be replaced by the gene-bits in G_K^{best} with a probability of $Brate_K^i$. That is to say, X_K^i took the initiative to narrow the Hamming distance with X_K^{best}.

If float-point encoding method was adopted, internal benchmarking learning was carried out by X_K^i means that with a probability of $Brate_K^i$, G_K^i would be updated according to rule (4) as mentioned below. That is to say, X_K^i took the initiative to diminish the Euclidean distance with X_K^{best}.

$$G_K^i = G_K^i + \lambda(G_K^{best} - G_K^i) \quad (4)$$

whereas, $\lambda \in [0,1]$, stands for the shift step length of individual X_K^i.

2.3 Pseudocode

Let $E = \{P_1, P_2 \dots P_{np}\}$ be the whole ecological system consists of np niche populations, N_i be the number of individuals in P_i, P_i^j be the jth individual in P_i, P_i^{best} be the best individual in P_i. Let f_i^j be the evaluation function value of P_i^j, $\tilde{f}_i$ be the average value of P_i at current generation, $\tilde{f}_E$ be the average value of E at current generation, P_{best} be the best individual in E. Let $Grate'$ be the external learning rate, $Brate'$ be the internal learning rate, $Srate'$ be the self-learning rate. Let max_gen be the maximum iteration times. Then the pseudocode for BLA can be given as mentioned below.

1. Initialize the np, N_i, $Grate'$, $Brate'$, $Srate'$ and other parameters if necessary.
2. $for \quad gen = 1 : max\_gen, \quad do$
   - a. $for \quad i = 1 : np, \quad do$
     - i. Evaluate f_i^j
     - ii. Evaluate $\tilde{f}_i$
     - iii. Find and record P_i^{best}
   - b. Find out and record P_{best}
   - c. Find out, record and update the best individual so far in E.
   - d. Evaluate $\tilde{f}_E = \left(\sum \tilde{f}_i\right) / np$
   - e. $for \quad i = 1 : np, \quad do$
     - i. P_i^j conduct external benchmarking learning
     - ii. If f_i^j does not be improved, then, P_i^j will conduct internal benchmarking learning

iii. If f_i^j does not be improved yet, then, P_i^j will carry out self-learning

f. *if* $\tilde{f}_E$ does not be improved or the best individual in E does not be replaced *do* P_i will exchange its best individual with other niche populations

3. Output the global optimal solution.

3 EXPERIMENT AND SIMULATION

It is still rarely seen some scientific literatures about structuring dynamic function, Angeline has proposed a kind of mobile parabolic problem [7], which involved in a single hump function based on float-point encoding. Brank et al. have put up a method structuring dynamic function [8]. Namely, the goal is to find out the maximum of all the peaks, whose position, height and width would change with slight changes in the environments. It is easy to construct all kind of mobile peak functions in this way, yet its process is rather trivial.

To compare with BLA, several well-known optimization methods including Simulated Annealing [4] (SA), Taboo Search [9] (TS), Primal Dual Genetic Algorithm [10] (PDGA), have been put to use to optimize Schwefel's function. In this experiment, binary encoding was adopted in all the optimization methods and the length of individual's gene expression was set as 50, namely, a control variable was represented by 25 binary bits. The purpose of this experiment is to test each method's ability for tracking the extreme point of Schwefel's function in the dynamic environments. The controls parameters in these methods were set as below.

SA: the iteration number was set as 80 for each phase of thermal equilibrium to compensate for the disadvantages of single individual's serial search; each individual's neighbor solutions were created by turning some bit in the gene expression into its opposite value; the number of iterations was set as the current annealing temperature. TS: each iteration, 10 neighbor candidate solutions were created in the same way as SA; Tabu list consisted of the optimal solution after each iteration. PDGA: tournament selection strategy, anti-XOR crossover and single-point mutation were adopted; crossover rate and mutation rate were set as 0.7 and 0.2 respectively. BLA: at the initialization stage, 10 niche populations and 10 individuals in each niche population, the initial value of the external learning rate, internal learning rate and self-learning rate were all set as 0.5.

The maximum iteration times in each method was set as 250, a template would be created every 50 iteration and all of the individuals would conduct bit XOR operation with the template, so environments went through 5 times of oscillation. r stands for the ratio between the number of bits whose value is 1 and the length of the template, namely, the intensity of environments changes. When r is equal to 0.1, 0.5 and 0.9 respectively, each method's ability for tracking the extreme point of Schwefel's function in the dynamic environments is shown below.

From Figure 1, it can be observed that the search performance of SA and TS was unstable, they could not find out the global optimal solution, namely they could not track the trajectory

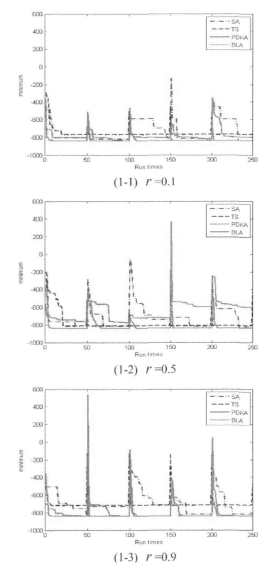

(1-1) r =0.1

(1-2) r =0.5

(1-3) r =0.9

Figure 1. Comparison of four methods' dynamic tracking ability.

27

of the maximum point after the environments changed, especially the environmental changes was acute. When the environments changed smoothly, as shown in figure (1-1) and figure (1-2), PDGA could not effectively track the trajectory of the maximum point, yet if the environments changed sharply, as shown in figure (1-3), PDGA had a very outstanding performance in tracking the trajectory of the maximum point, this is mainly because its dual-mapping strategy played an important role.

The great majority of individuals within the population had greatly deviated from the optimization purposes after the environments changed acutely, but the dual mapping was able to help these individuals bounce back to the high yielding regions. In addition, the dynamic environments was constructed by dynamic temples, so PDGA performed excellently; in fact, if the dynamic environments was constructed through other methods, for example, changing the scope of decision variables, the dual-mapping strategy would do nothing for PDGA. In the present experiment, the performance of SA was slightly better than TS on the whole, just because at each stage of thermal equilibrium, SA iterate 80 times, which was seven times more than TS. However, neither SA nor TS found out the global optimal solution in the new environments. In contrast, BLA had much more powerful search capabilities and adaptability. It would adjust its search direction and search step length to adapt to the new environments after detecting the smooth or acute changes of the environments, so BLA could search and find out the global optimal solution as soon as possible. From Figure 4, it is easy to find that there are a bit of slight bending in these value curves, this demonstrated that BLA is extremely sensitive to the changes of the environments and it could work friendly with the environments through transparent interaction, which could help BLA to seek out a better solution than the current one and improve it continuously. It is also easy to see that BLA was always able to seek out the global optimal solution at the beginning of each new cycle, and no matter how the intensity of environments changes was, these global optimal solutions were always the same, this indicated that the search performance of BLA is very stable.

4 CONCLUSIONS AND FUTURE WORK

In this paper, a competitive learning mechanism based on dynamic niches was set up according to the core values of benchmarking. Consequently, some defects of the Existing Intelligent Optimization Methods (EIOMs), for example, the running direction of the search process was indecisive, the EIOMs could not maintain the population diversity and could not be adapt to dynamic optimization problems, were all got overcome. As a result of the imitation and learning to the benchmark, individuals within the population are able to approach to the target regions in the solution space and seek out the optimal solutions quickly. The search behaviors of these individuals are no longer apt to be completely passive, self-adaptive and random, but active and direction-oriented. What is more, the formidable problem of maintaining the diversity of population was completely overcame through the self-organizing learning process of the niche system and its friendly interaction with the environments, thus, the exploration and exploitation of the BLA will be balanced self-adaptively. And then, BAL is able to accurately detect the slight changes of the environments and track the trajectory of the extreme points in the search space, and thus, BLA is naturally adaptable for the dynamic environments.

BLA, which originated from benchmarking theory of business management, is different from the EIOMs, which stem from the biological activities of nature. So BLA is brand new and it is a newborn member of the family comprising the modern intelligent optimization methods. However, as other optimization methods, BLA also involved a number of controls parameters, such as learning rates, etc. and how to set these controls parameters to optimize BLA to achieve the best effect, which itself is also a combinatorial optimization problem, is one of our next research topics.

ACKNOWLEDGMENT

According to "The length of the paper should not exceed 4 pages without extra pages fees", a high percentage of thesis was not included in this paper. The author thanks publishing house, my work, though only part of it, was published, after all.

REFERENCES

[1] Yang Dong-long, *Benchmarking: how to learn from benchmark enterprises*, China Social Sciences Press, Beijing, 2004. (in Chinese)

[2] Kennedy J., Eberhart R.C. "A discrete binary of the particle swarm optimization", *Proc. IEEE International Conference on systems, man and cybernetic*. Piscataway, NJ: IEEE Service Center, 1997: 4104–4209.

[3] J.H. Holland, *Adaptation in Natural and Artificial Systems*, University of Michigan Press, Arbor, 1975.

[4] Kirkpatrick S., Gelatt Jr C.D., Vecchi M.P. "*Optimization by simulated Annealing*". Science, 1983, 220: 671–680.

[5] Kennedy J., Eberhart R.C., "Particle Swarm Optimization", *Proc. IEEE International Conference on Neural Network*. Riscataway, NJ: IEEE Service Center, 1995: 1942–1948.

[6] Li Xiao-lei, Shao Zi-jiang, Qian, Ji-xin. "An optimizing method based on autonomous animals: fish-swarm algorithm". *System Engineering Theory and Practice*, 2002, 22(11): 32–38.(in Chinese)

[7] Angeline P.J. "Tracking extrema in dynamic environments" *Proceedings of the 6th international conference on Evolutionary Programming*, 1997, volume 1213 of LNCS.

[8] Brank J., Mattfeld D., Engelmann T. *A simple dynamic scheduling benchmark*. Evolutionary Optimization in Dynamic Environments Kluwer Academic Publishers, 2002.

[9] Glover F., "Tabu search—Part I". *ORSA Journal on Computing,* 1989, 1(3): 190–206.

[10] Yang S. "The primal-dual genetic algorithm", *Proc of the 3rd Int Conf on Hybrid Intelligent System*. IOS Press, 2003.

Information Technology and Applications – Li (Ed.)
© 2015 Taylor & Francis Group, London, ISBN 978-1-138-02677-3

The establishment of public opinion forecasting and early-warning through the methods of grey forecasting and pattern recognition

Hong Fu, Xiaogang Ren & Hui Li
Beijing Institute of Science and Technology Information, Beijing, China

Zhitao Du
Department of Journalism and Mass Communication, China Youth University for Political Science, Beijing, China

ABSTRACT: As the network public opinion features various topics, complex content and large amount of data, the authors have constructed a multi-level index system of network public opinion analysis which consists of three levels of indicators. And then, network public opinion forecasting model is established through the method of grey forecasting, and the early-warning model is established through the method of pattern recognition as well. In this paper, the authors have selected 60 network public opinion events as training sample, established the classification rules of network public opinion level, and also tested the usability of the forecasting and early-warning model. Eventually, the authors have put forward the whole frame of constructing the system of network public opinion forecasting and early-warning due to large amount of data and complicated calculation required by forecasting and early-warning.

Keywords: grey forecasting; pattern recognition; network public opinion; forecasting; early-warning

1 INTRODUCTION

Nowadays, network public opinion has brought about huge influence on politics, economy and life for the society because of its following characteristics such as spreading swift, large numbers of participants, and the intense interactivity among the participants. And the characteristics of the network users being spontaneous, irrational and emotional make it difficult to tell the difference between network public opinion and rumors. Therefore, the monitoring and management of network public opinion, which play an important role in the field of society management and political democracy, are facing a severe test. Therefore, it will be very important to grasp the development tendency of network public opinion and recognize the extent of its influence through strengthening real-time monitoring.

2 LITERATURE REVIEW

At present, the related research achievements focus on the following three aspects. First, there is the research about the analysis index system of network public opinion such as the index system established by Qing Wang from the heat extent, intensity, gradient and evolution tendency about the public opinion [1] and another index system established by Run-xi Zeng from the source, omens and the situations of public opinion through AHP (Analytic Hierarchy Process) [2]. Second, there is the research about the model of forecasting and early-warning and the related computing methods such as the early-warning model put forward by Chen Lin et al., through the method of fuzzy inference rule [3], the monitoring and early-warning model established by Eirinaki et al., through the method of data-mining about Web Log [4], and the monitoring and early-warning model established by Martens et al., through the methods of opinion-mining about single granularity information [5]. Third, there is the research about the information system construction of network public opinion. And there are several monitoring, forecasting and early-warning systems of network public opinion outstanding from the same products, such as Founder®, MZhiXun®, TRS®, Autonomy® network public opinion systems, and so on. These systems could help the users to grasp the public opinion hotspot, and depict the evolution tendency about the public opinion through monitoring the information sources of web portals, news website, BBS, Blog, and Micro-blog real-timely.

This paper tries to establish a forecasting and early-warning model of network public opinion, based on the characteristics of topic variety, content complexity, and large amount of data of public opinion through the methods of grey forecasting and pattern recognition.

3 THE CONSTRUCTION OF ANALYSIS INDEX SYSTEM OF NETWORK PUBLIC OPINION

There are three dimensions to measure the situation about network public opinion index system. The first dimension, which measures the heat extent, attention extent and influence extent from the total quantity of the public opinion, is named spread strength in this paper. The second dimension, which discusses the sentiment and opinion attitude of public opinion from the view of audience, is called network users' emotion attitude. The third dimension, which depicts the evolution path from the view of time and space, is called development degree. The analysis index system in this paper was established from the above mentioned three dimensions. And there are two methods to establish the whole index system. The authors put forward original index collection through literature review and summary. And then the final index is selected through combining the method of Delphi and gray statistic. As a consequence, the selected index is as following.

3.1 *Spread strength. It includes four lower-level indicators*

Degree of exposure. Generally speaking, the more the number of news release of related topics and the original theme posts release is, the higher degree of exposure of the theme and the stronger the public opinion will be.

Degree of attention. It measures the strength of public opinion from the perspective of the audience. The more the number of people who pay attention is, the stronger the public opinion will be and the greater the social influence will be.

Degree of coverage. It includes network coverage rate and area coverage rate.

Degree of authority. This includes website level, the guiding force of opinion leaders and extract rate.

3.2 *Emotion attitude*

Sensitivity of the theme. The standard of measurement has two indexes: theme classification and the number of sensitive words.

Content tendency. This refers to the attitude of the audience towards the content. The attitude can be classified into positive, neutral, and negative.

Turnout. This is a more explicit and direct behavior of network involvement, better able to reflect the public's attitude.

View dispersion. The formation of public opinion is the process of gradual convergence and unification of the opinion, attitude, and mood of the majority. The more the number of the people holding the same opinion is, the more powerful the influence of the public opinion will be. The measurement standard calculates the variance of sample attitude, based on the analysis of data for the audience emotion tendency of orientation. The greater its value is, the smaller the strength of public opinion will be.

3.3 *Development degree*

Diffusion velocity. It describes the speed of the development of public opinion, measured by the acceleration. There are three aspects: news and the main post acceleration, reply and comment acceleration, and browse acceleration.

Migration degree. It is used to measure the transfer degree of the public attention from one theme event to other topics.

Mutation degree. Network public opinion characterizes self-organizing, which allows for many uncertainties during its development process. The higher the mutation degree is, the less the validity and reliability of public opinion forecasting will be. The dispersion degree of opinion leader's view can be used to measure the mutation degree. First, opinion leader samples can be dug out. Second, opinion tendency of the opinion leader is evaluated. Finally, variance yields of opinion leader's view can be calculated. The greater the value is, the higher the mutation degree will be. Then it will be more difficult to forecast the development tendency of public opinion and the strength of public opinion will be less, and vice versa.

4 PUBLIC OPINION FORECASTING PATTERN

Network public opinion forecasting and early-warning include two periods: the first period is to forecast future trends through current data; the second period is to determine the level of warning according to certain determinant rules and forecasting values. This paper is to accomplish the forecasting task in the first period through the method of grey forecasting and accomplish the warning task in the second period through the method of pattern recognition [6,7].

4.1 *Constructing the network public opinion predictive model through the method of grey prediction*

Network public opinion possesses the characteristics of uncertainty, incomplete information and multiple indexes. Therefore, the construction of the model should use the method of grey prediction. As the system grey prediction model need to use the method of nested to construct the model, this paper embedded the model of GM (1, 1) into GM (1, N) with a view to obtain the predicted value about the behavior variables.

4.1.1 Constructing the system grey prediction model

In the system about N variables, the system grey prediction model could be expressed as equation (1) as follow:

$$x^{(0)}(k) + AZ^{(1)}(k) = Bx^{(1)}(k) + b, k = 1, 2, ..., m \quad (1)$$

In the equation (1), ① $x^{(0)}(k)$ stands for the observation sequence about the different variables in different time, $x^{(0)}(k) = [x_1^{(0)}(k), x_2^{(0)}(k), ..., x_N^{(0)}(k)]^T$ ② $x^{(1)}(k)$ stands for cumulative generated sequence (I-AGO) about $x^{(0)}(k)$, $x^{(1)}(k) = [x_1^{(1)}(k), x_2^{(1)}(k), ..., x_N^{(1)}(k)]^T$ ③ $z^{(1)}(k)$ stands for the mean sequence about $x^{(1)}(k)$, $z^{(1)}(k) = [z_1^{(1)}(k), z_2^{(1)}(k), ..., z_N^{(1)}(k)]^T$, $z_i^{(1)}(k) = 0.5(x^{(1)}(k) + x^{(1)}(k-1))$ ④ parameters includes A and B, b stands for the t development coefficient matrix and gray matrix respectively.

$$A = \begin{bmatrix} a_1 & & & 0 \\ & a_2 & & \\ & & \ddots & \\ 0 & & & a_N \end{bmatrix}$$

$$B = \begin{bmatrix} 0 & b_{12} & b_{13} & \cdots & b_{1N} \\ 0 & 0 & b_{23} & \cdots & b_{2N} \\ \vdots & \vdots & \ddots & & \vdots \\ 0 & 0 & 0 & \cdots & b_{N-1,N} \\ 0 & 0 & 0 & \ddots & 0 \end{bmatrix} \quad b = [0,0,...,b_{N0}]^T$$

⑤ parameters vector is $P_{GM(1,N)} = [a_1, b_{12}, b_{13}, ..., b_{1N}]^T$, and its identification formula use least squares $P_{GM(1,N)} = (B^T B)^{-1} B^T y_N$, B is high matrix, the number of lines is bigger than its columns

$$B = \begin{bmatrix} -z_1^{(1)}(2) & x_2^{(1)}(2) & \cdots & x_N^{(1)}(2) \\ -z_1^{(1)}(3) & x_2^{(1)}(3) & \cdots & x_N^{(1)}(3) \\ \vdots & \vdots & \vdots & \vdots \\ -z_1^{(1)}(k) & x_2^{(1)}(k) & \cdots & x_N^{(1)}(k) \end{bmatrix}$$

$$y_N = [x_1^{(0)}(2), x_1^{(0)}(3), ..., x_1^{(0)}(k)]^T$$

4.1.2 The nested solving steps for system grey prediction model

Step 1, calculating the predictive value of x_N, namely $\hat{x}_N^{(0)}(k) = [\hat{x}_N^{(0)}(m+1), \hat{x}_N^{(0)}(m+2), ..., \hat{x}_N^{(0)}(m+\xi)]$
Solving method: $\hat{x}_N^{(0)}(2) = \beta_{N0} - \alpha_N x_N^{(0)}(1)$

$$\hat{x}_N^{(0)}(k) = (1 - \alpha_N)x_N^{(0)}(k-1),$$
$$k = 3, 4, ..., m, m+1, ..., m+\xi \quad (2)$$

In equation (2), $\beta_{N0} = \dfrac{b_{N0}}{1 + 0.5a_N}$ $\alpha_N = \dfrac{a_N}{1 + 0.5a_N}$

Step 2, calculating the predictive value of x_{N-1}, namely $\hat{x}_{N-1}^{(0)}(k) = [\hat{x}_{N-1}^{(0)}(m+1), \hat{x}_{N-1}^{(0)}(m+2), ..., \hat{x}_{N-1}^{(0)}(m+\xi)]$

Solving method: $\hat{x}_{N-1}^{(0)}(k) = \beta_{N-1,N} - x_N^{(0)}(k)$
$$+ (1 - \alpha_{N-1})x_{N-1}^{(0)}(k-1) \quad (3)$$
In equation (3), $\beta_{N-1,N} = \dfrac{b_{N-1,N}}{1 + 0.5a_{N-1}}$
$\alpha_{N-1} = \dfrac{a_{N-1}}{1 + 0.5a_{N-1}}$.

Step 3, calculating the predictive value of x_{N-2}, namely

$$\hat{x}_{N-2}^{(0)}(k) = [\hat{x}_{N-2}^{(0)}(m+1), \hat{x}_{N-2}^{(0)}(m+2), ...,$$
$$\hat{x}_{N-2}^{(0)}(m+\xi)]$$

Solving method: $\hat{x}_{N-2}^{(0)}(k) = \beta_{N-2,N-1}x_{N-1}^{(0)}(k)$
$$+ \beta_{N-2,N}x_N^{(0)}(k) + (1 - \alpha_{N-2})x_{N-2}^{(0)}(k-1) \quad (4)$$

In equation (4), $\beta_{N-2,N-1} = \dfrac{b_{N-2,N-1}}{1 + 0.5a_{N-2}}$,
$\beta_{N-2,N} = \dfrac{b_{N-2,N}}{1 + 0.5a_{N-2}}$, $\alpha_{N-2} = \dfrac{a_{N-2}}{1 + 0.5a_{N-2}}$

Step N, calculating the predictive value of x_1, namely $\hat{x}_1^{(0)}(k) = [\hat{x}_1^{(0)}(m+1), \hat{x}_1^{(0)}(m+2), ..., \hat{x}_1^{(0)}(m+\xi)]$

Solving method: $\hat{x}_1^{(0)}(k) = \displaystyle\sum_{i=1}^{N} \beta_{1i}x_i^{(0)}(k)$
$$+ (1 - \alpha_1)x_1^{(0)}(k-1), i = 2, 3, ..., N \quad (5)$$

In equation (5), the solving method about α_1 and β_{1i} are the same to the methods of above equations.

4.2 Constructing the network public opinion predictive model through the method of pattern recognition

4.2.1 Determining the classification rules

According to the predicted value, we could realize early-warning about network public opinion through a certain kind of discrimination rule. The authors established early-warning model through the method of pattern recognition. Determined the classification rules for the predicted value, and then recognized its level based on the probability of the crisis. The level of warning-situation was classified to "Low, Medium and High" according to the influence of public opinion. "Low level" refers to the subject of public opinion in the stage of steady beginning, although there are network users participate in the discussion through posting message, replying, commenting and forwarding, but has only formed weak influence. "Medium level" refers to the subject of public opinion in the stage of steady development, already has a certain size, but has not yet formed strong influence. "High level" refers to subject of public opinion in the stage of big size, and has formed strong influence both on the cyberspace and the real society.

In this paper, w stands for the situation of warning level. $w = w_1$ refers to low level, $w = w_2$ refers to medium level, $w = w_3$ refers to high level. This paper will analysis the N kinds of observation value, which are $x_1, x_2, ..., x_N$, about network public opinion. The probably value about these observation value consist the N-dimensional feature space, $x = [x_1, x_2, ..., x_N]^T$ refers to N dimensions feature vectors. As the secondary indicators can reflected the characters about network public opinion completely in Table 1, we take on the secondary indicators as observation value, which is $N = 11$. And the secondary indicators should be obtained through the method about weight determine and add on the tertiary indicators collected by system.

If there is a vector x, which is a certain point in N-dimensional feature space, in the feature space. You should consider is what kind of w is appropriated for x. That is to say, you should confirm the classification rules for network public opinion level. Therefore, the authors minimize the classification error through the method of Bayes decision according to minimum error rate as determine basis.

Through Bayes

formula $P(w_i \mid x) = \dfrac{p(x \mid w_i)P(w_i)}{\sum\limits_{j=1}^{3} p(x \mid w_j)P(w_j)}$, $i, j = 1, 2, 3$ (6)

The conditional probability $P(w_i \mid x)$, which is called posterior probability could be obtained. Therefore, the real purpose of using Bayes formula

is to transfer the prior probability of $P(w_i)$ to the posterior probability $P(w_i \mid x)$ through observing the value of x. Therefore, the Bayes decision rule according to the minimum error rate is: if $P(w_i \mid x) = \max P(w_j \mid x)$, then $x \in w_i$.

Defining a discrimination function $g_i(x)$, $i = 1,2,3$. If given $g_i(x) > g(x)$ which matches the condition j≠i, x should be classified into w_i.

Here, $g_i(x) = P(w_i \mid x)$ or $g_i(x) = p(x \mid w_i)P(w_i)$ or $g_i(x) = \ln[p(x \mid w_i)P(w_i)]$ (7)

The classifier system could be designed according to the discrimination function mentioned above as Figure 1. First, calculating the value of the function about $g_i(x)$, and then choosing the maximum class, which is correspond to discrimination function, as decision-making result.

4.2.2 Pattern recognition through multivariate normal distribution

The distribution of N feature observed value x can be estimated through a lot of statistics data. As the normal distribution has the rationality of the physical and mathematical simplicity, here consider the algorithm about x in the situation of normal distribution. And the probability density function for multivariate normal distribution is as the equation below.

$$P(x \mid w_i) = \frac{1}{(2\pi)^{\frac{N}{2}} |\Sigma|} \exp\left\{ -\frac{1}{2}(x - \mu)^T \Sigma^{-1}(x - \mu) \right\}$$
(8)

In equation (8), $\mu = E\{x\}$ refers to N-dimensional mean vector for each kind of x. That is to say $\mu = [\mu_1, \mu_2, ..., \mu_N]^T$. Σ refers to N × N-dimensional covariance matrix, $\Sigma = E\{(x-\mu)(x-\mu)^T\}$. Σ^{-1} refers to the inverse matrix for Σ, $|\Sigma|$ is the determinant for Σ.

According to normal distribution, give minimum error rate discrimination equation is as follows:

Table 1. A part of data about the predicted value of public opinion for "Zhou Jiu-geng events".

	Date	Degree of exposure $\hat{x}_1^{(0)}(k)$	Degree of attention $\hat{x}_2^{(0)}(k)$	......	Migration speed $\hat{x}_{10}^{(0)}(k)$	Mutation speed $\hat{x}_{11}^{(0)}(k)$
Observe value	08.12.12	0.5631	0.4532	......	0.2364	0.2334
	08.12.13	0.6592	0.5561	......	0.2515	0.3316
	08.12.14	0.4431	0.6692	......	0.3310	0.2017
	08.12.15	0.5102	0.6310	......	0.2714	0.1095
	08.12.16	0.3314	0.7439	......	0.3926	0.0691
	08.12.17	0.6135	0.8235	......	0.2253	0.0563
Predicted value	08.12.18	0.5783	0.8541	......	0.2617	0.0541
	08.12.19	0.6562	0.8842	......	0.3148	0.0413

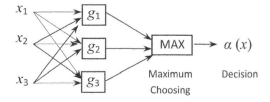

Figure 1. The components of classifier system.

$$g_i(x) = \ln\left[p(x|w_i)P(w_i)\right] \quad (9)$$

Through equations (8) and (9), in multivariate normal probability model $(p(x|w_i){\sim}N(\mu_i, \Sigma_i)$, $i = 1, 2, 3)$, obtain the discrimination function as follows:

$$g_i(x) = -\frac{1}{2}(x - \mu_i)^T \Sigma_i^{-1}(x - \mu_i)$$
$$-\frac{N}{2}\ln 2\pi - \frac{1}{2}\ln|\Sigma_i| + \ln P(w_i) \quad (10)$$

As $(N/2)\ln 2\pi$ has nothing to do with i, it is negligible in equation (10). Therefore, equation (11) can be obtained through simplifying (10).

$$g_i(x) = -\frac{1}{2}(x - \mu_i)^T \Sigma_i^{-1}(x - \mu_i) - \frac{1}{2}\ln|\Sigma_i|$$
$$+ \ln P(w_i) = x^T W_i x + w_i^T x + w_{io} \quad (11)$$

In equation (11), $W_i = -1/2\,\Sigma_i^{-1}$ ($n{\times}n$ matrix), $w_i = \Sigma_i^{-1}\mu_i$ (n-dimensional vector), $w_{i0} = -1/2\mu_i^T\Sigma_i^{-1}\mu_i - 1/2\ln|\Sigma_i| + \ln P(w_i)$.

4.2.3 Estimation for error rate

In this paper, the authors tried to use error rate to measure the merits and shortcomings about the classifier's performance during the course of classifying according to pattern recognition. As the warning level about network public opinion was classified into "low, medium and high", the feature space can be divided into three parts such as $\Re_1, \Re_2, \Re_3$. And $P(e)$ is the average error rate as there may be a lot of possible fault points during the course of classifying.

$$P(e) = [P(x \in \Re_2|w_1) + P(x \in \Re_3|w_1)]P(w_1)$$
$$+ [P(x \in \Re_1|w_2) + P(x \in \Re_3|w_2)]P(w_2)$$
$$+ [P(x \in \Re_1|w_3) + P(x \in \Re_2|w_3)]P(w_3)$$
$$= \sum_{i=1}^{3}\sum_{\substack{j=1 \\ j\neq i}}^{3}[P(x \in \Re_j|w_i)]P(w_i) \quad (12)$$

As there will be large amount of calculation solving $P(e)$ directly, this paper will calculate the average correct rate probability $P(c)$. As a consequence,

$$P(c) = \sum_{j=1}^{3}[P(x \in \Re_j|w_j)]P(w_j)$$
$$= \sum_{j=1}^{3}\int_{\Re_j} P(x|w_j)P(w_j)dx \quad (13)$$

$$P(e) = 1 - P(c) \quad (14)$$

4.3 Model application through cases study

4.3.1 Choosing the classificatory training ample and establishing its parameters and discrimination function

In this paper, 20 events, which have strong influence belonging to high-level warning about public opinion, were chosen as training sample. There are 20 tertiary indicators and 11 secondary indicators about network public opinion assessment. First, the tertiary indicators should be unified dimension of standardization. This paper use extemum method to make the original value be unified into the field of [0,1] through the formula of (original value-min. value)/(max. value-min. value). Second, calculate the value of 11 secondary indicators through the value of tertiary indicators and relatively weight. Third, through parameters estimation, the secondary indicators for the 60 training samples obey the normal distribution, and the prior probabilities are as following respectively: low-level warning public opinion $P(w_1) = 0.3$; medium-level warning public opinion $P(w_2) = 0.4$; high-level warning public opinion $P(w_3) = 0.3$. The parameters Σ_1, Σ_2 and Σ_3 for conditional probability density function $P(x|w_i)$ are matrix of 11×11 respectively, μ_1, μ_2, μ_3 are 11-dimension column vector. As the data about the training sample are big, in this paper only a part of the data was mentioned here. The discrimination function can be obtained according to the equation (11).

$$g_1(x) = \begin{vmatrix} x_1 \\ x_2 \\ \vdots \\ x_{11} \end{vmatrix}^T \begin{vmatrix} 0.310 & -0.905 & \cdots & -0.649 \\ -0.905 & 0.237 & \cdots & -0.255 \\ \vdots & \vdots & \cdots & \vdots \\ -0.649 & -0.255 & \cdots & 0.347 \end{vmatrix}$$
$$\times \begin{vmatrix} x_1 \\ x_2 \\ \vdots \\ x_{11} \end{vmatrix} + \begin{vmatrix} 0.218 \\ 0.093 \\ \vdots \\ 0.552 \end{vmatrix}^T \begin{vmatrix} x_1 \\ x_2 \\ \vdots \\ x_{11} \end{vmatrix} + 1.981 \quad (15)$$

$$g_2(x) = \begin{vmatrix} x_1 \\ x_2 \\ \vdots \\ x_{11} \end{vmatrix}^T \begin{vmatrix} 0.463 & -0.252 & \cdots & 0.133 \\ -0.252 & 0.370 & \cdots & 0.163 \\ \vdots & \vdots & \cdots & \vdots \\ 0.133 & 0.163 & \cdots & -0.209 \end{vmatrix}$$

$$\times \begin{vmatrix} x_1 \\ x_2 \\ \vdots \\ x_{11} \end{vmatrix} + \begin{vmatrix} 0.343 \\ 0.416 \\ \vdots \\ 0.071 \end{vmatrix}^T \begin{vmatrix} x_1 \\ x_2 \\ \vdots \\ x_{11} \end{vmatrix} + 3.525 \quad (16)$$

$$g_3(x) = \begin{vmatrix} x_1 \\ x_2 \\ \vdots \\ x_{11} \end{vmatrix}^T \begin{vmatrix} 0.571 & 0.239 & \cdots & 0.138 \\ 0.239 & -0.192 & \cdots & 0.093 \\ \vdots & \vdots & \cdots & \vdots \\ 0.138 & 0.093 & \cdots & 0.364 \end{vmatrix}$$

$$\times \begin{vmatrix} x_1 \\ x_2 \\ \vdots \\ x_{11} \end{vmatrix} + \begin{vmatrix} 0.963 \\ 0.885 \\ \vdots \\ 0.471 \end{vmatrix}^T \begin{vmatrix} x_1 \\ x_2 \\ \vdots \\ x_{11} \end{vmatrix} + 6.041 \quad (17)$$

4.3.2 *Calculating the predictive value for the indicators*

In this paper, the authors took "the events of Zhou Jiu-geng" as sample to test the feasibility about the model. And the information source covered 310 websites such as www.sina.com, www.163.com, www.mop.com and bbs.tianya.cn, and so on. Through data collection and statistics and unified indicators dimension using standardized method of Min-Max, the situation value about the public opinion can be calculated. And then depict the evolution path according to the values. Meanwhile, in order to show the effect of the model, the authors compared the data evolution path with "Network public opinion indicators evolution path about Zhou Jiu-geng events" depicted by IRI. As the standard about the indicators' value in this paper are different from the IRI public opinion indicators' measurement, the authors transferred the indicators value to centesimal system data with a view of helping comparing with IRI system. As a consequence, public opinion evolution path for "Zhou Jiu-geng events" about IRI and this paper are as shown in Figure 2.

It can be observed from Figure 2, the evolution tendency depicted in this paper is the similar to IRI, especially on the important turning point. Such as the date of December 19, the discipline inspection commission began investigating Zhou Jiu-geng under the pressure from netizens which made the network public opinion have a real impact on virtual society. In addition, the development curve

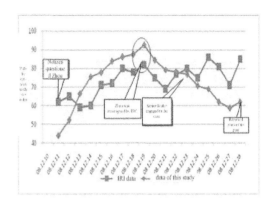

Figure 2. Evolution path about network public opinion of "Zhou Jiu-geng events".

described in this paper, which make the predictive about the overall trends of the public opinion, is smoother.

In order to test the effect of the model, given the situation about public opinion of "Zhou Jiu-geng events" is uncertain during December 18 to 19, taking the data from December 12 to 17 as a observe value to predict the public opinion trend and discriminate its warning level about December 18 and 19. Put the data of each indicator collected into equations (1) and (5), and part of the data can be expressed as Table 1.

4.3.3 *Calculating the discrimination function*

Putting predicted value into equations (15), (16) and (17), comparing the values of discrimination functions. If the value of $g_1(x)$ is maximum, the warning level of public opinion will be low-level. If the value of $g_2(x)$ is maximum which will refers to the medium-level. And $g_3(x)$ is maximum refers to the high-level. Through concrete calculating on this events, the level about the date of December 18 is $g_1(x) = 2.21$, $g_2(x) = 3.05$, $g_3(x) = 3.89$; and December 19 is $g_1(x) = 3.36$, $g_2(x) = 3.27$, $g_3(x) = 4.15$, which meant that the warning level about these two days are high-level and the influence became stronger and stronger. The situation fits for Figure 2 which indicate that the model is accurate.

4.3.4 *Tell the error rate about pattern recognition*

Putting the values obtained from the calculation above into equations (13) and (14) can acquire the error rate $P(e) = 1.391\%$. In this paper, the maximum error probability less than 5% is acceptable. Therefore, the conclusion about forecasting and early-warning of network public opinion according to the model mentioned above is credible.

5 CONCLUSION

In this paper, the network public opinion analysis index system was constructed. And the forecasting and early-warning model about network public opinion was established through the methods of grey forecasting and pattern recognition based on the analysis system. The authors tell the feasibility about the model taking "Zhou Jiu-geng events" as example. Three aspects need to be explained. (1) To simplify the processing, this paper defines the index variables as multivariate normal distribution. If which is evenly distributed, chi-square distributed or other function distributed, it can also be solved similarly as well. (2) As the classified rules and the accuracy about classified recognition during the course of early-warning depends to the choosing of training samples, the training sample number should be as much as possible, and the distribution should be as balance as possible and as representative as possible as well. (3) The purpose of forecasting and early-warning about network public opinion is better to predict, preview and control the crisis in advance. Therefore, it is very important to establish crisis plan base, knowledge base and related crisis management mechanism which should be the research direction after this paper.

ACKNOWLEDGEMENTS

This research work was supported by National Social Science Foundation of China Under Grant number 12CTQ046, the Philosophy and Social Science Foundation of Beijing Under Grant number 11 KDC0228, Humanities and Social Science Foundation of China Ministry of Education Under Grant number 12THCZG036 and the Financial Foundation of Beijing Under Grant number PXM2014_178214_000005.

REFERENCES

[1] Qing Wang, Ying Cheng, Nai-peng Cao et al. Construction about monitoring and early-warning index system of network public opinion[J]. Library and information service, 2011,55(8):54–57. *In Chinese.*

[2] Run-xi Zeng. Construction about early-warning index system for network public opinion of emergency[J]. The intelligence theory and practice, 2010,(1):77–80. *In Chinese.*

[3] Chen Lin, Bi-cheng Li, Jin Wang et al. Ealy-warning method about network public opinion based on fuzzy inference[J]. Journal of Information Engineering University, 2011,12(1):72–76. *In Chinese.*

[4] Eirinaki M., Vazirgiannis M. Web Mining for Personalization[J]. ACM Transactions on Internet Technology, 2003,3(1):12–13.

[5] Martens D., Bruynseels L., Baesens B., et al. Predicting going concern opinion with data mining[J]. Decision Support Systems, 2008,45(4):756–777.

[6] Ju-long Deng. Grey forecasting and decisioin[M]. Wuhan: Press of Huazhong university of Science and Technology, 2002:9. *In Chinese.*

[7] Zhao-qi Bian, Xue-gong Zhang. Pattern recognition[M]. Beijing: Tsinghua university press, 2000:1. *In Chinese.*

Information Technology and Applications – Li (Ed.)
© 2015 Taylor & Francis Group, London, ISBN 978-1-138-02677-3

The development of the anti-plagiarism software and its application in the teaching practice

Zhengheng Xie
Shanghai Jianqiao College, Shanghai, China

Wei Song
School of Business, Black Hills State University, USA

ABSTRACT: Academic dishonesty or cheating becomes a major issue which directly affects the quality of the learning outcomes and teaching process. This paper analyzes student plagiarism in a higher education setting. In order to detect the occurrence of the academic dishonesty, the anti-plagiarism software was developed to check the student project reports submission in the electronic formats. The software is able to scan and compare the attributes on electronic documents quickly and effectively. It can correctly identify plagiarists and form the investigation report. The software was used in the "Engineering Design Drawing Course", "MCU project", "3D Computer Drawing" etc. The software can identify the student who has committed plagiarism, in turn; it can improve the effectiveness of study and the quality of teaching. The software is available for use in other disciplines as well. It can standardize and optimize the quality of teaching curriculum designed to provide a solution against student plagiarism.

Keywords: anti-plagiarism; course design; digital fingerprint; MD5

1 INTRODUCTION

The purpose of student cheating is actually very simple, mostly in order to obtain better marks or be able to pass the exam. Plagiarism is a form of cheating behavior in academia. This behavior exists in entire campus. Number of cheating incidences has been escalated each year. It has also affected different disciplines in the schools. Especially in the large size class, due to the massive use of computers in the form of electronic document instead of original paper, the more serious plagiarism would be taken place. This is not only because of cheating on electronic documents more subtle than paper ones, but also this type of the cheating is much easier operated as long as the other students to do the work with U stick copy, then this person can change the file name and become legitimately to claim as his own work. However, to identify this kind of plagiarism, instructors' workload would be huge and it is really time-consuming. Therefore, many instructors could not catch all these incidences which left more academic dishonesty incidences unidentified.

Student plagiarism has many reasons. Some because of the uninteresting contents in learning or it is insignificant to learn. Also some of these contents are abstract or obsolete but the students must cope with the instructors. For these students, the objective of the cheating is to get the course passed while the objective of the instructors is to stop this unethical practice and build strong moral in the classroom.

2 ANTI-PLAGIARISM SOFTWARE DESIGN CONCEPT

As we all know, everyone has own unique fingerprint, which often becomes the identity of the offender. Similarly, any electronic document (regardless of its size, format, and quantity) also has an equally unique "digital fingerprint" which can be generated by the MD5 algorithm. If anyone made any changes on the electronic document, its corresponding MD5 value as a "digital fingerprint" will change. Conversely, if two electronic documents have identical MD5 value, they can be identified for the same file. This property is often used for security checks after transferring files to prevent them from being damaged or intentionally modified. Other properties of electronic documents such as file size (SIZE) and the Last Modification Time (LMT) can also be used to compare two files whether they are same. However, the conclusion is negative as two files with same SIZE or LMT

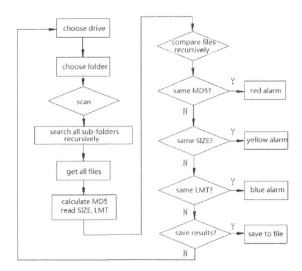

Figure 1. Anti- plagiarism software program block diagram.

cannot guarantee they are identical. But as the probability of occurrence of this scenario is very small, we can use it for further investigation.

The anti-plagiarism software identifies whether there is plagiarism cheating by comparing the MD5 value, SIZE and LMT of the electronic documents submitted by students. Its block diagram is shown in Figure 1.

3 IMPLEMENTATION OF ANTI-PLAGIARISM SOFTWARE AND ITS APPLICATION

Figure 2 shows the anti-plagiarism software user interface. The software is compiled into Microsoft ClickOnce installer which can be easily installed on other user machines for distribution. Prior to use all electronic documents submitted by students to be detected are put in a folder, run the program, the program interface appears as shown in Figure 2. Select the folder, and then click on "Scan Files" button, the program will search all sub-folders recursively and find all files with their path, file name, SIZE, LMT, and MD5 attributes which will be displayed in the list control of the form while the number of files found and the time elapsed will be shown on top of the list control. Click on "Compare Files" button, the program will conduct comparison among all files. If two files are of the same size, the background appears as yellow; if LMT are identical, the background appears blue; if MD5 values are the same, the background is displayed in red. Finally, click "Save" button, the document information with background color (i.e.

Figure 2. Anti-plagiarism software interface.

plagiarism is suspected of cheating) can be save to a text file for further confirmation.

The main code snippets and comments are as follows:

3.1 MD5 calculation subroutine

'Routine input file path variable, the output of 128 byte MD5 string

Public Function GetMD5Checksum (ByVal filePath As String) As String Dim md5 As MD5CryptoServiceProvider new MD5Crypto-toServiceProvider' Read the file in the file stream

Dim fs As FileStream

fs = New FileStream (filePath, FileMode.Open, FileAccess.Read, FileShare.Read, 8192)

'Calculate MD5 value

40

```
md5.ComputeHash (fs)
fs.Close ()
Dim hash As Byte () = md5.Hash
Dim sb As StringBuilder = New StringBuilder
Dim hByte As Byte
'Convert a hexadecimal string
For Each hByte In hash
    sb.Append (String.Format ("{0:   X2}",
hByte))
    Next
    Return sb.ToString ()
End Function
```

3.2 *Code for comparing of file attributes*

```
For i = 0 To List1.Items.Count - 1 'traverse all
records in list control
    For j = i + 1 To List1.Items.Count – 1
    Label1.Text = " being compared ..." + List1.
Items (j)
    SubItems (0) Text +.. "-" & List1.Items (i)
SubItems
(0) Text..
    If List1.Items (i). SubItems (1). Text = List1.
Items (j). SubItems (1). Text Then
    'If two files with the same size, the background
is shown as yellow
    List1.Items (i). BackColor = Color.Yellow
    List1.Items (j). BackColor = Color.Yellow
        End If
    If List1.Items (i). SubItems (2). Text = List1.
Items (j). SubItems (2). Text Then
    'If two file with identical LTM, the background
appears blue
            List1.Items (i). BackColor = Color.Blue
            List1.Items (j). BackColor = Color.Blue
        End If
    If List1.Items (i). SubItems (3). Text = List1.
Items (j). SubItems (3). Text Then
    'If two files with same MD5 values, the back-
ground is displayed in red
            List1.Items (i). BackColor = Color.Red
            List1.Items (j). BackColor = Color.Red
        End If
    Next
    Next
'Stop the clock, and when the display comparison
    stopwatch.Stop ()
    Label1.Text = " comparison with the end of
time:." + Stopwatch.Elapsed.ToString
    Button4.Text = "compare file attributes"
    stopwatch.Reset ()
    ...
```

3.3 *Code for save results*

```
For i As Integer = 0 To List1.Items.Count - 1 'LIST
control traverse all files
```

```
    If List1.Items (i). BackColor = Color.Red Then
'will write to the file with same MD5
    My.Computer.FileSystem.WriteAllText (Saved.
FileName, List1.Items (i). SubItems (0). Text +
ControlChars.Tab + List1.Items (i). SubItems
(1). Text + ControlChars.Tab + List1.Items (i).
SubItems (2). Text + ControlChars.Tab + List1.
Items (i). SubItems (3). Text + "(Md5)" + Control-
Chars.NewLine, True)
    ElseIf List1.Items (i). BackColor = Color.
Yellow Then 'will write to the file with same SIZE
information
    My.Computer.FileSystem.WriteAllText (Saved.
FileName, List1.Items (i). SubItems (0). Text +
ControlChars.Tab + List1.Items (i). SubItems
(1). Text + ControlChars.Tab + List1.Items (i).
SubItems (2). Text + ControlChars.Tab + List1.
Items (i). SubItems (3). Text + "(Size)" + Control-
Chars.NewLine, True)
    ElseIf List1.Items (i). BackColor = Color.
Blue Then 'will write to the file with same LMT
information
    My.Computer.FileSystem.WriteAllText (Saved.
FileName, List1.Items (i). SubItems (0). Text+
ControlChars.Tab + List1.Items (i). SubItems
(1). Text + ControlChars.Tab + List1.Items (i).
SubItems (2). Text + ControlChars.Tab + List1.
Items (i). SubItems (3). Text + "(Mod time)" +
ControlChars.NewLine, True)
    End If
    Next
    ...
```

3.4 *The effect of anti-plagiarism software*

We used anti-plagiarism software to check the archived electronic documents of last term's "Engineering Design Drawing" course, for a total of 50 students with a total of 773 files, Scan and Compare spend 7.87 seconds. We found almost every student had committed some degree of cheating which prompted us it is a very serious problem. The instructor in the current semester has deployed the anti-plagiarism software to detect in the classroom with a student demonstration turned quiz as a sample, who copied from others. The student admitted his wrong doing before the fact. Due to the concerns of plagiarism to be caught, each student finished his/her work on his/her own and ensured it was not copied from others. Thus, the students from Engineering Drawing Design Course generally do better because of the deployment of this software.

We also used this software in the National CAD Certification Exam held by Education Management Information Center of National Education Department with excellent results.

4 CONCLUSION

Study shows that the design and implementation of anti-plagiarism software can check student electronic document submissions to ensure they are not copied from each other in a timely fashion. Practice has proved that the software can effectively curb the occurrence of students' plagiarism when using the electronic document for their assignment. Hence, the overall quality of the learning and teaching has been strengthened.

REFERENCES

[1] Zhang Kangnian. VC++ in the use of the disk serial number encryption software [J]. Jiangxi Vocational and Technical College, 2009 (03), in Chinese.

[2] Wang Qianli. Application of MD5 encryption algorithm Based on C# technologies [J]. Xi'an Aero Technical College, 2010 (05), in Chinese.

[3] Pang Qining. A software registration key encryption algorithm [J]. Communications and TV Broadcasting, 2008 (02), in Chinese.

[4] Wang Gang. Research on MD5 encryption algorithm used in J2ME [J]. Software Guide, 2005 (16), in Chinese.

[5] Pan Qingfang. Design of using MD5 encrypted database system [J]. Yangtze University (Natural Science), 2006 (0), in Chinese.

[6] Che Zihui, Jing Shuangyan. Using of MD5 encryption algorithm in VFP [J]. Baoding Teachers College. 2006 (04), in Chinese.

[7] Zhang Guoxiang, Shu Yukun. Application and Research on IC card cash register management security information system [J]. Hubei Normal University (Natural Science Edition), 2008 (04), in Chinese.

[8] Tian Wenyan. MD5 encryption technology in ASP site user account [J]. Computer Knowledge and Technology (Academic Exchange), 2007 (01), in Chinese.

[9] Chang Yizhi, Zhao Yi, Tang Xiaobin. MD5 algorithm study [J]. Computer Science. 2008 (07), in Chinese.

[10] Zhang Shumin, Han Wenhong. Software encryption technology and its implementation [J]. Henan Agriculture 2009 (12), in Chinese.

[11] Simha, A.; Armstrong, J.; Albert, J. Attitudes and Behaviors of Academic Dishonesty and Cheating Do Ethics Education and Ethics Training Affect Either Attitudes or Behaviors? [J]. Journal of Business Ethics Education. Vol. 9 (2012) (1), 129–144.

Information Technology and Applications – Li (Ed.)
© 2015 Taylor & Francis Group, London, ISBN 978-1-138-02677-3

Analysis and design method of the Cyber-Physical Systems based on SCADE

Lichen Zhang & Tao Peng
Guangdong University of Technology, Guangzhou, Guangdong, China

ABSTRACT: Cyber-Physical Systems (CPS) is considered as the third information technology after computer and Internet. It integrates the computing system and physical world, and will change the way in which we interact with the physical world just like the Internet changed how we interacted with one another. Safety-Critical Application Development Environment (SCADE) is a high-security application development environment. This paper proposes about using the SCADE to analyze and design the CPS, and then from three aspects to illustrate the advantages of this operation. It can provide a reference for further research.

Keywords: CPS; SCADE; software design; high security; Cyber Physical Systems (CPS)

1 INTRODUCTION

1.1 *Structure of CPS*

CPS is a product of integration of computing process and the physical world, it is the next generation of intelligent system which implements the coordination and closely combine with the computing resources and physical resources. Academician He Jifeng pointed out that "CPS, broadly understood, is a controllable, credible and scalable networked physical device system, it deeply fuse computing, communications and control capabilities based on environmental perception. It using the feedback loop between calculation process and physical process to realize the depth of the fusion of real-time interactions to increase or extend new functions, in the form of safe, reliable, efficient and real-time monitoring or control of a physical entity. The ultimate goal is to realize the total integration of information world and the physical world, building a controllable, reliable, scalable and secure efficient CPS network, and eventually fundamentally changing the way of human build engineering physics system." [1].

The abstract structure of CPS is shown in Figure 1, figure illustrates the CPS implement the close contacts between physical world and the calculation process through the "3C" (Computer, Communication and Control). CPS use sensors embedded in physical device and the Wireless Sensor Network (WSN) to communicate with physical device and computer; Actuators control physical entity according to the calculated control command; decision control unit generate control

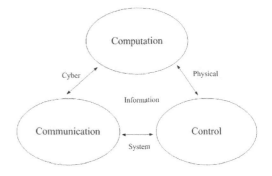

Figure 1. Abstract structure of CPS.

commands according to the input of sensors and the rules defined by user. CPS, with the integration of "3C", can monitor and control physical entity with a safety, real-time and reliable way. CPS not only emphasizes the physical sensing function, but also highlighted emphasizes control function. In essence, CPS is a network system with control attributes, and it contains wireless network in which traditional control rarely used, by the use of CPS, transmission, sharing, integration of information is more clear, and both open loop and closed loop controls are used [2].

1.2 *Characteristic of CPS*

CPS includes environmental perception ubiquitous in the future, embedded computing, network communication and network control systems engineering, and make the physical system to have calculation, communication, precise control, remote collaboration and autonomous five functions [3]. CPS is different from the traditional real-time embedded systems of some special properties:

1. Deep embedded: embedded sensors and actuators were deeply embedded in each of the physical components and may even be embedded into the material [4], it made the calculation becomes more common, computing process is closely connected with the physical world.
2. Complexity: CPS is commonly serving a large and complex system, with different types of devices and communication networks, and in time and space level, CPS also contains a high level of complexity.
3. High security: compared with the traditional software system, the system scale and complexity of the CPS put forward higher requirement for information system security [5].
4. Heterogeneity: CPS contains many subsystems with different functions and structures, each subsystem through wired or wireless network communication mode to coordinated work. Therefore, CPS is also called as the Systems of Systems [6].
5. Adaptability: As the future intelligent system, the CPS should have adaptive and the ability of re-configuration when the environment changes.

2 (SAFETY-CRITICAL APPLICATION DEVELOPMENT ENVIRONMENT) SCADE INTRODUCTION

2.1 Development process of SCADE

SCADE is a model-driven development environment, it was developed by Esterel Technologies Company. SCADE uses the concept of correct by construction, covering the whole process of the embedded software development from software requirements to the embedded code: requirements modeling, simulation, coverage analysis, formal verification and code generation [7]. Model-Driven Development (MDD) is used in SCADE to carry out software development. Through the graphical modeling to create a system model, then adopt coverage analysis and coverage analysis formal verification technology ensure the security of the system. After make sure the safety of the model, SCADE can automatically generate engineering oriented ANSI C language code, and the generated code can be directly applied to engineering practice without modification. The development process of SCADE is shown in Figure 2.

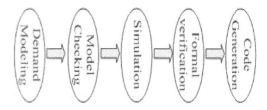

Figure 2. The development process of SCADE.

SCADE provides two graphical modeling mechanism: data flow graph and finite state machine, both are based on strict mathematical model and the mathematical semantics. It can guarantee the accuracy, completeness, consistency and no ambiguous of the design model [8]; formal verification is the use of strict mathematical reasoning to prove the correctness of the design and whether system is in line with the intention of the designers. Formal verification methods that SCADE provide use descriptive characteristics of the language LUSTRE to express the relevant features of the system; SCADE use of international quality certification standard D0178B code generator, the established model can be transformed to direct engineering oriented ANSI C code, simplifies the development mode of "coding" process. SCADE modeling mechanism is based on strict mathematical theory, it can create a clear and unambiguous model, SCADE code generator using the method of mathematical reasoning and automatically generate code from the model, ensuring the consistency of models and code [9].

2.2 Development feature of SCADE

Traditional software development process is a typical "V" type development process. It includes seven processes: software requirements, preliminary design, detailed design, coding, unit testing, integration testing and system testing. The entire software life cycle process is based on software code. Software requirements, preliminary design and detailed design are aim to carry out the coding; while unit testing, integration testing and system testing are to verify the correctness of the code. SCADE is based on the MDD, compared with traditional software development, SCADE has the following features:

1. SCADE based on strict mathematical theory, through rigorous modeling to requirements specification, cleared the ambiguity and vagueness of requirements in the early stages of system development;
2. Requirements document and design document can automatically generated by the model, the

document will adjust and update according to the change of the model;

3. Automatically generates the high-quality code, which can be applied directly to the project;
4. Formal verification techniques can find hidden error in software design and ensure the security of the software;
5. KCG can generate more than 50% code, the generated code is slightly smaller than the traditional hand-written code and operating efficiency is roughly equal, SCADE can greatly increase the efficiency of development, shorten the development cycle in designing high security software system.

3 RELATED WORK

3.1 *Reactive system and synchronous assuming*

Reactive system and synchronous assuming are the two basic concepts related with the model design. The reactive system refers to the constant interaction with the external environment of the system. Most real-time, embedded systems are reactive systems, the behavior of the reaction system can be abstracted as a constantly receive input and produce output of the black box. When the reactive system running, environment ready to provide new input and use output generated by the system and at the same time system can read and write to these inputs and outputs. Systems and environments constantly interacting, the behavior between system and its environments is concurrent rather than sequential, therefore concurrency is one of the fundamental characteristics of reactive system. Synchronous assumption is to assume that the reactive system has an infinite fast processing speed, namely the system response time input, and produces output and then wait for the next input at a minimum, indivisible particles. Due to the time of the particles is inseparable, therefore, in time particles of response to system input and generate output, system has no other changes, namely the system only responds to one input within a time particles. Synchronous assumption greatly simplifies the design of reactive system, suitable for many applications [10]. CPS is a highly autonomous intelligent system, it must have information interact between its environment, not only perceive the environment, but also affect the environment, Therefore, CPS is a reactive system, using SCADE to design CPS can simplify system design and improve efficiency.

3.2 *MDD*

SCADE using the model oriented software design method to analysis and design system. MDD is based on the model and its expression language, the model provides such a capability that can consistently display different views of the systems. The model can correctly describe software specifications and reflect the requirements of the software. Model can be quickly and quality translated or interpreted to executable object code. Compared with traditional software development methods, MMD has the following characteristics:

1. Faster: The application is designated as a higher level of abstraction model. Each element of the model (or other symbol pseudo programming language) may represent multiple lines of code. In this way, we can achieve more functionality at the same time.
2. Lower the costs: Using MDD developing software can use less engineers and non-professional to complete and ensure the software quality. In addition, use MDD to change is the development and maintenance projects can also save costs. In the aspect of maintenance, reading high abstract model of the behavior of the application is easier.
3. Improve quality: In the process of software development using MDD, the application uses advanced abstract model, while the model will be interpreted as the code by an engine, therefore the quality of the generated code is assured. Using SCADE to analysis and design CPS can make use of to the advantage of MMD.

3.3 *High security of SCADE*

SCADE is a high-security development environment which generally suitable for embedded applications in various fields software development, especially in aviation, aerospace, defense and other fields has been widely applied. SCADE modeling mechanism based on strict mathematical theory, in the initial stages of systems development it ruled out the system security risks. After establishing the model, we can use the Prover plug-in to carry out formal verification to test whether the model meets security. If the model is safe, it can give a security proof, if the model is unsafe, it can give a counter-example to help us perform error correction [11]. Formal verification can largely ensure the security of the target system. The code generated by SCADE has many security features: each variable assignment in its scope only once, no recursion, no endless loop, no dynamic pointer, and no dynamic memory allocation. SCADE provides all the mechanisms to ensure that the structure of the software with certainty, the user can not introduce uncertainty because SCADE compiler checks the software certainty, thus ensuring high security of systems. CPS general application some areas

45

required high security, such as health care facilities and assisted living, intelligent traffic control and safety, advanced automotive systems, energy reserves, environmental monitoring, defense systems, infrastructure construction and other fields, SCADE can fully guarantee the security requirements of CPS.

3.4 *Decision*

Using SCADE as a development environment to analyses and design the CPS can satisfy the characteristics of the CPS, and can also give full play to the characteristics of SCADE, combination of these two has a great advantage. CPS is the next generation of intelligent system after embedded system and SCADE covers the whole process of embedded system development. SCADE can also cover the overall process of development of the CPS. The advantages of this method are: improve the development efficiency of CPS, reduce development costs, and meet the high security requirements of CPS.

REFERENCES

[1] He Jifeng. Cyber-physical systems [J]. Communication of the China Computer Federation, 2010, 6(1): 25–29.

[2] Peng Yu. The development of networking technology and its direction of industrial applications[J]. Automation Instrumentation, 2011, 32(1):1–12.

[3] Halbwachs, N. A synchronous language at work: the story of lustre Proceedings, Third ACM & IEEE International Conference on Formal Methods and Models for Co-Design (IEEE Cat. No. 05EX1093), 2005:3–11.

[4] Kim J.E., Mosse D. Generic Framework for Design, Modeling and Simulation of Cyber Physical Systems[C]//Position Paper for NSF Workshop on Cyber-Physical Systems. 2006.

[5] Akella R., McMillin B.M. Model-checking BNDC propertiesin Cyber-physical systems[C]//Proceedings of International Computer Software and Applications Conference. 2009:660–663.

[6] Woo H., Yi Jian-liang. A Simulation Framework for PSoC Based Cyber Physical Systems[C]//Proceedings of International Conference on Distributed Computing Systems. 2008:525–528.

[7] Ouassila. Mode—Automata Based Methodology for Scade [Z]. Lecture Notes in Computer Science, 2005(3414):386–401.

[8] Peter Amey. Correctness by Construction: better can also be cheaper. Journal of Defense Software Engineering. 2002.3:24–30.

[9] Amar Bouali, Bernard Dion, Kosuke Konishi, Using Formal Verification in Real-time Embedded Software Development, JSAE Annual Congress, 2005:5430–5434.

[10] Ouassila Labbani, Jean-Luc Dekeyser, Pierre Boulet. Mode—Automata Based Methodology For SCADE. Hybrid Systems Computation and Control-8th International Workshop. HSCC 2005.2005: 386–400.

[11] Amar Bouali, Bernard Dion, Formal Verification for Model-Based Development, Esterel Technologies, Paper Number 05 AE-235.

Information Technology and Applications – Li (Ed.)
© 2015 Taylor & Francis Group, London, ISBN 978-1-138-02677-3

Performance of incremental-best-relay technique for two-way relaying systems

Jing Guo, Changxing Pei & Hong Yang
Department of State Key Laboratory of Integrated Services Networks, Xidian University, Xi'an, China

ABSTRACT: This letter investigates incremental-best-relay two-way Amplify-and-Forward (AF) relaying system. We studied the end-to-end performance of the incremental-best-relay cooperative-diversity networks over independent non-identical Rayleigh fading channels. Closed-form expressions for the lower bound of outage probability, asymptotic outage behavior and the finite-signal-to-noise ratio Diversity-Multiplexing Trade-off (DMT) are determined. Results show that the considered scheme not only realizes full diversity order (K+1), but also achieves improved expected spectral efficiency.

Keywords: two-way relaying network; Amplify-and-Forward (AF); incremental-best-relay; outage probability; Diversity-Multiplexing Trade off (DMT)

1 INTRODUCTION

Two-way relaying system has attracted much research interest because of its potential in achieving higher spectral efficiency and throughput for wireless networks [1]. Many efficient protocols have been proposed for two-way relay networks. For example, for Amplify-and-Forward (AF) based relaying in bidirectional networks, there are two major protocols, namely, Analog Network Coding (ANC) protocol and Time Division Broadcast (TDBC) protocol [2]–[4]. Under a half-duplex constraint, although using two time slot, the ANC protocol achieves diversity order just one, because it cannot utilize the direct-path between the two sources even if such direct-path physically exists. In contrast, TDBC can efficiently utilize the direct link but has the need of three time slot transmission, which yields considerable loss of spectral efficiency. Up to the date of the literature, there is little attention found on the incremental relaying protocol. Further to increase the spectral efficiency of the cooperative-diversity networks, the TDBC protocol considering the utilization of the direct-path for AF-based relaying network has motivated our work.

In this letter, we focus on a two-way AF relaying system, where two sources exchange information with the help of K cooperating AF relay nodes, as recently proposed by [5] and [6]. An incremental-best-relay two-way relaying scheme is proposed and its outage probability is analyzed. We derive a tight lower bound of the outage probability of the proposed protocol in closed form. The bound is very close to the exact outage probability obtained by simulation, irrespective of the values of channel variances. Asymptotic outage behavior is investigated in high Signal-to-Noise Ratio (SNR) regime. Moreover, useful insights into practical two-way AF relaying system are obtained by quantifying the finite-SNR Diversity-Multiplexing Trade-off (DMT).

2 SYSTEM MODEL

Consider a two-way AF relaying system, where two sources S_1 and S_2 exchange information with the help of K cooperating AF relay nodes R_k, ($k = 1, \ldots, K$). All terminals are single-antenna devices and operate in a half-duplex mode. We assume the direct link between S_1 and S_2 physically exist and can be utilized to transmit information. The channels associated with each link are independent but not necessarily identically distributed (i.n.i.d.) flat Rayleigh fading. Letting h_0, h_1 and h_2 denote the channel fading coefficient of $S_1 \leftrightarrow S_2$, $S_1 \leftrightarrow R_k$ and $S_2 \leftrightarrow R_k$, the channel gains associated with these links conform to exponential distributions with σ_0^2, σ_{1k}^2 and σ_{2k}^2, respectively. We assume that each terminal possesses the same transmit power P. We also assume here the Additive White Gaussian Noise (AWGN) at all nodes is independent and identically distributed (i.i.d.) with zero mean and unit variance.

A Time-Division Multiple Access (TDMA) scheme is employed for orthogonal channel access. In the first time slot, S_1 broadcasts its signal to all the K relays and S_2. In the second time slot, S_2 broadcasts its signal to all the K relays and S_1. Once the direct transmissions between S_1 and S_2 complete, both of them try to decode their respective message. Recalling that the channel is reciprocal and the assumption that transmit power and data rates are the same for the two sources, $S_1 \rightarrow S_2$ and $S_1 \leftarrow S_2$ links therefore have the same state, i.e., they are either successful or failed in transmission. Thus, if S_1 can decode the information from S_2 correctly, it sends a one-bit Acknowledgement (ACK), indicating the success of transmission, to S_2 and all the K relays. Otherwise, it sends back a one-bit Non-Acknowledgement (NACK), indicating the failure of transmission. It is assumed that the ACK and NACK can be received successfully by the corresponding nodes and consume negligible time for transmission. If relays receive ACK, a new information exchange process will start in the next time slot. Otherwise, in the third time slot, the selected relay R_k will forward a combined signal to both end sources S_1 and S_2. After S_1 and S_2 perform maximal ratio combining at the end of the third time slot, the combined SNRs at S_1 and S_2 can be formulated by following the similar derivation procedure of [3], respectively,

$$\gamma_{1k} \approx P|h_0|^2 + \frac{(1-\xi)P|h_{1k}|^2|h_{2k}|^2}{(1+\xi)|h_{2k}|^2+(1-\xi)|h_{1k}|^2}, \qquad (1)$$

$$\gamma_{2k} \approx P|h_0|^2 + \frac{\xi P|h_{1k}|^2|h_{2k}|^2}{(2-\xi)|h_{1k}|^2+\xi|h_{2k}|^2}, \qquad (2)$$

where ξ indicates the power allocation factor at relay for the normalized received signals from S_1 and S_2, $0 < \xi < 1$.

Usually, the optimal relay is chosen to maximize the minimum of the end-to-end SNRs, so the conventional optimal relay selection criterion can be expressed as follows:

$$k^* = arg \max_{k \in \{1, ..., K\}} \min \{\gamma_{1k}, \gamma_{2k}\}.$$

The mutual information of the direct link transmission (at the end of the second time slot) can be expressed as $I_0 = 1/2\log(1+P|h_0|^2)$, whereas the maximum mutual information for the two unidirectional relayed transmission (at the end of the third time slot) can be expressed as $I_{1k^*} = 1/3\log(1+\gamma_{1k^*})$ and $I_{2k^*} = 1/3\log(1+\gamma_{2k^*})$.

3 PERFORMANCE ANALYSIS

3.1 Outage probability

We assume that each block of transmission includes r bits information and each time slot has a

length of t seconds. The two sources thus have $r/2$ bits information to transmit to each other in one block. Define $R_s = r/(tB)$ bits/s/Hz as the baseline data rate, where B is the bandwidth of the occupied channel.

An outage event occurs if neither the direct transmission nor the relaying transmission succeeds. Using the concepts of probability theory, the outage probability of the proposed scheme can be expressed as follows:

$$P_{out} = \mathbf{Pr}\left(I_0 < \frac{R_s}{2}, \min\{I_{1k^*}, I_{2k^*}\} < \frac{R_s}{3}\right), \qquad (3)$$

It is very difficult to obtain the exact expression of P_{out}. In this letter, we derive a tight lower bound.

By using the inequality $xy/(x+y) \leq \min(x,y)$, γ_{1k^*} and γ_{2k^*} can be upper bounded by the following equations:

$$\gamma_{1k^*} \leq P\left(|h_0|^2 + \min\left\{\frac{1-\xi}{1+\xi}|h_{1k^*}|^2, |h_{2k^*}|^2\right\}\right), \qquad (4)$$

$$\gamma_{2k^*} \leq P\left(|h_0|^2 + \min\left\{\frac{\xi}{2-\xi}|h_{2k^*}|^2, |h_{1k^*}|^2\right\}\right), \qquad (5)$$

From above section, the lower bound for the outage probability can be derived in closed-form as follows:

$$P_{out} \geq \tilde{P}_{out}$$
$$= \mathbf{Pr}\left(\max_k \left\{|h_0|^2 + \min\left\{a|h_{1k}|^2, b|h_{2k}|^2\right\}\right\}\right.$$
$$\left. < \frac{2^{R_s}-1}{P} \triangleq \frac{\rho}{P}\right) \qquad (6a)$$
$$= \mathbf{Pr}\left(\max_k \min\left\{a|h_{1k}|^2, b|h_{2k}|^2\right\}\right.$$
$$\left. < \frac{\rho}{P} - |h_0|^2\right)$$
$$= \int_0^1 \frac{\rho}{P\sigma_0^2} e^{-\frac{\rho(1-x')}{P\sigma_0^2}} \mathbf{Pr}\left(\max_k \min\left\{a|h_{1k}|^2, b|h_{2k}|^2\right\}\right.$$
$$\left. < \frac{\rho x'}{P}\right) dx' \qquad (6b)$$

where $a = \min\left(\frac{1-\xi}{1+\xi}, 1\right)$ and $b = \min\left(\frac{\xi}{2-\xi}, 1\right)$.

By substituting (4) and (5) into (3), (6a) is obtained. Since $|h_0|^2$ is exponentially distributed variable with parameter σ_0^2 and letting $x = \frac{\rho}{P}(1-x')$, we have (6b)

$$J = \Pr\left(\max_k \min\left\{a|h_{1k}|^2, b|h_{2k}|^2\right\} < \frac{\rho x'}{P}\right)$$

$$= \prod_{k=1}^{K}\left[1 - \Pr\left(a|h_{1k}|^2 > \frac{\rho x'}{P}\right)\Pr\left(b|h_{2k}|^2 > \frac{\rho x'}{P}\right)\right]$$

$$= \prod_{k=1}^{K}\left[1 - e^{-\left(\frac{1}{a\sigma_{1k}^2} + \frac{1}{b\sigma_{2k}^2}\right)\frac{\rho x'}{P}}\right]$$

$$= 1 + \sum_{i=1}^{K}(-1)^i \sum_{k_1=1}^{K-i+1}\sum_{k_2=k_1+1}^{K-i+2}\cdots\sum_{k_i=k_{i-1}+1}^{K}\prod_{j=1}^{i} e^{-\left(\frac{1}{a\sigma_{1k_j}^2} + \frac{1}{b\sigma_{2k_j}^2}\right)\frac{\rho x'}{P}}$$

$$(7)$$

The desired result $\tilde{P}_{out}$ can then be calculated by substituting (7) into (6), we get

$$\tilde{P}_{out} = 1 - e^{-\frac{\rho}{P\sigma_0^2}} + \frac{1}{\sigma_0^2}e^{-\frac{\rho}{P\sigma_0^2}}\sum_{i=1}^{K}(-1)^i\sum_{k_1=1}^{K-i+1}\sum_{k_2=k_1+1}^{K-i+2}\cdots\sum_{k_i=k_{i-1}+1}^{K}\psi_i,$$

$$(8)$$

where

$$\psi_i = \begin{cases} \dfrac{\rho}{P}, & \phi_i = 0 \\[2ex] \dfrac{1}{\phi_i}\left[1 - e^{-\frac{\rho\phi_i}{P}}\right], & \phi_i \neq 0 \end{cases}$$

and $\phi_i = \displaystyle\sum_{j=1}^{i}\left(\frac{1}{a\sigma_{1k_j}^2} + \frac{1}{b\sigma_{2k_j}^2}\right) - \frac{1}{\sigma_0^2}$.

3.2 Asymptotic analysis

Now, based on the preceding results, we perform asymptotic analysis of the outage probability in high SNR regime. In this letter, since the noise is assumed to have unit variance and the gains of the channels associated with each link can be different, we simply assume that $SNR = P$. For sufficiently high SNR, by using the fact that $e^x = 1 - \sum_{k=1}^{K} x^k/k!$, $J = \prod_{k=1}^{K}\left(1/a\sigma_{1k}^2 + 1/b\sigma_{2k}^2\right)\frac{\rho x'}{P}$, we have

$$\tilde{P}_{out} \approx \int_0^1 \frac{\rho}{P\sigma_0^2}\prod_{k=1}^{K}\left(\frac{1}{a\sigma_{1k}^2} + \frac{1}{b\sigma_{2k}^2}\right)\frac{\rho x'}{P}dx$$

$$= \frac{1}{(K+1)\sigma_0^2}\prod_{k=1}^{K}\left(\frac{1}{a\sigma_{1k}^2} + \frac{1}{b\sigma_{2k}^2}\right)\left(\frac{\rho}{P}\right)^{K+1}, \quad (9)$$

From (9), it can be observed that $\tilde{P}_{out}$ decays proportional to $SNR^{-(K+1)}$, which indicates that the considered scheme achieves the diversity order of $K+1$. This means that the diversity order increases linearly with the number of relays although we use one relay only.

3.3 DMT analysis

In this section, we present a no asymptotic, finite-SNR DMT for realistic propagation conditions. The proportionality constant, which can be interpreted as a finite-SNR spatial multiplexing gain, dictates the sensitivity of the rate adaptation policy to SNR. From [7], for given $R_s = r\log_2(1+P)$, the finite-SNR DMT can be derived as follows:

$$d(r,P) = -\frac{P}{\tilde{P}_{out}(r,P)}\frac{\partial \tilde{P}_{out}(r,P)}{\partial P}$$

$$= \frac{r(1+P)^{r-1}P - (1+P)^r + 1}{P\sigma_0^2\tilde{P}_{out}(r,P)}e^{-\frac{(1+P)^r-1}{P\sigma_0^2}}$$

$$\times\left[1 + \sum_{i=1}^{K}(-1)^i\sum_{k_1=1}^{K-i+1}\sum_{k_2=k_1+1}^{K-i+2}\cdots\sum_{k_i=k_{i-1}+1}^{K}\right.$$

$$\times\left.\left(\frac{\psi_i}{\sigma_0^2} + e^{-\left(\sum_{j=1}^{i}\left(\frac{1}{a\sigma_{1k_j}^2} + \frac{1}{b\sigma_{2k_j}^2}\right) - \frac{1}{\sigma_0^2}\right)\frac{(1+P)^r-1}{P}}\right)\right],$$

$$(10)$$

Now, we investigate the asymptotic DMT of the proposed system. As $P \to \infty$, from (10) and the definition of DMT, we can arrive at $d(r,P) = (K+1)(1-r)$, $0 \leq r \leq 1$. It is observed that the full diversity order, $K+1$, can be achieved by the proposed scheme. The maximum multiplexing gain of the proposed scheme is unity, whereas the counterpart of TDBC is only 2/3. This indicates that, by using incremental relaying into typical TDBC, we can achieve the full diversity order and an improved spectral-efficiency.

4 NUMERICAL RESULTS

In this section, we demonstrate the performance of the proposed scheme through numerical simulation and Monte Carlo simulations. For illustration purposes, we assume i.i.d. fading statistics for the links $S_1 \leftrightarrow R_k$ and $S_2 \leftrightarrow R_k$, respectively (i.e., $\sigma_{1k}^2 = \sigma_1^2$ and $\sigma_{2k}^2 = \sigma_2^2$, for $k = 1, ..., K$). And also we assume that $\xi = 0.5$.

Figure 1 depicts the outage performance versus system SNR (P) for the proposed scheme. First of all, an excellent agreement between the analytical results and simulations is observed. This confirms the accuracy of our analysis. The outage probability approximations are accurate for large P. This confirms the validity of our analysis in diversity order and array gain. With an increase of K, the outage performance is improved in the form of decreased outage probability and increased system

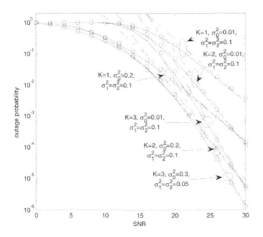

Figure 1. Outage probability against SNR P.

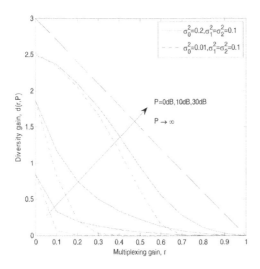

Figure 2. Finite-SNR DMT performance of the proposed scheme for different SNRs P ($K = 2$).

diversity order, as expected. In addition, for a given value of K, the outage behavior improves significantly when the average channel quality of the direct links becomes better, even if the average channel quality of the relaying links tends to be worse. Moreover, the outage performance improves significantly when the average channel quality of the direct links becomes better with the less number of relay. This phenomenon highlights the importance of the direct links and the necessity of exploiting it.

Figure 2 plots the estimates of finite-SNR DMT with different parameter settings. Here,

it is observed that the discrepancy between the finite-SNR DMT and the asymptotic DMT can be significant, particularly for low multiplexing gains. This clearly shows that the asymptotic DMT may overestimate the achievable diversity in practice. In the figure, it is also observed that the DMT performance of $\sigma_0^2 = 0.2, \sigma_1^2 = \sigma_2^2 = 0.1$ is better than that of $\sigma_0^2 = 0.01, \sigma_1^2 = \sigma_2^2 = 0.1$, for any given multiplexing gain r. This is because the proposed scheme could fully utilize the ACK feedback to improve the spectral efficiency when the average channel quality of the direct links becomes better.

5 CONCLUSIONS

In this letter, we have studied the AF-based incremental-best-relay protocol involving the direct-path between two sources for two-way relaying network. We have first derived a tight lower bound of the outage probability of the proposed scheme in closed-form when each terminal consumed the same transmission energy. Then, based on the lower bound, we have obtained the asymptotic outage behavior and the finite-SNR DMT of the proposed scheme. Results show that the proposed scheme can achieve significant spatial diversity, which highlights the importance of the direct links and the necessity of exploiting it.

REFERENCES

[1] B. Rankov, A. Wittneben., Spectral efficient protocols for half-duplex fading relay channels. *IEEE J. Select. Areas Commun.*, 25(2): 379–389, 2007.

[2] M. Ju, I.-M. Kim. Relay selection with ANC and TDBC protocol in bidirectional relay networks. *IEEE Trans. Commun.*, 58(12): 3500–3511, 2010.

[3] Z. Yi, M. Ju, I.-M. Kim. Outage probability and optimum combining for time division broadcast protocol. *IEEE Trans. Wireless Commun.*, 10(5): 1362–1367, 2011.

[4] Raymond H.Y. Louie, Yonghui Li, Branka Vucetic. Practical Physical Layer Network Coding for Two-Way Relay Channels: Performance Analysis and Comparison. *IEEE Trans. Wireless Commun.*, 9(2): 764–777, 2010.

[5] Lingyang Song. Relay Selection for Two-Way Relaying With Amplify-and-Forward Protocols. *IEEE Trans. Veh. Technol.*, 60(4): 1954–1959, 2011.

[6] E.Y. Li, S.Z. Yang. Simple relay selection criterion for general two-way opportunistic relaying networks. *Electron. Lett.*, 48(14): 881–882, 2012.

[7] R. Narasimhan. Finite-SNR diversity-multiplexing tradeoff for correlated Rayleigh and Rician MIMO channels. *IEEE Trans. Inf. Theory*, 52: 3965–3979, 2006.

Information Technology and Applications – Li (Ed.)
© 2015 Taylor & Francis Group, London, ISBN 978-1-138-02677-3

The explore of teaching model in mobile learning environment

Xiao-hui Guan
Department of Computer and Information Engineering, Zhejiang University of Water Resources and Electric Power, Hangzhou, Zhejiang, China

Ya-guan Qian
College of Science, Zhejiang University of Science and Technology, Hangzhou, Zhejiang, China

ABSTRACT: With the development of mobile networks and the popularity of intelligent mobile terminals, mobile learning enters into our vision. This paper firstly analyzes the characteristics of mobile learning and then proposes a new teaching model suitable for mobile learning environment. Then some measures are proposed to reform the existing teaching model.

Keywords: mobile learning; intelligent terminal; teaching model

1 INTRODUCTION

With the development of wireless network and the popularity of mobile terminals, such as ipad, iphone, etc., mobile learning, a new learning method, is entering into our vision. Mobile learning is a new learning method using digital mobile devices in wireless communication environment. It can be implemented anytime anywhere. The learning time is scattered. It can be very short or very long. The learning place is not limited in classroom, library and other fixed places. Now, students begin to pay more and more attention to their mobile phone. They prefer mobile phones to computers. So we should make full use of wireless network and mobile terminals to build a new teaching environment and introduce it into the traditional teaching system. We can customize special teaching and learning environment for personalized needs of students.

2 THE CHARACTERISTICS OF MOBILE LEARNING

Mobile learning is a new learning method by means of portable wireless network devices. It is out of control of time and place. It has unique features:

2.1 *It is informal and transient*

Mostly the mobile learner acquires knowledge by using the sporadic free time. The time span is short and irregular.

2.2 *Learning device is portable and movable*

In general, learning devices include smartphone, ipad and other mobile terminals. They have small body and light weight and easy to carry. The learning process is conducted by using wireless network to connect course platform. Because the mobile device has small screen size and storage capacity, we should avoid these drawbacks in teaching process.

2.3 *Learning object is young students*

The objects of mobile learning almost are young students. They like to pursue fashion and novel things. In order to attract the attention of students, some interaction segment should be integrated in the teaching process.

2.4 *It is autonomous*

The mobile learning is different from centralized teaching. It needs students to study consciously and actively. The teacher should shift from the transmitters of knowledge to the assistant of students, and then become the partner or collaborator.

3 THE TEACHING MODE

According to the characteristics of mobile learning, we design a suitable teaching mode for computer courses. The new mode makes full use of the advantages of wireless mobile technology and wireless network technology to meets the learner needs. The mode includes four parts of course analysis, course resources design and development, teaching organization and teaching evaluation.

3.1 Course analysis

The explicit course content with reasonable structure is the basis to achieve teaching objectives. So course analysis should be implemented before course design. The following content should be confirmed:

3.1.1 Making a teaching plan
We firstly analyze the status and role of course in curriculum system, formulate the teaching goal of course, and then constitute the idea and framework according to the goal.

3.1.2 Confirming the teaching content
According to the teaching goal, we confirm the teaching content. And then we divide the content into several relatively independent modules, and designate the key points and difficult points and their orders.

3.1.3 Planning the style to show the teaching content
In general, we use multimedia, video, animation and other style which youth like. We also design some theme discussions and surveys to make professional knowledge to be easily accepted and make classroom atmosphere active.

3.1.4 Designing a small project basing on teaching content
The inquiry learning from small projects is a suitable way for mobile teaching. We allow students to participate in small-scale projects to know well the knowledge by group collaboration.

3.2 Designing and developing teaching resources

The mobile learning resources are a series of digital resources which are customized for mobile learning. They must reflect teaching content and achieve teaching goal. The teaching resources should provide servers for teaching strategies. The following principles should be followed:

3.2.1 The teaching unit should be small
Because students study anytime anywhere, the learning environment is complex and the learning process maybe be disturbed and break at any time. So teaching unit should be small.

3.2.2 The knowledge should be encapsulated well
The correlation of different units should be little. The students can achieve some teaching goals only

by an individual unit without referencing other resources. The teaching designer should elaborate teaching content from different perspectives. It can make learners understand knowledge points in multi-aspects.

3.2.3 The learning resources should be diversity
We should use various forms to express the teaching content, such as text, audio, video flash and other animation. The learner can select suitable resource according to the environment. We also can combine several forms to make noise immunity.

3.2.4 The interface should be friendly and simple
Because mobile phone has small screen, the interface of learning platform should be simple, intuitive and easy to operate. The function of platform should include chatting room and real-time communication. They can provide one-stop teaching environment for students and teachers to interact each other in real-time.

As an important resource, the mobile teaching platform offers an efficient and practical entrance to study anytime and anywhere. So the platform should have the following function modules (Fig. 1).

3.3 Teaching organization and implementation

The teaching process is implementing a series of activities, such as resolving inquiry question, collaborative team, group discussions and FAQs. The purpose is creating specific learning contexts for students and teachers to strengthen the interaction of them.

The teaching activities are related to course objective and teaching content. They reflect specific knowledge points. The following is a teaching

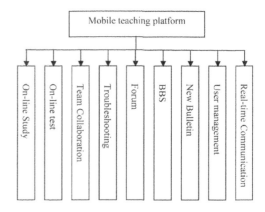

Figure 1. The function modules of mobile teaching platform.

52

strategy which exploratory method is used to resolve problems.

- Teacher creates a new, open and uncertain problem situation to stimulate and sustain students' attention.
- Teacher brings forward a question related with course content. He properly gives some hints and reference materials and encourages students to find relevant information.
- Students then discuss the questions they met by BBS or on-line chatting room. The teamwork activities, such as discussions, role playing, collaboration, are proper for mobile learning.

3.4 *Teaching evaluation*

Teaching evaluation includes two parts. One is evaluating students, which can measure how much knowledge students have by mobile learning. Through the evaluation, teachers can have a comprehensive understanding of students. At the same time students can look back at what they did. Two is evaluating the teaching process, teaching result and teaching resources.

When evaluating students we should focus on the teaching process. We need collect the amount of students browsing course content and course resource, whether students are active to speak, how many materials they devote, whether they answer questions, whether they finish their homework, and result of periodic examination, etc. These data becomes the evidence of evaluations for students. The evaluation system pays close attention to developing many abilities of students. It can help students get a thorough understanding of them and build self-confidence.

Evaluating the teaching process is determining whether the teaching content is proper, whether the teaching content can achieve the teaching goal. It also considers the consistency of teaching content and representation form and whether they match different mobile devices. Evaluating learning resources is also an important task. It focus on whether the knowledge structure is clear, whether the operation is fast and efficient, whether the interface is friendly, whether the layout is reasonable and whether the system is stable and have high efficiency. Through the evaluation teacher can adjust the instructional strategies for better achievement.

4 REFORM THE EXISTING TEACHING MODEL

Currently, we adopt the "teaching learning and doing" teaching model in most courses. We should integrate the above model into the existing teaching model. It includes the following contents:

4.1 *Building a comprehensive teaching guidance system by perfecting the existing teaching model*

Currently in most courses we have implemented the "teaching, learning and doing" teaching model. It greatly improves the teaching quality. As students is beginning got absorbed in mobile terminals, we should draw the interest of students from games through decentralized and situational teaching. So we integrate the mobile teaching into the existing teaching model to give students a comprehensive guidance in daily life.

4.2 *Reorganizing existing digital curriculum resources and constructing curriculum resources suitable for mobile learning*

In our college, most courses have many digital resources. But they are only applied in classroom teaching. With the implementation of mobile teaching, we should transfer these resources to wireless network. Because of the restrictions of terminal devices we should extract and recombine existing resources to build diverse resources with scattered knowledge. They are very easy to operate and suitable for mobile learning. With the development of cloud computing, we also can transplant the resources into cloud.

4.3 *Constructing mobile learning platform supporting 3G technology for each course*

In order to promote and facilitate mobile learning, we should build mobile learning platform for each course. It puts the spare time of students together and becomes a "pocket classroom" to let students learn at any time and in any place.

Also the platform needs to provide functions for convenient management and evaluation. Teachers and students can interact with each other through the platform.

5 CONCLUSION

With the development of mobile Internet and the popularity of intelligent mobile terminal, mobile teaching is becoming an important part of modern teaching model. The students can study anytime anywhere. In this paper, we firstly analyzed the character of mobile learning and then constructed a teaching model for mobile learning. Finally, we reformed our existing teaching model to fit for mobile learning.

ACKNOWLEDGMENT

This work is supported by the Project of national education information technology. NO (136241136).

REFERENCES

Cui Jian. Research on mobile teaching model [D]. University Of Hebei, 2009.

Ju Chunhua, Peng Jianliang. Research on the fusion of teaching and practice basing on mobile learning – The electronic commerce and logistics as an example. Chinese higher education research, 2012(09).

Yu Yuanbo, Chen Mingrui, Li Xiaoling. The teaching process management of computer course in mobile learning environment[J]. Journal of Hainan University, 2013(02).

Information Technology and Applications – Li (Ed.)
© 2015 Taylor & Francis Group, London, ISBN 978-1-138-02677-3

Empirical study on the reasons of IPO underpricing

Ping Zhang

Business School, Hohai University, Nanjing, China

ABSTRACT: The phenomenon of IPO underpricing prevalently exists in the worldwide stock market. It is found that the IPO underpricing in China's stock market is the most severe around the world. Small and medium-sized enterprises board occupies a very important position in the capital market system, and there are unique reasons for the underpricing phenomenon. This paper selects initial public offering stocks in small and medium-sized market (2006 to 2013) as samples, analyzing factors that influence IPO underpricing, degree of influence, and reasons account for such phenomenon.

Keywords: small and medium-sized enterprises board; pyramid hierarchy; IPO underpricing

1 INTRODUCTION

IPO Underpricing is the increase in stock value from the initial offering price to the first-day closing price. IPO underpricing is always shown in the great increase of new stock on the first day of listing. Researches find that such phenomenon prevalently exists in the stock market around the world, in which mature markets are lower than emerging markets [2]. China has the highest underpricing rate in the world, especially in small and medium-sized market [3]. Although the IPO underpricing brings considerable excess return to some investors, this phenomenon contradicts market rules. In the long run, it is not conducive to the healthy development of capital market and would even distort the concept of correct investment, resulting in the widespread of speculation.

2 LITERATURE REVIEW AND THEORETICAL ANALYSIS

Asymmetric Information theory: The asymmetric Information in IPO underpricing is mainly manifested in the information asymmetry between issuers and investors, investors and investors, as well as issuers and underwriters. Compared with investors, issuers have more information. To obtain such necessary information, investors need to spend more time and costs for the search of authentic information. Issuers need to lower their issuing price to compensate for the these searching costs of investors [6]. With regard to issuers and underwriters, underwriters have more information about the market and know market demands better. At the same time, underwriters will perform underpricing for successful issuance and thus more profits [7]. In addition, investors can be further divided into information possessor and non-information possessor, in which information asymmetry exists. Information possessors will purchase undervalued stocks and sell overvalued stocks to investors who don't have information superiority. This situation is called as "Winner's Curse" [1].

3 THE RESEARCH HYPOTHESIS

China's stock market suspended the issuance of new stocks during the implementation of equity division reform. IPO was restarted in June, 2006, mainly concentrating on small and medium-sized market in Shenzhen. As a result, the selected data are from IPO companies in small and medium-sized board market from 2006 to 2013.

3.1 *Sample selection and data sources*

This paper selects all the listing IPO corporation from 2006 to 2013 in small and medium-sized board market in Shenzhen, removing parts of incomplete data as well as companies that do not meet the requirements. After screening, 583 data are finally obtained.

All the financial data involved in this paper are from Guotai'an database, in which Pyramid hierarchical data are calculated from the "company, shareholders and actual controllers" section in the IPO announcement of companies.

3.2 Definitions and measurements of variables

1. Explained variables—IPO underpricing.
   In this paper, excess return rate represents the extent to which IPO pricing has been underestimated. The formula to calculate excess return rate is:

   $$AR = (P1-P0)/P0 \qquad (1)$$

   AR stands for the excess return rate on the first day of listing (underpricing rate of issuance). P1 refers to the closing price on the first day of listing. P0 means the issuing price.

2. Explanatory variables

   1. Pyramid hierarchy. Pyramid hierarchy is the number of company levels between the listed corporation and the ultimate controller [2].
      Hypothesis 1: Pyramid hierarchy and underpricing are negatively related.
   2. Hypothesis 2: Issuing size and underpricing are negatively related.
   3. Hypothesis 3: Issuance Price-earnings ratio and underpricing rate are positively related.

3.3 Assumption of model

In this paper, Multiple Regression Model is used to test various factors that impact the IPO underpricing in China's small and medium-sized stock market. The regression model is as follows:

$$AR = \beta0 + \beta1 \text{ LEVEL} + \beta2 \text{ SIZE} + \beta3 \text{ IPO PE} + \beta4 \text{ Lotrate}$$

4 EMPIRICAL ANALYSIS OF THE TEST RESULTS

4.1 Descriptive statistics

In this paper, data are analyzed in aspects of excess return rate, pyramid hierarchy, issuing size, issuance Price-earnings ratio, issuance lottery rate, turnover rate on the first day of listing, asset profit rate before issuance, etc. See Table 1.

1. From data in Table 1, we can see that the average excess return rate in small and medium-sized market is close to 88%. Compared with the rate of 20% to 30% in mature markets, this number shows that small and medium-sized stock markets are still immature in China, in which a high rate of underpricing exists.
2. In terms of issuance Price-earnings ratio data, the mean is approximately 65 times. With the pricing and issuing mode reform of stocks, the government releases stock pricing. Stock prices gradually rise by way of bidding. This situation also manifests that the stock market has gotten rid of the control of government and is achieving marketization.

4.2 Regression analysis

1. $\beta1 < 0$, namely Pyramid hierarchy and underpricing rate are negatively correlated, but not significantly. But through regression analysis, Pyramid hierarchy of small and medium-sized market are found to be negatively correlated to underpricing rate, which is decided by the characteristic of the small and medium-sized market.
2. $\beta2 < 0$, issuing scale and underpricing rate are negatively related, and the significance is strong. This analysis result means that the larger the listing corporation is, the lower the IPO underpricing rate is. Large scale companies have more information disclosure, attract more attentions, and have more transparent internal operations.
3. $\beta3 > 0$, issuance Price-earnings ratio and underpricing rate have a positive correlation. From the result of analysis, price-earnings ratio and underpricing rate are positively related, namely the higher the price-earnings ratio is, the higher the underpricing rate would be. High price-earnings ratio will attract more investors, especially in small and medium-sized market where speculations and bubbles exist.

Table 1. Descriptive statistics.

	N	Minimum	Maximum	Mean	Std. deviation
Lotrate	583	0.01294100	10.10997500	0.6137714940	0.89291182397
SIZE	583	3870.0000	593480.0000	67343.682535	57535.2650232
PE	583	9.45	223.11	64.7443	31.45893
AR	583	0.000532	6.267442	0.87979987	0.936273579
LEVEL	583	0	4	1.55	0.563
Valid N (listwise)	583				

Table 2. Coefficients[a].

Model	Unstandardized coefficients		Standardized coefficients	t	Sig.	Collinearity statistics	
	B	Std. error	Beta			Tolerance	VIF
1							
(Constant)	−0.485	0.647		−0.750	0.454		
Lotrate	−0.133	0.036	−0.127	−3.722	0.000	0.786	1.272
SIZE	−4.701E-6	0.000	−0.289	−8.740	0.000	0.836	1.196
PE	−0.017	0.001	0.571	18.338	0.000	0.943	1.060
LEVEL	−0.041	0.051	−0.025	−0.813	0.417	0.975	1.026

a. Dependent variable: AR.

5 CONCLUSIONS AND SUGGESTIONS

5.1 *Reasons for the IPO underpricing in small and medium-sized market*

1. The difference in industries makes it impossible for investors to have particular understandings of each one of them. This effects on the pricing of different industries, leading to different underpricing rates [5]. In China, small and medium-sized market is a special system of our country's regime. Since it has only developed for a short time, the market is immature compared to the main board market.

2. The reason that there are more serious speculation behaviors in small and medium-sized market than in the main board market is for its newly development. In addition, Shenzhen Stock Exchange stops issuing new stocks for the following 3 years to set up the growth enterprises market. The board trading in mall and medium-sized enterprises on June 25, 2004 also adds to its uncertainty, so speculation prevails.

5.2 *Recommendations for the policy*

1. Improving the distribution mechanism of the market, and releasing market risks caused by high IPO underpricing. The split share structure reform lays a system foundation for an effective price discovery mechanism, but there is still a long way to go for the achievement of complete and effective price discovery mechanism [4].

2. Promoting rapid development of emerging capital market in China, enriching financial products, and broadening investment channels. The development of emerging markets, great enthusiasm for investment, and the large number of idle funds result in serious market bubbles. By creating more financial products, widening investment channels, and squeezing the market bubble, market equilibrium would be achieved.

REFERENCES

[1] Claessenss, Djankovs, Langlhp. 2000. The Separation of Ownership and Control in East Asian Corporations [J]. Journal of Financial Economics (58): 81–112.
[2] Fan, J.P.H, Wong, T.J. and Zhang Tianyu. The Emergence of Corporate Pyramids in China [R]. Working Paper. The Chinese University of HongKong.
[3] Qian Liu, Congtao Song. An Empirical Study of the pyramid structure influence on the performance of private listed companies [J]. Friends of Accounting, 2012(10).
[4] Lili Shao. The impact of private enterprise organizational structure underpricing of initial public offerings and related transactions [J]. Economic Survey, 2008(5).
[5] Xuemin Zhuang. Our plates IPO underpricing reason research [J]. Research on Economics and Management, 2009(11).
[6] Jinhua Li, Haibin Huang. Our SME Issues of IPO underpricing [J]. Market Modernization, 2007(3).
[7] Jianchao Li, Zhaohua Zhou. An Empirical Study of SME IPO underpricing phenomenon [J], Soft Science, 2005(5).

Information Technology and Applications – Li (Ed.)
© *2015 Taylor & Francis Group, London, ISBN 978-1-138-02677-3*

Simulation model of micro-blogging opinion dissemination under sudden natural disasters

Hongliang Wang & Jie Wang
Business School, Hohai University, Nanjing, China

ABSTRACT: The micro-blogging opinion generated by the sudden natural disasters disseminates rapidly and has great impact. We build the SIR model to simulate the dissemination of micro-blogging opinion caused by The Ya'an Earthquake in Sichuan province in April 20 by MATLAB. We adopt the empirical research to study the dissemination law of micro-blogging opinion under sudden natural disasters. Then we conclude the characteristics of the dissemination of micro-blogging opinion from both the perspective of intelligence science and epidemiology, which is ground-breaking. One is scattering distribution principle, small world principal and the least effort principle, the other is pathogen, infectivity, immunity. At last we put forward the controlling strategy of the dissemination law of micro-blogging opinion for the government to implement the public opinion persuasion micro-blogging from both the perspective of intelligence science and epidemiology.

Keywords: micro-blogging opinion; SIR model; dissemination law; controlling strategy; sudden natural disasters

1 INTRODUCTION

Micro-blogging, a kind of unique, brief, real-time, interactive, free social network, has been spurting and increasingly becoming the eye of the public opinion storm of Internet since Sina micro-blogging was provided in August 2009. According to the data in "31th China Internet Development Survey Report" released by the China Internet Network Information Center (CNNIC), the number of netizens is 564 million as of march 2012 in China (2013). The number of micro-blogging users has reached 309 million, which has increased 58.73 million compared with the end of 2011. And the proportion of micro-blogging users in netizens has reached 54.7%.

Sudden natural disaster events are one kind of natural anomalies which occur suddenly and unexpectedly (Cai Meizhu 2012). They may cause or threaten to cause serious economic and social harm, which needed the emergency measures to cope with. In recent years, we have met with the snowstorms in southern China, Wenchuan Earthquake, Zhouqu large debris flow and so on, which have led serious losses to the social, economic development and properties. Now the micro-blogging has been becoming the most influential communication tools. When we meet with the sudden natural disaster, the fraudulent, partial micro-blogging might lead to the risk of possible terror strikes, might harm the public interest, undermine state economic plans or disrupt social economic order. Cost-minimization, most economical and most successful approach of crisis management is to avoid crises or destroy the bud of the crisis (Zhang Xiaoming 2006). Consequently, research on the micro-blogging opinion dissemination law and controlling strategy under sudden natural disasters is of great significance for the government's emergency management decision makings.

2 RESEARCH STATUS

Emergencies micro-blogging public opinion refers that the initial event spreads through the micro-blogging in a very short period and then disseminates, spreads and rages. Contagion is a random process, which is essentially a dynamic phenomenon. When correlation changes in the sudden unpredictable case, the contagion occurs (Focardi & Fabozzi 2005). The initial event spreads in the interactive and sharing micro-blogging network, which easily forms a result of the development from a single event to a multi-level events, than led the unpredictable and uncontrollable changes. The root causes of the risk of internet emergencies are the contagion and the variation (Ye Jinzhu & She Lian 2012).

Currently, some scholars have tried to study emergencies public opinion from different perspectives. Ma Shoushuai (2011) analyzed the spread of emergencies micro-blogging rumor (Ma Shoushuai 2011); Sun Wei (2011) analyzed the characteristics and reasonable expressions of internet public opinion, then conclude the channel of the expressions and guidance of internet public opinion (Sun Wei & Zhang Xiaolin); Tang Xiaobo (2012) expanded the ideas and methods of co-word network and complex network analysis to the micro-blogging public opinion analysis, and designed a micro-blogging public opinion analysis model based on network visualization (Tang Xiaobo & Song Chengwei 2012); Jiang Xin (2012) had proved that communication network in micro-blogging community by using complex network theory and social network analysis method (Jiang Xin & Tian Zhiwei 2012); Qian Ying (2012) found SIR model simulation result fitted the historical data well, which was an evidence that the model could be used to study the public opinion spreading on micro-blogging, then based on the model simulation, they analyzed how public opinion spread under different scenarios (Qian Ying et al. 2012). Lan Yuexin (2013) established the differential equation model of micro-blog opinion diffusion law base on research of emergency micro-blog public opinion influence trend (Lan Yuexin 2013)

The disadvantages of the study on emergencies internet public opinion are following: ①Research started late, and the number of papers is not large, moreover, the majority of the scholars studied the emergencies internet public opinion from the perspective of news public opinion, fewer studied from the perspective of intelligence; ②Existing studies were mostly qualitative analysis, and lack of quantitative analysis and model lacks support; ③The studies adopted the method of empirical analysis lacked theoretical depth, which were not deep enough. This paper studies the law of micro-blogging opinion dissemination law under sudden natural disasters. We build the SIR model, and then we simulate the dissemination of micro-blogging opinion caused by The Ya'an Earthquake in Sichuan province in April 20 by MATLAB. We study the law of micro-blogging opinion dissemination innovatively from the perspectives of both intelligence and epidemiology so that the government could implement micro-blogging control, guidance adopted quantitative models.

3 THE SPREAD PROCESS OF MICRO-BLOGGING PUBLIC OPINION AND MODEL CONSTRUCTION

3.1 *The spread process of micro-blogging public opinion*

The classical epidemic model (SIR model) was proposed by Kermack (Kermack & McKendrick 1927)

in 1927. This paper builds the SIR model to simulate the dissemination of micro-blogging opinion in the cyberspace under sudden natural disasters. We assume that the micro-blogging network has N nodes which can broadcast to its followers. The information can only pass along the directed edges due to the information asymmetry in the micro-blogging network. The users are divided into three categories: ①health users, also S-state users, are the one who haven't received the information flow; ②infected users, also I-state users, are the one who have received the information flow and then have forwarded the information; ③immune user, also R-state users, are the one who have received the information flow and then haven't forwarded the information. In the entire network, I-state users infect the S-state users by α. On the other hand, I-state users become the R-state users by β if the S-state users have previously known the message and I-state users may think the information flow worthless. Some I-state users broadcast to its followers in the micro-blogging network when natural disaster occurs. Along with the evolution of the system, all users of the micro-blogging no longer broadcast the information flow, then I-state users disappear. Only do the S-state users and R-state users exist in the system. The micro-blogging public opinion doesn't dissemination and system will soon stabilize. The process of the micro-blogging opinion dissemination is shown in Figure 1.

In the micro-blogging network, we assume N as all the number of nodes, also the total users of the micro-blogging network. We assume S as the healthy users, assume I as the infected users, assume R as the immune ones. Only do the S-state users, I-state users and R-state users exist in the system. So does this equitation: N = S + I + R. Only do the S-state users and I-state users exist in the system firstly and the number of R-state users is zero. All the I-state users become the R-state users, then only do the S-state users and R-state users exist in the system at last. Therefore the number of R-state users is able to reflect the extent of the micro-blogging opinion dissemination under sudden natural disasters.

3.2 *The model of the micro-blogging opinion dissemination*

We build the differential equations on the basis of the research on the SIR model of Zanette D

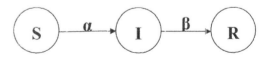

Figure 1. The process of the micro-blogging opinion dissemination.

(Zanette 2002) H and Moreno Y (Moreno et al. 2004). In the following equations, $s(t)$, $i(t)$, $r(t)$ separately means the number of the S-state users, I-state users, R-state users at t point.

$$\begin{cases} \dfrac{di(t)}{dt} = \alpha * s(t) * \left(i(t)/N\right) \\ \dfrac{dr(t)}{dt} = \beta * i(t) * \left(i(t) + r(t)\right)/N \\ N = s(t) + i(t) + r(t) \end{cases} \quad (1)$$

4 THE CASE OF MICRO-BLOGGING PUBLIC OPINION DISSEMINATION UNDER SUDDEN NATURAL DISASTERS

At 8:02 on April 20, 2013, the magnitude-7 earthquake occurred in Ya'an Lushan County (latitude 30.3, longitude 103.0) Sichuan Province. Tremors were felt as far away as Chendu, Chongqing, Baoji, Hanzhong, Ankang. As of 28 of April, more than 200 million people were affected, 196 people loses their lives. Ya'an Earthquake had brought Sichuan the immeasurable loss.

Sichuan Ya'an earthquake was a major sudden natural disaster emergency. Ya'an became the eye of the hurricane of all nations in very short time. The social media, represented by Sina Webo, highlighted the powerful platform for influence and appeal as the earthquake "information center". The micro-blogging network became the most convenient communication platform between civil and government which enhanced the efficiency of rescue consumedly. This paper aims to study the micro-blogging opinion dissemination law under sudden natural disasters in the case of the information flow dissemination of Ya'an Earthquake in Sina Webo.

4.1 Data collection

This paper collected the micro-blogging data from 8:00 on 19 April, 2013 to 8:00 on 4 May using the in the "advanced search" function of Sina Webo. The data collected in this study were all open micro-blogs except the private micro-blogs and the collection micro-blogs. The private micro-blogs and the collection micro-blogs couldn't infect the S-state users that also could not we collect. The missing data didn't affect the construction of the SIR model and the analysis of the collection data.

We adopted the different key words that we could collect the different data. The popular keywords related to this earthquake are, "Ya'an Earthquake", "Lushan Earthquake", "earthquake", "aid", "blessing", "Ya'an people search" and so on.

The number of micro-blogs, taken the key words as "Ya'an" were nearly 37 234 622, which was much more than the sum of results taken the key words. Also did the other results usually include the key words "Ya'an". What's more, by the comparison with the data we collected, we also collected the data taken the key words "Ya'an", then we acquired 620 items of mico-blogs, which was nearly negligible. 5.3 earthquake occurred on April 22nd at the junction of the Inner Mongolia Autonomous Region and Fuxin, Tongliao and Liaoning provinces. So do we take "earthquake" as the key words might bias the monitoring results. For this reason, this paper took "Ya'an" as the key words which could depict precisely the process of micro-blogging public opinion dissemination under sudden natural disasters.

This paper counted the number of the daily total micro-blogs, the original micro-blogs, the forward micro-blogs, pictures micro-blogs, video micro-blogs, links micro-blogs. Table 1 is shown a collection of the micro-blogs data of Ya'an Earthquake from 8:00, April 20 to 8:00 4th, May.

4.2 Data analysis

Certain generalizations can be derived from the data in Table 1:

1. We could draw the trend charts of the amount of total micro-blogs, origin micro-blogs, forward micro-blogs (Fig. 2). We found that micro-blogging opinion dissemination regarding this earthquake grew exponentially. What's more, total micro-blogs change tendency were almost the same as forward micro-blogs. On the first day (24 h), the event had become the center of public opinion in the whole microblogging network, and being disseminated, then the number of the original micro-blogs reached the peak and the total micro-blogs, forward micro-blogs continued to disseminate. On the second day, the total micro-blogs, forward micro-blogs reached the peak. In the third day, the two fell sharply, then the micro-blogging opinion subsided further in the next three days. On the eighth day, the total micro-blogs, forward micro-blogs fell sharply again. In the next week, the micro-blogging opinion subsided further, but it was also the eye of the public opinion storm of Internet. In the whole process, the original micro-blogs subsided on the second day. In contrast, the original micro-blogs is the source of the micro-blogging opinion, and which promote the dissemination of the microblogging opinion. At the same time, we notice that the micro-blogging opinion was being at a high level in the first three days, which is called the "golden 72 hours". A large number

Table 1. The micro-blogs data of Ya'an earthquake.

Time 8:00 19th	Total micro-blogs	Pictures micro-blogs	Original micro-blogs	Forward micro-blogs	Video micro-blogs	Links micro-blogs
20/4	145	124	108	37	2	101
21/4	7 211 648	4 551 776	1 621 632	5 347 760	898	583 392
22/4	8 744 288	5 913 024	1 298 624	7 445 664	8240	626 240
23/4	4 059 360	3 849 728	774 560	3 284 800	3296	398 816
24/4	3 319 072	2 383 008	375 744	2 943 328	658	174 688
25/4	2 900 480	1 163 488	224 128	2 676 352	366	98 880
26/4	2 746 844	1 069 619	194 464	2 552 380	308	49 543
27/4	2 596 112	2 062 160	166 693	2 429 419	177	103 048
28/4	199 794	125 738	43 250	156 544	26	228
29/4	128 841	105 592	32 897	95 944	37	336
30/4	75 442	53 198	17 956	57 486	42	433
1/5	110 483	88 239	20 770	89 713	67	429
2/5	44 756	37 231	14 783	29 973	64	263
3/5	43 014	36 247	13 534	29 480	69	474
4/5	35 376	29 413	13 132	22 244	75	711

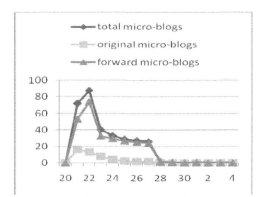

Figure 2. The change of total micro-blogs, original micro-blogs, the forward micro-blogs.

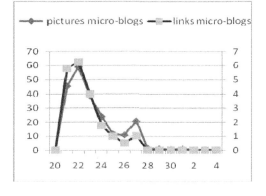

Figure 3. The change of pictures micro-blogs, links micro-blogs.

of rescue workers searched for survivors day and night, hoped to rescue the prime time after the earthquake, "72 hours", they rescued more injured victims.

2. We could draw the trend charts of pictures micro-blogs, links micro-blogs (Fig. 3). We found the dissemination of pictures micro-blogs is almost same as the dissemination of links micro-blogs. On the first day, the two grew exponentially. On the second day, the two reached the peak. In the next four days, pictures micro-blogs, and links micro-blogs subsided. On the 7th day, the number grew dramatically. Since Sichuan Province government had decided to the 7th day for a province-wide day of mourning, accompany with the people of

Sichuan, internet users cherished the memory of the victims. In the second week, micro-blogging opinion subsided.

3. In Table 1, we found that the number of the links micro-blogs was much bigger than the number of the video micro-blogs. We may have thought the video micro-blogs disseminate much easier than the links micro-blogs, but the fact was that I-state users preferred the links micro-blogs. There were two reasons: ①the internet traffic is expensive, most of the mobile users could not afford the cost of the video micro-blogs; ②the mobile users could easily understand the fact according to the links micro-blogs, and not have to take several minutes to buffer or ads.

4.3 *Model simulation*

According the statistics of Sina in fourth quarter of 2012, the number of registered users was nearly 503 million, and the number of the daily active users had reached 46.2 million. There were 456.8 million inactive users, who had little effect on the dissemination of micro-blogging opinion. So we took the 46.2 million active users into consideration.

We found that Chendu High-tech Institute of Care-life sent out the first government micro-blog and announced the earthquake on 8:02 a.m. The first private micro-blog sent out on 8:04 and "Ya'an feel strongly". Several minutes after the earthquake, "Ya'an", "earthquake" became the top two entries in Sina. In the time of we-media, every network node is possible to be the origin of news. We collect 42 824 the original micro-blogs between 8:02 to 9:00 on 20th April. In this model, we take the original micro-blogs in 9:00 a.m. $I(0)$, the I-state users. We take $S(0)$ as S-state users and $R(0)$ as R-state users in that time. SO we list the parameters as follow:

$$N = S + I + R$$
$$N = 46\ 200\ 000;$$
$$S(0) = 46\ 157\ 152;$$
$$I(0) = 42\ 848;$$

We fit the Model on MATLAB program and the result is as follows in Figure 4.

In Figure 4, we found the fitting data in general accord with the fact data in Ya'an Earthquake. After the incident, the micro-blogging opinion grew exponentially, then reached the peak on the second day. Then the micro-blogs subsided sharply utile the "golden 72 hours". After that, the micro-blogs subsided continuously. The micro-blogging opinion again slumped then subsided after 8 days. The dissemination process of the micro-blogging opinion in general accord with the dissemination of the disease. The fitting data could demonstrate the correctness of this analysis.

5 CONCLUSION

This paper analyses the dissemination of the micro-blogging opinion under sudden natural disasters. We build the epidemic model (SIR model), then we simulate the dissemination of micro-blog opinion caused by The Ya'an Earthquake in Sichuan province in April 20 by MATLAB. Then we conclude the characteristics of the dissemination of micro-blogging opinion from both the perspective of intelligence science and epidemiology, which is ground-breaking. At last we put forward the controlling strategy of the dissemination law of micro-blogging opinion for the government to implement the public opinion persuasion microblogging from both the perspective of intelligence science and epidemiology. The shortcoming of this research is that we applied just one case. In the further study, we will try to analyse the other cases of sudden natural disasters so that might examine in more depth.

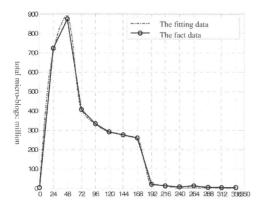

Figure 4. SIR model of micro-blogging opinion dissemination and the model fitting.

REFERENCES

Cai Meizhu. 2012. Research on properties of internet opinion of sudden natural calamity. *News Research Guide* (10):13–17.

China Internet Network Information Center. 2013. The 31st China Internet Development Statistics Report.

Focardi, S.M. & Fabozzi, F.J. 2005. An Autoregressive Conditional Duration Model of Credit—Risk Contagion. *Journal of Risk Finance* 6(3):208–225.

Jiang Xin & Tian Zhiwei. 2012. An Empirical Study on Information Dissemination in Microblog Community from the Perspective of Small-world Property. *Information Science* 30(8):1139–1142.

Kermack, W.O. & McKendrick, A.G. 1927. Contributions to the mathematical theory of epidemics, part I. *Proceedings of the Royal Society of London A* 115:700–721.

Lan Yuexin. 2013. Research on Microblog Opinion Diffusion Model of Emergent Events. *Information Science* (03):31–34.

Liu Yi. 2007. Research on the concept, properties, expression and dissemination of the internet opinion. *Theory Horizon* (1):11–12.

Ma Shoushuai. 2011. Analysis of the dissemination of the rumors of sudden natural calamity in the context of micro-blog. *Journalism Knowledge* (9):63–65.

Moreno, Y. & Nekovee, M. & Pacheco, A.F. 2004. Dynamics of rumor spreading in complex networks. *Physical Review E* 69(6):066130.

Qian Ying et al. 2012. The spread of public sentiment on micro-blogging under emergencies. *Journal of the China Society for Scientific and Technical Information* 31(12):1299–1304.

Sun Wei & Zhang Xiaolin. The Expression and Guidance of Public Opinions on Internet concerning Abrupt Natural Disasters—Taking the Earthquake Tsunami in Eastern Japan for Example. *Academic Exploration* 06:115–118.

Tang Xiaobo & Song Chengwei. Microblog Public Opinion Analysis Based on Complex Network. *Journal of the China Society for Scientific and Technical Information* 31(11):1153–1162.

Ye Jinzhu & She Lian. The Mechanism of Internet Emergency Spread. *Journal of Intelligence* 31(3):1–5.

Zanette, D.H. Dynamics of rumor propagation on small-world networks. *Physical Review E* 65(4):041908.

Zhang Xiaoming. 2006 Public warning mechanism design and construction of index system. *Chinese Public Administration* (7):14–19.

Zhu Huaxin & Shan Xuegang & Hu Jiangchun. 2012 China Internet public opinion analysis report.

Information Technology and Applications – Li (Ed.)
© 2015 Taylor & Francis Group, London, ISBN 978-1-138-02677-3

The performance evaluation of IT industry: Based on exploratory factor analysis

Yang Wu & Yongjun Tang
School of Business, Hohai University, Nanjing, China

ABSTRACT: Performance evaluation is an important part of modern enterprise management. This paper selects 13 financial indicators of performance evaluation based on the studies of corporate performance evaluation and relative theories and uses SPSS software to do exploratory factor analysis of listed companies in the IT industry. We try to reduce the dimension of complex indicators to few comprehensive several main factors and name the main factors, calculate the scores of every company. The result of analysis provides a simple and clear evaluation system for enterprises to evaluate its operating results.

Keywords: corporate performance; IT industry; EFA

1 INTRODUCTION

Corporate performance objectively reflects the financial position and operating results brought by business activities during certain period. Operators and other stakeholders can make effective decisions according to corporate performance evaluation. And corporate performance evaluation guides operators to take actions to improve management. Therefore, how to build a comprehensive and effective corporate performance evaluation system becomes a focal problem at home and abroad.

IT industry is a highly specialized industry with low profits and rapid development. After years of development, the electronic information industry, including IT industry, has become a pillar industry in China. Therefore, this paper evaluates the performance of listed companies in the IT industry by exploratory factor analysis is conducive to the healthy and rapid development of the IT industry. Besides, the evaluation system we get in this paper provides a basis for listed companies. It is easy to find out the existing problems and take actions to make improvement and development of enterprises.

2 LITERATURE REVIEW

2.1 *Foreign research*

The research in aboard of corporate performance evaluation can be divided into three stages:

The first stage is the stage of cost performance, which produced at the beginning of capitalism. Enterprises in this period paid attention to the cost and the system of corporate performance is very simple.

The second stage flourished in the early 20th century called financial performance evaluation. During this period, DuPont designed many important business and budget targets. And professor Wall and Denning proposed Wall comprehensive score method.

The third stage is the comprehensive evaluation stage. Due to the rapid development of economy, cost and financial performance evaluation did not meet the needs of enterprises, so a large number of new evaluation methods emerged. Such as: American consulting firm Stern Stewart proposed the concept of economic value added, and Kelvin Cross and Richard Lynch combined the financial information and non-financial information of overall corporate strategy to establish a performance pyramid model. In 1992, Professor Kaplan and Norton presented Balanced Score Card and then, Pinches, Mingo and Caruthers adopted the method of principal component analysis to corporate performance evaluation.

2.2 *Domestic research*

The research of corporate performance evaluation is relatively late in our country. In recent years, our studies of corporate performance evaluation are mostly concentrated in the factor analysis.

Chen Jingpu and Li Yanping (2014) used factor analysis to do research of financial performance of listed financial companies. They evaluated the financial performance from four aspects, including profitability, capital expansion, growth capacity and solvency in order to explore the advantages and disadvantages of corporate financial governance. Li Junmei and Sun Gongmei (2013) applied factor analysis method to 20 listed household appliances companies. As a result, they got the main factor score model and total score model by the analysis method, and at last, they gave their rankings and conclusions.

Therefore, based on previous studies, this paper tries to use EFA to do evaluation of IT listed companies to provide a simple and clear evaluation system for the IT industry, and gives reasonable suggestions to improve corporate performance.

3 THE EMPIRICAL RESEARCH

3.1 *The selection of companies*

IT industry is one of the fastest growing industries in recent years. IT industry use tools and technology to collect, collate, store and transfer the information to providing information services, appropriate information tools and information technology services. IT industry has become the cornerstone of today's economic activities and social life.

This paper selects the Chinese listed companies of IT industry in 2013 as a sample. The data are gathered from the GTA database, and we eliminate the ST companies, non A-share companies and the companies with incomplete information. The remaining 129 companies are the research object in this paper.

3.2 *Index design*

We select the 13 indicators in four dimensions to measure the corporate performance after

Table 1. Evaluation indicators of IT industry.

Assessment angle	Indicator name	Index numbers
Profitability	Earnings Per Share (EPS)	X_1
	Operating margin	X_2
	Return On Assets (ROA)	X_3
	Return On Equity (ROE)	X_4
Operation ability	Accounts receivable turnover ratio	X_5
	Inventory turnover rate	X_6
Solvency	Current ratio	X_7
	Quick ratio	X_8
	Asset-liability ratio	X_9
	Equity ratio	X_{10}
Development capability	Net profit growth rate	X_{11}
	Operating margin growth rate	X_{12}
	Operating income growth rate	X_{13}

synthesizing the former studies of performance evaluation and theories at home and abroad.

3.3 *The introduction of EFA*

Exploratory Factor Analysis (EFA) is a method which can locate the few comprehensive indexes from several relevant indicators to reflect the key messages contained in the original index. Therefore, EFA can integrate variables with complex relationships to a few specific integrated core factors and draw factor score coefficient matrix at last.

Exploratory factor analysis mainly consists of seven steps: ①collect variable values; ②construct the correlation matrix; ③determine the number of main factors; ④extract the main factors; ⑤do rotation of factors; ⑥explain the structure of factors; ⑦calculate the main factors score.

3.4 *The application of EFA*

3.4.1 *The pretreatment of the indicators*
Among the 13 indicators selected in this article, the four indicators, the current ratio, quick ratio, equity ratio, asset-liability ratio are moderate indicators, while other indicators are positive indicators. Thus this article needs to turn moderate indicators to positive indicators. In this paper, we selected the approach: $x' = 1/(1 + |x–A|)$, "A" represents the optimal values of x, current ratio is 2, quick ratio is 1, equity ratio is 1.2 and asset-liability ratio is 50%.

3.4.2 *Applicability test of EFA*
Before using exploratory factor analysis, we should test the correlation between indicators because one prerequisite of using EFA is to have a strong correlation between the original indicators. This paper uses KMO and Bartlett test to show the correlation between the primary indicators:

As we can seen from the KMO and Bartlett test table (Table 2), the value of KMO test is 0.733 which shows that the original indicators are suitable for factor analysis according to the KMO standards. Besides, the score of Bartlett test is 1236.18, Sig value is 0, and the variables can be considered that the correlation coefficient matrix between indicators has a significant difference.

Table 2. KMO and Bartlett test.

Kaiser-Meyer-Olkin measure of sampling adequacy.		0.733
Bartlett's test of sphericity	Approx. chi-square	1236.180
	df	78
	Sig.	0.000

3.4.3 *Determine the number of main factors*

According to the total variance contribution of each factor, this paper uses principal component analysis of EFA to extract 4 main factors from the 13 original indicators. Then, we use the 4 main factors to replace the original indicators to reflect the whole information. The variance contribution was shown in Table 3.

As can be seen from Table 3, this paper extracted 4 main factors from the 13 indicators after factor analysis. The variance contribution of the 4 main factors is 76.211%, which means that four main factors reflect the 76.211% of the overall information, there is less loss of information. So it is feasible to use the factors to replace the original variables to do performance evaluation of listed company.

3.4.4 *Name the main factors*

In the analysis, due to the comprehensive of original factors is strong, sometimes it is difficult to work out their practical significance. So this paper took the varimax rotation of these factors, which shown in Table 4. And we can mark those high load values in yellow so that it would be more intuitive.

The rotated component matrix shown above indicates that different indicators show high load in different main factors. Therefore, we integrated the factors which shows high load in same main factors to name the main factors.

From Table 5, firstly, there are 4 indicators showing high load on the first main factor. The current ratio, quick ratio represents the short-term solvency

Table 3. Total variance explanation and cumulative sum.

Component	Initial eigenvalues			Rotation sums of squared loadings		
	Total	% of variance	Cumulative %	Total	% of variance	Cumulative %
1	4.197	32.287	32.287	3.569	27.455	27.455
2	2.897	22.281	54.568	3.376	25.971	53.426
3	1.760	13.538	68.106	1.880	14.464	67.891
4	1.054	8.104	76.211	1.082	8.320	76.211
5	0.970	7.459	83.669			
6	0.744	5.724	89.393			
7	0.418	3.212	92.605			
8	0.311	2.394	95.000			
9	0.196	1.511	96.510			
10	0.170	1.306	97.816			
11	0.159	1.226	99.042			
12	0.071	0.549	99.591			
13	0.053	0.409	100.000			

Table 4. The factor loading matrix after rotation.

	Component			
	1	2	3	4
VAR00001	−0.124	0.870	−0.003	0.104
VAR00002	−0.419	0.761	0.062	0.110
VAR00003	−0.077	0.948	0.066	0.000
VAR00004	0.079	0.947	0.103	−0.032
VAR00005	0.150	−0.049	0.036	0.618
VAR00006	−0.140	0.149	−0.056	0.778
VAR00007	0.844	0.004	−0.052	0.087
VAR00008	0.921	−0.078	−0.002	0.005
VAR00009	0.963	−0.099	0.008	−0.014
VAR00010	0.884	−0.118	−0.013	−0.048
VAR00011	−0.077	0.184	0.931	−0.019
VAR00012	−0.045	−0.050	0.918	0.048
VAR00013	0.217	0.393	0.379	−0.239

Table 5. Name the main factor.

Sequence	Indicator with high load	Name
1	X_7 current ratio	Solvency
	X_8 quick ratio	
	X_9 asset-liability ratio	
	X_{10} equity ratio	
2	X_1 Earnings Per Share (EPS)	Profitability
	X_2 operating margin	
	X_3 Return On Assets (ROA)	
	X_4 Return On Equity (ROE)	
3	X_{11} net profit growth rate	Development capability
	X_{12} operating margin growth rate	
	X_{13} operating income growth rate	
4	X_5 accounts receivable turnover ratio	Operation ability
	X_6 inventory turnover rate	

of a company, and asset-liability ratio and equity ratio indicates the long-term solvency of a company. There, the first major factor is defined as solvency.

Secondly, 4 factors show higher load on the second main factor, including EPS, operating margin, ROA, ROE. EPS reflects the ability of shareholder benefit, while operating margin, ROA and ROE report the capacity of business benefit. Therefore, the second main factor is profitability.

Thirdly, there are three indicators having high load on the third main factor, net profit growth rate, operating margin and operating income growth rate, the three indicators reflect the growth in corporate profitability, so you can define the third main factor as development capacity.

Fourthly, on the fourth main factor, two indicators show high load. High accounts receivable turnover ratio means the recovery speed of accounts receivable is fast and lower bad debt losses. Inventory turnover rate reflects the speed

of inventory turnover. Therefore, the last factor is operating capacity.

3.4.5 *Calculate the score of main factors*

In this paper, we use the regression coefficients to estimate factor scores, according to Table 6 above; we can draw all the main factor score function to calculate the main factor scores:

$$F_1 = 0.014X_1 - 0.078X_2 + 0.03X_3 + 0.074X_4 \\ + 0.055X_5 - 0.015X_6 + 0.247X_7 + 0.264X_8 \\ + 0.274X_9 + 0.25X_{10} - 0.012X_{11} - 0.014X_{12} \\ + 0.08X_{13};$$

$$F_2 = 0.267X_1 + 0.211X_2 + 0.293X_3 + 0.3X_4 - 0.036X_5 \\ + 0.012X_6 + 0.048X_7 + 0.026X_8 + 0.021X_9 \\ + 0.014X_{10} - 0.018X_{11} - 0.093X_{12} + 0.117X_{13};$$

$$F_3 = -0.065X_1 - 0.02X_2 - 0.039X_3 - 0.02X_4 \\ + 0.058X_5 - 0.023X_7 + 0.006X_8 + 0.012X_9 \\ + 0.001X_{10} + 0.5X_{11} + 0.515X_{12} + 0.166X_{13};$$

$$F_4 = 0.058X_1 + 0.065X_2 - 0.037X_3 - 0.061X_4 \\ + 0.586X_5 + 0.716X_6 + 0.097X_7 \\ + 0.028X_8 + 0.013X_9 - 0.021X_{10} + 0.024X_{11} \\ + 0.096X_{12} - 0.214X_{13};$$

We get total score function by using the variance contribution rate:

$$G = 0.27455F_1 + 0.25971F_2 + 0.14465F_3 + 0.0832F_4$$

We can see that the solvency does the largest contribution to corporate performance and profitability followed according to the score function. Finally, we calculate the scores of each listed company in the IT industry and make a ranking of the 129 companies.

3.5 *Suggestions*

Due to limited space, this article only lists the companies ranked in the top five and last five, which

Table 6. The component score matrix.

Indicator	Component			
	1	2	3	4
VAR00001	0.014	0.267	−0.065	0.058
VAR00002	−0.078	0.211	−0.020	0.065
VAR00003	0.030	0.293	−0.039	−0.037
VAR00004	0.074	0.300	−0.020	−0.061
VAR00005	0.055	−0.036	0.058	0.586
VAR00006	−0.015	0.012	0.000	0.716
VAR00007	0.247	0.048	−0.023	0.097
VAR00008	0.264	0.026	0.006	0.028
VAR00009	0.274	0.021	0.012	0.013
VAR00010	0.250	0.014	0.001	−0.021
VAR00011	−0.012	−0.018	0.500	0.024
VAR00012	−0.014	−0.093	0.515	0.096
VAR00013	0.080	0.117	0.166	−0.214

Table 7. Companies ranking.

Rank	Name	F_1	F_2	F_3	F_4	G
1	Ganglian	46.94	−29.65	48.29	507.85	54.43
2	Topway Video	−2.43	3.07	2.01	255.3	21.66
3	Focus Technology	2.38	−0.43	4.48	171.55	15.46
4	Zhongqingbao	−2.66	2.81	2.85	178.39	15.26
5	People.cn	−2.84	3.03	1.07	167.96	14.14
125	GQY Video	0.56	0.32	−2.48	2.46	0.08
126	Telecom Electronic	0.45	0.54	−3.93	1.43	−0.18
127	Allwin Telecommunication	0.73	0.71	−6.07	0.77	−0.43
128	Keybridge Communication	1.16	1.08	−9.11	0.91	−0.64
129	Geeya Technology	1.62	3.45	−28.98	0.58	−2.8

can be seen from Table 7. The top three companies, including Shanghai Ganglian, Topway Video, Focus Technology, get higher scores in corporate performance. But compared with other companies, most of the top companies have large amount of accounts receivable turnover or inventory turnover, and even hundreds of times more than others. Thus, it is difficult to explain the corporate performance only by comprehensive ranking. The impact of higher accounts receivable turnover ratio and inventory turnover rate of the corporate is also a double-sided problem. Shanghai Ganglian gets the highest score in solvency owing to the rapid asset recovery. At the same time, Shanghai Ganglian does well in development capacity and operational capabilities which means the benefits of adequate liquidity is very large. But Shanghai Ganglian gets low scores in profitability, so Shanghai Ganglian should pay more attention to profitability and take actions to make improvement.

The problem of the last ten companies focuses on the development capacity, almost every company get negative scores in the third main factor. This means these companies are going downhill, they should timely innovate new products to saving the declining trend of the enterprise.

4 BRIEF CONCLUSION

From the analysis above, we can see that factor analysis plays an important role in corporate performance evaluation. By using the method of exploratory factor research, complex evaluation index can be integrated into several main factors. And EFA provide us a clear and intuitive evaluation system of corporate performance. Based on the factor score model and comprehensive score model from EFA, we can get the score of every company and find the existing problems of each company. So that listed company of IT industry can develop appropriate strategy, which plays an important role to better development in the future.

ACKNOWLEDGEMENTS

We would like to thank support of Guidance Project of Jiangsu Philosophy and Social Science Research Foundation for University (number: 2012SJD630069).

REFERENCES

Chen Jingpu, Li Yanping. The financial performance evaluation of financial listed companies based on factor analysis [J]. *Friends of Accounting*, 2014 (05):102–105.

Drunker P. Harvard business review on measuring corporate performance [J]. *Harvard Business Review*, 1998(2).

Gong Guangming, Zhang Liuliang. Performance Evaluation of listed companies in Hunan Province based on Factor Analysis [J]. *Friends of Accounting*, 2012 (01):36–39.

Lijun Mei, Sun Hongmei. The performance evaluation of listed companies based on factor analysis—take household electrical appliance enterprises as a example [J]. *Accounting Communications*, 2012 (11):30–31.

Lv Zhentong, Zhang Lingyun. The statistical analysis and application of SPSS [M]. *China Machine Press*, 2009.

Peter F. Drunker. Harvard Business Review on Measuring Corporate Performance [M]. *Harvard Business Review Paperback*, 1998.

Information Technology and Applications – Li (Ed.)
© 2015 Taylor & Francis Group, London, ISBN 978-1-138-02677-3

An empirical study on relationships between intangible assets and profitability of IT industry

Yuqian Shi & Wenlin Gu
School of Business, Hohai University, Nanjing, China

ABSTRACT: The connotation and nature of the intangible assets play a very important role in the profitability of the business, especially for the new high-tech enterprises. Intangible assets are not the only key to gain a competitive edge, but also an important reference point of research and development capabilities of companies. In our article, 162 listed IT companies in China are researched with three aspects: their represented liabilities scale, solvency and risk level. The results show that there is a close correlation between the intangible assets and the profitability of the business in the IT industry.

Keywords: intangible assets; profitability; correlativity

1 INTRODUCTION

With the advancement of technology and the advent of the information age, the role of intangible assets of the companies is becoming more and more important. The proportion of intangible assets owned by the companies is increasing. Especially for the IT industry, intangible assets have become an important indicator to measure the value of a company. In order to enhance the competitiveness of companies and gain more comprehensive benefits, companies consume tangible assets to form intangible assets in the process of creation of value and wealth.

IT industry in China has maintained a momentum of rapid development, which has a close relationship with the value and promotion of the national government. According to statistics, in the ranking of IT industry competitiveness index in 2011, China ranked the world's 38th, while Hong Kong ranking the 19th and Taiwan ranking the 13th. In the aspect of human capital, China ranked the second in the world, for it had a huge number of l engineering students and labor forces in IT industry. The rapid development of IT industry also has contributing to the development of China's economy, and always plays an important role for rising China's GDP. As a measurement of corporate value, intangible asset is one important factor of gaining the rapid development of IT industry. Therefore, it is very important to research intangible assets' boost to corporate profitability.

2 BACKGROUND AND OBJECTIVE

2.1 *IT industry overview*

Since from the 1990s, as a symbol of the IT revolution, the Internet and multimedia technology made, emerging industries in some developed countries gradually replacing the dominant position of the automobile, construction and other traditional industries in the national economy, which became a new economic growth point. IT refers to the information-related technologies, mainly composed by sensor technology, communications technology and computer technology. Broadly speaking, IT industry consists of electronics products such as computers, mobile phones, projectors, printers and all computer peripherals, and the technologies to operate these products including networking, software and so on. Today, computer products have become a major force in stimulating industry economic growth.

Rapid development of information technology brings the trend of socio-economic "online" globalization accompanied with the information network, its deeper meaning is wealth transfer and redistribution of resources worldwide based on information technology architecture. As the deepening globalization of the IT industry as well as the government's attention and promotion of technical information, China's information technology industry has leapt to the largest industries of the national economy. IT market size has become one of the most dynamic areas of the national economy, where the rapid development of Internet is an important factor in expansion of China's IT industry in recent years.

2.2 Research purpose

With the transition from industrial economy to knowledge economy in the society, scientific and technological content in the economic activity is increasing, the focus of enterprise transits from physical assets to intangible assets gradually. The current intangible assets have become an important economic resource and played an increasingly important role in the enterprise value creation. Since most of intangible assets have a higher technological content (excluding land use rights), particularly playing a significant role in value creation in high-tech enterprises. Currently, the key support of high-tech industry mainly focuses on: information technology industry, biological technology and new pharmaceutical industry, new materials and new energy industry. Therefore, this article is intended to put listed companies of China's information technology industry as research objects, analyzed the correlation between intangible assets and corporate profitability, and obtained the degree of intangible assets contribution to companies. Thus, reveal the disclosure status quo, existing problems and solutions of intangible assets information of IT industry in China's current economic environment, making the IT industry to better serve the economic development.

3 PROCESS OF EMPIRICAL RESEARCH

3.1 Introductions of research methods and models

This article chooses panel data analysis. Panel data refers to taking multiple sections in time sequence, in order to analyze the sample data in three dimensions (time, individuals and indicators), which is analyzing the differences between individuals by taking the time factor into account. Panel data analysis can not only overcome the multicollinearity problems troubled by time series, but also provide more degrees of freedom and greater estimate efficiency. In addition, it could reflect some combined effects of the ignored factors of time and individual differences. As this article is to examine the influence of intangible assets on profitability in the IT industry through panel data, it can be assumed that the slope coefficient is constant. Then, select the optimal model by F test and Hausman test among mixed regression models, fixed effects and random effects.

3.2 Selection of variables

What the sample selected is five financial indicators of listed companies in China's IT industry

from 2011 to 2013: return on total assets ratio, intangible assets ratio, debt asset ratio, current ratio, and comprehensive leverage. Where return on total assets ratio is as the dependent variable Y, intangible assets ratio as independent variable X1, and the remaining three indicators were used as control variables X2, X2, and X3.

Return on total assets ratio is the ratio of EBIT (Earnings Before Interest and Tax) to average total assets, which means the level of income obtained by all the assets of the company. It fully reflects the profitability and input-output state of the business, the higher the ratio, the more effective the operations of the company's assets. If the index is greater than the market rate, it means that companies can take full advantage of financial leverage to conduct debt management and obtain profits as much as possible. The return on total assets ratio is an important measure of profitability of corporate assets, this article use it as an index to measure the profitability of the IT industry.

Intangible assets ratio refers to intangible assets accounting for the proportion of total assets. IT industry is a high-tech industry based on the research and development. Development and production of electronic products require a lot of scientists and huge capital investment. As the relevant disclosure of the IT industry in the domestic market is insufficiency, the data of R & D expenditure could not be gained, so this article uses the data of intangible assets ratio to represent R & D capabilities of companies.

Debt asset ratio is the ratio of total liabilities to total assets, is a comprehensive index to evaluate company's debt level and a measure of the operation ability of companies by using the funds of creditors. In general, an appropriate level of debt asset ratio is from 40 to 60%. However, there are different expectations in different positions, for example, creditors want the ratio as low as possible, investors or shareholders consider that a higher ratio will bring greater benefits, while the operators concern about mostly is to make full use of the borrowed funds to bring profits for companies. This article will use debt asset ratio as a measure of the scale of corporate debt, belonging to control variables.

Current ratio is defined as the ratio of current assets to current liabilities, a measure of the solvency of companies. The higher the ratio, the stronger the solvency of companies, the higher the credit rating of companies, creditors are more willing to lend more money to companies, which could maintain the capital structure of a high debt ratio. However, the current ratio is too high to keep the efficiency of capital turnover and profitability. Generally, the reasonable minimum ratio is considered as 2.

Comprehensive leverage refers to the combined effects of the operating leverage and financial leverage, is the ratio of the gradient of net profit to the gradient of the main business income and a measure of the degree of influence on changes in sales to earnings per common share. In this article, comprehensive leverage reflects the risk level of the enterprise, as a control variable.

3.3 *To establish models*

3.3.1 *Choose the best model*

This article uses econometric software Eviews to analyze samples by panel data. The key of establishing models is to select the best one from mixed effects regression model, fixed effects regression models, and random effects regression models.

First, perform Hausman test and put forward a hypothesis:

H_0: the individual effects have nothing to do with the independent variables (random effects regression model should be established).

H_1: the individual effects are associated with the independent variables (fixed effects regression model should be established).

The results obtained by using the software is that Chi-Square Statistic = 16.103691, Probability = 0.0029. Thus, the chi-square test value is too high, and the P value is too low. Therefore, reject the null hypothesis that the random effects model should be established.

Then, by Hausman test, eliminate the random effects model. Now the F test can choose the best model between the mixed effects regression model and the fixed effects regression model. Similarly to present the hypothesis:

H_0: the intercept terms of different individuals in the models are the same (true model is mixed effects regression model).

H_1: the intercept terms of different individuals in the models are not the same (true model is fixed effects regression model).

F statistic is defined as: $F = [(SSEr−SSEf)/(N−1)]/[SSEf/(NT−NK)]$, Table 1 lists the values needed.

According to the formula of F statistic, $F = 4.806945$, and calculate the F critical value at the 5% significance level by Stata, F. INV (0.05, 161, 320) = 0.794400685. Significantly, F value is much greater than the F critical value at the 5% significance level, which illustrates that the probability of rejecting the null hypothesis reaches 95%. Therefore, the most suitable model of the sample is the fixed effects regression model.

3.3.2 *Establish the optimal model*

According to the conclusions of 3.3.1, establish a fixed effects regression model, obtain the regression equation of intangible assets ratio and return

Table 1. The values needed.

Sum squared resid		
None (SSEr)	1.654893	
Fixed (SSEf)	0.4841	
N	T	K
The number of individual section	The number of times	The number of explanatory variables
162	3	4

on total assets ratio under the three control variables in the IT industry:

$$Y = -0.303927X1 - 0.076417X2 - 0.0019X3 - 0.506EX4 + 0.102025 \ (R^2 = 0.979724)$$

R^2 value is very high, which indicates a high goodness of fit of the regression equation. There is a strong correlation between intangible assets and profitability in the IT industry. According to the coefficient of each variable in the regression equation above, in the IT industry, the impact of comprehensive leverage on behalf of companies risk level on the corporate profitability is least, the solvency and the debt scale followed, and shows negative correlation. But the most noteworthy is that the intangible assets ratio on behalf of the R & D capabilities and the return on total assets ratio on behalf of profitability shows negative correlation.

4 ANALYSIS OF THE EMPIRICAL RESULTS

The empirical results show that intangible assets ratio and profitability was negatively correlated in the IT industry, which seems contrary to common sense. But in fact, there are following reasons:

4.1 *The disclosure of intangible assets contains excessive land use rights*

For the IT industry of the high-tech industry, the benefits brought by the land use rights are actually not significant. Nowadays, accounting disclosure is to divide the land use rights as intangible assets, but did not subdivide the intangible asset in the financial statements or in the notes. Some IT companies continue to invest its assets to the land use rights vested in intangible assets, in order to get some international certification. Then, intangible assets do not achieve the practical effect on the economic, although the proportion of intangible assets of companies is increasing continuously.

4.2 The value of technological intangible assets is less

The value of technological intangible assets mainly incarnates in the high-tech industry, therefore the investment in technological intangible assets will bring the IT companies with substantial benefits. However, the domestic IT companies have focused on research and development, but they could not get results, the value of technological intangible assets is less, resulting that intangible assets owned by companies could not well represent their R & D capabilities. The reason for this is the lack of high-tech personnel, uncomprehensive information technology theory, inadequate machinery and equipment.

4.3 The most of intangible assets are purchased

The value of technological intangible assets in the IT industry is not only low, but also purchased in large part, instead of their own R & D. Although this can temporarily bring economic benefits to the company, it could not enhance corporate value, or reach the target of sustainable development, it either could not help to enhance the capabilities of research and development. IT companies only enhance their capabilities of research and development, in order to lay a solid foundation for the long-term development of companies.

5 ADVICES

Based on the above problems, IT industry and national policies need to be further improved in the following areas:

5.1 Improve the definition and disclosure of intangible assets in accounting standards

The definition of intangible assets is currently not unified; it's also not clear in the division of investment costs and capitalization of intangible assets. Thus, accounting standards require a clear definition of intangible assets, to expand the scope of intangible assets. The "controllability", "measurable" intangible assets should be confirmed and revealed appropriately, making the carrying value of intangible assets reflect its true value better. Because the company owns a large part of the broader intangible assets which are not recognized by accounting standards and these intangible assets play an important role in corporate earnings growth.

Moreover, it is lack of disclosure of intangible assets; the requirements for disclosure of intangible assets are too simple in accounting standards.

Therefore, it needs to increase the disclosure of intangible assets in financial statements; the specific constituent name and total amount of intangible assets need disclosure in the notes, just as fixed assets in the notes divided into houses, buildings, machinery and equipment, and transport equipment. In addition, it needs to strengthen the behaviors of disclosure of intangible assets of listed companies, making the names of the specific items of intangible assets to be unified.

5.2 To increase investment in human resources and training of high-tech enterprises

High-tech enterprise talents are the soul of the IT industry, the country should focus on training high-tech enterprise personnel and increase investment, while companies should cultivate technical personnel belonging to the corporate. It is a key to prevent the loss of talents from the enterprise. According to statistics, the total attrition rate is highest in the electronic information companies, because researchers have expertise and more options of employment opportunities in scientific research departments, expectations of realizing their values are also relatively higher, so the attrition rate would be higher. The essential reason is that there is a huge gap in the standard of living and level of social development between countries and regions in the world, China is still in the developing countries and far less than Western developed countries. Therefore, the national high-tech enterprise personnel study abroad, in order to seek better development space. For high-tech companies, to avoid the loss of talents from the companies not only relies on their own strength, but also needs abundant material wealth and higher level of development of the country for the foundation. The businesses can build an attractive salary system and corporate culture with the sense of community, to retain technical personnel from the material and value talents from the emotion.

5.3 To increase the protection of intangible assets

The science and technology of the knowledge economy era are changing fast; a lot of intangible assets have strong timeliness, especially technological intangible assets. And the decision of the level of corporate profits not only relies on the amount of intangible assets, but also depends on the ability of enterprises to use them. Therefore, enterprises should make full use of existing intangible assets within the validity period, to improve efficiency in the use of intangible assets. The government should set some appropriate supportive policies to help companies use intangible assets effectively; it is an important prerequisite of achieving systematic and comprehensive relevant laws and regulations to maximize the value of intangible assets. As long as the government improves relevant laws and regulations of intangible assets' management and increase efforts to protect intangible assets of the enterprises, it can encourage companies to increase investment in intangible assets and technological intangible assets, and improving the degree of their use.

REFERENCES

Abrahams, Tony & K. Sidhu. The Role of R&D Capitalizations in Firm Valuation and Performance Measurement [J]. Australian Journal of Management, 1998, (23):169–184.

Gupta, Neeraj J. Do stock prices reflect the value of intangible investments in customer assets? [J]. Dissertation Abstracts International Section A: Humanities and Social Sciences, 2008, (68): 4808.

Huang Tongcheng, Yang Jian. Quantitative Analysis of High-tech Listed Company's Profitability Factors [J]. China Management Science, 2002(8):13–17.

Lev, B. & P. Zarowin. The boundaries of financial reporting and how to extend them. [J]. Journal of Accounting Research, 1999, (37):353–385.

Ma Yu. Correlation Study between Intangible Assets and Business Performance based on the IT Industry [D]. Tianjin University of Finance and Economics, 2009.

Qi Wangyue. Correlation Study between Intangible Assets and Enterprise Value—Based on Further Testing of Subsidiary Ledger of Intangible Assets of IT Industry [D]. Southwestern University of Finance and Economics, 2009.

Shivaram Rajgopal, Mohan Venkatachalam & Suresh Kotha. The Value Relevance of Network Advantages: The Case of E-Commerce Firms [J]. Journal of Accounting Research, 2002, (41):135–162.

Sun Yanxia. Scoping and Measurement of Intangible Assets [N], Technology Consulting Herald, 2006(2): 77–79.

Wu Ruizhi, Luo e'xiang, Qian Shengsan. Study on the Relationship between Intangible Assets and Profitability of Listed Companies [N]. Journal of University of Shanghai for Science and Technology, 2013 (2): 135–139.

Wu Shangrong. Analysis of the Main Drivers and Development Trends of China's IT Industry [J]. Electronic Commerce, 2007 (8):57–58.

Information Technology and Applications – Li (Ed.)
© 2015 Taylor & Francis Group, London, ISBN 978-1-138-02677-3

The research on growth evaluation of Chinese IT listed companies

Yu-jin Fu

School of Business, Hohai University, Nanjing, P.R. China

ABSTRACT: On the basis of the data of Chinese IT listed companies which can be widely obtained, this paper tries to extract the key factors that affect the growth capacity of IT listed companies, and seek out the internal relationship between the growth ability and its main influential elements by using factor analysis. The five key factors can be defined as innovation; growth speed; growth efficiency, and growth equality. Next, comprehensive evaluation of Chinese IT listed companies is gotten by using entropy method creatively. Finally, some suggestions are presented to optimize the growth ability of IT listed companies in China based on the above evaluation.

Keywords: IT listed companies; growth evaluation; factor analysis; entropy method

1 INTRODUCTION

According to Miller and Modigliani, growth means value creation. Today, growth is more and more important. Growth is the core of a listed company, the life in the stock market, the motive force of the sustainable development of national economy and an important indicator which measures the management abilities and development prospect of the company. Scholars such as Manigart, S., Zhou Zhidan etc. have put a lot of effort on studying the principal elements that affect the growth ability most, and built up evaluating systems aiming to strengthen companies' growth ability. On the basis of the previous study, this paper will evaluate the growth ability of Chinese IT listed companies below.

2 EXPERIMENTAL SECTION

2.1 *Introduction of factor analysis and entropy method*

2.1.1 *Introduction of factor analysis*
Factor analysis is a multivariate statistical method for reducing a large number of complex variables to fewer underlying dimensions.

It is a proper method for evaluating the growth of listed companies because the method eliminates the overlapped information of primitive indexes. It makes the result more rational and reliable.

2.1.2 *Introduction of entropy method*
Entropy is a measurement of the disorder degree in a system. It serves as a criterion for the degree of difference among the samples.

2.2 *Sample selection and data sources*

This paper selected 116 Chinese IT listed companies from Shanghai Stock Exchange and Shenzhen Stock Exchange, eliminating those having incomplete information, changing main business or having a record of ST or *ST. 15 index data from 2011 to 2013 are used to evaluate the growth capacity of Chinese IT listed companies comprehensively.

2.3 *Variable descriptions*

Growth ability of a company can be divided into two main parts: one is a condition of high-speed development on the purpose of maximizing enterprise value based on existing resources; another is the room for further development and progress.

Growth ability of Chinese IT listed companies is evaluated comprehensively from five dimensions, including innovation, risk, growth speed, growth efficiency, and growth equality in this paper. Table 1 shows 13 specific variables used to describe the growth capacity of a company from those five dimensions.

2.4 *Factor analyses*

2.4.1 *Feasibility test of factor analysis*
First, this paper uses KMO and Bartlett's Test to inspect whether the correlation between indicators are feasible to factor analysis (Table 2). The KMO is 0.675, more than criteria 0.6, while Bartlett sphere test value is 2220.176 (the significance level of 0). As a result, it is possible to do factor analysis due to the certain correlation between indicators.

Table 1. Variable description of growth indicators.

Variable type	Symbol	Variable name
Innovation	X1	Intangible assets/total assets
Risk	X2	Equity ratio
	X3	Current ratio
	X4	Quick ratio
	X5	Cash ratio
Growth speed	X6	Total assets growth rate
	X7	Net assets growth rate
	X8	Net profit growth rate
Growth efficiency	X9	Ratio of profits to cost
	X10	Accounts receivable turnover
	X11	Inventory turnover
Growth equality	X12	Total assets payment growth rate
	X13	Net assets profit growth rate

Table 2. KMO and Bartlett's test.

Kaiser-Meyer-Olkin measure of sampling adequacy.		0.675
Bartlett's test of sphericity	Approx. chi-square	2220.176
	Sig.	0.000

Table 3. Communalities.

	Initial	Extraction
Intangible assets/total assets (X1)	1.000	0.785
Equity ratio (X2)	1.000	0.654
Current ratio (X3)	1.000	0.950
Quick ratio (X4)	1.000	0.952
Cash ratio (X5)	1.000	0.884
Total assets growth rate (X6)	1.000	0.969
Net assets growth rate (X7)	1.000	0.951
Net profit growth rate (X8)	1.000	0.990
Ratio of profits to cost (X9)	1.000	0.668
Accounts receivable turnover (X10)	1.000	0.858
Inventory turnover (X11)	1.000	0.694
Total assets payment growth Rate (X12)	1.000	0.990
Net assets profit growth rate (X13)	1.000	0.994

2.4.2 Establish factor variables

According to Table 3, the communalities of all the variables are higher than 60% or even 90%, indicating that factors been extracted have covered most information of initial variables.

By analyzing Table 4, we could merge all the 13 variables into 5 kinds of common factors with the Initial Eigen values larger than 1. Their contribution rates to the growth capacity of Chinese IT listed companies are 30.737%, 22.974%, 15.473%, 8.665 and 8.603%. The cumulative variance is about 86%.

2.4.3 Factor component matrix

From the rotated factor component matrix in Table 5, we can find that the first common factor has big load on the indicators of X4, X3, X5, X2, and X9 which can be defined as risk factor. The second common factor that has big load on the indicators of X12, X3, and X8 can be defined as growth equality factor. Also, the third common factor has big load on the indicator of X6 and X7 which can be defined as growth speed factor; while the forth common factor that has big load on the indicators of X10 and X11 can be defined as growth efficiency factor. Finally, the fifth common factor has big load on the indicators of X1 which can be defined as innovation factor.

2.4.4 Scoring of factors

Factor scoring is the ultimate expression of factor analysis. The scores about 5 common factors are cited in Table 6. We can calculate the scores of the factors for each observation based on the coefficient and initial variables. After factor scoring, we can achieve the effect of dimensionality reduction by replacing initial variables with the scores of factors.

2.5 Comprehensive scoring by the entropy method

Finally, we used the entropy method to determine the weight of evaluating indicators, calculate the composite score reflecting each sample company's growth ability on the basis of the results of factor analysis, and then ranking the sample companies according to the scores.

$$P_{ij} = \frac{F_{ij}}{\sum_{i=1}^{n} F_{ij}} \tag{1}$$

$$S_j = -\frac{1}{\ln n} \sum_{i=1}^{n} P_{ij} \, Ln P_{ij} \tag{2}$$

Following the above formulas, the entropy of 5 common factors (S_j) is (0.2231, 0.1898, 0.1900, 0.3543, and 0.2928).

$$D_j = 1 - S_j \tag{3}$$

$$W_j = \frac{D_j}{\sum_{j=1}^{m} D_j} \tag{4}$$

Table 4. Total variance explained.

	Initial eigen values			Rotation sums of squared loadings		
	Total	% of variance	Cumulative %	Total	% of variance	Cumulative %
1	4.241	32.625	32.625	3.996	30.737	30.737
2	2.960	22.771	55.396	2.987	22.974	53.711
3	1.837	14.133	69.529	2.011	15.473	69.184
4	1.186	9.126	78.655	1.126	8.665	77.849
5	1.014	7.797	86.452	1.118	8.603	86.452
6	0.811	6.236	92.687			
7	0.482	3.707	96.395			
8	0.310	2.386	98.781			
9	0.078	0.601	99.382			
10	0.058	0.444	99.826			
11	0.015	0.118	99.945			
12	0.006	0.046	99.991			
13	0.001	0.009	100.00			

Extraction method: principal component analysis.

Table 5. Rotated component matrixes.

	1	2	3	4	5
X4	0.974	0.019	0.041	−0.030	0.009
X3	0.973	0.009	0.037	−0.046	−0.008
X5	0.934	0.015	0.102	0.018	0.035
X2	−0.801	0.034	0.030	−0.098	−0.030
X9	0.724	0.130	0.316	0.158	−0.036
X13	0.016	0.996	−0.001	0.017	−0.023
X12	0.046	0.994	0.007	0.014	0.020
X8	0.014	0.988	0.120	0.002	−0.002
X6	0.099	0.069	0.973	−0.046	0.071
X7	0.124	0.028	0.966	−0.019	−0.042
X10	0.036	0.001	−0.054	0.923	−0.049
X11	0.168	0.068	0.016	0.592	0.459
X1	−0.075	−0.043	0.012	−0.154	0.868

Extraction method: principal component analysis.
Rotation method: varimax with Kaiser normalization.
a. Rotation converged in 4 iterations.

Table 6. Scores of common factors.

	Component				
	1	2	3	4	5
X1	−0.018	−0.006	−0.011	−0.195	0.793
X2	−0.210	0.021	0.073	−0.036	−0.014
X3	0.259	−0.007	−0.060	−0.101	−0.012
X4	0.258	−0.004	−0.058	−0.087	0.002
X5	0.239	−0.009	−0.020	−0.039	0.021
X6	−0.049	−0.017	0.499	−0.005	0.048
X7	−0.042	−0.033	0.499	0.027	−0.057
X8	−0.016	0.331	0.020	−0.018	0.000
X9	0.157	0.019	0.113	0.114	−0.055
X10	−0.043	−0.022	0.017	0.840	−0.108
X11	0.010	0.012	0.004	0.366	0.500
X12	0.001	0.337	−0.042	−0.017	0.022
X13	−0.006	0.338	−0.043	−0.010	−0.017

Extraction method: principal component analysis.
Rotation method: varimax with Kaiser normalization.
Component scores.

Following the above formulas, the weight of 5 common factors (W_j) is (0.2072, 0.2161, 0.2160, 0.1722, and 0.1886).

$$Z_i = \sum_{j=1}^{n} W_j P_{ij} \qquad (5)$$

The final composite score calculated by adopting factor analysis and entropy method implies that GHYX (600037), YGRJ (002063), SWTX (002405), GLD (002410) and GDWL (600831) score higher than other companies. However, if we only use entropy method to determine the weight the evaluating indicators, companies with higher scores will be GHYX (600037), YZJ (300085), GLD (002410), YGRJ (002063) and DRZB (300183). From the above comparison it can be observed that ranking results will be different depending on different evaluating methods. Using entropy method directly neglects the overlapped information of primitive indexes, thus resulting in irrationality in the ranking of growth ability. By using factor analysis combined with entropy method, we can remove both application correlations between indexes and repeatability, at the same time, these two methods both reflect the effect of variation of each index to comprehensive

Table 7. Composite scores.

Entropy method combined with factor analysis		Entropy method only	
Item	Score	Item	Score
GHYX	16.8844	GHYX	10.4366
YGRJ	12.4734	YZJ	7.4531
SWTX	10.2448	GLD	7.0665
GLD	10.1989	YGRJ	6.8066
GDWL	9.1385	DRZB	6.6450

evaluation. The result will be more rational and reliable.

3 CONCLUSION AND SUGGESTION

3.1 Conclusion

By factor analyzing, we could merge all the 13 variables that affect the growth capacity of a company into 5 kinds of common factors, namely, risk, growth equality, growth speed, growth efficiency and innovation. Their contribution rates to the growth of Chinese IT listed companies are 30.737, 22.974, 15.473, 8.665, and 8.603%.

The empirical research has shown that, as a whole, Chinese IT listed companies have good growth propensity and bright growing prospect. The average net profit growth rate reached 0.1280. The final composite score implies that GHYX (600037), YGRJ (002063), SWTX (002405), GLD (002410) and GDWL (600831) score higher than other companies.

Using entropy method directly to estimate the growth ability of Chinese IT listed companies will neglect the overlapped information of primitive indexes, thus resulting in irrationality in the ranking of growth ability. By using factor analysis combined with entropy method, we can remove both application correlations between indexes and repeatability, at the same time; these two methods both reflect the effect of variation of each index to comprehensive evaluation. The result will be more rational and reliable.

3.2 Suggestions

3.2.1 Improve the specialization level of cooperation in IT industry through asset reorganization
Facing the growth problems brought by low industry concentration rate and poor resources utilization efficiency, IT industry should strengthen asset restructuring and the integration of resources to reduce the cost of competition, cut transaction costs, increase profits and improve anti-risk ability.

3.2.2 Vigorously promote technological progress in IT industry
The constraints analysis about the growth ability of IT companies shows that IT companies widely have the problem of weak R & D efforts and low scientific and technological achievements conversion rate. Vigorously promote scientific and technological progress, will become an essential way to improve the growth ability of IT listed companies. On the one hand, improving the scientific research ability of IT listed companies needs the support from the government. On the other hand, Chinese IT listed companies should strengthen the international technology cooperation to improve their scientific research ability.

3.2.3 Improve the competitiveness of IT cooperation comprehensively
The most prominent feature of the listed company's competitiveness is the combination of human resource and advanced products. The strategies used to promote the competitiveness of products and innovations have been discussed before. Four measures promoting the competitiveness of human resource will be proposed below. First, cultivate a unique brand culture. Second, focus on the innovation of marketing and service. Third, build the core capability of human resource. Finally, build a unique corporate culture.

REFERENCES

[1] Manigart, S., Wright, M., and Robbie, K., et. al, Venture Capitalists' Appraisal of Investment Projects: An Empirical European Study [J]. Entrepreneurship Theory and Practice, Summer. 1997:29–43.
[2] Bottazzi, Giulio and Angelo Secchi, A. Stochastic model of firm growth. Physical A, 2003, 3(24): 213–219.
[3] Liu Zhaode. A Study of Evaluation Theory about Enterprise Growth of High-tech SMEs [J]. Science & Technology Progress and Policy, 2009(24): 98–101.
[4] Zhou Zhidan. Study on the growth evaluation of growing high-tech enterprises—an empirical analysis based on Ningbo [J]. Science Research Management, 2010(4):9–16.
[5] Qu Wenbin. Research on the growth of listed companies of high and new technology in Hubei based on catastrophe progression method [J]. Science & Technology Progress and Policy, 2012(10):104–108.

Information Technology and Applications – Li (Ed.)
© *2015 Taylor & Francis Group, London, ISBN 978-1-138-02677-3*

Case study on EIT industry-university-research collaborative innovation platform

Jie Wang & Hongliang Wang
Hohai University, Nanjing, China

ABSTRACT: Industry-university-research collaborative innovation is the key to carry out innovation-driven growth strategy. China is implementing the "2011 Plan", trying to build the national level industry-university-research collaborative innovation platform. EIT (European Institute of Innovation and Technology) started earlier and more successful industry-university-research collaborative innovation platform aimed at revitalizing the EU economy. In this case, we studied EIT platform case from the analysis of structure and collaborative innovation path, digging its internal mechanisms to achieve collaborative innovation. On this basis, we will make recommendations for the implementation of China collaborative innovation centers and industry-university-research collaborative innovation platform.

Keywords: EIT; industry-university-research; collaborative innovation; platform

1 INTRODUCTION

The industry-university-research collaborative innovation is an important part of the world's construction of national innovation system. China has started the "2011 plan" in 2012, aims to collaborative innovation center as the carrier, to establish a national collaborative innovation platform. In the Lisbon strategy (2005), the EU proposed the innovation economy and planed to establish EIT (European Institute of Innovation and Technology) to realize the innovation of EU level. After 5 years, the platform has made remarkable achievements in industry-university-research collaborative innovation, which has important implications for the establishment of China's collaborative innovation platform.

2 BACKGROUND

In recent years, Europe has recognized that although Europe owes the world famous universities and research bases, filled with vigor and vitality of enterprises and talents, but these subjects did not form a joint force, which result the lack of innovation culture, fewer ideas can be transformed into a new product or service. In order to promote social and economic sustainable development and national innovation capability, the EU launched EIT under the guidance of the goal in Lisbon Strategy in April, 2008. The mission is to solve the problem of innovation, promote sustainable economic growth in Europe and improve all aspects of the competition in Europe. EIT also creates a statute explicitly to the KIC (Knowledge and Innovation Communities) operation of the whole platform model. But until 2010, EIT identified three KICs in key areas.

3 THEORY

3.1 *Industry-university-research collaborative innovation*

At present, the definition of the domestic and foreign industry-university-research collaborative innovation is basically the same. However, the present research is mainly from the causes (Lee 1996, Geuna & Nesta 2006), mechanism (Wang Yi & Wu Guisheng 2001, Perkmann & Walsh 2007), models (Fontana & Geuna 2006, Yu Hebing 2012), technical characteristics and knowledge management (Jensen et al. 2003, Wu Yue & Gu Xin 2012) and efficiency evaluation (Xie Zhiyu 2004, Bercovitz & Feldman 2008) perspective, which emphasize the synergistic factors have the parties to achieve innovation research, but lacking of research on how to make body to release its own elements. Only breaking the barriers between the main institutional mechanisms, we can establish a collaborative innovation platform that the main body of industry-university-research collaborative innovation can share elements.

Industry-university-research collaborative innovation can be divided into three levels: micro

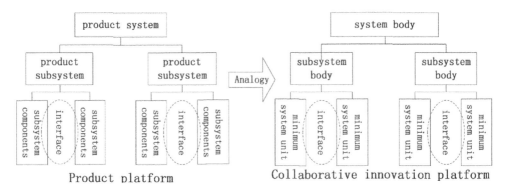

Figure 1. Collaborative innovation platform architecture and product platform analogy.

level—taking enterprise as the object of study of individual cooperative; meso level—in a particular industry as the object of study of alliance cooperation; macro level—with national collaborative research between objects. Collaborative innovation platform is the foundation of nation level. It can be for collaborative innovation laying structure of meso and micro level, to guide the direction of. In this paper, EIT university industry collaborative innovation platform belongs to the macro level.

3.2 Collaborative innovation platform

In the study of the platform, the most mature one is the research of product platform. America Northwestern University professor Meyer defined that product platform is composed of a collection of subsystems and interfaces of public structures. By means of this collection, it can develop and manufacture a range of products which have a competitive advantage smoothly and efficiently (Meyer 1997). Baldwin & Clark (1997) puts forward three features of product platform: (1) the modular structure; (2) interface (module interaction and communication architecture and interface); (3) standard (design principle modules follow). Analogous to the definition and characteristics of the product platform, we define the collaborative innovation platform as a hierarchy of subsystems subject and structure of the interface, which can break the barriers of institutional mechanisms between the minimum system unit, releasing factor, and enabling collaborative innovation. The basic architecture of collaborative innovation platform is shown in Figure 1.

The ultimate goal of collaborative innovation platform is to make the body can break the system barriers of the minimum system unit—between industry, university and research and release factors. But how to break the system barrier requires longitudinal interface management, as well as internal minimum system unit for horizontal interface management. Therefore, we will study the EIT case from the following two aspects: first, analyze the structure of the EIT platform. Second, analyze the interface management of EIT, and research the realization path of collaborative innovation.

4 STRUCTURAL ANALYSIS OF EIT PLATFORM

According to the definition of collaborative innovation platform, we will analyze the EIT platform structure from the three part—module, interface and standard.

4.1 The EIT platform module

Module structure of EIT platform includes EIT Governing Board, KICs, and the partnerships under the co-location in the center of the KIC.

4.1.1 EIT governing board

The Board is a permanent management institution of EIT. It is entrusted with the overall command of the headquarters of the implementation of the EIT strategic leadership and operational activities. It is an independent agency. It can independently make decisions and be responsible for the evaluation and selection, sponsorship of KIC.

Presently, EIT Governing Board brings together 22 high-level professionals. They all have outstanding expertise in the field of higher education, research, business and innovation. There are 18 appointed members and 4 representative members in it. The appointed members are European experts. They can provide experience in enterprises, higher education and research; representative members responsible for providing the policy recommendations for the Board from KIC's perspective. The list of candidates is nominated by the KIC itself from the higher education, research institutions and innovation partner institutions. This constitutes of the Board is to ensure that all important decisions are able to consider the needs of balancing the interests of industry-university-research body.

4.1.2 *KIC*

KIC refers to the independent cooperation relationship between universities, research institutions, enterprises and other partner organizations with the form of strategic network formed in the process of innovation. The theme of KIC is usually based on long-term innovation planning, which in order to promote the establishment of EIT regulations target.

In 2010, according to the 2020 strategic goal of sustainable social and economic development, EIT has launched three KICs. They are Climate-KIC, EIT ICT Labs, and KIC InnoEnergy. KIC has launched a series of activities cover the whole chain of innovation, effectively promoted the process of the research to the market, innovation projects and enterprise incubation.

4.1.3 *Co-location centers and internal partner organizations*

KIC partner organizations means members of KIC, including higher education institutions, research institutions, public or private enterprises, financial institutions, regional and local authority departments, foundations and non-profit organizations (2014). Partner organizations are divided into two categories. Among them, the core partner organizations are excellent organization of world-class, fully committed to the KIC operation and to raise funds for the KIC. Affiliated partner organizations participate only in the KIC project activities. These partner organizations can communicate and share innovative ideas through activities in the co-location center. At present, there are 17 co-location centers spread all over the EU, they are activity bases of the partner organizations.

4.2 *EIT platform interface*

The EIT platform is composed of three levels of system structure, that is, macro-, meso- and micro-systems. Macro-system consists of EIT board and three KICs. Meso-system consists of each KIC and its affiliated co-location center. Micro-system consists of inner parts of each co-location center. Similarly, the EIT platform also contains two longitudinal crossing level interfaces and a parallel hierarchical interface (Fig. 2). Two longitudinal crossing level interfaces are respectively EIT-KICs macro-system interface (as the interface ① shown in Fig. 2), the meso-system interface (as the interface

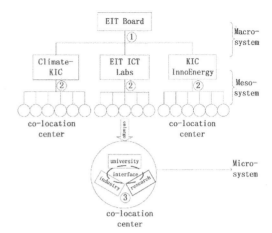

Figure 2. The EIT platform structure model.

② shown in Fig. 2), and the micro-system interface (as the interface ③ shown in Fig. 2).

5 ANALYSIS OF COOPERATIVE INNOVATION PATH TO THE EIT PLATFORM

If the EIT platform wants to achieve collaborative innovation, it needs to get rid of the system and mechanism system barriers between all the parts with the interface management. Then, the parts will release their innovative elements. Longitudinal cross level interface in EIT platform is macro- and meso-system interface, and parallel level interface is the micro-system interface. This paper will analyze the cooperative innovation path to the EIT platform from the management of three interfaces.

5.1 *Interface management of EIT-KICs*

Interfacial activity of longitudinal and cross level is the interaction between the main vertical relationship bodies. Two types of body are in a non equal status in authority, scale, nature and the allocation of resources. The EIT-KICs macro-system interface is a typical vertical interface. Not only the KIC theme directly designed by EIT governing board, but also each KIC must be in accordance with the requirements of EIT to carry out the work. However, the information communication in this level of interface is not one-way. Because the EIT board is not responsible for industry-university-research collaboration behavior, if they cannot get feedback from KIC, they are unable to control the micro-system, which leads to the final result that collaborative innovation strategy deviation high intention.

Thus, EIT board has carried out monitoring and management. Annually, KIC will report to EIT board the monitoring content of KIC level and cross-KIC level in the performance appraisal system. EIT board will have a grasp on the whole for the performance of KIC, which is easy for the EU to make performance appraisal on EIT platform. Each KIC will also give help which the micro-system need to the EIT board, the way is representative members of EIT. Representative members of EIT are chosen from the KIC, which is an exemplification of the KIC feedback mechanism in EIT. This interface management in macro-system breaks the barriers between EIT board and KIC from the macro level and achieves smooth operation in macro-system.

5.2 *Interface management of KIC- co-location center*

KIC-co-location meso-system interface also belongs to the cross-layer interface, and the two sides exist affiliation. However, the interface management needs bidirectional information flow at the interface between the two sides, so as to realize the goal of interface management. Therefore, information feedback mechanism must also be set up in the vertical interface management. From KIC's point of view, partner organizations must obey the KIC targets and achieve KIC levels of performance appraisal content. Each partner organization is able to share resources, information in the co-location center, and obtain funds or other elements from KIC. The benefit mechanism makes the number of KIC partners climbed steadily in recent years. Management activities in meso-system interface realize the coordination of KIC and micro-system unit.

5.3 *Interface management of industry-university-research*

The main parts of micro-system are partners in the co-location centers. The system management of this interface includes regulations, the exchange of information channels and promote cooperation atmosphere. The micro-system of rules and regulations is mainly EIT rules, EIT rules for the management of production, learning and research, the parties from the system level, the university industry partnership cooperation can have a support system. The rules of micro-system is EIT rules, EIT rules manage all the parts of industry-university-research from the system level. The main way to achieve information exchange is the activities, related education and training in co-location centers. Promote the atmosphere of cooperation including the establishment of a unified goal,

cultivating culture of innovation, increasing the trust and improving the mechanism of interests. The goal of micro-system is the mutual cooperation of partner organizations, working together to solve the main problems of specific KIC field, to realize the innovation in this field. Interest mechanism is reflected in the benefits that the university provides guidance classes for the enterprises and research institutions, and the enterprise is able to get the research achievements of the university and research institutes, and so on.

6 CONCLUSIONS AND SUGGESTIONS

6.1 *Case summary*

In the interface management of EIT-KICs, EIT board and KIC realize the coordination in macro-system level by monitoring management and feedback mechanism. In the interface management of KIC- co-location center, KIC and partners in co-location center realize the coordination in meso-system level by bidirectional interaction. In the interface management of industry-university-research, partner organizations in the co-location center realize the coordination. Finally, EIT realizes the coordination from the EIT board to KIC to co-location center (Fig. 3).

6.2 *Suggestions advices for the construction of collaborative innovation platform in China*

China's "2011 plan" collaborative construction mode of collaborative innovation platform and the internal mechanism of EIT collaborative

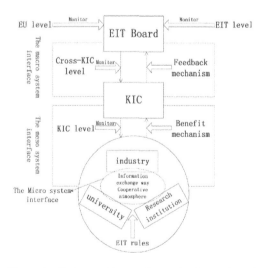

Figure 3. Path to achieve collaborative innovation of EIT platform.

innovation platform are very similar. According to the overall operation mechanism of EIT platform, we make the following suggestions for the implementation of China's "2011 Plan" collaborative innovation center and building innovative collaborative industry-university-research platform.

To make the duties of the main body clear, as well as the relationship between the main parts of each system level. China's "2011 plan" formulates that identified units of the collaborative innovation center is the Ministry of education and the Ministry of Finance. From the present situation, there are no complete regulations to make the duties of Ministry of education, Ministry of finance, and the main body clear, as well as the relationship between the main parts. 2011 Collaborative Innovation Center in our country is divided into four aspects in the collaborative innovation model oriented, but with no specific department responsible for every aspect, which leads to a control from national level directly to center level. This kind of management is not scientific. Therefore, our country must be combined with the actual situation, introduced corresponding policy to clear the duties of main parts, including coordination mechanism, evaluation mechanism, and so on.

To establish monitoring and feedback mechanisms between competent ministries and collaborative innovation centers. Every level of EIT has a corresponding monitoring and feedback mechanisms, making the entire platform keep an efficient dynamic management. Our country has no monitoring and feedback mechanism. It cannot be check that whether the qualification process is benefit for the main part of industry-university-research collaboration. China should learn from EIT platform model, develop monitoring and feedback mechanism.

To establish a cultural system of innovation. EIT platform often promotes the study for college students to enterprises and research institutions, which make them full of pioneering and innovative spirit. Similarly, the university will also provide training courses for entrepreneurs, companies and research institutions. Besides, KIC also provide business support services for entrepreneurs with innovative ideas outside the platform. These practices all guarantee the formation of innovation culture in EIT platform. China also has best industries, universities and research institutions. Therefore, China can also promote the cultural exchange and innovation between the main part of industry-university-research, so that the whole platform to build the cultural system of innovation.

REFERENCES

Baldwin, C.Y. & Clar, K.B. 1997. Managing in an age of modularity. *Harvard Business Review* 75(5):84–93.

Bercovitz, J. & Feldman, M. 2008. Entrepreneurial universities and technology transfer: a conceptual framework for understanding knowledge-based economic development. *Journal of Technology Transfer* 31(1):175–188.

Du Lanying. & Chen Xin. 2012. Study on Scientific Research Collaborative Innovation Mechanism and Pattern—An Example of Small and Medium Enterprises. Science & Technology Progress and Policy 29(22):103–107.

EIT Regulation [EB/OL]. [2014-4-13] http://eur-lex. europa.eu/LexUriServ/LexUriServ.do?uri=OJ:L:2013: 347:0174:0184:EN:PDF.

EUROPEAN COMMISSION. Working together for growth and jobs: a new start for the Lisbon strategy, European Commission, Brussels, 2005(24):234–345.

Fontana, R. & Geuna, A. & Matt, M. 2006. Factors affecting university-industry R&D projects: the importance of searching, screening and signaling. *Research Policy* 35(2):309–323.

Geuna, A. & Nesta, L. 2006. University patenting and its effects on academic research: the emerging European evidence. *Research Policy* 35(6):790–807.

He Yu-bing. 2012. The theoretical model of I-U-R collaborative innovation. *Studies in Science of Science* 30(2):165–174.

Jensen, R.A. et al. 2003. Disclosure and licensing of university inventions. *International Journal of Industrial Organization* 21(9):1271–1300.

Jiang Shi-mei. et al. 2012. Industrial cluster I-U-R synergy innovation mechanism—case study of Baoding new energy & electric power transmission and distribution cluster. *Studies in Science of Science* 30(2): 207–212.

Lee, Y.S. 1996. Technology transfer and the research university: a search for the boundaries of university-industry collaboration. *Research Policy* 25 (6):843–863.

Meyer, M.H. & Lehnerd, A.P. 1997. *The power of product platform: building value and cost leadership.* New York: Free Press.

Perkmann, M. & Walsh, K. 2007. University-industry relationships and open innovation: towards a research agenda. *International Journal of Management Reviews* (9):259–280.

Wu Guis-heng. & Wang Yi. 2001. Sticky Knowledge Transfer in University-Industry Collaboration. *Science Research Management* 22(6):114–121.

Wu Yue. & Gu Xin. 2012. Knowledge Collaboration Process in Industry-university-research Institute Collaborative Innovation. *Forum on Science and Technology in China* (10):17–23.

Zhou Pei. & Zhang Daoyun. & Yao Shiwu. 2013. Research on Multiple Interactive Collaborative Innovation and Enterprises. *Management World* (8):181–182.

Information Technology and Applications – Li (Ed.)
© 2015 Taylor & Francis Group, London, ISBN 978-1-138-02677-3

Empirical study on relationship between equity incentive and R&D expenditure of information technology listed companies

Yuxuan Yin & Wenlin Gu
School of Business, Hohai University, Nanjing, China

ABSTRACT: Based on the principal-agent theory and incentive theory, this article contrasts a panel data model by using the financial data of information technology listed companies from 2011 to 2013 to discuss the relationship between equity incentive and R&D expenditure. Using the method of grouped regression, the essay has reached the following conclusion: (1) Equity incentive can be conducive to increase the spending on research and development. (2) Profitability is positively correlated to the expenditure on research and development. (3) Development ability is positively related to the R&D expenditure. (4) The asset-liability ratio and R&D spending have a negative relation. The conclusion of this article can further verify the necessity of increasing the spending on research and development of enterprises and the positive role of implementing equity incentive. It also provides the theory basis for improving the mechanism of R&D input and optimizing the equity incentive system.

Keywords: equity incentive; R&D expenditure; information technology listed company

1 INTRODUCTION

As the first productive power, science and technology play an important role in the process of promoting China's economic development. Followed by the strategy of being an innovative country, China improves the government R&D spending as a share of GDP step by step recently. In order to form a unique competitive advantage and maintain a sustained growth momentum in the fierce competition, increasing R&D expenditure has become the main method in today's high technology industry. Especially for information technology listed companies, most of which are technology intensive enterprises and have the characteristic of high growth and high liquidity. By being shareholders through means of technology, they have the double identities as core executive and core technical staff, therefore equity incentive will be more frequent in these companies. However, whether the R&D expenditure will be conducive to the enterprise performance need to be discussed. The study on this issue will be helpful to optimize equity incentive system, improve the R&D mechanism and improve the core competitiveness, thus realizing high-speed development of the enterprise.

2 LITERATURE REVIEW

Most scholars take the listed companies in Shanghai and Shenzhen stock market as the samples to study the influence imposed by the executive equity incentive on the corporate R&D spending. Among them, Chen Xiaodong conducted the survey from perspective of different motivation of equity incentive, he found that motivational equity incentive can significantly improve the level of performance and welfare equity incentive does not have obvious effect. According to the different nature of controlling shareholders, Xie Shangwei found that the promoting effect in state-owned company is more obvious. Studying from different sectors, Liu wei found that equity incentive plays a positive role in high technology enterprise and there is no obvious correlation between them in the non-high technology enterprise. Hsiang-Lan Chen takes the high technology enterprises in Taiwan as research object and reaches the same conclusion. By constructing a panel data, Jiangfeng Wu found that the positive correlation is more significant when spare resources are sufficient or corporate performance is high. However, from the endogenous perspective, Tang Qingquan found that the correlation will turn to be negative when the manager ownership reaches to a certain level. Moreover, Balkin take the high-tech enterprises in the United States as the research object and also found that there is no obvious correlation between them.

3 RESEARCH DESIGN

3.1 *Research hypothesis*

According to the agency cost theory, based on the principal agent relationship, corporate management can obtain more information compared

Table 1. Definition and measurement of the variables.

Variable type	Code	Definition	Measurement
Dependent	RDI	R&D intensity	R&D expenditure/total assets
Independent	MSR	Executives shareholding ratio	Executive share/total share capital
	ROE	Financial performance	Net profit/net assets
	TBQ	Market performance	Enterprise value/total equity
	DAR	Asset-liability ratio	Total debt/total assets
	GROW	Development ability	Growth rate of total revenue
Controlled	SIZE	Enterprise scale	LN (total assets)

with outside investors when there exists conflicts of interest between shareholders and managers. Facing the risk of moral hazard and adverse selection, the management may cause damage to shareholder interests. Achievements in scientific research by increasing the spending on research and development can be converted to realistic productivity, thus effectively improving the production efficiency of enterprises and bringing excess profits. However, with high investment, big risk and lagged yield, managers are unwilling to increase spending on research and development. Executive shareholding can not only motivate the managers, but also realize the identity transformation from operators to owner. This approach can reduce the agency cost effectively and ease the conflict of shareholders and managers by producing interest assimilation. It's also helpful to reduce the adverse effects of asymmetric information, thus being the main power to motivate executive to increase R&D spending. Considering the fact that executives may hold shares in a certain proportion as being an original shareholders. Double identities of both shareholders and managers may drive them to pursue short-term interests, thus producing the entrenchment effect. Therefore, this article further divides the samples into groups according to whether implementing the equity incentive or not.

Based on the above analysis, this article puts forward the following hypothesis:

H1: Managerial ownership is significantly positively related to the enterprise research and development spending.
H2: The stock option incentive has a positive effect on the R&D expenditure.

3.2 Data sources and sample selection

Through consulting the annual financial report of information technology listed companies from 2011 to 2013. This article selected 203 companies. By excluding 58 companies with incomplete disclosure and relevant data, it finally chose 145

listed companies as the research samples. All the financial data referred comes from CSMAR and annual reports published by Shenzhen and Shanghai stock market. This article chooses Excel 2013 and EVIEWS7.2 to sort the data for statistical analysis.

3.3 Variable definition and model construction

3.3.1 Dependent variable
The former essays generally choose proportion of R&D expenditure and the companies' revenue to measure the intensity of enterprise research and development. Considering the fluctuation of revenue due to the characteristic of better growth and high profitability of information technology companies, this article chooses the proportion of R&D expenditure and the companies' total assets to measure the R&D intensity.

3.3.2 Independent variable
Learning from the former scholars, this essay sets the executive shareholding ratio as index of equity incentive. Then, this article selects ROE and Tobin Q value to measure business performance. The indicator of ROE reflects financial achievement in the short-term and the value of Tobin Q shows the market performance of the enterprise in the long-term. In addition, the article takes asset-liability ratio as the indicator of debt constraint and uses the growth rate of business income to measure the development ability.

3.3.3 Controlled variable
Considering the influence of other factors, the article sets the size of enterprise as controlled variables. Enterprise with larger scale will acquire more capital and ability to carry on the research and development activities.

4 EMPIRICAL ANALYSIS

In order to verify the relationship between the equity incentive and the R&D expenditure, this article learned from Tang Qingquan (2009) to construct model as follows:

$$RDI_{it} = \beta_0 + \beta_1 MSR_{it} + \beta_2 ROE_{it} + \beta_3 TBQ_{it} + \beta_4 DAR_{it} + \beta_5 GROW_{it} + \beta_6 SIZE_{it} + \gamma$$

4.1 *Descriptive statistics*

It can be observed from the descriptive statistics, average R&D input proportion of overall samples has reached 2.57%. It is generally believed that enterprise can maintain basic survival when the spending intensity reached 2%. It will be conducive to form competitiveness when the ratio comes up to 5%. Therefore, the level of R&D input from information technology listed companies still remains at lower level. From the group result, the average R&D input level of enterprises which the executives hold share has reached 2.97%. It's higher than the average level of companies whose executives without shareholding. It means that managerial ownership will need to be conducive to increase the R&D expenditure.

4.2 *Regression analysis*

Under the panel data model, this article finally chooses the individual fixed effect model after

Hausman test and *F* test. Using the panel data model to estimate can increase the amount of data and the degree of freedom. Meanwhile, it can reduce the multicollinearity of explanatory variables so as to improve the estimation accuracy. It selects the cross-section weighs to do the generalized least squares to eliminate the influence of interface heteroscedasticity.

As the original shareholders, both of identities offer them greater management autonomy. It will reduce the supervision of the management and manipulate the board of directors, thus reducing R&D spending to pursue short-term interests. Therefore, the essay further divides the samples with shareholding into two groups: samples with and without stock option incentive.

The results show that executive share ratio is positively related with R&D spending at significant level of 5% under the all samples. It means that managers tend to increase the R&D expenditure for better performance when their interests and enterprise's interests converge. According to the grouped regression result, in the samples with shareholding, the relation between R&D expenditure and enterprise performance is more obvious. It testifies the promotion of R&D spending by the shareholding. In addition, development

Table 2. Descriptive statistics result.

Variable	MSR	RDI		
	All	All	Executive with stock ownership	Executive without stock ownership
Sample capacity	145	145	93	52
Mean	0.294844	0.025694	0.029656	0.0132039
Maximum	0.651188	0.106939	0.817839	0.377008
Minimum	0.014903	0.005150	0.000043	0.000002
Std. Dev.	0.024211	0.201242	0.089464	0.015378

Table 3. Test result of the model selection.

	F test		Hausman test		Optimal model
	F value	Critical value	Chi-Sq. statistic	Prob.	
1	4.6034	0.78348	8.5417	0.0201	Fixed
	Reject		Reject		
2	10.2533	0.734728	9.7071	0.0273	
	Reject		Reject		Fixed
3	–	–	1.8596	0.0517	
	–		Accept		Random
4	3.6917	0.5537	7.6548	0.0019	
	Reject		Reject		Fixed
5	11.4952	0.6886	8.9770	0.0075	
	Reject		Reject		Fixed

Table 4. Regression result.

Variable	(1) All	(2) With shareholding	(3) Without	(4) With stock incentive	(5) Without
C	−0.011	−0.082	−0.045	−0.184	−0.048
MSR?	0.006**	–	–	0.036**	0.009*
ROE?	0.154***	0.332**	0.297*	1.336***	0.997*
TBQ?	0.005	0.217*	0.163	0.734*	0.347
DAR?	−0.010*	−0.012**	−0.009	−0.014*	−0.003*
GROW?	0.002***	0.009***	0.003***	0.004	0.003
SIZE?	0.003**	0.026***	0.332*	0.265**	0.798*
RS	0.983	0.978	0.976	0.8721	0.796
Ad-R2	0.972	0.966	0.964	0.856	0.714
F	84.384	82.618	78.805	76.119	71.380
DW	2.639	2.962	2.644	2.123	2.345
Prob.	0.0000	0.0000	0.0000	0.0000	0.0000

(*, **, *** means the significant level of 10%, 5% and 1%).

ability is positively related to the R&D input. With greater growth, executives will have more ability and be more confident for the prosperity of enterprise. Meanwhile, companies will have more retained earnings for R&D input when the operating income growth improves gradually. However, debt constrain is negatively correlated to the R&D spending under significance level. It shows that the higher level of asset-liability ratio is, the greater pressure the enterprise has to repay the debt, thus reducing the motivation of executives to increase spending on research and development. From the point of controlled variables, the regression results of five groups all shows that managers in larger companies prefer to take part in R&D activities.

5 SUGGESTION

1. Attach great importance to the research and development activities

    R&D activities have become the source of enterprises to form core competitiveness. However, China's spending on research and development remains low. The enterprise should look far ahead from a high plane to increase the R&D input actively and continuously, thus promoting research and development level.

2. Perfect the executive equity incentive system

    Due to the information asymmetry between shareholders and executives, implementation of equity incentive system can better alleviate the agency problem than supervision. It's imperative for enterprises to improve the executive shareholding. In addition, developers also play an important role in the enterprises, the agency problem between developers and management also needs attention. By implementing the long-term incentive system such as equity incentive, on the one hand, it can reduce the accident loss result from attrition of R&D personnel; on the other hand, it can also better motivate developers to focus on research and development to turn the R&D input into realistic productivity effectively, thus ultimately achieving the purpose of improving enterprise performance.

3. Regulate information disclosure of enterprise R&D spending

    Present accounting standards did not make specific provision. There is a lack of transparency in classifying research phase and development phase. The names of detailed items of research and development are far from united. All of the defects make the information disclosure lack of sufficiency and comparability. In addition, the different manners and positions of disclosure for R&D spending also reduce the reliability and availability of data. Therefore, the relevant department should make clearer regulation, specify the rules and unify the standard. By strengthening the integrity, reliability and comparability of information disclosure, it will further promote the theoretical research and better guide the practice.

REFERENCES

[1] Chen Xiaodong, Zhou Jianan. Study on the relationship between equity incentive and R&D spending—based on the empirical data from A share market [J]. Securities Market Herald, 2014(2):33–41.

[2] Xie Shangwei, Liao Baoli. Study on the relevance of equity incentive and corporate R&D input—based on the empirical data from manufacturing industry [J]. Journal of Business Economics, 2012, (9):100–102.

[3] Xia Yun, Tang Qingquan. Analysis on the equity incentive and R&D expendtirue of high-tech industry [J]. Securities Market Herald, 2008, (10):29–34.

[4] Hsiang-Lan Chen, Yen-sheng Huang. Employee stock ownership and corporate R&D expenditure: evidence from Taiwan's information-technology industry [J]. Asia Pacific Journal of Management, 2006, 23(3):369–384.

[5] Jiangfeng Wu, Rungting Tu. Ceo stock option pay and R&D spending: a behavioral agency explanation [J]. Journal of Business Research, 2007, 60(5):482–492.

[6] Wang Qinrong. Ceo equity incentive and R&D input—based on the empirical data from Chinese listed companies [D]. Xi'an Electronic and Science University, 2010.

Information Technology and Applications – Li (Ed.)
© 2015 Taylor & Francis Group, London, ISBN 978-1-138-02677-3

Empirical research on capital structure and corporate performance of listed companies in information technology industry

Dandan Zhang

School of Business, Hohai University, Nanjing, P.R. China

ABSTRACT: Based on the domestic and international studies about the capital structure and corporate performance, this paper first introduces the general situation of the information technology industry. Then the concept of capital structure and corporate performance are defined. Ninety-five listed companies in information technology industry from Shanghai stock exchange and Shenzhen stock exchange are chosen as the research subjects. The comprehensive evaluation of the companies' corporate performance from 2011 to 2013 is calculated by using factor analysis. Then this paper use regression analysis methods to study the impact of capital structure on corporate performance. Finally, combined with the above analysis, some suggestions are presented to improve the corporate performance of listed companies in information technology industry.

Keywords: capital structure; corporate performance; factor analysis; regression analysis

1 INTRODUCTION

In 1958, Modigliani and Miller proposed the MM Theory, the capital structure has gradually become a research subject of theoretical value and practical significance. The capital structure not only relates to the company's capital cost, but also affects the operating performance and sustainable development of the company. Reasonable capital structure will help to improve the company's performance. With the rapid development of the information technology, information technology has become the highlight of the national economy. Advanced information technology industry will lead to the development of the whole country, which is the consensus of the world. In order to promote the healthy and sustainable development of the information technology industry, the relationship between capital structure and corporate performance of the listed companies in information technology industry from Shanghai stock exchange and Shenzhen stock exchange is studied empirically. Therefore, the study of capital structure and corporate performance is of important practical significance.

2 EXPERIMENTAL SECTION

2.1 *Definition of capital structure and corporate performance*

2.1.1 *Definition of capital structure*
Capital structure refers to the proportional relationship of the owner's equities and the interest of creditors. Debt asset ratio and short-term debt asset ratio are selected to measure the capital structure in this paper.

2.1.2 *Definition of corporate performance*
Corporate performance refers to the business efficiency and operator performance during a certain period. In this paper, the corporate performance is only refers to the listed companies' business efficiency. The business efficiency is evaluated from four dimensions: profitability, debt-paying ability, operational capacity, and development capability.

2.2 *Sample selection and data sources*

This paper selected listed companies in information technology industry from Shanghai stock exchange and Shenzhen stock exchange, to calculate indicators and evaluate the corporate performance of 95 listed companies from 2011 to 2013.

2.3 *Index selection*

2.3.1 *The dependent variable*
To calculate the corporate performance of the listed companies in information technology industry, this paper selected 13 financial indicators: ROE (X_1), net profit margin on assets (X_2), net profit margin on sales (X_3), EPS (X_4), current ratio (X_5), quick ratio (X_6), debt asset ratio (X_7), receivables turnover ratio (X_8), inventory turnover (X_9), total assets turnover

(X_{10}), total assets growth rate (X_{11}), net profit growth rate (X_{12}), and revenue growth rate (X_{13}).

2.3.2 *The explaining variable*

Debt Asset Ratio (DRA) and short-term debt ratio (SDAR) are selected to measure the capital structure in this paper. The specific variables are shown in Table 2.

2.3.3 *The control variable*

The company size (SIZE) and growth (GROW) are selected as the control variables, which expressed by the logarithm of total assets and total assets growth rate.

Table 1. Description of index (1).

Index type	Symbol	Formula
Profitability	X_1	Net profit/average net assets
	X_2	Net profit/average total assets
	X_3	Net profit/revenue
	X_4	Net profit/equity
Debt-paying ability	X_5	Current assets/current liabilities
	X_6	Liquid assets/current liabilities
	X_7	Liabilities/total assets
Operational capacity	X_8	Main business income/ average balance of accounts receivable
	X_9	Main business costs/ average balance of inventories
	X_{10}	Main business income/ average total assets
Development capability	X_{11}	Assets increase in current/ total assets at first
	X_{12}	Net profit growth in this year/net profit in the previous year
	X_{13}	Revenue growth in this year/revenue in the previous year

Table 2. Description of index (2).

Index type	Symbol	Formula
Explanatory variables	DAR	Liabilities/total assets
	SDAR	Short-term liabilities/ total assets
Control variables	SIZE	LN (total assets)
	GROW	Assets increase in current/ total assets at first

2.4 *Research hypothesis*

Based on the previous studies on the capital structure and corporate performance of listed companies in information technology industry, the hypotheses are proposed as follows:

Hypothesis 1: The Debt Asset Ratio (DAR) and corporate performance of listed companies in information technology industry are negatively related.

Hypothesis 2: The Short-term Debt Asset Ratio (SDAR) and corporate performance of listed companies in information technology industry are negatively related.

Hypothesis 3: The company size (SIZE) and corporate performance of listed companies in information technology industry are positively related.

Hypothesis 4: The company's growth (GROW) and corporate performance of listed companies in information technology industry are positively related.

According to the hypotheses, establish the regression model:

$$F = \varepsilon + \alpha_1 DAR + \alpha_2 SDAR + \alpha_3 SIZE + \alpha_4 GROW$$

2.5 *Factor analysis*

2.5.1 *KMO and Bartlett's test*

Use SPSS19.0 to calculate the corporate performance of the listed companies in information technology industry. According to KMO and Bartlett's Test, the KMO is 0.607, which illustrate the correlation between indicators. Bartlett's test value is 3341.730 (Sig. = 0.000), it get through the significant inspection.

2.5.2 *Total variance explained*

From the factor analysis, five factors explained 74.068% of the variance among the 13 variables. The first characteristic factor is 3.177, the variance of contribution value is 24.438%. The second characteristic factor is 2.938, the variance of contribution value is 22.598%. The third characteristic factor is 1.233, the variance of contribution value is 9.484%. The forth characteristic factor is 1.162, the variance of contribution value is 8.939%. The fifth characteristic factor is 1.119, the variance of contribution value is 8.609%. The contribution of public factor showed that the five factors reflect original indexes.

Extraction Method: Principal Component Analysis.

2.5.3 *Rotated component matrix*

According to rotated component matrixes, the first characteristic factor's positive load is larger on

Table 3. Total variance explained.

	Initial eigen values			Rotation sums of squared loadings		
	Total	% of variance	Cumulative %	Total	% of variance	Cumulative %
1	4.011	30.854	30.854	3.177	24.438	24.438
2	2.276	17.506	48.360	2.938	22.598	47.037
3	1.207	9.282	57.642	1.233	9.484	56.520
4	1.094	8.412	66.054	1.162	8.939	65.460
5	1.042	8.014	74.068	1.119	8.609	74.068
6	0.942	7.249	81.317			
7	0.846	6.510	87.827			
8	0.692	5.320	93.147			
9	0.456	3.508	96.655			
10	0.267	2.052	98.708			
11	0.133	1.020	99.728			
12	0.034	0.260	99.987			
13	0.002	0.013	100.00			

Table 4. Rotated component matrixes.

	Component				
	1	2	3	4	5
X_5	0.944	0.057	0.009	0.064	0.016
X_6	0.942	0.064	0.018	0.056	0.013
X_7	−0.790	−0.174	0.117	0.107	0.048
X_1	0.635	0.615	−0.197	0.039	0.052
X_3	−0.115	0.953	0.036	−0.014	0.061
X_2	0.196	0.943	−0.001	−0.069	0.018
X_4	0.226	0.781	0.071	0.319	−0.065
X_8	0.042	−0.053	0.885	0.002	0.042
X_{10}	−0.483	0.159	0.612	−0.095	−0.072
X_{11}	0.140	0.226	0.065	0.736	−0.156
X_9	0.108	0.098	0.106	−0.618	−0.197
X_{12}	−0.015	−0.077	0.000	−0.156	0.756
X_{13}	0.040	0.142	0.010	0.281	0.681

Table 5. Scores of common factors.

	Component				
	1	2	3	4	5
X_1	0.140	0.174	−0.117	−0.047	0.041
X_2	−0.025	0.345	−0.021	−0.153	0.019
X_3	−0.135	0.375	−0.026	−0.097	0.055
X_4	−0.001	0.243	0.057	0.213	−0.074
X_5	0.334	−0.083	0.124	0.022	0.014
X_6	0.334	−0.079	0.130	0.014	0.013
X_7	−0.256	0.001	0.014	0.136	0.037
X_8	0.130	−0.066	0.766	0.021	0.062
X_9	0.061	0.077	0.082	−0.552	−0.139
X_{10}	−0.113	0.091	0.452	−0.067	−0.045
X_{11}	0.011	0.003	0.073	0.644	−0.179
X_{12}	0.010	−0.015	0.019	−0.174	0.688
X_{13}	−0.005	0.023	0.031	0.199	0.596

current ratio (X_5), quick ratio (X_6) and ROE (X_1), the negative load is smaller on debt asset ratio (X_7), which indicate that the first factor represents the company's debt-paying ability. The positive load of second characteristic factor's is larger on net profit margin on sales (X_3), net profit margin on assets (X_2) and EPS (X_4), indicating that the second factor represents the company's profitability. The third characteristic factor's load on receivables turnover ratio (X_8) and total assets turnover (X_{10}) is large, which means that the third factor and the forth factor represent the company's operational capacity. For the fifth characteristic factor, the net profit growth rate (X_{12}) and revenue growth rate (X_{13}) is larger, indicating that the fifth factor represent the development capability of listed companies in information technology industry.

2.5.4 *Component matrix*

Use orthogonal solution to get the five factors as follows:

$$F_1 = \mathbf{0.140}X_1 - 0.025X_2 - \mathbf{0.135}X_3 - 0.001X_4$$
$$+ 0.334X_5 + \mathbf{0.334}X_6 - 0.256X_7 + 0.130X_8$$
$$+ \mathbf{0.061}X_9 - 0.113X_{10} + \mathbf{0.011}X_{11} + 0.010X_{12}$$
$$- 0.005X_{13}$$

$$F_2 = 0.174X_1 + 0.345X_2 + 0.375X_3 + 0.243X_4$$
$$- 0.083X_5 - 0.079X_6 + 0.001X_7 - 0.066X_8$$
$$+ 0.077X_9 + 0.091X_{10} + 0.003X_{11} - 0.015X_{12}$$
$$+ 0.023X_{13}$$

Table 6. Model summary.

Model	R	R square	Adjusted R square	Std error of estimate
1	0.404	0.163	0.151	1.8755874

Table 7. Anova[b].

Model		Sum of squares	df	Mean square	F	Sig.
1	Regression	192.199	4	48.050	13.659	0.000
	Residual	984.992	280	3.518		
	Total	1177.191	284			

Table 8. Coefficient.

Model		Unstandardized coefficients		Standardized coefficients		
		B	Std error	Beta	t	Sig.
1	Constant	3.764	2.969		1.267	0.206
	DAR	−3.863	0.749	−0.328	−5.156	0.000
	SDAR	−3.562	0.816	−0.248	−4.366	0.000
	SIZE	0.106	0.131	0.052	0.807	0.421
	GROW	0.634	0.213	0.164	2.973	0.003

$$F_3 = -0.117X_1 - 0.021X_2 - 0.026X_3 + 0.057X_4$$
$$+ 0.124X_5 + 0.130X_6 + 0.014X_7 + 0.766X_8$$
$$+ 0.082X_9 + 0.452X_{10} + 0.073X_{11} + 0.019X_{12}$$
$$+ 0.031X_{13}$$

$$F_4 = -0.047X_1 - 0.153X_2 - 0.097X_3 + 0.213X_4$$
$$+ 0.022X_5 + 0.014X_6 + 0.136X_7 + 0.021X_8$$
$$- 0.552X_9 - 0.067X_{10} + 0.644X_{11} - 0.174X_{12}$$
$$+ 0.199X_{13}$$

$$F_5 = 0.041X_1 + 0.019X_2 + 0.055X_3 - 0.074X_4$$
$$+ 0.014X_5 + 0.013X_6 + 0.037X_7 + 0.062X_8$$
$$- 0.139X_9 - 0.045X_{10} - 0.179X_{11} + 0.688X_{12}$$
$$+ 0.596X_{13}$$

According to the component score coefficient matrix, the comprehensive score model can be established as follows:

$$F = (24.438\%F_1 + 22.598\%F_2 + 9.484\%F_3$$
$$+ 8.939\%F_4 + 8.609\%F_5)/74.068\%$$

2.6 Regression analysis

Before regression analysis, test the correlation between each variable. The result shows that there is no multicollinearity, as the coefficient between independent variables is very small.

Use SPSS19.0 for multiple linear regression, the results are shown in Table 6, Table 7 and Table 8. It is shown in Table 6 that R-Squared is 0.163, adjusted R-Squared is 0.151, which indicated that about 15% of corporate performance is explained by explaining variables. This paper only studies on the impact of capital structure on corporate performance, not on all factors affecting the performance, so a relatively low goodness of fit is reasonable.

It is shown in Table 8 that the standardized coefficients of Debt Asset Ratio (DAR) is 0.328 (Sig. = 0.000). Therefore, the hypothesis 1 is established. The standardized coefficients of Short-term Debt Asset Ratio (SDAR) is −0.248 (Sig. = 0.000). The hypothesis 2 is established. The standardized coefficients between company size (SIZE) and corporate performance is positive (0.052), but non-significant (Sig. = 0.421), verifying the hypothesis 3. The standardized coefficients between company's growth (GROW) and corporate performance is 0.164 (Sig. = 0.003). The hypothesis 4 is established.

3 CONCLUSION AND SUGGESTION

3.1 Research conclusions

1. The debt asset ratio and short-term debt asset ratio are negatively related with corporate performance of listed companies in information technology industry. To some extent, the debt financing of listed companies in information technology industry does not have the financial leverage effect. The high debt ratio leads to the increase of financial risk and probability of bankruptcy.

2. The company size and corporate performance of listed companies in information technology industry are positively related, but the significance is not strong. The larger the company size is, the more the investment opportunity generate.
3. The company's growth and corporate performance of listed companies in information technology industry are positively related. A higher level of growth means better prospects for development, indicating a high value of enterprise market.

3.2 *Policy suggestions*

1. Innovate the financing method of listed companies in information technology industry
   In order to maintain the sound development of the information technology industry and prevent the financial risk, it is necessary to improve the financing channel. Listed companies in information technology industry need to preclude the use of new financing methods, such as insurance funds, financial leasing, issuing bonds, and backdoor listing. By optimizing its capital structure, companies can reduce financial risk and improve corporate performance.
2. Promote the scale development of listed companies in information technology industry
   Due to the limited market share in the current, competition among enterprises is very intense. Corporate mergers can promote the growth of enterprise, achieve the purpose of scale economies, and conducive to the long-term development of enterprise.
3. Concern about business growth of listed companies in information technology industry
   Because of the different industrial characters, information technology companies have been part of high-growth business. With the support of national policy, the development of information technology industry has good prospects. Due to rapid growth, the funding gap is large. Enterprises should meet the required capital to avoid losing investment opportunities. In the fierce competition, the high-growth of information technology industry is destined to become the driving force of China's economic development.

REFERENCES

[1] Frank. M.Z., Goyal. V.K. Capital structure decisions [J]. Working Paper, 2003:56–98.
[2] Berger. A.N., E.B. Patti. Capital structure and firm performance: a new approach to testing agency theory and an application to the banking industry [J]. Journal of Banking & Finance, 2006, (30):1065–1102.
[3] Zhang zhaoguo, He Weifeng, Liang Zhigang. Capital structure and corporate performance-the empirical evidence from state holding listed companies and private listed companies [J]. China soft science, 2007, (12):147–151.
[4] Bai Jiyang. An empirical analysis of capital structure and corporate performance of listed companies in China's cultural industries [J]. Journal of Shanghai Business School, 2012, 13(5):26–38.
[5] Xu wei, Gao Ying, Xing Ying. Capital structure, stock ownership structure and corporate performance [J]. Journal of Shanxi Finance and Economics University, 2005, 27(4):116–120.
[6] Chen Deping, Zeng Zhihai. The interactive relation between capital structure and corporate performance-empirical test based on the listed companies on the growth enterprise market [J]. Accounting Research, 2012, (8):66–71.

Information Technology and Applications – Li (Ed.)
© *2015 Taylor & Francis Group, London, ISBN 978-1-138-02677-3*

Ecosystem health assessment of Japanese larch in Liaodong area based on grey relation projection method

Qiang Liu
College of Information and Electrical Engineering, Shenyang Agricultural University, China

Shu-rong Hui, Cong Wei & Ling Fang
College of Science, Shenyang Agricultural University, China

Chen Liang
China National Aeronautical Radio Electronics Research Institute, China

Hui Liu
Shanghai Environmental Engineering Design Science Research Institute, China

ABSTRACT: A grey relation projection model for ecosystem health assessment is established based on grey system theory and vector projection principle. The Analytic Hierarchy Process is used to calculate the index weight and analyze the health of Larix Kaempferi forests plantation of the eastern region of Liaoning Province. In this model, evaluated samples and classified standards are treated as vectors, and project on the same ideal sample. Then, evaluated samples are classified and arranged in the projective value. Hence, the theory can provide a reference value for the health management of ecosystem.

Keywords: grey relation projection method; Japanese larch plantation of the eastern region of Liaoning Province; ecosystem health assessment

1 INTRODUCTION

Japanese larch is an excellent fast-growing timber species. It is widely used all over the word because of liking adequate light, warm and humid climate, strongly branches sprout, and considerable resistance to cold. The eastern region of Liaoning Province as one of the region which has a large area of Japanese larch plantation and the main distribution area of the forest resources plays an important role in the health management of ecosystem[2]. In order to manage Japanese larch plantation well and improve forest productivity. This study used grey theory and starting from the point of view of the vector projection to propose grey relational degree projection model. The health of Japanese larch forests plantation of the eastern region of Liaoning Province with this model is analyzed. In order to understand the health status of Japanese larch in the region. This paper provides decision support and scientific basis for the rational utilization and protection of resources, make it toward the direction of healthy and sustainable growth, and enhance the economic, social and ecological benefits.

2 THE NATURAL CONDITIONS OF THE STUDY AREA

The study area is located in Fushun City, Liaoning Province, BeiWang Qing forest. Between latitude 41°14′10″ to 41°58′50″ and longitude 124°15′56″ to 125°27′46″. Belonging to the extension part of the Changbai Mountain offshoot, the terrain is tilted from northeast to southwest. Xinbin has annual frost-free period of 128 days, the surface is rich in water resources, a total of 1.44 billion cubic meters. The landform types of Xinbin are tectonic erosion low-mountain. The average elevation is 492 meters. The territory of Xinbin belongs to the seasonal continental climate of the northern temperate zone. It has distinct seasons, the pleasant climate, and abundant rainfall. The time of the annual average rainfall in the frost-free period is about 150 days. The average of annual rainfall is 750–850 mm. The main soil type of Xinbin is brown. Accounted for 85% of the total land area, the average of soil thickness is from 20 to 40 cm. There are all kinds of natural vegetations. Such as Oak woods, broad leaved mixed shaw, mixed coniferous stick shrubbery, Lespedeza shrub meadow vegetation and aquatic vegetation, and so on. Artificial vegetation such as larch, Korean pine forest, fruit trees and so on. The vegetation of the county has 175 species, 35 families and 79 genera. The kind of birds has 12 orders, 27 families, and 95 species.

3 FIELD INVESTIGATION

This study selected 23 sublots to investigate in Xinbin, Liaoning Province, Beiwangqing forest. According to the different stand age and aspect select representative lots. The arbor specifications within 20 m × 20 m quadrats are set in each plot, including 5 m × 5 m shrub quadrats and 1 m × 1 m herbaceous quadrats, the investigation of factors such as average height, DBH, forest age, Canopy density, Soil thickness, Litter depth, vegetation coverage and so on.

4 BUILDING THE HEALTHY EVALUATION INDEX SYSTEM OF JAPANESE LARCH

4.1 *Construction principle[8]*

In order to make the Japanese larch plantation more scientific, reasonable, comprehensive, the theory developed three principles as follows:

Scientific Principles. It can reflect the development of ecosystem health exactly. The method of assessment is easy to grasp.

Representative Principles. It can adequately represent the inherent characteristics of the ecosystem as well as the natural and socio-economic status.

Operability Principle. The data of evaluation accessibility, the way of consideration and measuring is simple and operability. It can reasonably unify theoretical science and practical feasibility.

4.2 *The selection of the evaluation*

According to the above mentioned three principles, we surveyed and sorted out the basal data for Japanese larch plantation, combining with its own situation. In order to make Japanese larch plantation better economic and social development, meeting the growth of site conditions and the dominant factor, we selected 12 indicators based on site factors, stand structure, resistance indicators, sustainability indicators, four major categories in many indicators[3]. Hence, it can provide the basis for Japanese larch forests plantation ecosystem health assessment of the eastern region of Liaoning Province.

5 SETTING UP THE GREY RELATION PROJECTION METHOD

5.1 *The basic principle of grey relation projection method*

Deng julong, who is the professor of Science and Engineering in Huazhong university, put forward the grey system in 1982[1]. It aims to solve system problems which exist part of the information explicitly and part of the information is not clear in life and production. Such as forest ecosystem system. It has uncertainty and incomplete information. The system has both been known of the white information and have not yet fount the black information. But more general qualitative understanding of the grey information. In fact, comprehensive evaluation of this system is actually a grey multi-objective decision making[4].

Grey correlation projection method, in essence, is a simple weighted addend method. Weighting coefficient is proportional to the square of the weighting coefficients of the original in the method a new set of weighting coefficients. Important indicator of weighting coefficients to be further strengthened by the new algorithm. It makes the method simple, additive weighting method has a completely different connotation.

In grey relation projection method, combined the size of the die size and the cosine of the angle, reflected comprehensively and accurately the closeness between the policy program with the ideal solution, the ecosystem health assessment will be closer to the objective reality.

5.2 *Setting up the grey relation projection model*

5.2.1 *Setting up multi-criteria decision making domain collection A = {Case 1, Case 2,..., Case n} evaluation collections V = {Index 1, Index 2,..., Index m} Decision-making programs on indicators attribute value*

$Y_{ij} = \{i = 0,1,2,...,n; j = 1,2,3,...,m\}$. Indicators are usually divided into two categories. One indicator is the greater the better property values, the other indicator is opposed to the former. Decision-making matrix is case of A to V

$$Y = \left(Y_{ij}\right)_{(n+1)\times m} \quad (i = 0,1,2,...,n; j = 1,2,3,...,m) \quad (1)$$

5.2.2 *Initialization process of the data*

$$Y' = \left(Y'_{ij}\right)_{(n+1)\times m} \quad (i = 0,1,2,...,n; j = 1,2,3,...,m)$$

At the same time

$$Y'_{ij} = Y_{ij}/Y_{0_j} \quad (i = 0,1,2,\ldots,n; j = 1,2,3,\ldots,m) \quad (2)$$

Y_{0j} is the initialized matrix in the ideal solution for property values.

Y_{0j} is initialized after $Y_{0j} = 1(j = 1,2,3,\ldots,m)$, which is the ideal case.

r_{ij} is the sub-factor Y'_{ij} $(i = 1,2,\ldots,n; j = 1,2,3,\ldots,m)$ on the parent factor.

$Y_{0j} = 1(j = 1,2,\ldots,m)$ of the correlation degree.

$$r_{ij} = \frac{\min\limits_{n}\min\limits_{m}|Y'_{0j} - Y'_{ij}| + \lambda\max\limits_{n}\max\limits_{m}|Y'_{0j} - Y'_{ij}|}{|Y'_{0j} - Y'_{ij}| + \lambda\max\limits_{n}\max\limits_{m}|Y'_{0j} - Y'_{ij}|} \quad (3)$$

$\lambda = 0.5$

5.2.3 *Building the Judgment matrix F, which is composed by (n+1)m × r*

$$F = \left(r_{ij}\right)_{nxm}, (i = 1,2,\ldots,n; j = 1,2,3,\ldots,m) \quad (4)$$

5.2.4 *Evaluation weighted vector is*

$$W = \{W_1, W_2, \ldots, W_m\}^T > 0 \quad (5)$$

During the effect of the weighting vector

$$F' = F \cdot W = (F'_1, F'_2, \ldots, F'_m)$$

Augmented matrix each decision scheme as a row vector (vector). Then angle between each decision-making program A_i with the ideal solution A_0 for the grey relation projection angle θ_i. Cosine greater, indicating that change the direction of the more consistent decision-making program A_i between the ideal solution A_0. Cosine of the angle is as follows:

$$\cos\theta_i = \frac{A_i A_0}{\|A_i\|\|A_0\|} = \frac{\sum\limits_{j=1}^{m} W_j F_{ij} W_j}{\sqrt{\sum\limits_{j=1}^{m}\left(W_j F_{ij}\right)^2}\sqrt{\sum\limits_{j=1}^{m} W_j^2}}, (i = 0,1,2,\ldots,n) \quad (6)$$

The cosine of the angle $0 < \cos\theta_i \leq 1$ is always the greater the better. The greater the cosine of that change direction more consistent decision-making program between the ideal solution d_i. As a modulus of the decision-making program A_i is as follows:

$$d_i = \sqrt{\sum_{j=1}^{m}\left(W_j F_{ij}\right)^2} \quad (7)$$

$\overline{W_j}$ as a grey relation projection weight vector, is

$$\overline{W_j} = \frac{W_j^2}{\sqrt{\sum\limits_{j=1}^{m} W_j^2}} \quad (8)$$

5.2.5 *The calculation of Grey related projection value*

$$D_j = d_i \cos\theta_i = \sum_{j=1}^{m} F_{ij}\overline{W_j}, (j = 1,2,3,\ldots,m) \quad (9)$$

According to the size of the projection, the relationship between the decision-making program with the ideal solution evaluated is determined. Projection values obtained by the Euclidean distance in SPSS cluster analysis are divided into the forest ecosystem health, sub health, and unhealthy three levels.

6 THE RESULTS AND ANALYSIS

6.1 *The establishment of the evaluation*

Considering the existing achievements at home and abroad[5–8], and combined with the the BeiWang Qing forest actual situation, the study chose 12 indicators. Indicators of site factors, such as altitude, slope gradient, and indicators of Stand structure, such as average height, DBH, canopy density, forest age, volume, and vegetation coverage. Resistance indicators, such as Fire danger index and Pests rate. Sustainability indicators, such as soil thickness and litter depth.

Exsiting plan $A = \{plot1, plot2, \ldots, plot24\}$, Index set $V = \{$forest age, altitude, slope gradient, average height, DBH, canopy density, soil thickness, litter depth, vegetation coverage, volume, fire danger index, and pests rate$\}$.

6.2 *Determining the health standards*

According to the health of evaluation criteria as well as the characteristics of Japanese larch, maxima selected indicators including average height, DBH, fire danger index, litter depth, vegetation coverage, soil thickness, and volume. Moderate selected indicators including forest age, altitude, vegetation coverage; Minimum selected indicators including slope gradient, and pests rate. Moderate type indicators take the average of each plot. The ideal plan $A_0 = (25, 605, 3, 19.60, 20.10, 0.8, 90, 5.30, 85, 341, 0.9, 0.11)$.

6.3 *The evaluation of standing forest*

1. Initialization of the data According to the formula (2) shown in Table 1, we initialized the data.
2. According to the formula (3) to get the judgment matrix F
3. Obtaining a set of weighted vector W by the Analytic Hierarchy Process, $W = (0.03154, 0.02126, 0.06362, 0.03301, 0.02986, 0.03453, 0.08799, 0.11475, 0.06412, 0.08106, 0.19635, 0.24194)$
4. According to the formula (8), we can obtain to the grey relation projection weight vector. $W_j = (0.0027, 0.0012, 0.0109, 0.0029, 0.0024, 0.0032, 0.0209, 0.0355, 0.0111, 0.0177, 0.104, 0.1579)$ then from the formula (9)
   $W_j = (0.3609, 0.3371, 0.3215, 0.3303, 0.3378, 0.2956, 0.3433, 0.3130, 0.2764, 0.3002, 0.3144, 0.3318, 0.3047, 0.3667, 0.3293, 0.2925, 0.3116, 0.3205, 0.2938, 0.3542, 0.3382, 0.3207, 0.2860)$
5. Clustering analysis for each small class projection value, differential square method for clustering using Euclidean distance to calculate the distance between the sample results are shown in Figure 1. According to the results of cluster analysis, 23 small classes can be divided into health, sub-health, in-health, and unhealthy four categories.

Figure 1 shows that the health of plots includes plots number 1, 14, 20 accounting for 13% of all plots. They are characterized by abundant species, growing well, high species diversity, projection is larger. The sub-health small class from number 2, 4, 5, 7, 12, 15, 21, accounting for 30.4% of all small class. In healthy small classes from number 3, 8, 11, 17, 18, 22 small class, accounting for 26%. The unhealthy small class from number 6, 9, 10, 13, 16, 19, 23 small classes, accounting for 30.4% of all small classes, smaller projection values.

7 CONCLUSION

1. This study establishes a forest ecosystem health evaluation model based on grey relation projection method. Comprehensive evaluation of the health status of the 23 artificial Japanese larch forest is in small classes with this model. The results show that the health, sub-health, healthy, unhealthy larch small class accounted for 13%, 30.4%, 26%, and 30.4% of the total number of samples. Evaluation results are consistent with the reality that the area of Japanese larch in overall is better health status, should be continue to maintain to the development of a healthy direction.
2. Grey relation projection method to small classes in health standards projection values comparable standard size as the state of health, forest health assessment closer to the objective reality. It makes forest health assessment closer to the objective reality, the process is simple, more comprehensive evaluation, more truly reflect the health status of forests. But its greatest weakness is not strong, discrete projection values tend to homogenize, and it is not an easy classification.
3. The main use of cluster analysis in this study is to designate the health and sub-health, unhealthy boundaries, with the change in the number of cluster sample, the results may be different, however it has a certain relative.

Source: the provincial key project of promoting science and technology in comprehensive Agricultural Development.

Topic title: Evaluation of the performance appraisal in promotion projects.

REFERENCES

[1] Deng J.L. The gray theoretical system tutorial [M]. Wuhan: Huazhong University of Science and Technology Press, 1990:89–103.
[2] Xiao F.J., Ouyang H. Forest ecosystem health assessment indicators and application in China [J]. Acta Geographica Sinica, 2003, 8(6):793–807. (In Chinese)
[3] Lu S.W., Liu F.Q., Yu X.X. Health assessment of forest ecosystem in Badaling forest center [J]. Journal of Soil and Water Conservation, 2006, 20(3):39–45. (In Chinese)
[4] Lü F., Cui X.H. Multi-Criteria decision grey relation projection method and its application [J]. Systems Engineering-Theory & Practice, 2002, (1):103–107. (In Chinese)
[5] Li X.Y. Pilot study and application on the indicators for forest health assessment [D]. Beijing: Chinese Academy of Forestry, 2006. (In Chinese)
[6] Gu J.C. Health analysis and evaluation of the main forest types in the typical Soil-gravel Area in north China [D]. Beijing: Beijing Forestry University, 2006. (In Chinese)
[7] Kong H.M., Zhao J.Z. Method of ecosystem health assessment [J]. Journal of Applied Ecology, 2002, 13(4):486–490.
[8] Kong H.M. Theory of forest ecosystem health and Evaluation Index System [J]. Beijing: Ecological Environment Research Center, Chinese Academy of Sciences, 2002:32–43.

Information Technology and Applications – Li (Ed.)
© 2015 Taylor & Francis Group, London, ISBN 978-1-138-02677-3

Research on mean-variance approach and CAPM approach for determining the expected risk and return

Lingfen Zhang & Jianfei Leng
School of Management, Hohai University of Business School, Nanjing, China

ABSTRACT: The purpose of this paper is to examine mean-variance approach and CAPM approach for determining the expected risk and return of a two stock portfolio which are Seven Network Limited (SEV) stock and CSR group stock listed on the Australian Stock Exchange. Based on the calculation and analysis, we gave some suggestions about which portfolio to invest in. In addition, we critique the two approaches based on academic literature.

Keywords: SEV; CSR; risk and return

1 INTRODUCTION: SEV AND CSR

Seven Network Limited (SEV) is a commercial television broadcaster in Australia with principal activities in television station, magazine publishing, and investments. According to company category, SEV belong to media sector in ASX. CSR Limited (CSR) is one of the Australia's leading manufacturing companies. The principal activities of entities in the CSR group included the manufacture and supply of building products in Australia, China, Malaysia, New Zealand, Singapore and Thailand. The group mills raw sugar from sugarcane as well as producing renewable electricity and ethanol in Australia, and manufactures and distributes refined sugar products in Australia and New Zealand. In Australia, the group has an interest in the smelting of aluminium. CSR belong to Capital goods sector in ASX.

2 CHOICE AND JUSTIFICATION OF DATA

2.1 *Time period chosen between December 2003 and August 2008*

Brailsford, Faff & Oliver (1997) suggested that it is generally accepted that around 50 data points are required to obtain reliable OLS estimates, and beta estimates appear to be reasonably stable over a four-to five-year period. Therefore, the time period between December 2003 and August 2008 is chosen in our research.

2.2 *Sampling frequency: Monthly*

When it comes to choose sample frequency, the most commonly used sampling intervals are daily, weekly and monthly (Brailfords, Faff & Oliver, 1997). However, Hawawini (1983) indicated that, data of different intervals may provide dissimilar betas. Furthermore, Carleton and Lakonishok (1985) showed that, trading pattern may cause beta to be biased when beta is computed basing on daily or weekly data, and also in their study of NYSE securities, monthly data provided no bias in the estimated beta. In addition, monthly data have been historically more commonly available than either weekly or daily data. (Brailfords, Faff & Oliver, 1997). Therefore, monthly interval has been used in our study as it probably to provide more reliable rate of return.

2.3 *Market indices: All Ordinaries Accumulation Index*

In Australia, All Ordinaries and the S&P/ASX 200 are the most commonly quoted indices. However, these indices have some limitations. One limitation is that the indices do not fully reflect the change of shareholders' wealth, because of the calculations do not take dividends paid into consideration. (Frino et al., 2006). Due to this limitation, an alternative to the All Ords, the All Ordinaries Accumulation Index (AOAI) is preferred as it captures the stock returns. However, owing to the fact that the All Ordinaries Accumulation Index can only be found during December 2003 and August 2008, our research can only use this time period.

2.4 *The risk-free rate: Australian government 10 year bond*

The 10-year Australian Government 10 year bond is regarded as risk-free rate in this study. First, the 10-year bonds are government debt securities, which the return is guaranteed to the investors by the federal government. Second, Treasury bond rates are usually facing less random variation than Treasury-bill rate (Clinebell et al. 1994), and most importantly, the long-term investment horizon reflects long-term inflation expectation such as 10-year government bonds is the best choice for measurement.

3 CALCULATION

3.1 *Mean-variance approach*

Mean-Variance analysis was derived as a systematic approach to measure risks in financial markets. This approach established a criterion for the comparison of different securities and portfolios as a trade-off between their return- as measured by expected return- and risk- as measured by the variance of the return. (Jaksa Cvitanic, Fernando Zapatero, 2004)

3.1.1 *Each company*

In our study, first, rate of return for SEV and CSR are calculated respectively. However, as mentioned earlier, CSR pays dividend twice a year as at July and December. The next step is to calculate the expected return which is the mean value of the stock returns and variance, and the standard deviation is calculated by squaring root of Variance.

The expected return of SEV is 0.009978397, whereas expected return of CSR is 0.01546892. The standard deviation of SEV's return is 0.078549393 and CSR's is 0.0891492. The result shows that comparing SEV with CSR, CSR has a higher return, and more risk.

3.2 *CAPM approach*

CAPM is built on the theory that investors face two types of risks in market:

- A Market Risk—These risks are systematic and non-diversifiable. For example, interest is decreasing several times recently. This is a systematic risk.
- Unsystematic Risk—unique risk, this risk is specific to individual stocks and can be diversified away if the investor to hold the stock in a portfolio. In more technical terms, it represents the component of a stock's return that is not correlated with general market moves.

Modern portfolio theory indicates that unsystematic risk can be diversified away by including more shares into a portfolio. However, diversification still does not solve the problem of systematic risk; even a portfolio of all the shares in the stock market cannot eliminate that risk. Therefore, when calculating a expected return, systematic risk is what plagues investors most. CAPM, therefore, evolved as a way to measure this systematic risk.

3.3 *The formula*

The CAPM is a model for pricing an individual security or a portfolio. The standard formula of CAPM describes the relationship between risk and expected return of a risky security or portfolio.

We can observe it is important to define a risk-free rate first in CAPM. Usually a 10-year government bond yield is used. To this is added a premium that equity investors demand to compensate them for the extra risk they accept. This equity market premium equal the expected return from the market minus the risk-free rate of return. It is the price of risk. Then the equity risk premium is multiplied by a coefficient "beta".

According to CAPM, beta is the only relevant measure of a stock's risk. It measures a stock's relative volatility—that is, it shows the relationship between return on a stock and return on the market. As below table x.x illustrates, if a share price moves exactly in line with the market, then the stock's beta is 1. A stock with a beta of 1.5 would rise by 15% if the market rose by 10%, and fall by 15% if the market fell by 10%.

Beta is found by statistical analysis of individual, daily share price returns, in comparison with the market's daily returns over precisely the same period. In their classic 1972 study titled "The Capital Asset Pricing Model: Some Empirical Tests", financial economists Fischer Black, Michael C. Jensen and Myron Scholes confirmed a linear relationship between the financial returns of stock portfolios and their betas. They studied the price movements of the stocks on the New York Stock Exchange between 1931 and 1965.

In this assignment, SEV and CSR are selected to setup a stock portfolio. First, their beta will be calculated separately.

To estimate beta of one stock, one needs a list of returns for the stock and returns for the index; these returns can be daily, weekly or any period. In this assignment, monthly data including adjusted share price, dividend and market index are collected to get the stock and returns for the index. 10-year bond yield is used and be converted to a monthly rate as risk-free rate of return.

The formula is as follows:

$$Y_t = (r_{i,t} - r_{f,t}) \qquad (1)$$

$$X_t = (r_{m,t} - r_{f,t}) \qquad (2)$$

where $r_{i,t}$ = the return on stock I,
earned over period t (3)

$r_{f,t}$ = the risk-free rate of return,
earned over period t (4)

$r_{m,t}$ = the market rate of return,
earned over period t (5)

After calculation (detail steps refer to Appendix xx), the result is as follows:

For the stock portfolio, the beta is:

P_{SEV} = proportion of the stock SEV in
the portfolio.   (6)

P_{CSR} = proportion of the stock CSR in
the portfolio.   (7)

To calculate the expected return of the portfolio, we use the mean of monthly risk-free rate of return as r_f and mean of monthly index return as r_m. Then the expected return is:

After construct a series of portfolios of the two stocks by varying the weights of each at 2.5% intervals from 0 to 100%, a series of beta and expected returns are calculated (detail data refer to Appendix xx). Based on that, the characteristic line of the stock portfolios is shown in figure x.x.

4 ANALYSIS AND RECOMMENDATION

We can define the market beta of 1, and beta as a standardized measure of systematic risk. The higher the beta, the more systematic risk has been taken. According to historical data analysis and literature review, we suggest that media sector is a low beta sector and the risk of media sector usually lower than average market risk, and the beta of News CORP (NWS) in NYSE is 0.9847 as an illustration. Capital goods sector will have a slight higher beta than market beta. Hence, we expect the estimated beta of SEV will close to 1 and the estimated beta of CSR will more than 1 and higher than SEV's beta.

We calculated the beta of SEV is 1.215 and the beta of CSR is 1.224 from January 2004 to August 2008. Though SEV's beta is larger than market beta, but it is still close to 1. CSR's beta slight higher than SEV's beta and market beta. Our assumption has been proven correct.

As was previously stated, the expected return of SEV is 1.37% and the expected return of CSR is 1.38% from January 2004 to August 2008. CSR have higher rewards than SEV, because CSR with a higher beta than SEV and took more systematic risk.

The covariance of returns between SEV and CSR is 0.002645 from January 2004 to August

2008, which close to zero but showing positively correlated. Hence, an increase in returns on SEV was probably to be associated with an increase in returns on CSR during this period, and vice versa. Meanwhile, the correlation coefficient of returns between SEV and CSR is 0.377716. Correlation coefficient gives a more readily interpretable measure of the degree of association between two securities. Correlation coefficient equal to 1 is perfectly positively correlated, means no diversification between two securities. Correlation coefficient equal to -1 is perfectly negatively correlated, which implies perfect diversification between two securities. A correlation coefficient of 0 means there is no relationship between two securities. A correlation coefficient of 0.377716 indicated that there is a mild positive relationship between the returns of SEV and CSR.

Figure 1 demonstrates that the point P is the best portfolio choice for the investor wishes to minimize the risk of their investment, because it has the minimum standard deviation.

We construct a series of portfolios of the two stocks (SEV & CSR) by varying the weights of each at 2.5% intervals, and then calculated the expected return and standard deviation of each

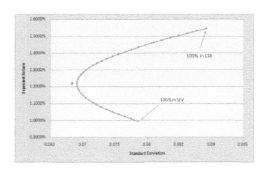

Figure 1. Risk and return TradeOff between for difference combinations of SEV and CSR, from Jan. 2004 to Aug. 2008.

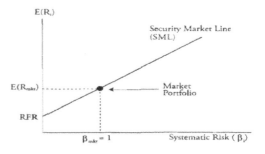

Figure 2. Capital Asset Pricing Model.

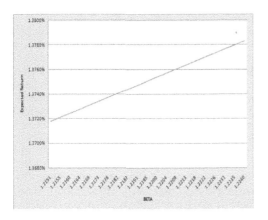

Figure 3. Possible return and beta outcomes that could have been achieved from combining SEV and CSR, from Jan. 2004 to Aug. 2008.

portfolio. Among these 41 portfolios, we recommend the portfolio combining with 60% of SEV and 40% of CSR, which has the minimum standard deviation of 0.069 and the expected return of 1.2175% at this risk level.

We also estimated expected returns and betas of these 41 portfolios by using CAPM approach. The portfolio combining with 100% of SEV and 0% of CSR has lowest beta (1.2151) and lowest expected return (1.3718%). The portfolio structured by 100% CSR securities has the highest expected return but highest beta, which means we need taken highest risk to achieve more rewards than other portfolio choices.

Figure 3 illustrated that the possible returns and beta outcomes between for difference combinations of SEV and CSR is positive linear relationship. It is also highlight that the increasing beta augments expected return, when SEV amount in portfolio decreased from 100% to 0% at 2.5% intervals and CSR increased proportion in portfolio correspondingly.

Combination all above of we discussed, we recommend the portfolio combining with 60% of SEV and 40% of CSR.

REFERENCES

[1] Brailsford, T., Faff, R. and Oliver, B. 1997. *Research Design Issues in the Estimation of Beta* (McGraw-Hill, Sydney).

[2] Carleton W.T. and Lakonishok J., *Risk and Return on Equity: The use and misuse of historical estimates*, Financial Analyst Journal, January–February 1985.

[3] Clinebell, John M., Kahl, Douglas R. and Stevens, Jerry L., 1994. Time-Series properties of the Equity Risk Premium, *The Journal of Finance Research*, v17 n1 p105.

[4] D. Johannes Juttner and Kim M. Hawtrey, *Financial Markets/Money and Risk*, Fourth Edition, Longman, Copyright ©1997.

Information Technology and Applications – Li (Ed.)
© 2015 Taylor & Francis Group, London, ISBN 978-1-138-02677-3

Research on the impact of IM technology on college students' intimacy

Xinrui Chen & Jingrong Sha
College of Modern Educational Technology, Northwest University for Nationalities, Lanzhou, China

Shiliang Lv
China's National Linguistic Information Technology, Northwest University for Nationalities, Lanzhou, China

ABSTRACT: As the rapid growth of information technology and delivery of multiple service of the Internet, Instant Messaging (IM) has become college students' main communication medium. To examine the relationship between IM use and intimacy, a survey was conducted and responses (109) were collected. This paper adopts the statistical method of study, through the parameter analysis, the result showed the total amount of IM correlated with verbal intimacy positively; the amount of IM communication correlated with verbal and affective intimacy positively. In addition, IM communication could be taken as an indicator for predicting intimacy, which is stronger than IM amount. The result implied that IM brought the positive influence on college students, and fostered college students' interpersonal relationship.

Keywords: IM; college students; intimacy

1 INTRODUCTION

The traditional theories points out that intimate relationship can occur only through face-to-face interaction. Along with the development of computers and network technologies, intimate relationship may occur and develop in IM, which is a kind of computer mediated communication (Walther & Burgoon, 1992). As a new type of communication medium, IM differs from the traditional ones. IM eliminates the limitation of time or spaces (Ramire, Dimmick, & Lin, 2004), and provides a new space for users to build and strengthen intimate relationships (Rheingold, 1983).

1.1 *IM characteristics*

IM is one of text-based form of online communication that allows synchronous communication via the media of internet (Hameed, Mellor, & Badii, 2006). In contrast with other communication media, IM has the unique characteristics of instantaneity, presence, and spontaneity (Garrett, & Danziger, 2007). Instantaneity means it provides access to exchange information synchronously; presence indicates, in contrast with other communication media (e.g., phone and E-mail), IM users can easily be aware of whether friends on their contact list are online and available to chat before the action of communication; spontaneity refers that IM allows the users to select communicator and to take them in IM contact list, which defines

their friendships in order to communicate freely (Huang & Yen, 2003).

1.2 *College students' IM usage*

As a new type of communication media, IM eliminates the limitation of waiting time and space associated with other communication media and provides a new space for interpersonal relationship. As IM makes communication easy and convenient (Ramire et al., 2004), IM can satisfies the psychological needs of college students, such as sustain relationship with others. In addition, when college students use IM, it supplies access to feel strong communication atmosphere (Nicholson, 2002), which is important for them to build and sustain relationships with family and friends, and promote closeness in interpersonal relationship (Leung, 2001).

For college students, IM is the most popular online activity. According to the findings presented in the study of Pew Internet and American Life Project (Jones & Madden, 2002), college students are twice as likely to use IM compared with others on a given day, with 72% of respondents to the survey among college students reported that they mainly use IM for communicating with friends. Quan-Haase (2005) conducted a survey with 268 college students and found that 67% of college students use IM every day. Junco and Mastrodicasa (2007) carried out a survey with 7705 college students and found that 75.5% of the participants use IM every day. In China, college students is the largest population and IM has become the main communication media for college students (Nie, Ding, Jiang, & Liu, 2007). As the percentage of college students who use IM gets higher, it is clear that IM use is dominated by college students. More and more college students are begin taking IM as the main communication media due to the development of network technology, the continuous improvement of the IM technology and the increase of the level of accepting new communication medium.

1.3 The relationship between IM use and intimacy

Intimacy is a multidimensional concept that can be defined along different dimension. Previous researchers conceptualized intimacy as a kind of very close relationship that form and develop through long association (Perlman & Fehr, 1987). Generally speaking, verbal intimacy and affective intimacy are widely accepted as the components of intimacy (Tolstedt & Stokes, 1983). Verbal intimacy is regarded as the essential condition that must occur for intimacy to form and the important component of intimacy process, it stands for the apparent verbal information exchange just as self-disclosure (Kjeldskov et al., 2004). Affective intimacy "stands for the individual's feelings of closeness and emotional bonding, including degree of liking, moral support, and the ability to tolerate drawback in the significant other" (Tolstedt & Stokes, 1983). It reflects the share of emotion and the awareness of intimacy between individuals and the significant other, including the perception of closeness (Sinclair & Dowdy, 2005).

Hu, Smith, Westbrook and Wood (2004) studied the relationship between intimacy and IM use amount and found IM use related to intimacy positively, indicated that the intimacy increase as IM use amount increasing. However, recent studies show that in predicting psychological well-being besides the frequency of internet use online activities is better (Gordon, Juang & Syed, 2007; Morgan & Cotton, 2003). This finding indicates that when predict the psychological well-being online activities matter (but not the total amount that including amount of just having IM open). Up to this point, published studies on the IM use (total) amount include amount of just having IM open and online activities amount. Only online activities supply the chance to associate with others for college students and affect the experience of interpersonal communication. Among all the types of online activities (communicate with others, send and receive E-mail, entertainment), communicating with others is the main online activity. The current study attempts to analyze IM communication and provides a better understanding of the mechanism that underlying college students' intimacy in IM.

The current study aims to extend previous study by examining the relationship between IM use and intimacy. The study measures the amount of IM use and the intimacy by using a developed intimacy questionnaire. This study also examines the relationship between IM use and intimacy: verbal intimacy and affective intimacy.

The study hypotheses examined were:

Hypotheses 1: the amount of IM communication correlated with verbal intimacy significantly.
Hypotheses 2: the amount of IM communication correlated with affective intimacy significantly.
Hypotheses 3: the amount of IM communication is the indicator of intimacy, which is better than IM total amount.

2 METHODS

2.1 Study tools

2.1.1 Intimacy questionnaire

The current study used a 17-item intimacy questionnaire by selecting items from the intimacy questionnaire that developed by Hu et al. (2004) used to measure intimacy (Appendix), which include two dimensions of intimacy: verbal intimacy (section A) and affective intimacy (section B).

In section A, all items measure the verbal intimacy by depth and breadth of interpersonal communication, items 1–9 were encoded into ten-point Likert scale, anchored by never (1) and always (10), item 9 was reverse coded. After factor analyses, all items were kept. Students were asked to answer question based on their feelings when they use IM, such as: "when you talk to others on IM, to what extent your conversation are about your family?" The higher the score, the higher the verbal intimacy will be.

In section B, all items measure the affective intimacy by the connection that users fell in the process of interpersonal communication, items 1–8 were encoded into ten-point Likert scale, anchored by disagree (1) and strongly agree (10), items 3 and 7 were reverse coded. After factor analyses, all items were kept. Students were asked to answer question based on their feeling when they use IM, such as: "when you communicate with others by IM, to what extent you feel they are close to you?" The higher the score, the higher affective intimacy will be.

In current study, Cronbach's alphas were 0.852 for verbal intimacy and 0.756 for effective intimacy was, indicating such questionnaire has a good level of internal consistency in accordance with previous studies (e.g., Hu et al., 2004).

2.1.2 *IM use*

Three items were used to measure of IM use. Participants were asked to respond to the following questions: (1) Do you use IM everyday?; (2) How long do you use it totally?; (3) How long do you use it for communicating with others?

2.1.3 *Sociodemographics*

Participants indicated their demographic information that include age and gender.

2.2 *Procedure and date analyses*

All date analyzed by SPSS 16.0. Descriptive statistics were conducted to illustrate the demographic characteristics of the participants and their use of IM. Correlations were conducted to evaluate the relationship between IM amount and intimacy. To examine the hypotheses, 2 hierarchical linear regression analyses were done to examine which factor affect verbal intimacy and effective intimacy.

3 RESULTS

3.1 *IM usage*

College students in current study, spending a mean of 6.95 hours (SD 1.05) totally. Spending a mean of 4.88 hours (SD 0.87) for communicating with others, indicating college students mainly use IM to communicate with others, and IM has become the important way for them to communicate.

3.2 *Relationship between IM total amount, IM communication amount and intimacy*

Correlation analyses were conducted to examine the relationship between intimacy and IM use. The result shows that IM total amount positively correlated with verbal intimacy ($r = 0.266$, $p < 0.001$), did not correlate with affectively significantly ($r = 0.04$, $p > 05$); both the verbal intimacy ($r = 0.710$, $p < 0.001$) and the affective intimacy ($r = 0.710$, $p < 0.001$) positively correlated with IM communication amount.

Previous studies found that gender and age may affect intimacy (Hu et al., 2004; McNelles & Connolly, 1999), thus gender and age were taken as control variables when conduct regression analyses. Hierarchical linear regression analyses were conducted to examine which factor (IM total amount, IM communication amount) predict intimacy. Statistical analysis has three steps.

A hierarchical linear regression was conducted to build a model for predicting verbal intimacy. In the first step, two indicators were added: gender (coded into dummy variables) and age

(continuous variable). This model was not statistically significant, $F(2, 101) = 0.44$, $p > 0.05$, $R2 = 0.01$, $\Delta R2 = -0.01$. IM total amount entered in the second step. Addition of this indicator significantly increased the fit of the model, $B = 0.28$, $SEB = 0.26$, $F(3, 100) = 2.79$, $p < 0.05$, $R2 = 0.08$, $\Delta R2 = 0.05$. Though IM total amount had a significant effect, it only accounts for 5% of the variances. The third step consisted of IM communication amount. Addition of this indicator increased the fit of the model significantly, $B = 0.74$, $SEB = 0.10$, $F(4, 99) = 15.45$, $p < 0.05$, $R2 = 0.38$, $\Delta R2 = 0.36$. For IM total amount, B changed from ($B = 0.28$, $SEB = 0.10$, $p < 0.05$) to ($B = 0.10$, $SEB = 0.09$, $p > 0.05$). These results indicate that IM communication has a significant effect, accounts for 36% of the variances, and verbal intimacy increases with the increasing of amount of IM communication.

A hierarchical linear regression was conducted to build a model for predicting affective intimacy. In the first step two indicators were added: gender (coded into dummy variables) and age (continuous variable). This model was not statistically significant, $F(2, 101) = 1.31$, $p > 0.05$, $R2 = 0.03$, $\Delta R2 = 0.01$. The total amount of IM entered in the second step. Addition of this indicator did not significantly increase the fit of the model, $B = 0.28$, $SEB = 0.26$, $F(3, 100) = 0.96$, $p > 0.05$, $R2 = 0.03$, $\Delta R2 = -0.01$. The third step consisted of IM communication amount, addition of this indicator increased the fit of the model significantly, $B = 0.74$, $SEB = 0.10$, $F(4, 99) = 30.58$, $p < 0.05$, $R2 = 0.55$, $\Delta R2 = 0.54$, For IM total amount, B changed from ($B = 0.06$, $SEB = 0.12$, $p > 0.05$) to ($B = -0.22$, $SEB = 0.09$, $p > 0.05$). These results indicate that IM communication has a significant effect, accounts for 54% of the variances, and affective intimacy increases with the increasing of amount of IM communication.

4 DISCUSSIONS

4.1 *College students' use of IM*

With a high rate of use of IM (95.4%), results of this study indicates that college students use IM positively and IM has become an important communication media for college students. There was positive relationship between IM communication amount and intimacy. Moreover, IM communication amount was positive predictive of intimacy, indicates that as college students' IM communication amount increase, therefore the level of perceived intimacy also increases.

IM becomes the main communication medium may due to the characteristics of IM and college students' unique needs. As the important media of online communication, IM breaks limitation of traditional communication media, and satisfies college students' needs of constant connectivity and real-time conversation in such an information age (Frand, 2000), strengthens the bonds between friends and enriches social networks by extending the real community into the cyberspace (Ellison, Steinfield & Lampe, 2007). By breaking the limitation of traditional media, IM facilitates emotional information sharing and exchange, also supplies new space for college students for fostering emotional support (Rovers & Essen, 2006). Moreover, according to McGuire's Psychological Motivations theory, college students have the need for stimulation and expression, and IM satisfies needs and wants of college students.

4.2 The relationship between IM communication and verbal, affective intimacy

The results of two hierarchical liner regressions shows that IM communication amount is stronger indicator of intimacy, this finding is congruent with studies on uses of internet that found when predicting outcomes online activities were as important, even more important than time spend (Gordon et al., 2007; Junco & Cotton, 2010). These findings indicate that IM communication amount is the strong indicator of intimacy, that is to say the perceived verbal, affective intimacy increased along with the increase of IM communication amount.

Previous studies showed that self-disclosure fostered intimacy and promoted interdependence between communicators (Altman & Taylor, 1973), and disclosure in the internet fostered the development of interpersonal friendship (Henderson & Gilding, 2004). Related study showed that self-disclosure in internet was more intense and more likely compared with face-to-face conversation (Pornsakulvanich, 2005). Similar to traditional communication medium, internet users also expect to reduce the uncertainty of interpersonal relationship and develop affinity (Walther & Parks, 2002). Along with the increasing of online communication amount, individuals would reveal more intimate information about one's self that regarded as essential for forming and sustaining personal relationship and would affect perceived intimacy (Hiltz & Turoff, 1978). Therefore, as the IM communication amount increased, the level of perceived verbal intimacy increased as well.

Affective intimacy refers to the sharing of emotional feelings and the offer of emotional support (Shaughnessy, 1995), which was related to the social cues such as physical distance, posture, and eyes contact (Argyle & Dean, 1965). Though IM communication is limited in non-verbal cues just as emotions and intonation, the appearance of network language compensated for social cues. Network language is a type of communication way that has the function of vice language cues, and can be used to assist language. Among all types of network language, emoticons are distinctive and have the characteristics of vice cues. IM users could enrich their conversation with social meaning via the use of Emoticons (Walther & D' Addario, 2001). With the continuous improvement of the IM technology, IM manufacturers enriched the diversity of emoticons via adding dynamic, magic icon. When College students use these emoticons to communicate with friends, it could not only simulate face to face communication via compensating the lack of non-verbal cues, but also construct verbal communication situation that supply access to express the user' s emotion and accurately perceive other's emotion. Along with the increase of IM communication amount, college students will reform their unique communication style (how to express themselves and how to recognize other's emotion). Because college students internalize the emotions of others over time, the perceived affective intimacy increase along with the increase of IM communication amount.

4.3 Limitations

The major limitation of the current study is that it is impossible to confirm the causal mechanisms between IM use and intimacy by such cross-sectional study. Though the results show that IM communication amount and intimacy are related, it is hard to determine the direction of the effect. For instance, college students who spent more time communicating with others on IM also scored higher in the level of perceived intimacy; however, it may be that college students who perceived higher level of intimacy were more likely to spend more time communication on IM. Future longitudinal studies would be needed to determine the mechanism of causation.

REFERENCES

Altman, I., & Taylor, D., (1973). *Social penetration: the development of interpersonal relationships.* New York: Holt, Rinehart & Winston.

Argyle, M., & Dean, J. (1965). Eye contact, distance and affiliation. *Sociometry, 28,* 289–304.

Ellison, N.B., Steinfield, C., & Lampe, C. (2007). The benefits of facebook "friends:" social capital and college students' use of online social network sites. *Journal of Computer-Mediated Communication, 12(4),* 1143–1168.

Frand, J.L. (2000). The information-age mindset: changes in students and implications for higher education. *Educause Review, 35(5),* 14–19.

Garrett, R.K., & Danziger, J.N. (2007). IM = interruption management? Instant messaging and disruption in the workplace. *Journal of Computer-mediated Communication, 13*, 1.

Gordon, C.F., Juang, L.P., & Syed, M. (2007). Internet use and well-being among college students: beyond frequency of use. *Journal of College Student Development, 48(6)*, 674–688.

Grinter, R., & Palen, L. (2002). Instant messaging in teen life. *Proceedings from computer supported cooperative work 2002* (pp, 21–30). ACM.

Hameed, S., Mellor, J & Badii, A. (2006). The impact of the increasing use of instant messaging (IM) on user's real social communication and integration. *International Journal on Transactions on Internet Research, 2(20)*, 38–44.

Henderson, S., & Gilding, M. (2004). I' ve never clicked this much with anyone in my life: trust and hyperpersonal communication in online friendships. *New Media & Society, 6(4)*, 487–508.

Henggang, N., Li, D., Pei, J., & Yi, l. (2007). On the scale making and characteristics of young people s internet communication. *Journal of Guangzhou University (social sciences), 6(5)*, 3–8.

Hiltz, S.R., & Turoff, M. (1978). *The Network Nation: Human Communication via Computer.* New York: Addison-Wesley.

Huang, A.H., & Yen, D.C. (2003). Usefulness of instant messaging among young users: social vs. work perspective. *Human Systems Management, 22(2)*, 63–72.

Hu, Y., Smith, V., Westbrook, N., & Wood, J.F. (2004). Friendships through IM: examining the relationship between instant messaging and intimacy. *Journal of Computer-Mediated Communication, 10(1)*, 38–48.

Junco, R., & Cotten, S.R. (2010). Perceived academic effects of instant messaging use. *Computers & Education, 56*, 370–378.

Junco, R., & Mastrodicasa, J. (2007). *Connecting to the net.generation: what higher education professionals need to know about today's students.* Washington, D.C.: NASPA.

Junco, R., Merson, D., & Salter, D.W. (2010). The effect of gender, ethnicity, and income on college students' use of communication technologies. *CyberPsychology, Behavior, and Social Networking, 13(6)*, 37–53.

Jones, S., & Madden, M. (2002). *The internet goes to college: How students are living in the future with today's technology.* Washington, DC: Pew Internet and American Life Project.

Kjeldskov, J., Gibbs, M., Vetere, F., Howard, S., Pedell, S., Mecoles, K., & Bunyan M. (2004). Using cultural probes to explore mediated intimacy. *Australasian Journal of Information Systems, 12(1)*, 102–115.

Leung, L. (2001). College student motives for chatting on ICQ. *New Media & Society, 3(4)*, 483–500.

McNelles, L.R., Connolly, J.A. (1999). Intimacy between adolescent friends: age and gender differences in intimate affect and intimate behaviors. *Journal of Research on Adolescence, 9(2)*, 143–159.

Morgan, C., & Cotten, S.R. (2003). The relationship between internet activities and depressive symptoms in a sample of college freshmen. *CyberPsychology & Behavior, 6(2)*,133–142.

Nicholson, S. (2002). Socialization in the virtual hallway: instant messaging in the asynchronous web-based distance education classroom. *Internet and Higher Education, 5*, 363–372.

Perlman, D., & Fehr, B. (1987). The development of intimate relationships, In Perlman, D., & Duck, S (Eds). *Intimate Relationships* (pp, 13–42). Newbury Park, CA: Sage.

Pornsakulvanich, V. (2005). *Testing a uses and gratifications model of online relationships.* Doctoral dissertation. College of Communication and Information, Kent State University.

Quan-Haase, Anabel, Joseph Cothrel & Barry Wellman (2005). Instant messaging for collaboration: a case study of a High-Tech firm. *Journal of Computer-Mediated Communication, 10(4)*.

Ramirez, A.R., Dimmick, J., Feaster, J., & Lin, S. (2004). The gratification niches of instant messaging, e-mail, and telephone. *Communication Research, 35(4)*, 529–547.

Rheingold, H. (1993). *The virtual community: homesteading on the electronic frontier.* Reading, MA: Addison-Wesley.

Rovers, A.F., van Essen, H.A. (2006). Guidelines for haptic interpersonal communication applications: an exploration of foot interaction styles. *Virtual Reality, 9*, 177–191.

Shaughnessy, M.F. (1995). Sexual intimacy and emotional intimacy. *Sexological Review, 3(1)*, 81–95.

Sinclair, V.G., & Dowdy, S.W. (2005). Development and validation pf the emotional intimacy scale. *Journal of Nursing Measurement, 13*, 193–206.

Tolstedt, B.E., & Stokes, J.P. (1983). Relation of verbal, affective, and physical intimacy to marital satisfaction. *Journal of Counseling Psychology, 30(4)*, 573–580.

Walther, J.B., & Burgoon, J.K. (1992). Relational communication in computer-mediated interaction. *Human Communication Research, 19*, 50–88.

Walther, J., & D' Addario, K.P. (2001). The impacts of emoticons on message interpretation in computer-mediated communication. *Social Science Computer Review, 19*, 323–345.

Walther, J.B., & Parks, M.R. (2002). Cues filtered out, cues filtered in: Computer-mediated communication and relationships. In M.L. Knapp & J.A. Daly (eds.), *Handbook of interpersonal communication* (3rd ed, 529–563). Thousand Oaks, CA: Sage.

Information Technology and Applications – Li (Ed.)
© 2015 Taylor & Francis Group, London, ISBN 978-1-138-02677-3

Research on accounting conservatism and investment efficiency of IT enterprises: A perspective of impairment of assets

Chenyue Yu & Jixiao Xu
Business School of Hohai University, Nanjing, Jiangsu, China

ABSTRACT: This paper studies the influence of accounting conservatism on investment efficiency of information technology enterprises from the perspective of impairment of assets. Through regression analysis, we find that accounting conservatism, which is measured by the impairment of assets, can curb over-investment of IT enterprises, while it can also intensify under-investment of IT enterprises.

Keywords: accounting conservatism; investment efficiency; impairment of assets; IT enterprises

1 INTRODUCTION

In today's increasingly competitive market environment, investment activity is an important foundation for enterprises to achieve maximum value target and it is directly related to the survival and development of enterprises. It will also affect the efficiency of resource allocation and distribution of society as a whole. However, as a result of the imperfection of China's capital market, the dominance of state-owned shares and the mismanagement of mechanism, agency problems and information asymmetry problems of our enterprises are more outstanding, that make the inefficient investment phenomenon relatively common. The typical inefficient investment behaviors mainly include under-investment and over-investment. Thus how to availably improve the investment efficiency is not only becoming a difficult problem faced by all the enterprises, but also becoming a hot research topic.

Accounting conservatism, also known as cautiousness, means that the enterprise accounting measurement should follow the principle of prudential, do not overstate assets or revenue, and do not underestimate liabilities or losses in the case of uncertainty. Research on accounting conservatism at home and abroad has been a hot topic. Since Basu (1997) proposed measures of the model, research on accounting conservatism has attracted many scholars. With the deepening of research on the formation and influence of the accounting conservatism, and Watts (2003) proposed it as a corporate governance mechanism, more and more scholars began to concentrate on the economic consequences of accounting conservatism, research findings mainly focus on accounting conservatism of the financing and investment effects. As accounting conventions, the conservatism inevitably plays a relief role in information asymmetry of contradiction, which is between the owner and the agent. Therefore, accounting conservatism can impact on the investment efficiency of enterprises in a certain degree.

The impairment of assets, which is one of the concrete embodiments of accounting conservatism in accounting standards, reflects that enterprises confirm all the possible economic loss in time. Therefore, from this perspective, we attempt to choose the impairment of assets as a measure of accounting conservatism, then select the information technology enterprises (IT enterprises) as the research objects, and finally use the regression analysis method to study the influence of accounting conservatism on investment efficiency of IT enterprises.

2 DATA AND RESEARCH DESIGN

2.1 Hypothesis

A review of previous studies has founded that accounting conservatism can curb over-investment and intensify under-investment. Thus, in this paper, we generate the following hypotheses:

H1: Accounting conservatism which is measured by the impairment of assets can curb over-investment of IT enterprises.
H2: Accounting conservatism which is measured by the impairment of assets can intensify under-investment of IT enterprises.

2.2 Data and sample

We select the data of information technology listed companies from Shanghai and Shenzhen stock markets during the period of 2011 to 2013 as samples and delete companies which are ST or PT and issued A shares, B shares and H shares at the same time. Then we winsorize all continuous variables at 1% level to eliminate the effects of extreme values. All the data come from the CSMAR and the RESSET.

2.3 Variable selection and model construction

2.3.1 Model construction of investment efficiency
We use the Richardson (2006) investment expectation model to measure the investment level of IT enterprises. The model is set up as follows:

$$Invest_{t+1}/A_t = \chi_0 + \chi_1 Growth_t + \chi_2 Size_t + \chi_3 Age_t$$
$$+ \chi_4 Return_t + \chi_5 Lev_t + \chi_6 Cash_t/A_t$$
$$+ \chi_7 Invest_t/A_t + \varepsilon_0 \tag{1}$$

where $Invest$ = new investments; A = total assets; $Growth$ = growth rate of operating revenue; $Size$ = natural log of total assets; Age = number of years the firm has been listed on stock market; $Return$ = stock returns; Lev = debt ratio; $Cash$ = cash flow divided by total assets; and ε_0 is the residual, which is on behalf of inefficient investment.

2.3.2 Model construction of accounting conservatism
In this part, we draw on the Khan and Watts (2009) model to build accounting conservatism model which is measured by the impairment of assets.

$$WD_t = \alpha_0 + \alpha_1 D_t + \beta_0 R_t + \beta_1 D_t \times R_t + \varepsilon_1 \tag{2}$$

$$\beta_0 = \mu_1 + \mu_2 Size_t + \mu_3 PTB_t + \mu_4 Lev_t \tag{3}$$

$$\beta_1 = \lambda_1 + \lambda_2 Size_t + \lambda_3 PTB_t + \lambda_4 Lev_t \tag{4}$$

Replace equation (2) with (3) and (4), then the result is as follows:

$$WD_t = \alpha_0 + \alpha_1 D_t + (\mu_1 + \mu_2 Size_t + \mu_3 PTB_t$$
$$+ \mu_4 Lev_t) R_t + (\lambda_1 + \lambda_2 Size_t + \lambda_3 PTB_t$$
$$+ \lambda_4 Lev_t) D_t \times R_t + \varepsilon_1 \tag{5}$$

where WD = impairment of assets divided by total assets; R = stock returns; D = dummy variable, equal to 1 when R is negative and equal to 0 otherwise; $Size$ = natural log of total assets; PTB = price-to-book ratio; Lev = debt ratio; and ε_1 is residuals.

In the operation, firstly put all the samples' data into the equation (3) and (4), and then calculate the values of μ_1, μ_2, μ_3, μ_4 and λ_1, λ_2, λ_3, λ_4. Finally, calculate the value of β_0 and β_1. The β_0, which is measured by regression, shows the sensitive degree

of impairment of assets for good news, while $(\beta_0 + \beta_1)$ represents the sensitive degree of impairment of assets for bad news. Therefore,

$$C\text{-}WD = (\beta_0 + \beta_1)/\beta_0 \tag{6}$$

We choose the equation (6) to stand for conservatism. The greater $C\text{-}WD$ is, the higher the degree of accounting conservatism.

2.3.3 Model construction of accounting conservatism and investment efficiency
Many research findings show that internal cash flow and executive compensation can also influence the investment efficiency of enterprises. Therefore, the final research models are as following:

The model of influence of accounting conservatism on over-investment of IT enterprises:

$$OI_{t+1} = \eta_0 + \eta_1 C\text{-}WD_t + \eta_2 CF_t/A_t + \eta_3 Pay_t/A_t + \varepsilon_2 \tag{7}$$

The model of influence of accounting conservatism on under-investment of IT enterprises:

$$UI_{t+1} = \omega_0 + \omega_1 C\text{-}WD_t + \omega_2 CF_t/A_t + \omega_3 Pay_t/A_t + \varepsilon_3 \tag{8}$$

where OI = over-investment, equal to ε_0 when ε_0 is positive; UI = under-investment, equal to ε_0 when ε_0 is negative; $C\text{-}WD$ = the coefficient of accounting conservatism; A = total assets; CF = internal cash flow; Pay = top three executive pay; and ε_2, ε_3 is residual.

3 EMPIRICAL TEST AND RESULTS

3.1 Descriptive statistics

Table 1 provides the descriptive statistics of the samples of investment efficiency of IT enterprises.

Table 1. Descriptive statistics of the samples of investment efficiency of IT enterprises.

Variables	Mean	Median	Max	Min	St. dev.
$Invest_{t+1}/A$	0.112	0.062	3.089	−0.119	0.220
$Growth$	0.384	0.162	103.811	−0.953	3.766
$Size$	22.031	21.953	26.949	18.380	1.170
Age	11.600	12.000	26.000	2.000	4.343
$Return$	−0.298	−0.323	1.430	−0.732	0.194
Lev	0.524	0.529	2.362	0.021	0.212
$Cash/A$	0.175	0.141	0.852	0.000	0.126
$Invest_t/A$	0.091	0.061	1.778	−0.504	0.141

The number of the samples is 369.

The mean of $Invest_{t-1}/A$ and $Invest_t/A$ are 0.112 and 0.091, and the median are 0.062 and 0.061, both of them are positive. This demonstrates that the scale of investment in IT enterprises is expanding and the prospect is bright.

The maximum of *Growth*, which is measured by growth rate of operating revenue, is 103.811 and the minimum is −0.953, while the mean is 0.384. This shows that the operating performance of IT enterprises has a lot of difference, but they still have large investment opportunities as a whole.

The maximum and the minimum *Size* are 18.380 to 26.949, however, *Age* are 2 to 26. This suggests that the scales of IT enterprises keep little difference between each other but the ages differ too much.

The mean and median values of *Return* are negative, in other words, the yield of IT enterprises in stock market is not ideal.

The difference between maximum and minimum value of debt ratio (*Lev*) is large, the mean and median are more than 0.5 at the same time. This demonstrates that the debts of IT enterprises are keeping at a high level.

At last, the mean of *Cash* is 0.175 and the standard deviation is 0.126, which means IT enterprises have maintained cash in a certain amount.

Table 2 and 3 show that the difference between the group of over-investment and under-investment of new investment is large, which is determined by the positive and negative residuals. The difference of the coefficient of conservatism (*C-WD*) is small.

From the perspective of internal cash flow, the maximum, mean and median of *CF/A* in over-investment group are 0.950, 0.034 and 0.043, which are larger than the value in under-investment group. That means over-investment group's internal cash flow is higher than under-investment group among IT enterprises, which can make inefficient investment.

From the perspective of executive pay, *Pay/A* has the mean of 0.001 and the median of 0.000 in over-investment group; which are similar to the value in under-investment group.

3.2 *Regression analysis*

In the Table 4, we can see that the regression coefficient of accounting conservatism (*C-WD*) and over-investment (*OI*) is −0.306, *T*-value is −1.727, and it is significant at 0.01 level. This means that accounting conservatism which is measured by the impairment of assets has a significantly negative correlation with over-investment. In other words, the impairment of assets can curb over-investment of IT enterprises. Hypothesis 1 is set up.

Table 5 shows that the regression coefficient of accounting conservatism (*C-WD*) and under-investment (*UI*) is 0.182, *T*-value is 2.950, which is also significant at 0.01 level, and this is to say that accounting conservatism which is measured by the

Table 2. Descriptive statistics of the samples of accounting conservatism and over-investment of IT enterprises.

Variables	Mean	Median	Max	Min	St. dev.
OI	0.142	0.072	7.089	0.001	0.328
C-WD	0.054	0.043	0.912	−0.608	0.026
CF/A	0.034	0.043	0.950	0.001	0.317
Pay/A	0.001	0.000	0.019	0.000	0.002

The number of the samples is 148.

Table 3. Descriptive statistics of the samples of accounting conservatism and under-investment of IT enterprises.

Variables	Mean	Median	Max	Min	St. dev.
UI	−0.051	−0.042	−0.000	−0.330	0.028
C-WD	0.053	0.043	0.190	−0.019	0.023
CF/A	0.027	0.001	0.550	−0.001	0.217
Pay/A	0.001	0.000	0.019	0.000	0.002

The number of the samples is 221.

Table 4. Regressions with both accounting conservatism and over-investment of IT enterprises.

Variables	Coeff.	T	Sig.	Multi-co linearity Tole.	VIF
Constant	0.074	3.360	0.001		
C-WD	−0.306	−1.727	0.001	0.976	1.024
CF/A	0.939	22.22	0.000	0.947	1.056
Pay/A	−5.068	−2.387	0.017	0.966	1.306

Dependent variable: *OI*.

Table 5. Regressions with both accounting conservatism and under-investment of IT enterprises.

Variables	Coeff.	T	Sig.	Multi-co linearity Tole.	VIF
Constant	0.189	1.351	0.177		
C-WD	0.182	2.950	0.003	0.887	1.127
CF/A	−0.047	−3.880	0.000	0.831	1.203
Pay/A	−3.031	−2.225	0.022	0.931	1.074

Dependent variable: *UI*.

impairment of assets has a significantly positive correlation with under-investment. In other words, the impairment of assets can intensify under-investment of IT enterprises. Hypothesis 2 is set up.

4 CONCLUSION

This paper uses the empirical study to inspect the relationship between accounting conservatism, which is measured by the impairment of assets, and investment efficiency of IT enterprises. The main conclusions are laid below:

1. Accounting conservatism which is measured by the impairment of assets can curb over-investment of IT enterprises.

   Through the provision for impairment of assets, the expected future losses can be reflected in the financial statements without delay, which will timely help IT enterprises' managers to give up the investment project that future cash flow are negative. Meanwhile, the loss of information can attract the attention of managers, and prompt managers to maintain a more cautious attitude to treat future investment. In a word, it is conducive to curb over-investment of IT enterprises.

2. Accounting conservatism which is measured by the impairment of assets can intensify under-investment of IT enterprises.

The reason of this finding is that the conservative accounting policies have made the financial report data more pessimistic and even hidden the real strength of IT enterprises. It will let IT enterprises' managers maintain a more cautious attitude when making investment choices. For example, it may lead to a lack of investment in the way that the managers blindly give up an investment project which has risks but the net present value is positive.

REFERENCES

Basu, S. 1997. The conservatism principle and the asymmetric timeliness of earnings. *Journal of Accounting and Economics* 24(1): 3–37.

Givoly, D., Hayn, C.K. & Natarajan, A. 2007. Measuring reporting conservatism. *The Accounting Review* 82(1): 65–106.

Khan, M. & Watts, R.L. 2007. Estimation and validation of a firm-year measure of conservatism. *Available at SSRN* 967348.

Watts, R.L. 2003. Conservatism in accounting part I: Explanations and implications. *Accounting Horizons* 17(3): 207–221.

Yang Dan, Wang Ning & Ye Jianming. 2011. Accounting conservatism and investment behavior of listed firms. *Accounting Research* 3: 27–33.

Information Technology and Applications – Li (Ed.)
© *2015 Taylor & Francis Group, London, ISBN 978-1-138-02677-3*

Information technology security patent application analysis in China

Yuangang Yao
China Information Technology Security Evaluation, Beijing, China

Xiaoyu Ma
Patent Examination Cooperation Center of the Patent Office, SIPO, Beijing, China

Xianghui Zhao, Runpu Wu, Lin Liu & Yanzhao Liu
China Information Technology Security Evaluation, Beijing, China

ABSTRACT: With the development and extensive application of information technology, information security has become increasingly important, and information security technology gets unprecedented attention. Therefore, information security industry has been developing rapidly. Intellectual property protection of information security technology is important for the development of the information security industry, and it also reflects the current developments in information security. In this article, we use China patent retrieval system to analyze information security technology patents in China according to the domestic patent applications, the applicants, and the comparison of Chinese and foreign patent applications. We summarize the situations of information security industry and the trends of information security in China, and provide some advices for information security technology research and industry development.

Keywords: information security; patent analysis; intellectual property

1 INTRODUCTION

With the development of information technology, the connotation of information technology security develops gradually from the initial information confidential to the information integrity, availability, controllability and non-repudiation, and forms the basic theory and implementation technology of security attack, prevention, detection, control, management, evaluation and so on. Currently, The contents of information technology security mainly include equipment security, data security, content security and behavior security (Feng 2010, Whitman 2010).

Information technology security is an interdisciplinary technology, involving knowledge of computer, communication, network, mathematics, physics, law, management, sociology and so on. Technically, information security technology can be broadly divided into basic security technologies and application security technologies. The basic security mainly includes encryption-based security, security integration, security management, security system architecture, security evaluation, and project management. The application security mainly includes electromagnetic leakage protection,

security operation platform, information detection, computer virus prevention, information security enhancement, security audit, intrusion detection, safety precaution, content regulation, and information security offensive and defensive (Shen 2007, Siponen 2007). Vulnerability is an important cause of information security issues. As a result, in the information technology security researches and practices, vulnerability analysis technology is increasingly important (Wu 2012).

The development of information technology security technology and the increasing market demands lead to the development of information technology security products. And the user requirement propositions are shifted from passive to active, from a single demand to diverse ones. This prompted the types of information security products from a single network security product to diverse network security, application security, data security, security system products (Shen 2001).

The increase of market investment brings more competition. With the increasing awareness of property rights, the security industry pays more attention on product intellectual property protection while strengthening technological innovation. As a result, information technology security patents reflect the state of art of industry and research trends. In this paper, we carry out an analysis for patent application situation of information technology security in China, describe and summarize current development of information technology security industry from multiple dimensions, and identify the trends to provide a reference for security industry and the intellectual property protection.

2 INFORMATION TECHNOLOGY SECURITY PATENT SEARCHING STRATEGIES

In our research, data are retrieved from China Patent Retrieval System (CPRS) (Yan 2013). CPRS contains all patent application information in China since the year 1985, which has a full data coverage and rich data items. We choose specialized domain vocabulary in information technology security area as keywords and their extension combing with the support of related classification numbers in application to search patent documents. The data retrieved are up to May 31, 2014.

After the submission of patent applications, it usually takes about 18 months for patent disclosures before they can be retrieved. So far, some patent applications submitted in 2012, 2013 and 2014 haven't been retrieved yet. Therefore, the statistics of last two years are lower than actual ones, and they are only reference data in the following analysis.

3 INFORMATION TECHNOLOGY SECURITY PATENT ANALYSIS

We totally retrieve 3615 patent applications in information technology security areas after searching and filtering from CPRS, in which 47.77% of the applications are authorized. In the rest of this paper, we will analyze these patent applications by multiple dimensions of patent application quantity, applicants, inventors, locations, and keywords.

3.1 *Analysis of application quantity*

To overview the information technology security patent application in China, we first analyze the application quantity through statistical analysis according to application dates. Figure 1 shows the number of patent applications over the years since the year 1985.

Judging from the number of applications over the years, patent applications in information technology security can be divided into 3 stages.

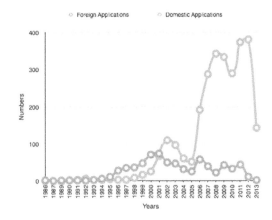

Figure 1. Information technology security patent applications in China.

The first stage is from the year 1985 to 1994, in which patent applications were less than 10 per year. That is because the threshold for information technology security research is high, and the market is immature and the growth is slow. This is the initial stage of the information technology security industry.

The second stage is from the year 1995 to 2005, in which the number of applications had significant growth year after year. In 1995, there were only 16 applications, and by 2002 the number grew to 160. With the rise and popularity of the Internet, information technology security industry had been growing rapidly for the first time. The important value of security was well recognized. This is the growth stage of information technology security industry.

The third stage is from the year 2006 to now, in which the number of applications increased rapidly with an average of more than 300 applications each year. It reached to 418 applications in 2011. During this phase, security events caused by viruses, trojans, phishing attacks, spam, botnets, spyware, malware, attacks against vulnerabilities is increasingly serious. Sampling data published by China Information Security Evaluation Center (CNITSEC) show that more than 68.76 million virus samples, 0.92 million web trojan samples, 0.12 million phishing sites, and 0.15 million hidden links websites were found in 2012 (Wu 2013), which reflects the great demands for security technologies and products. In this situation, the information technology security industry has been sustained rapid growth and made a large number of technology and product developments and innovations based on the breakthroughs achieved by companies, universities and institutes in security theory and technology. In particular, the support of national policies also plays a significant role to promote the development of information technology security in China.

3.2 Analysis of patent applicants

In the retrieved 3615 patent applications, enterprise applications are 2528, accounting for 69.93% of total applications; universities and research institutes applications are 652, accounting for 18.03% of total applications; individual applications are 435, account for 12.03% of total applications. Enterprises, who are the main participants in security industry, play a key role for the development of information technology security. Considering the fierce competition in market, they have the strongest demands for innovation, and intellectual property protection. Universities and research institutes, who focus on advanced technology research, are also important component of innovation and intellectual property protection; the number of individual applications is small.

These patent applications involve a total of 527 patent applicants. We focus on the enterprises, universities and research institutions to analyze the applicants.

Figure 2 shows the main 23 enterprises of the patent applications and authorizations, including 13 domestic enterprises and 10 foreign enterprises. The top 10 domestic enterprises are Feitian, Huawei, Qihoo, ZTE, Lenovo, Tencent, Kingsoft, Venustech, Antiy, Senselock, Qinchuan. And the top 10 foreign enterprises are Sony, Microsoft, Panasonic, IBM, Samsung, Intel, Kaspersky, Hitachi, Fujitsu, Toshiba. Less than half of the 23 enterprises are traditional security technology ones; many of them are hardware or software vendors, or Internet companies. This shows that security is a challenge for the entire information technology industry that needs to guarantee the security of products and services. Domestic enterprises Feitian, Huawei, and Qihoo attach great importance to the protection of information technology security innovation and intellectual property. Meanwhile foreign applications are mainly from Japan, the U.S. and South Korea enterprises, and Japanese enterprises apply half of foreign applications. From the point of patent application authorization, Fujitsu, Lenovo, Sony, Toshiba, Panasonic, Feitian, Hitachi, IBM, Venustech and Huawei have authorization rate more than 60% with strong innovation. In the top 10 foreign enterprises, Japanese enterprises account for half, which reflects the importance of intellectual property protection in Japan. China Internet enterprise Qihoo also need to be concerned. As a leading Internet security and information services company, it applies a total of 75 security related patents, and 2 of them are authorized. Although the authorization is only 2.67%, there are nearly 70 patent applications submitted in the year 2012 and 2013, which are still in patent examination process. This also reflects the fact that Qihoo is increasing the investment in the field of information technology security.

Figure 3 shows the major universities and research institutes for information technology security patent applications and authorizations in

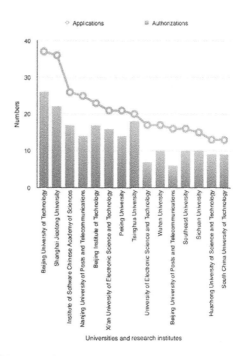

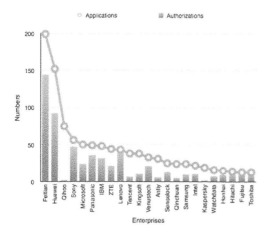

Figure 2. Patent applications and authorizations by enterprises.

Figure 3. Patent applications and authorizations by universities and research institutes.

China. The top 15 applications in order are Beijing University of Technology, Shanghai Jiaotong University, Institute of Software Chinese Academy of Sciences, Nanjing University of Posts and Telecommunications, Beijing Institute of Technology, Xi'an University of Electronic Science and Technology, Peking University, Tsinghua University, University of Electronic Science and Technology, Wuhan University, Beijing University of Posts and Telecommunications, Southeast University, Sichuan University, Huazhong University of Science and Technology, South China University of Technology. These universities and research institutes all have strong background of traditional information security technology research and product development, while others have less patent applications. Compared with enterprises, the authorization rate is slightly higher, but the number is smaller. Furthermore, universities and institutes have fewer security products related to these patents, so the influence to information technology security industry is indirect.

3.3 Analysis of Chinese and foreign patent applications

The geographical distribution of information technology security patent applications in China is wide, including 33 domestic provinces, municipalities and autonomous regions and 26 foreign countries and regions. There are 2877 domestic applications and 738 foreign applications. Domestic applications concentrate in economically developed regions, where have not only the high level of information but also high incidence of security events. Beijing, Shenzhen, Shanghai are the top 3 cities in domestic applications, and the top 10 cities account for 78.21% of total domestic applications. Figure 4 shows the distribution of domestic applications in different cities.

Foreign patent applications are mainly from the U.S. and Japan, accounting for 38.21% and 29.95% of total foreign applications respectively, followed by South Korea, France, and Russia. Figure 5 shows distribution of foreign applications. The foreign patent applications entered into growth stage since the year 1995, reached the peak in the year 2000, and then have no increasing trend after that. While Chinese applications started relatively late, which began to grow around the year 2000, and exceeded foreign applications in 2005. After 2005, domestic applications grow rapidly. Although the numbers of applications between the domestic and foreign are quite different, the overall trend is similar. This indicates that on one side, the China's information technology security industry is developing rapidly, and it has a strong capability of independent innovation; on the other side,

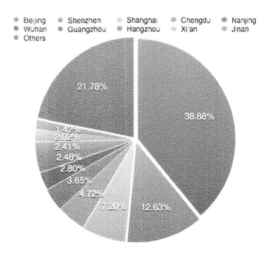

Figure 4. The domestic city distribution of patent applications.

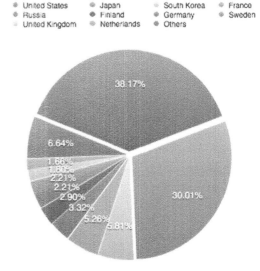

Figure 5. The foreign distribution of patent applications.

China has already been in hand among globalization under the influence of international situation, and the security industry and market has become part of the global ones.

4 CONCLUSION

Compared with the information technology security researches abroad, there is a late start in China. But because of the national attention and continuous investment, China's information technology security industry is experiencing a stage of rapid development. As the market continues to mature and expand, more and more research institutes and manufacturers join to the competition in the market.

Based on the analysis and summaries above, we list some trends and suggestions for information technology security development and intellectual property protection in China. 1. The industry attaches great importance to technology innovation and intellectual property protection. Whether in China or in the world, a variety of patent litigations are there that we can't ignore. Strategic layout in information technology security technology patents should include not only the art of intellectual property protection but also the key technologies in the future. 2. Enterprises strengthen the cooperation with universities and research institutes. Universities and research institutes have strong background knowledge for technology research, and they have applied 609 patent applications for security technology. To promote the further development of information technology security technology in China, enterprises need to strengthen cooperation with universities and research institutions. 3. The governments and industry increase investment to support and expand the scope of applications of information security. Information

technology security as the protection of national economic and social development needs continued guidance and support.

REFERENCES

Feng, D.G., 2010. The current information security technology situation in China and abroad. E-government, 7: 27–33.

Shen, C.X., 2001. Thoughts and suggestions on Information Security Industrialization. China Information Security, 2: 12–16.

Shen, C.X., Zhang, H.G., Feng, D.G., 2007. Information security review. Science in China, 37(2): 129–150.

Siponen, M.T., Oinas-Kukkonen, H. 2007. A review of information security issues and respective research contributions. ACM Sigmis Database, 38(1): 60–80.

Whitman, M.E., Mattord, H.J. 2010. Principles of information security. Cengage Learning.

Wu, S.Z., Guo, T., Dong, G.W., 2012. Software vulnerability analyses: A road map. Journal of Tsinghua University, 10: 1309–1319.

Wu, S.Z., 2013. 2012 Annual national information security situation assessment. Beijing: Current affairs press.

Yan, X.S., 2013. Introduction to intranet China patent retrieval system (CPRS). http://www.sipo.gov.cn/wxfw/mfzljzyjspx/mfpxkj/200804/P020080402390606898112.ppt.

Information Technology and Applications – Li (Ed.)
© 2015 Taylor & Francis Group, London, ISBN 978-1-138-02677-3

Ownership concentration, balance degree and corporate performance: An empirical study on Chinese listed companies of IT industries

Hairong Lu
Hohai University, Nanjing, China

ABSTRACT: As the information technology of Chinese industries and enterprises has been developed further, the IT industry has a good prospect. As the most important factor in the formation of corporate governance mechanism, the conditions of ownership structure directly affect the company operational efficiency. This paper empirically analyses the relationship between ownership structure and corporate performance, using listed companies of IT industry from 2010 to 2012 as research samples. It can be found that ownership concentration has a U-shaped curve with corporate performance, and equity restriction is significantly and positively correlated with corporate performance.

Keywords: IT industry; ownership concentration; equity balance degree; corporate performance

1 INTRODUCTION

Since 1930s, the effect of ownership structure on corporate governance efficiency has become one of the researches focus in financial field for a long period. Scholars have carried out fruitful researches on this issue, but most of the researches are aimed at the whole or only a certain industry. Few scholars focus on the large number of IT industries listed enterprises in the current market.

In the western, equity structure is highly decentralized and the main corporate governance problem is the principal-agent conflict between shareholders and managers. But in the IT industries listed companies in China, equity is highly centralized and the corporate governance problem gradually evolved into the interest conflicts between large shareholders and small shareholders and the controlling shareholders make an invasion upon the inferior dormant partner. It is urgent to explore how to establish an effective corporate governance mode to suppress the big shareholder looting by researching on the relationship of ownership concentration, equity balance degree and corporate performance.

2 RESEARCH HYPOTHESIS

2.1 *Ownership concentration and corporate performance present a U-shaped quadratic curve relationship*

Research results show that big shareholders of listed company behave differently when the ownership concentration is very low or very high. When ownership concentration is low, as the holdings rise, major shareholders will have a strong incentive to pursue their own utility maximization, thus to encroach on the interests of small shareholders. At this stage, the more concentrated the equity is, the worse company performance will be. This is called "Entrenchment Effect". When ownership concentration is high, big shareholders' occupy behavior would be weakened, since large shareholders accounted for a large part of interest in the listed companies and they can gain more by improving company performance than encroaching on minority shareholders interests. Besides, as the proportion of large shareholders is increased to a certain degree, big shareholders will have stronger power to constraint management, to gradually reduce the agency cost and to improve the company's performance. This is called "Alignment Effect".

This paper attempts to study, in listed companies of IT industries, whether or not "Entrenchment Effect" and "Alignment Effect" exist at the same time. Thus, in this paper, it is assumed that in listed companies of IT industries, there is a nonlinear relationship between ownership concentration and corporate performance. In particular, it is a relatively stable U-shaped relationship: When ownership concentration is low, the company performance decreases with the increase of large shareholder's stake; when ownership concentration is high, the company performance increases with the increase of large shareholder's stake. Therefore, this article puts forward hypothesis 1.

Hypothesis 1: ownership concentration and corporate performance present a U-shaped quadratic curve relationship.

2.2 Equity balance degree has a positive correlation with corporate performance

By analyzing the research findings, this paper finds that as the ownership balance degree of listed companies becomes higher, compared to controlling shareholder, outside shareholders will have relatively stronger power, and the motivation and ability of external shareholders' supervision also become stronger. As the ability of controlling shareholders abuse become weaker, equity checks and balances can better improve the company performance. And the true sense of equity checks and balances means that shareholders have the ability and motivation to suppress the depredations of large shareholders from inside and a mutual supervision is formed, which is an effective method to improve the governance structure of listed companies. Therefore, in a narrow sense, the equity checks and balances is the balance of power for largest shareholder, the purpose of which is to protect the interests of minority shareholders and to maintain the healthy development of securities market. Therefore, to a certain extent, a reasonable and effective equity checks and balances structure can protect the minority shareholders, and corporate performance will increase with the increase of degree of ownership balance. Therefore, this article puts forward hypothesis 2.

Hypothesis 2: equity balance degree has a positive correlation with corporate performance.

3 RESEARCH DESIGN

3.1 Sample selection

The study samples are listed companies from the IT industry. In order to guarantee the validity of the data, this article uses some selection criteria to filter the original samples. The screening criteria include: the sample companies are all publicly listed before December 31st, 2009; rejecting companies that are *ST or ST, and companies with negative profit in any of the three years; excluding listed companies with B shares or H shares. Through the above screening steps, this article finally determines 204 sample data from 68 listed companies in three years. The financial index data, annual report data and equity structure indexes of listed companies in this paper are all from CSMAR database. The statistical software used is SPSS19.0.

3.2 Variable definitions

In this paper, the variables we study include:

The dependent variable: it is the company performance indicators. Company performance indicators can generally be divided into two categories, one is accounting indicators that are obtained based on the enterprise's own statements; the other is market indicators, for example, Tobin's Q performance measurement. As in recent years, regulators gradually strengthen the supervision of earnings management for listed companies; ROE has become more effective, thus becoming an important indicator of the operating performance of listed companies. Tobin's Q is the ratio of market value and replacement cost of an enterprise. Its economic implications are to compare the value created by the use of resources and the cost of investment. Tobin Q value reflects the ratio relationship between the input and output of the companies from the angle of market. This article selects the ROE and Tobin's Q to measure the accounting performance and market performance of the company respectively.

The independent variables: the independent variables include ownership concentration index and equity balance degree index. First, this paper chooses the first big shareholder's shareholding, the sum of the top five shareholders' shareholding and the sum of the top ten shareholders' holding to measure equity concentration. Next, this paper chooses the ratio of first big shareholder's shareholding and the sum of the proportions of the second to the fifth largest shareholders' shareholding, the ratio of the first big shareholder's stake and the sum of the proportions of the second to the tenth largest shareholders' shareholding to measure equity balance degree.

Control variables: many scholars have found that a number of other factors are closely related to the company's performance, which are called the control variable. In order to accurately describe and characterize the effects of ownership concentration and equity restriction on the performance of the company, the paper selected the followings as control variables: Company Size, as the market value of the company cannot reflect company size well, this paper use the natural logarithm of the company's total assets to measure company size;

Table 1. Variable definitions.

Variable	Names and measurement
ROE	Return on equity
Tobin Q	Tobin Q value
PFIRS	First big shareholder's shareholding
PFIVE	Sum of the top five shareholders' shares
PTENS	Sum of the top t shareholders' shares
Z	Sum of top five/sum of second to fifth
S	Sum of top five/sum of second to fifth
SIZE	Natural logarithm of company's assets
LEV	Total debts/total assets

Capital Structure, the paper uses the asset-liability ratio to measure, which is defined by the ratio of total liabilities to total assets. Asset-liability ratio tends to reflect a company's financial risk and capital cost.

Many scholars have found that a number of other factors are closely related to the company's performance, which are called as the control variable. In order to describe and characterize the effects of ownership concentration and equity restriction on the performance of the company, this paper selected the followings as control variables: Company Size, as the market value of the company cannot reflect company size well, this paper use the natural logarithm of the company's total assets to measure company size; Capital Structure, the paper uses the asset-liability ratio to measure, which is defined by the ratio of total liabilities to total assets. Asset-liability ratio tends to reflect a company's financial risk and capital cost.

4 EMPIRICAL ANALYSIS

4.1 Descriptive statistics

With SPSS software, this paper concludes the average, maximum and minimum figures for the sample data in the individual year and the overall three years.

It should be noted that average proportion of the top five shareholders is 0.5617, and average proportion of the holdings of the top ten shareholders is 0.6215, which means that big shareholders have actually controlled the company and is consistent with the assumptions of "Entrenchment Effect" and "Alignment Effect".

4.2 Multiple regression models

According to the assumption, this paper establishes multiple regression models as follows.

Ownership Concentration Models:

$$ROE = \alpha_{11}(PFIRS) + \alpha_{12}(PFIRS^2) \\ + \alpha_{13}(SIZE) + \alpha_{14}(LEV) + C_1 + e \quad (1)$$

$$ROE = \alpha_{21}(PFIVE) + \alpha_{22}(PFIVE^2) \\ + \alpha_{23}(SIZE) + \alpha_{24}(LEV) + C_2 + e \quad (2)$$

$$ROE = \alpha_{31}(PTENS) + \alpha_{32}(PTENS^2) \\ + \alpha_{33}(SIZE) + \alpha_{34}(LEV) + C_3 + e \quad (3)$$

$$Tobin\,Q = \alpha_{41}(PFIRS) + \alpha_{42}(PFIRS^2) \\ + \alpha_{43}(SIZE) + \alpha_{44}(LEV) + C_4 + e \quad (4)$$

$$Tobin\,Q = \alpha_{51}(PFIVE) + \alpha_{52}(PFIVE^2) \\ + \alpha_{53}(SIZE) + \alpha_{54}(LEV) + C_5 + e \quad (5)$$

$$Tobin\,Q = \alpha_{61}(PTENS) + \alpha_{62}(PTENS^2) \\ + \alpha_{63}(SIZE) + \alpha_{64}(LEV) + C_6 + e \quad (6)$$

Equity Balance Degree Models:

$$ROE = b_{11}(Z) + b_{12}(SIZE) + b_{13}(LEV) + C_1 + e \quad (7)$$

$$ROE = b_{21}(S) + b_{22}(SIZE) + b_{23}(LEV) + C_2 + e \quad (8)$$

$$Tobin\,Q = b_{31}(Z) + b_{32}(SIZE) + b_{33}(LEV) \\ + C_3 + e \quad (9)$$

$$Tobin\,Q = b_{41}(S) + b_{42}(SIZE) + b_{43}(LEV) \\ + C_4 + e \quad (10)$$

4.3 The empirical results

According to the regression model and using SPSS software, the empirical analysis results are shown in Table 2.

Data from Table 2 shows that the correlation of PFIRS and ROE is significant at 1% level, and they both show a highly significant positive U-shaped relationship. Data from Table 3 shows that the correlation of PFIRS and ROE is significant at 1% level, and they both show a highly significant positive U-shaped relationship. Meantime, the correlation of PFIRS and Tobin Q is significant at 1% level, and they show a significant positive U-shaped relationship. The correlation of PFIVE and ROE is significant at 5% level, and they also show a highly significant positive U-shaped relationship. PTENS and ROE cannot pass the significant correlation test, so the two are substantially irrelevant. However, the correlation of PFIRS and Tobin Q is significant at 5% level, and they also show a significant positive U-shaped relationship. Thus, on the whole, the results of the regression analysis can confirm the Hypothesis 1 made above: in the IT industry listed companies, ownership concentration and corporate performance present a U-shaped quadratic curve relationship.

Table 3 shows that the correlation coefficient of Z index and ROE is −0.041, which is significant at 10% level, and the correlation coefficient of Z index and Tobin Q is −0.041, which is also significant at 10% level. It means that there is a basic significant negative correlation between Z index, ROE and Tobin Q. The correlation coefficient of S index and ROE is −0.037, which is significant at 10% level, and the correlation coefficient of S index and Tobin Q is −0.055, which is significant at 5% level. So it can be concluded that and ROE show a basically significant negative correlation, S index and Tobin Q show a significant negative correlation. From the above empirical study we can find that as the Z index and the S index increase, equity balance degree decreases and corporate performance declines. So, on the whole, the results of the regression analysis can confirm the Hypothesis 2 made above: in the listed companies of IT industries, equity balance degree has a positive correlation with corporate performance.

Table 2. Regression results of ownership concentration on corporate performance.

Variable	ROE			Tobin Q		
PFIRS	−0.391 (−3.250)^a ***^{b}			−13.550 (−4.098)^a ***^{b}		
PFIRS^2	0.502 (3.403)^a ***^{b}			15.361 (3.791)^a ***^{b}		
PFIVE		−0.519 (−2.118)^a **^{b}			−26.348 (−3.989)^a ***^{b}	
PFIVE^2		0.458 (2.097) **^{b}			20.441 (3.474) ***^{b}	
PTENS			−0.479 (−1.579)^a			−20.460 (−2.472)^a **^{b}
PTENS^2			0.388 (1.554)^a			14.496 (2.127)^a **^{b}
SIZE	0.035 (7.965)^a ***^{b}	0.040 (10.067)^a ***^{b}	0.041 (10.414)^a ***^{b}	−0.195 (−1.621)	−0.063 (10.067)^a	0.030 (0.281)^a
LEV	−0.106 (−5.036)^a ***^{b}	−0.120 (−5.737)^a ***^{b}	−0.121 (−5.788)^a ***^{b}	−5.383 (−1.621)^a	−5.434 (−9.637)^a ***^{b}	−5.589 (−9.763)^a ***^{b}
R-squared	0.186	0.181	0.178	0.240	0.258	0.236
F value	30.721	28.506	27.909	40.706	44.941	39.941

^{a}Number in the parentheses is the statistic value of T-test.
^{b}***Indicates significant at 1% level; **indicates significant at the 5% level; *indicates significant at 1% level.

Table 3. Regression results of equity balance degree on corporate performance.

Variable	ROE		Tobin Q	
Z	−0.041 (−1.724)^a *^{b}		−0.032 (−1.684)^a *^{b}	
S		−0.037 (−1.689)^a *^{b}		−0.055 (−2.437)^a **^{b}
SIZE	0.040 (10.284)	0.041 (−6.058)^a **^{b}	0.032 (0.300)	0.025 (0.234)^a
LEV	−0.126 (−6.045)^a **^{b}	−0.126 (10.236)^a **^{b}	−6.099 (−10.628)^a ***^{b}	−6.108 (−10.672)^a ***^{b}
R-squared	0.175	0.176	0.217	0.214
F value	36.595	36.679	47.744	48.202

^{a}Number in the parentheses is the statistic value of T-test.
^{b}***Indicates significant at 1% level; **indicates significant at the 5% level; *indicates significant at 1% level.

From Table 2 and Table 3, it can be seen that there is a significant positive correlation between the listed companies of IT industries and their corporate performance, indicating the presence of scale effect. There is a significant negative correlation between the asset-liability ratio of listed companies of IT industries and their corporate performance, indicating that weak debt governance for listed companies of IT industries.

5 CONCLUSIONS AND RECOMMENDATIONS

Through the above empirical analysis, the main conclusions can be drawn as the following: In listed companies of IT industries, "Entrenchment Effect" and "Alignment Effect" exist at the same time and Ownership concentration and corporate performance present a U-shaped quadratic curve relationship; in the listed companies of IT industries of China, the equity restriction, the higher equity balance degree is, the better company performance can be.

Through the analysis of the research findings, the article comes up with the following suggestions to improve the ownership structure of listed companies of IT industries:

5.1 Improve the legislation and institution building of securities market, strengthening law enforcement

In listed companies of IT industries, the situation that big shareholders damage the interests of small shareholders is relatively common through occupation of funds, related party transactions and illegal guarantee. As a result, we find that the existing laws and regulations cannot effectively restrain large shareholders encroaching on the interests of minority shareholders. Therefore, the government not only must improve the legislation and system construction of the securities market, but also need to strengthen law enforcement.

5.2 Reduce the degree of equity concentration, especially the proportion of the largest shareholder

By the empirical result of the ownership concentration and corporate performance, we can see that the ownership concentration has a significant positive U-shaped relationship with corporate performance. Various indexes of Ownership concentration show that equity structure of listed companies of IT industries in our country is excessively concentrated. Over concentrated Ownership is not only likely to create prejudice to the rights and interests of minority shareholders, but also is an important

reason for the big shareholders' abuse of control in current system. So it is necessary to take appropriate way to gradually reduce the degree of equity concentration of listed companies of IT industries in China, to attract investors widely at the establishment of companies and to build a diversified ownership finally.

5.3 Improve equity restriction and build an equity checks and balances of large shareholder

The largest shareholders in listed companies of IT industries tend to show selfish behavior to a certain degree. When the proportion of other large shareholders is high, they can balance and supervised the power of the largest shareholder better, take part in company management, prevent the transfer of company's assets by the largest shareholder and enhance the company's operating performance. Therefore, nowadays, to improve equity balance degree can help to solve the governance problems of IT industries companies in China.

REFERENCES

Berle & Means. 1932. The Modern Corporation and Private Property. Journal of Guizhou University. New York HY: Macmillan Co.

Chen, Deping & Chen, Yongsheng. 2011. A Study on the Relationship between Equity Balance Degree and Corporate Performance. Accounting Research.

Demsetz, H. 1983. The Structure of Ownership and the Theory of the Firm. Journal of Guizhou University. Journal of Law and Economics.

Gao, Jianlong. 2011. A Research Review of the Relationship between Ownership Concentration and Corporate Performance. Cuide to Business.

Gu, Xiang & Zhu, Dan. 2010. An Empirical Research On Ownership Structure and Operating Performance of Listed Companies. Market Weekly Economics.

Shleifer & Vishny. 1986. Vishny. Large Shareholders and Corporate Control. Journal of Political Economy.

Zang, Weiguo. 2011. Equity Concentration, Balance Degree and Competitiveness of Listed Companies, an Empirical Study Based on Factor Analysis Method. The Accounting Issue.

Zhang, Hongjun. 2004. A theoretical and empirical analysis of ownership structure and corporate performance of listed companies in China. Economic Science.

Zhang, Liang & Wang, Ping & Mao, Daowei. 2011. The Impact of Ownership Concentration and Equity Balance Degree on Enterprise Performance. Market Weekly Economics.

Zhang, Zhiqian & Yan, Qihua. 2011. An Empirical Research on Ownership Concentration and Corporate Performance of Listed Companies. Journal of Nanjing Audit College.

Zhu, Jing. 2011. Equity Concentration, An Research Review of the Relationship Between corporate equity structure and Corporate Performance. Journal of Guizhou University.

Information Technology and Applications – Li (Ed.)
© 2015 Taylor & Francis Group, London, ISBN 978-1-138-02677-3

A novel Redundant Inertial Measurement Unit and error compensation technology

Kunpeng He, Yuping Shao, Xiaoxue Wang & Chenyang Wang
College of Automation, Harbin, Heilongjiang, China

ABSTRACT: Redundancy configuration is used to improve the precision and reliability of Inertial Measurement Unit (IMU), a tetrahedron-based skew emplacement of 3 IMUs forms Redundant Inertial Measurement Unit (RIMU), then make a RIMU error analysis and establish the precise mathematical model. Calibration and parameter identification method is designed, and this method is simple and reliable with many groups of experiments, all parameters of the RIMU can be estimated only by one-time calibration, and the calibration precision of accelerometer reaches 2 mG while the gyro precision is better than 0.05°/s.

Keywords: redundancy configuration; RIMU; calibration; parameter identification

1 INTRODUCTION

Micro Electro-Mechanical Systems (MEMS) inertial devices become the first choice gradually in low-cost micro inertial navigation system because of its micro power consumption, small size and other advantages. However, it has a poor precision, stability and reliability ascribe to the material and process constraints. In order to improve the reliability and precision of navigation control system, redundancy technology is developing [1] [2], and it becomes a research hotspot. Usually Inertial Measurement Unit (IMU) is composed of three orthogonal single-axis gyros and three orthogonal single-axis accelerometers, while Redundant Inertial Measurement Unit (RIMU) is composed of more IMU, and it's mainly divided into orthogonal redundancy and non-orthogonal redundancy [3]. The ordinary IMU is used as a basic unit of RIMU to construct a new RIMU, and then the error is analyzed and accurate error compensation model is established according to the RIMU, and a calibration method is designed, finally, make an experimental verification.

2 THE NEW REDUNDANT CONFIGURATION SCHEME

ADIS16405 in AD Corporation is used as a basic component of RIMU, and it's a complete inertial measurement system consisting of a three-axis gyro, magnetometer and accelerometer. The axis-axis alignment error of gyro and accelerometer is respectively ±0.05° and 0.2° while temperature stability is respectively ±40 ppm/C and ±40 ppm/C,

the angle random walk of gyro is less than 2°/√hr while the rate random walk of accelerometer reach 0.2 m/sec/√hr, and the sensor axial diagram is shown in Figure 1.

RIMU is based on tetrahedral form, and 3 ADIS16405 are respectively arranged on two sides and the bottom of tetrahedral, and the whole schematic diagram of RIMU is shown in Figure 2(a). ADIS16405 installed on the bottom surface is defined as IMU_1 while the one installed on two side center are defined as IMU_2 and IMU_3, and three IMUs are connected with the navigation computer through a line, and the real diagram is shown in Figure 2(b).

After calculation, the angular rate measurement precision of this redundancy configuration

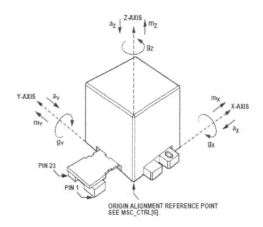

Figure 1. Axial diagram of ADIS16405.

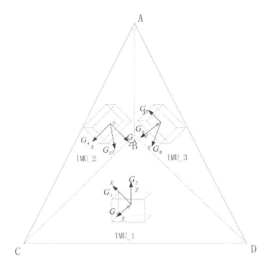

Figure 2(a).   Whole schematic diagram of RIMU.

Figure 2(b).   Real diagram of RIMU.

is nearly 2 times higher than the precision of orthogonal configuration [4]. Ensuring the angular rate measurement function, there are 455 kinds of measurement mode of the redundancy configuration to ensure fault tolerance performance, and there are enough fault reconstruction scheme when fault occurs to gyro or accelerometer of RIMU [4]. Therefore, the RIMU has high precision and reliability that conventional IMU can't reach. However, in order to improve the performance of RIMU further, calibration of device error is necessary.

Assuming measurement vector along the reference orthogonal coordinate is $\omega = [\omega_x\ \omega_y\ \omega_z]^T$, the measurement output for N inertial devices is $m = H\omega$, where, $m = [m_1\ m_2\ ...\ m_n]^T$, H is the measurement matrix. Installation position of one

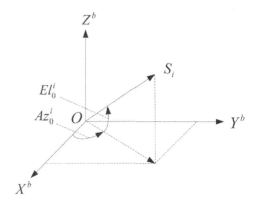

Figure 3.   Installation position of one gryo.

inertial device is shown in Figure 3, El_0^i and Az_0^i are the theory installation angle for ith inertial device, and S_i is the unit vector along installation direction of the ith sensor, therefore, S_i can be expressed as $S_i = \cos(El_0^i)\cos(Az_0^i)i + \cos(El_0^i)\sin(Az_0^i)j + \sin(El_0^i)k$ in carrier coordinate $OX^bY^bZ^b$. i, j, k are 3 unit vectors along the carrier coordinate. Therefore,

$$H = \begin{bmatrix} \cos El_0^1 \cos Az_0^1 & \cos El_0^1 \sin Az_0^1 & \sin El_0^1 \\ \cos El_0^2 \cos Az_0^2 & \cos El_0^2 \sin Az_0^2 & \sin El_0^2 \\ \vdots & \vdots & \vdots \\ \cos El_0^n \cos Az_0^n & \cos El_0^n \sin Az_0^n & \sin El_0^n \end{bmatrix}$$

3   REDUNDANT IMU ERROR MODEL

3.1   Constant error

The measurement error caused by constant error of RIMU composed of n inertial devices is $\Delta m_b = B = [b_1\ b_2\ ...\ b_n]^T$.

3.2   Scale factor error

The measurement error caused by scale factor error is due to inconsistency between the true scale factor and one by test [5], expressed as $\Delta m_k = K_s m$. Where, $K_s = diag[k_{s1}\ k_{s2}\ ...\ k_{sn}]$, k_{si} is scale factor error of the ith inertial device.

3.3   Installation error

The measurement error caused by installation error is due to the difference between the inertial device in the assembly and device design position [4]. When there is an installation error, the measurement matrix H perturbation, i.e. $H' = H + \Delta H$.

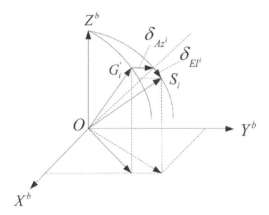

Figure 4. Schematic diagram of installation error angle.

Here, the pitch error δ_{El^i} and the azimuth error δ_{Az^i} is used to represent installation error of ith inertial device, which is shown in Figure 4, then the actual installation angle is $El^i = El_0^i + \delta_{El^i}$ and $Az^i = Az_0^i - \delta_{Az^i}$, after calculation, corresponding installation error matrix is

$$
\Delta H = \begin{bmatrix}
-\delta_{El^1}\sin El_0^1\cos Az_0^1 & -\delta_{El^1}\sin El_0^1\sin Az_0^1 & \\
+\delta_{Az^1}\cos El_0^1\sin Az_0^1 & -\delta_{Az^1}\cos El_0^1\cos Az_0^1 & \delta_{El^1}\cos El_0^1 \\
-\delta_{El^2}\sin El_0^2\cos Az_0^2 & -\delta_{El^2}\sin El_0^2\sin Az_0^2 & \\
+\delta_{Az^2}\cos El_0^2\sin Az_0^2 & -\delta_{Az^2}\cos El_0^2\cos Az_0^2 & \delta_{El^2}\cos El_0^2 \\
\vdots & & \vdots \\
-\delta_{El^n}\sin El_0^n\cos Az_0^n & -\delta_{El^n}\sin El_0^n\sin Az_0^n & \\
+\delta_{Az^n}\cos El_0^n\sin Az_0^n & -\delta_{Az^n}\cos El_0^n\cos Az_0^n & \delta_{El^n}\cos El_0^n
\end{bmatrix}
$$

The measurement error caused by installation error is $\Delta m_c = \Delta H(H^TH)^{-1}H^Tm$.

3.4 The error model of inertial device

Through the analysis above, inertial device measurement error is consists of constant error, scale factor error and installation error [6], and it's expressed as $\Delta m = \Delta m_b + \Delta m_k + \Delta m_c = B + K_sm + \Delta H(H^TH)^{-1}H^Tm$. The measurement error of ith inertial device is $\Delta m_i = b_i + k_{si}m_i + p_i\delta_{El^i} + q_i\delta_{Az^i}$ where, p_i and q_i are installation error coefficients corresponding.

4 REDUNDANCY CALIBRATION ALGORITHM

The gyro precision in RIMU is lower, and it's not sensitive to the angular velocity of rotation of the earth, however, angular rate provided by turntable is used to stimulate error and then applied in calibration. While the accelerometer can be calibrated by acceleration of gravity G [6]. The calibration path design is shown in Table 1.

Through calculation, the constant error of inertial devices is $b_i = \sum_{j=1}^{6}\Delta m_i^j/6$, Δm_i^j expresses the measurement error of the jth position of the ith inertial device. The scale factor error of the ith inertial device is

$$
\begin{aligned}
k_{si} =\ & \frac{-\Delta m_i^1 - \Delta m_i^2 + 2\Delta m_i^3 - \Delta m_i^4 - \Delta m_i^5 - 4\Delta m_i^6}{6\theta\sin El_0^i} \\
& + \frac{\left(\Delta m_i^3 - \Delta m_i^6\right)\sin El_0^i}{2\omega_0} \\
& - \frac{\left(\Delta m_i^1 - \Delta m_i^4\right)\cos Az_0^i\cos El_0^i}{2\theta} \\
& + \frac{\left(\Delta m_i^2 - \Delta m_i^5\right)\sin Az_0^i\cos El_0^i}{2\theta} \quad (i=3,5,6,8,9)
\end{aligned}
$$

where, $\theta = \begin{cases} \omega_0 & \text{gyro calibration} \\ g & \text{accelerometer calibration} \end{cases}$, the same below,

$$
\begin{aligned}
\delta_{El^i} =\ & \frac{\left(\Delta m_i^3 - \Delta m_i^6\right)\cos El_0^i - \left(\Delta m_i^1 - \Delta m_i^4\right)\cos Az_0^i\sin El_0^i}{2\theta} \\
& - \frac{\left(\Delta m_i^2 - \Delta m_i^5\right)\sin Az_0^i\sin El_0^i}{2\theta} \\
\delta_{Az^i} =\ & \frac{\left(\Delta m_i^1 - \Delta m_i^4\right)\sin Az_0^i - \left(\Delta m_i^2 - \Delta m_i^5\right)\cos Az_0^i}{2\theta\cos El_0^i} \quad (i\neq3)
\end{aligned}
$$

The installation angle of each inertial device can be calculated according to the spatial structure of IMU and the mechanical relationship among them, as seen in Table 2 in detail.

Because of $El_0^i = 0$ ($i = 1,2,4,7$), δ_{El^i} corresponding is unable to solve problem, and equations should be wrote out respectively aimed at the six positions, according to specific situation, we calculate

$$
k_{s1} = \frac{\Delta m_1^1 - \Delta m_1^4}{2\theta}, \quad k_{s2} = \frac{\Delta m_2^2 - \Delta m_2^5}{2\theta},
$$
$$
k_{s4} = \frac{\Delta m_4^2 - \Delta m_4^5}{2\theta}.
$$

In addition, because of $El_0^3 = 90°$, the third inertial device is a point in carrier coordinate, then δ_{Az^3} can't be calculated, and the error model is not suitable for it, the error model is needed to be established, i.e. $\Delta m_3^i = b_3 + k_{s3}\omega_z + \delta_{Az^3}\omega_x + \delta_{El^3}\omega_y$, the six-position calibration method is still used, formula derivation is

$$
k_{s3} = \frac{\Delta m_3^3 - \Delta m_3^6}{2\theta}, \quad \delta_{Az^3} = \frac{\Delta m_3^1 - \Delta m_3^4}{2\theta},
$$
$$
\delta_{El^3} = \frac{\Delta m_3^3 - \Delta m_3^5}{2\theta}.
$$

Table 1. Reference orthogonal axis direction and reference axis equivalent output.

	P 1	P 2	P 3	P 4	P 5	P 6
Input	x y z	x y z	x y z	x y z	x y z	x y z
Output of accelerometer	G 0 0	0 G 0	0 0 G	$-$G 0 0	0 $-$G 0	0 0 $-$G
Output of gryo	ω_0 0 0	0 ω_0 0	0 0 ω_0	$-\omega_0$ 0 0	0 $-\omega_0$0	0 0 $-\omega_0$

Table 2. The installation angle of each inertial device (°).

	1	2	3	4	5	6	7	8	9
El_0^i	0	0	90	0	289	341	0	290	341
Az_0^i	0	90	0	90	0	180	330	240	60

Table 3. MEMS gyro error calibration results.

Error category	Results	Error category	Results	Error category	Results	Error category	Results
k_{s1}	0.0051	$\delta_{El}{}^1$	$-$0.5789	$\delta_{AZ}{}^1$	0.1826	b_1	0.0912
k_{s2}	0.0066	$\delta_{El}{}^2$	$-$0.9179	$\delta_{AZ}{}^2$	$-$0.3462	b_2	0.3671
k_{s3}	0.0075	$\delta_{El}{}^3$	$-$0.5264	$\delta_{AZ}{}^3$	$-$0.4834	b_3	0.0322
k_{s4}	0.0051	$\delta_{El}{}^4$	0.9155	$\delta_{AZ}{}^4$	0.8391	b_4	0.4946
k_{s5}	0.0014	$\delta_{El}{}^5$	0.2199	$\delta_{AZ}{}^5$	$-$2.0020	b_5	0.5703
k_{s6}	$-$0.0109	$\delta_{El}{}^6$	$-$0.4501	$\delta_{AZ}{}^6$	0.9154	b_6	0.4026
k_{s7}	0.0056	$\delta_{El}{}^7$	0.5204	$\delta_{AZ}{}^7$	1.5240	b_7	1.0231
k_{s8}	$-$0.0001	$\delta_{El}{}^8$	0.0462	$\delta_{AZ}{}^8$	$-$2.9159	b_8	$-$0.0705
k_{s9}	$-$0.0082	$\delta_{El}{}^9$	0.2868	$\delta_{AZ}{}^9$	1.0967	b_9	$-$0.1062

Following we make a test according to the MEMS-IMU redundant inertial system, the specific steps of calibration test are: the system is mounted on a turntable, regarding x, y, z axis of the bottom IMU as the reference orthogonal coordinate axis, the position command makes x, y, z axis refer to the sky and land respectively with angular rate $\omega_0 = 10°$/s rotating turntable, and gyro and accelerometer data is collected for 10 s. Through simulation analysis, the gyro error calibration results are shown in Table 3. Experiments show that the calibration precision of accelerometer can reach $2mG$ while the one of gyro is 0.05 °/s, and can error compensate effectively.

5 CONCLUSIONS

IMU is the basic unit of this new redundant inertial measurement unit, which is different from the past in such a way that a single inertial device as the basic unit. And also it's integrated easily and easy to repair and low-cost, has high precision and reliability. According to the RIMU, error measurement model is established and redundant calibration method corresponding is designed, and experiment verifies the

method is simple and reliable, and can compensate the error effectively, the calibration precision is higher, and it's also a good reference for the actual engineering design of strap-down inertial navigation system.

REFERENCES

[1] Li Xuelian, Sun Yao, Mo Hongwei et al. A nine gyro redundant configuration based on MIMU [J]. Journal of Harbin Institute of Technology, 2009, 41(5):90–94.

[2] Xia Kehan, Xu Hualong. Study on optimal allocation of redundant inertial measurement unit [J]. Winged Missiles Journal, 2004(7):53–56.

[3] Xia Kehan, Xu Hualong. Research on application of redundant technology in trap-down inertial navigation system [J]. Shanghai aerospace. 2002(3):36–38.

[4] Liang Haibo. Key technique of micro inertial system based on redundant gyros [D]. Harbin: Harbin Engineering University, 2011.

[5] Lin Yurong, Deng Zhenglong. Calibration system level of inertial devices error in laser gyro strap-down inertial navigation system [J]. Journal of Harbin Institute of Technology, 2001, 01:112–115.

[6] Hua Bing, Liu Jianye, Xiong Zhi, Li Rongbing. Study on error calibration techniques for redundant strap-down inertial sensors [J]. Sensor technology, 2005, 05:31–33.

Information Technology and Applications – Li (Ed.)
© 2015 Taylor & Francis Group, London, ISBN 978-1-138-02677-3

Study of Coke Dry Quenching control system based on PROFIBUS Fieldbus

Zheng-shi Chen & Ya-xun Lan
Guangzhou Vocational College of Science and Technology, Guangzhou, China

Jun-zheng Song
Guangdong Institute of Metrology, Guangzhou, China

ABSTRACT: According to CDQ (Coke Dry Quenching) production process, CDQ Automatic Control System could be divided into Scorching Coke Control Subsystem, Body Control Subsystem and Recycle Gas Control Subsystem. As the introduction of PROFIBUS Fieldbus and intelligent PLC field apparatus, it realized the various subsystems' distributional control and the simplified installment. To satisfy control request, the whole system used three-layer network structure, and specifically discussed the CDQ control system hardware overall design and communication settings. It could provide some help on CDQ industry Modernization.

Keywords: PROFIBUS; PLC; CDQ

1 CDQ PRODUCTION PROCESS

The coke that produced by coke oven, its temperature is about 1000 degree, called red coke. Pushed out red coke from coking chamber is accepted by vehicle-mounted rotary coke cans. Coke cans trolley is pulled to the bottom of the CDQ raise derrick by electric locomotive, elevated to the top of the derrick by hoister, and then translated to coke dry quenching-boiler furnace roof. Red coke is loaded into coke dry quenching-boiler through the roof. Red coke carries on heat exchange with inert gas directly in the furnace, and coke cooled to 250 degree below is unloaded onto conveyor from the row of coke plant and then sent to Sieve Coke System[1].

The inert gas used to cool coke blew into Coke Dry Quenching-boiler from air-supply device by circulating fan, and the output gas temperature is about 850 degree. After primary dedusting, inert gas was sent into coke dry quenching-boiler to exchange heat and its temperature will drop to 200 degree below. Cold inert gas will be cooled to 150 degree below after secondary dedusting and heat exchanger cooling. Then the circulating fan will pressurize the cold inert gas and sent it into Coke Dry Quenching-boiler for recycle use. The steam produced by coke dry quenching-boiler could be sent into CDQ steam turbine power station for power generation or incorporated into the steam pipe.

2 HARDWARE OVERALL DESIGN OF CDQ CONTROL SYSTEM

According to CDQ (Coke Dry Quenching) production process, CDQ Automatic Control System could be divided into Scorching Coke Control Subsystem, Body Control Subsystem and Recycle Gas Control Subsystem.

2.1 Network architecture design of system hardware

To satisfy scene control requirements, this system uses three-tier network architecture: Manage Layer, Monitoring Layer and Filed Device Layer[2].

Field Device Layer mainly related to the sensor of production site, drives, I/O components, transmitter, solenoid valve and so on. And the main equipments include: electric locomotives, coke cans and delivery vehicles, hoisters, feeding equipments, row of coke plant, coke dry quenching-boiler, blower device, circulating fans, exhaust-heat boiler, primary and secondary dedusting equipments and so on. The diversity of equipments request Filed Device Layer to satisfy the open requirements. manufacturers should follow generally accepted standards to ensure products meet standards. So that devices from different manufacturers could be replaced by the same functional capabilities of other devices without affecting the devices' functionality, nor the expense of control system integration.

Monitor Layer is composed by IPC (Industrial Personal Computer). It is mainly for information transmission, such as monitoring, optimization and scheduling, and provides necessary control information for upper Manage Layer. Its characteristic is the information transmission has certain periodicity and timeliness, the data throughput is higher. Therefore it requires a larger bandwidth network. PROFIBUS-DP is a high-speed low-cost communication, could just meet the requirements of information transmission.

Manage Layer could not only be used to remote access control to Monitoring Layer, but also mainly used to transmit management information, such an business plans, sales, stocks, finance, human and corporation management. Data packets transmitted on Manage Layer are usually longer and data throughput is larger. The initiation of data communication is stochastic and non-rule-based, so it demands a network with larger bandwidth. The Manage Layer network is composed of the industry Ethernet. The system structure is shown in Figure 1.

2.2 PLC selection and module configuration

According to the analysis of various part of the control object and control task, decide to adopt CPU315-2DP, one of the SIMATIC S7-300 series PLC, as master station and intelligent slave station.

CPU315-2DP is a kind of Siemens serial PLC suitable for medium-sized control system. Its extendibility and communication ability is good. CPU315-2DP has two communication interfaces: A PROFIBUS-DP interface meet for EN5170 and an MPI interface. It could realize PROFIBUS, industrial Ethernet system and MPI communication through these two interfaces[3].

Taking into account the scale of PLC control system was larger; a rack can not accommodate all of the modules. It needs to add extension rack, and the rack with CPU is called main rack. The main and extension rack through the IM360 and

IM361 interface module to form a unified whole and Interface module of each rack through bus connector to connect to I/O modules.

Select CP443-1 as communication interface cards of S7-300 PLC and Industrial PC, and connect in MPI mode. The main functions of IPC are completing configuration of S7-300 PLC, programming, parameter setting, on-line monitoring, data acquisition and storage.

3 PLC CONTROL SYSTEM DESIGN BASED ON STEP 7

3.1 Program design of control system

According to its function, this system program could be divided into host monitoring program and lower PLC control program. Host IPC equipped with monitoring software WinCC and programming software SIMATIC STEP7.

3.1.1 Host monitoring program

Monitoring system uses Siemens WinCC configuration software. The monitoring station transfers and changes types of screen through an interactive form, achieves process control function. And it has fault alarm, real-time, historical trends, production reports and document archiving functions. This monitoring system also has the comprehensive opening characteristic, in addition to support DDE and OLE, but also in line with OPC industrial standard. It could carry on data access through ODBC and SQL mode[4].

3.1.2 Lower PLC control program

The designation of control system control program is based on structured programming ideas. Complex CDQ automatic control system could be divided into Scorching Coke Control Subsystem, Body Control Subsystem and Recycle Gas Control Subsystem. Each subsystem could also be divided into lots of smaller subtasks. If it goes on, more complex CDQ production control system could be divided into lots of easier small questions. The process of programming was exactly the opposite. Firstly write subroutine to solve subtask, then through layer upon layer transfers, finally finish the overall CDQ production control procedures. Part of the hardware configuration is shown in Figure 2.

3.2 Communication realization of control system

3.2.1 System communication settings

The system has two networks, industrial Ethernet and PROFIBUS-DP. Industrial Ethernet is for

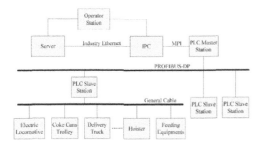

Figure 1. System structure.

Figure 2. Hardware configuration.

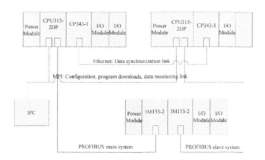

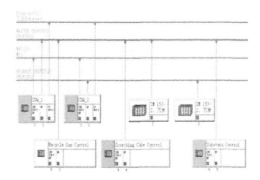

Figure 3. Software redundancy system.

the factory LAN, using twisted-pair connection. Mainly realizes picture browsing of local station for remote PC as well as the communication between PC of manage layer. Remote users take the way of user name password authorization to access local IPC, and dedicated firewall is also used to protect local IPC from external disturbance[5]. Local servers must have fixed IP to guarantee that release picture successfully.

PROFIBUS-DP is industry-specific communication network, primarily to realize communication between IPC and PLC. As the distance between IPC and PLC is far, a multi-mode fibre is used for connection. This ensures fast data transmission and data security, and at the same time it prevents other electromagnetic interference. In PROFIBUS-DP network, IPC and PLC supposes separately for the master station. This causes IPC could connect with all PLC process data through special software, realize various control and display. To ensure the stability of data transmission and high speed, network connection speed should be set to 1.5 M baud rate.

3.2.2 System redundancy settings

CDQ Body Control System, as the core part of the system, requires a stable, continuous production capacity. Therefore, it needs redundancy design to this system. Speaking of the redundancy technical of Siemens automation systems, it mainly consists of Hardware Redundancy and Software Redundancy. Hardware redundancy is aimed at S7-300/400 system, and the CPU itself had redundancy function. But it is much expansive to use CPU for redundancy. Software redundancy is referred to S7-300/400 system through the ways of MPI, PROFIBUS and Ethernet to realize redundancy. It needs the support of redundant software package and it is also a low-cost solution for realizing SIEMENS redundant functions. It is mainly used for control systems which demand less standby system switching time[6].

To form S7-300 soft-redundant systems, it needs: STEP 7 programming software, software

Figure 4. Software redundancy network configuration figure.

Redundancy package, two S7-300 CPU, two powers, three communication links (the RPOFBIUS communication link between the main system and the slave station, the RPOFBIUS communication link between the standby system and slave station, the MPI or PROFIBUS or Ethernet data synchronization communication link between the standby and the main system), Numbers of ET200M slave station, and each slave station includes one power and two IM153-2 interface modules (dedicated to redundant system, each module provides a PRPOFBIUS interface). Besides, it still needs programmable computer, download adapter, PROFIBUS or Ethernet connector wires, etc. In view of the CDQ Body Control System formed S7-300 software redundancy system architecture figure and network configuration figure is shown in Figures 3 and 4.

4 CONCLUSIONS

The introduction of PROFIBUS Fieldbus and industry Ethernet technology in CDQ control system greatly reduces the work load of field wire connection. And the installment, debugging and maintenance of equipment are very convenient. Through the communication between host monitoring machine and lower PLC, and other intelligent field equipments, realizes system's distributional control. The entire network's speed is fast, openness is good, and has a high performance-to-price ratio. Software redundancy plan adopted by the control system could effectively avoid plant shutdown caused by the failure of a control device, which reduces the factory loss greatly. Thus, the CDQ control system designed in this article may provide some significance for the modernization of CDQ production industry.

REFERENCES

[1] Miao Zhi-quan, Zhou Chun-lai, Zhang Bo. Study of Underground Railway Signal PLC Control System Based on PROFIBUS Fieldbus. Coal Mine Machinery, 2009, 30(2).

[2] Shuo Xiao, Xueye Wei, Yu Wang. A multipath routing protocol for wireless sensor network for mine security monitoring[J]. Mining Science and Technology (China). 2010 (1).

[3] Thomas A. Babbitt, Christopher Morrell, Boleslaw K. Szymanski, Joel W. Branch. Self-selecting reliable paths for wireless sensor network routing[J]. Computer Communications. 2008 (16).

[4] Alda J., González F.J. Fresnel zone antenna for dual-band detection at millimeter and infrared wavelengths. Optics Letters. 2009.

[5] Reid D.R., Smith G.S. A comparison of the focusing properties of a Fresnel zone plate with a doubly-hyperbolic lens for application in a free space, focused-beam measurement system. IEEE Transactions on Antennas and Propagation. 2009.

[6] Xue Hui. Brief Discussion on the Improving Measures for the Application of PLC Control System in Power Factory. Xinjiang Chemical. 2009.

[8] Fei Yuezhe,Liu Yu, Jiang Yi, Li Ziliang. Analysis of Interference to the PLC Control System. Computer Processing. 2011.

[9] Chang Jiangdong, Han Shuhong, Chen Yayu. Anti-Interference Technology for the PLC Control System. Coastal Engineering. 2009.

Information Technology and Applications – Li (Ed.)
© 2015 Taylor & Francis Group, London, ISBN 978-1-138-02677-3

Development of a wireless monitoring system based on ZigBee in chiller measurements

Yuhuang Zheng
Department of Physics, Guangdong University of Education, Guangzhou, China

ABSTRACT: In this paper, we have described a wireless monitoring system for chiller measurements. The system can allow faster reconfiguration of plant-floor networks as applications change. It can achieve higher throughput, lower average message delay and less average message dropping rate in wireless communication. The field test results show that the proposed method gives a good and robust performance in terms of transmission the data of chillers.

Keywords: ModBus; ZigBee; chiller; wireless networks

1 INTRODUCTION

Wireless networks have been under rapid development during recent years. Types of technologies being developed to wireless personal area network for short range, point-to multi-point communications, such as Bluetooth and ZigBee. The application of wireless technology for industrial communication and control systems has the potential to provide major benefits in terms of flexible installation and maintenance of field devices and reduction in costs and problems due to wire cabling (Phan R.C.W. 2012 and Zhan W., Porter J.R., Morgan J.A. 2014).

Nissan automobile assembly plants in Guangzhou, China, are using four chillers that have efficiency of 0.54 kw/ton. A water cooled chiller is a machine that removes heat from a liquid via a vapor-compression. This liquid can be circulated through a heat exchanger to cool the air. For many factories, chillers are the large energy users, and comprehensive maintenance is critical to ensure their reliability and efficient operation. Some factories use predictive maintenance to diagnose problems in advance, but a comprehensive Preventive Maintenance (PM) plan remains the key to ensuring the best performance and efficiency of chillers. Chillers now have tighter operational tolerances, and regular maintenance is more crucial than ever. When developing a PM plan for chillers, managers should consider a Supervisory Control and Data Acquisition (SCADA) system for chillers at first. The SCADA system document chiller performance daily with an accurate and detailed log, comparing this performance to design and start-up data to detect problems. This process allows managers to assemble a history of operating conditions, which can be reviewed and analyzed to determine trends and provide advanced warning of potential problems.

The water cooled chiller has a RS485 interface with ModBus RTU protocol. Allowing for up to 31 addressable chiller nodes, a RS485-equipped center computer with SCADA software can send commands to all chillers and each chiller can be controlled and monitored separately when assigned unique (Fovino I.N. 2012 and Liu X. 2012).

Because higher installation and maintenance costs of the wired chiller-monitoring network, wireless technology are required to overcome the limitations of wired networks, and benefit from mobility and design freedom it offers. The use of wireless networks in this chillers SCADA system has several advantages such as the reduction of time and cost to install new devices, since there is no need to provide a cabling infrastructure, along with the possibility of installing new devices in hard-to-reach or hazardous areas and the flexibility to alter existing designs (Alcaraz C. 2010 and Alcaraz C. 2011).

With adopting wireless technology for this chillers SCADA system, many important requirements should be considered regarding the solutions presented by the new standards, protocols, methodologies, and support tools. The most important requirements are interoperability, integration with existing systems and support tools for designing the network (Huang Y. 2012, Tseng C.H. 2013 and Yi P. 2011).

This paper presents a research effort towards the development of a wireless SCADA system for chillers based on ModBus protocol that is capable to collect and process data autonomously. The prototype system is designed to automatically detect chillers operating data based on ZigBee. ZigBee Gateway implementation approach is presented in detail. In addition, field tests validating the monitoring system are shown.

2 SYSTEM ARCHITECTURE

ZigBee provide the wireless network with the ability to reconfigure on the fly without being tied down by signal cables. The goal of the chiller SCADA system is to implement such a network using chiller controls connected by ZigBee gateway to a central computer that interfaces with a database accessible. The three major components consist of different kinds of chiller controls, ZigBee gateway, and the SCADA software hosted on the central computer. The system architecture is provided in Figure 1.

Four ZigBee gateways work as router. Each of them polls one chiller controller and collects the

equipment operating parameters in 100 ms. Each ZigBee gateway repacks the operating parameters in ModBus format and transmits them to ZigBee gateway works as coordinator. At the central computer, incoming data from the ZigBee coordinator are received and processed by the SCADA software which is developed in ForceControl6.0.

3 ZIGBEE GATEWAY DESIGN

ModBus is a type of local bus technology, which is similar to other local bus standard with open, intelligent, high environmental adaptability and other characteristics. Standard ModBus communication protocol uses two modes: ASCII (American Standard Code for Information Interchange) and RTU (Remote Terminal Unit). These two models are the same in the packet structure and functional command. Only the frame information expressed is different. ModBus has a query-response cycle characteristics, the master and slave devices can communicate individually, but also it uses broadcast to communicate to all slave devices. The slave device response message is also constituted of ModBus protocol. It includes confirm action domain and any data to be returned and error detection domain (Han Y., X. 2013). In this system, we use RTU mode.

The ZigBee gateway is based on the CC2530 System-on-Chip, which combines a RF transceiver with an industry-standard enhanced 8051 MCU, in-system programmable flash memory, 8-KB RAM, and other powerful peripherals. Figure 2 is a ZigBee coordinator gateway which consists of CC2530, SP3203E devices and some interfaces. And the ZigBee Router gateway consists of CC2530, SP483 device, which is shown in Figure 3.

Because the central computer has a UART interface and the interface is RS-232. But UART of CC2530 is TTL, so RS232-TTL conversion is done with a SP3203E chip. Every chiller control has a RS485 interface, and RS485-TTL conversion is done with a SP483 chip.

ZigBee gateway allows device containing UART to communicate via radio with other devices. Each device connects to ZigBee gateway. In this system,

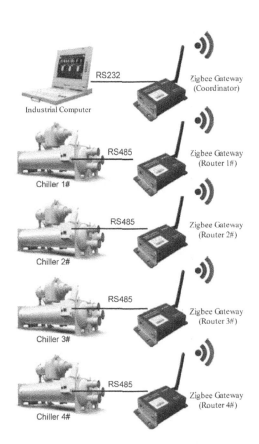

Figure 1.   System architecture.

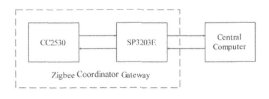

Figure 2.   ZigBee coordinator gateway.

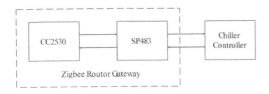

Figure 3. ZigBee Router gateway.

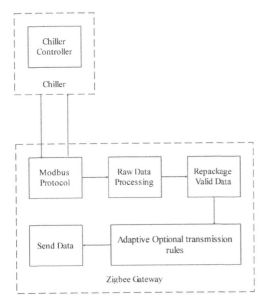

Figure 4. The flow chart of ZigBee Router gateway.

the five ZigBee gateways provide the radio communication link.

When the ZigBee gateway is used in an application, it is assumed that a permanent power source will be available at both ends of the wireless link. This means the on-chip radio can always be active, eliminating the need to synchronize the transmission/reception of data. The link is designed to operate at up to 19200 baud.

The ZigBee Coordinator is responsible for starting the network and allocating an address to the other gateway, which acts as a Router. Figure 4 is the program flow chart of ZigBee Router gateway.

A ZigBee Router gateway polls one chiller controller every 100 ms, and gets raw state data of the chillers. Because there are too many invalid data in raw data, the ZigBee Router gateway processes the raw data and gets all valid data and repackages them in ModBus frame. According to adaptive transmission cycle time rules, the ZigBee Router gateway selects the optimal transmission cycle and sends the data to the ZigBee coordinator gateway. The central computer reads the data that

retransmission by the ZigBee coordinator gateway and the SCADA software records the states of chillers.

4 EXPERIMENT

The SCADA system in central computer is made by ForceControl6.0. This software possesses many remarkable characteristics including completed functions, integrated management, modular development, high stability and easy to learn and use, etc. And it also provides abundant graphical development tools, stunning graphics elements and others which compose a facilitated developing environment. Figures 5–7 are some important monitoring interfaces about our system. Figure 5 is the whole chillers monitoring system, and Figure 6 is No.1 chiller monitoring interface. Figure 7 is the history data that water temperature in No.1 chiller. This system meets the real-time needs of the application.

In all measurements, we have observed Packet Error Rate less than 1% and response latency was always less than 50 ms. In periodic ModBus reporting case, median value of query and response

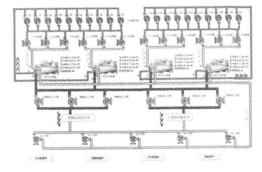

Figure 5. System monitoring interface.

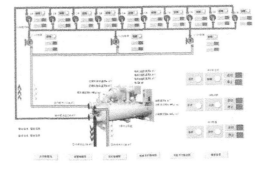

Figure 6. No.1 chiller monitoring interface.

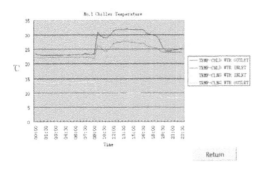

Figure 7. Water temperature history curves in No.1 chiller.

latencies were around 85 ms and 5 ms respectively. It is a reasonable strategy that a ZigBee Router gateway polls one chiller controller every 100 ms.

5 CONCLUSION

The design and implementation of a wireless monitoring system for chiller is discussed and presented in this paper. Tests are carried out to determine system performance for both the instrumentation and maintenance applications, and as the results are quite satisfactory. The results show the performance and interoperability for the wireless data acquisition system is good enough for some monitoring and non-critical instrument systems.

Further efforts are necessary to improve reliability of sensor nodes, security, and standardization of interfaces and interoperability. In addition, further studies are necessary to improve the protocol's functionality by checking the impact of the mobility of sensor nodes.

ACKNOWLEDGMENT

This project is funding by Appropriative Researching Fund for Professors and Doctors, Guangdong University of Education (No. 11ARF04). This project is funding by higher school science and technology innovation project of Guangdong Province (No. 2013LYM_0063).

REFERENCES

Alcaraz C. and Lopez J. 2010. A Security Analysis for Wireless Sensor Mesh Networks in Highly Critical Systems. *IEEE Transactions on Systems, Man, and Cybernetics, Part C: Applications and Reviews* 40(7): 419–428.

Alcaraz C., Fernandez-Gago C. and Lopez J. 2011. An Early Warning System Based on Reputation for Energy Control Systems. *IEEE Transactions on Smart Grid* 2(12):827–834.

Fovino I.N., Coletta A., Carcano A. and Masera M. 2012. Critical State-Based Filtering System for Securing SCADA Network Protocols. *IEEE Transactions on Industrial Electronics* 59(10):3943–3950.

Han Y., X. 2013. Research and Implementation of Collision Detection Based on ModBus Protocol. *Journal of Engineering Science and Technology Review* 6(1): 91–96.

Huang Y., Pang A., Hsiu P., Zhuang W. and Liu P. 2012. Distributed Throughput Optimization for ZigBee Cluster-Tree Networks. *IEEE Transactions on Parallel and Distributed Systems* 23(3):513–520.

Liu X. 2012. Quality of Optical Channels in Wireless SCADA for Offshore Wind Farms. *IEEE Transactions on Smart Grid* 3(3):225–232.

Phan R.C.W. 2012. Authenticated ModBus Protocol for Critical Infrastructure Protection. *IEEE Transactions on Power Delivery* 27(7):1687–1689.

Tseng C.H. 2013. Coordinator Traffic Diffusion for Data-Intensive ZigBee Transmission in Real-time Electrocardiography Monitoring. *IEEE Transactions on Biomedical Engineering* 60(12):3340–3346.

Yi P., Iwayemi A. and Zhou C. 2011. Developing ZigBee Deployment Guideline Under WiFi Interference for Smart Grid Applications. *IEEE Transactions on Smart Grid* 2(3):110–120.

Zhan W., Porter J.R. and Morgan J.A. 2014. Experiential Learning of Digital Communication Using LabVIEW. *IEEE Transactions on Education* 57(2):34–41.

Information Technology and Applications – Li (Ed.)

U.S. foreign aid management system and its inspiration for China

J.M. Hu, Y.C. Hua & J.M. Guo
School of Economics and Management, Hebei University of Technology, Tianjin, China

ABSTRACT: Foreign aid is an important part of U.S. foreign policy. After more than half a century's development, the United States has established a sound management system of foreign aid. China's foreign aid, although started earlier, still has some problems in its management system, which is not well adapted to the development need of China's foreign aid. Thus the experience of U.S. foreign aid management system may provide some inspiration for China's foreign aid management system reform.

Keywords: foreign aid; America; China; policy recommendation

The United States is the initiator of foreign aid, in the modern sense. Foreign aid in American foreign policy has always played a very important role. In recent years, foreign aid has become one of the three poles of U.S. foreign policy, together with foreign affairs and defense. U.S. has established a very sound foreign aid management system to manage huge amounts of fund from diverse sources. Since 1950, with improving foreign relations and foreign aid work, China has established foreign aid agencies at all levels, and has continuously strengthened their management. However, in view of the current situation, China's foreign aid management system cannot meet the need for its rapidly developing foreign aid cause. Studying and learning from the experience of other countries, especially of developed countries on foreign aid management, can help promote the development of China's foreign aid.

1 THE U.S. FOREIGN AID MANAGEMENT SYSTEM

1.1 *U.S. foreign aid management organizational model*

The organization and management of a country's foreign aid management system primarily include establishing agencies responsible for development & aid policy formulation and implementation, reorganizing and reforming them in accordance with changes in domestic and international political and economic situations, promoting coordination among sectors in the field of foreign assistance, strengthening human resources management, advancing foreign aid budget management, and strengthening international cooperation, etc.

According to the classification of foreign aid management organizational models by OECD-DAC, the U.S. foreign aid management system is "a department solely responsible for development aid policy-making, with separate independent bodies responsible for the implementation", that is, the Secretary of State in charge of the aid affairs at the cabinet level, with the USAID (United States Agency for International Development), the State Council, the Ministry of Finance, the Ministry of Health, the Millennium Challenge Corporation, and the U.S. peace corps etc., in charge of specific policy implementation.

1.2 *The management of U.S. foreign aid agency*

U.S. foreign aid structure initially formed from Marshall Plan. As mentioned earlier, there are a great number of U.S. foreign aid implementing agencies. In addition to the implementation of major federal governmental agencies associated with foreign aid, the United States has set up special foreign aid implementation agencies, also with some organizations engaged in foreign trade and economic exchanges undertaking part of foreign aid execution. By administrative level, all these agencies can be divided into Cabinet agencies, sub-Cabinet agencies and independent agencies. Cabinet agencies consist of 13 institutions including the State Cabinet, the Ministry of Agriculture, Defense, Commerce, etc. Sub-cabinet agencies include six agencies, viz. Centers for Disease Control (CDC), the National Institute of Standards and Technology, United States Patent and Trademark Office, the National Oceanic and Atmospheric Administration, the U.S. Fish and Wildlife Service and the U.S. Forest Service, etc. Separate external aid agencies include 11 institutions such as the U.S.

Agency for International Development (USAID), the Millennium Challenge Corporation, etc. The main foreign aid agencies number as many as 29, in addition to the U.S. embassy agencies responsible for liaison with foreign governments, such as the FBI, the Federal Drug Enforcement Administration and other agencies. It should be noted that the U.S. foreign assistance has always been decentralized. With domestic institutions having increasing capability in U.S. foreign relations and international affairs, every year there are agencies entering or withdrawing from the field of foreign aid. As of fiscal year 2001 there were 16 government departments and agencies providing foreign assistance; in fiscal year 2007 there were 24, and in FY2011 the number was 29.

1.3 *The functions of the U.S. foreign aid agency*

U.S. foreign aid implementing agencies have their own aid expertise, with types of assistance and operation fields crossed. USAID, the main implementer of long-term development assistance, is committed to promoting sustainable economic development in the recipient countries, and to enhancing the recipient government departments' and the private sector's capacity to meet the basic needs of their people. State Department's foreign aid activities focus on assistance to refugees, promoting democracy, and stability and reconstruction during post-conflict period, the international antidrug and law enforcement, voluntary contributions to multilateral aid agencies as well as providing financial assistance to support U.S. foreign policy objectives through strategic alliances. Ministry of Health provides international assistance related to global health research and training through the CDC and American national institutes of health, and provides assistance to refugees. Ministry of Agriculture provides emergency food aid and development food aid funding, and provides food security checks. Ministry of Finance is primarily responsible for the multilateral development assistance agencies' donations, including the World Bank, IMF and regional development banks. In addition, the Ministry of Finance is also responsible for providing technical assistance to developing countries for the implementation of large-scale economic reforms and debt relief. Founded in 2004, the Millennium Challenge Corporation is designed to provide targeted aid to the given countries which have good governance and the potential of effective use of aid funds to promote economic growth and poverty reduction. In addition, many other U.S. federal agencies also provide funding to foreign aid activities related to their own functions or manage of some foreign aid activities.

It should be noted that, although aid agencies have their own expertise, it does not mean that their perform activities only in their own areas of assistance, but with some overlapping. For example, although food aid is funded by the U.S. Department of Agriculture, the specific implementation of it is by the United States Agency for International Development (USAID), including the donated products for the overseas school food programs and mothers, children and infant nutrition program under the food-for-education Dole-McGovern bill.

1.4 *Coordination between the agencies*

Foreign aid policy makers and experts have no consensus on how to better promote the coordination of foreign aid agencies and on the need for more formal coordination. There are three different points of view. Some believe that a number of foreign assistance funds accounts and multiple actuator can well reflect the development assistance programs of various capacity requirements, as well as a wide range of U.S. foreign aid target; others hold that the lack of centralized coordination of foreign aid authorities has limited transparency, efficiency and effectiveness of U.S. foreign aid; still others think that there has been inadequate assessment of foreign aid programs to know whether inter-agency coordination is needed to provide the program efficiency and effectiveness. Proposals to solve the disputes more or less contain the following aspects, namely developing a national foreign assistance strategy, authorizing a new or existing agency to coordinate all foreign aid programs, coordinating through the National Security Council, improving the ability of U.S missions abroad to coordinate aid activities at the country level, separating strategic assistance programs from development assistance programs, and improving monitoring and reporting activities.

In fact, the drafters of the U.S. Foreign Assistance Act of 1961 had realized the need for coordinated foreign assistance, asking the U.S. delegation leaders to each country to coordinate the functions of U.S. government representatives. Currently, the coordination between U.S. foreign aid agencies are mainly carried out in two channels, namely the Coordinating Committee for Development Policy and the Director of Foreign Assistance. Given the large numbers of U.S. foreign aid implementing agencies, the National Security Council was established under Policy Co-ordination Committees on Development (PCC) as a forum of inter-organizational coordination, whose members include the representatives of major U.S. institutions managing U.S. foreign aid and foreign trade. Policy Coordination Committee on Development is under the leadership of the Director of Foreign Assistance. In 2006 the Director of Foreign Assistance was established under the State Department, responsible for coordinating U.S. foreign aid programs. The Director of Foreign Assistance has administrative rights over most of State Department and USAID foreign aid projects, and provides guidance to other foreign aid bodies, but does not control such major foreign aid projects as the Millennium Challenge Corporation, the Office of the Global AIDS Coordinators, etc. In addition, there are coordinators for assistance acts (such as Support for Freedom Act, Support for Eastern European Democracy Act, etc.), coordinators for reconstruction and stability responsible for their respective areas of aid coordination.

2 U.S. REVELATION TO FOREIGN AID MANAGEMENT SYSTEM REFORM IN CHINA

With the continuous development of China's foreign aid and accumulation of foreign aid experience, China's foreign aid management system began to take shape and gradually developed. The current China's foreign aid management agencies include, apart from the Ministry of Commerce, which is the competent authority, 23 other ministries among the central government agencies and departments, such as Ministry of Foreign Affairs, Ministry of Finance. In addition, the business sector of local governments and embassies and consulates abroad also participate in the management of foreign aid.

Because there are so many departments involved the division of functions between the various departments is relatively obscure. Take the Ministry of Foreign Affairs and the Ministry of Commerce for example. An evident conflict of interest exists on the management of foreign aid between the two. While Chinese embassies belong to the Ministry of Foreign Affairs, and the commercial offices of the embassies are the expatriate organizations of Ministry of Commerce. On the one hand, the offices are doing business under the ambassadors' guidance. On the other hand, domestically the Ministry of Commerce is the centralized management department of foreign aid, with Ministry of Foreign Affairs in a secondary position.

In addition, the regulatory aid agencies only report to their superiors, lacking contact necessary in the horizontal direction. Thus these agencies are not only isolated from the public, but from other agencies in the same department.

Sound communication and coordination between the U.S. foreign aid agencies deserve our notice. Two ways can be adopted o solve the current inter-department problems, namely establishing independent regulatory agencies, and improving foreign aid coordination mechanisms. Li Anshan (2009) also proposed similarly, that the establishment of high-level foreign aid commissions and specialized agencies can improve the effect of foreign aid. Such measures could make a more effective use of the limited funds, avoid mutual restraint between departments and aid overlap, and help form an integrated mechanism of decision making, management, and evaluation in the field of foreign aid.

ACKNOWLEDGEMENT

This article is funded by the MOE Humanities and Social Sciences Research Youth Fund project of 2012 Economic Analysis of Foreign Aid: Theoretical Research and China's strategy (Grant No. 12YJC790060), National Social Science Foundation of China (Grant No. 13BJL054).

REFERENCES

Cai, C.H. 2010. Sino-US relations at issue, regional security and global governance: the 10th "Sino-US relations", regional security and global governance dialogue "meeting Summary". *American Studies* (2): 180–197, in Chinese.

Kegley, C.W. & Hook S.W. 1991. U.S. Foreign Aid and U.N. Voting: Did Reagan's Linkage Strategy Buy Deference or Defiance? *International Studies Quarterly* 35(3):295–312.

Li, A.S. 2009. French history and current situation of aid to Africa—On Thoughts on Chinese aid to Africa. *West Africa* (11):13–21, in Chinese.

Wang, T.Y. 1999. U.S. Foreign Aid and UN Voting: An Analysis of Important Issues. *International Studies Quarterly* 43(1):199–210.

Zhou, H. 2008. China's foreign aid reform and opening up 30 years. *World Economics and Politics* (11):33–43, in Chinese.

Information Technology and Applications – Li (Ed.)
© *2015 Taylor & Francis Group, London, ISBN 978-1-138-02677-3*

A kind of characters segmentation method for text-based CAPTCHA attacking

Zhao Wang & Xiang Li
Key Laboratory of HCST (PKU), MoE, School of EECS, Institute of Software, Peking University, Beijing, China

ABSTRACT: The most widely deployed CAPTCHAs are text-based schemes. Attacking on the CAPTCHA can be used to find design flaws and weaknesses of CAPTCHA, and to improve the safety of CAPTCHA. The generation of text-based CAPTCHA typically uses the method of characters adhesion and characters segmentation is the most important issue for CAPTCHA cracker, as so far, there isn't a common and effective way to solve all the characters segmentation problems. In this paper we used CAPTCHAs from the educational administration site of Peking University as the samples, at first we used connected domain based de-noising, then used improved projection algorithm based segmentation method which selects the extreme, trough or crest points within a certain range as a dividing line, segmentation rate reached 80.8%, while the segmentation rate of basic projection algorithm is 12.5%.

Keywords: CAPTCHA; projection; segmentation; recognition

1 INTRODUCTION

CAPTCHA plays an important role in resisting malicious program attacks suffered by the website, for example, to prevent a malicious website registration, to prevent spam released in batches in the forum, to prevent password brute force, and so on. However, if CAPTCHA is designed to be rough, it is still vulnerable to attacker, therefore it is necessary to study the identification of CAPTCHA.

CAPTCHAs are divided into three types: text, image, and sound. Of which, text-based scheme is based on the character recognition has been widely used. The following methods are mainly used in the generation of text-based CAPTCHA: adding background noise to the picture, the rotation and deformation of the characters, the covering and adhesion of the characters. The cracking of CAPTCHA includes characters segmentation and characters recognition; however, segmentation is more difficult than recognition, once individual character is separated from each other, character recognition problem is not difficult to solve. As so far, there is no common and effective way to solve all the characters segmentation problems.

The CAPTCHA from the educational administration website of Peking University, a characters segmentation algorithm is presented in this paper. The features of this CAPTCHA are: there are background noises, characters adhesions and no character distortion.

2 STATUS OF CHARACTERS SEGMENTATION TECHNOLOGY FOR CAPTCHAS

Characters segmentation technology for CAPTCHA can be divided into projection and connected domain segmentation method. The basic projection method is projection of character to x-axis or y-axis, according to the position of adjacent troughs to determine the division position. Projection method is convenient for implementation, which has a fast operation speed, but it can only be used to characters with no adhesions, if there is adhesion, the projection method will treat adhesive characters as a character. Thus, various modifications have made to the projection method. The projection results were analyzed to determine an optimal segmentation. The advantage of connected domain segmentation method is that it does not use a horizontal or vertical line to characters segmentation, all areas connected will be separated. In the case of characters adhesion is not serious and is some of certain ways, connected domain segmentation algorithm can be used.

In 2008, Huang et al proposed a segmentation method for Yahoo and MSN's CAPTCHAs [1], it is based on the projection method, sliding window analysis are made to eliminate noise and achieve segmentation purposes. In 2010, he proposed projection and middle-axis point separation segmentation method to MSN's CAPTCHAs, the segmentation rate is about 75% [2]. In 2007, Yan et al used connected domain-based algorithm to CAPTCHAs with non-adhesion provided by CAPTCHA service website [3]. In 2008, Microsoft's CAPTCHAs are processed by using graphics knowledge, including the x-axis and y-axis projection, connected domain segmentation, characters segmentation rate is about 90% [4]. In 2012, according to the character color, size and positional relationship among the characters, Ling Chen used rough set based connected domain to make characters segmentation to non-adhesive CAPTCHAs [5]. Honggang Liu used the techniques of image preprocessing, projection technology, closed-loop detection principle and four-line spectrum, where the segmentation rate of Yahoo's CAPTCHAs was 78% [6].

3 PREPROCESSING OF CAPTCHAS

CAPTCHA recognition takes almost the same process: preprocessing of the picture, then segmentation and character identification. Preprocessing of CAPTCHA in this paper is as follows: input image, graying & binaryzation, image de-noising and output image.

3.1 *Image graying and binaryzation*

Image graying is for RGB graph, the CAPTCHA is a single frame GIF image. There is a clear distinction between the color of characters and background, therefore the background color is just set to 255, that is white; the rest position is set to 0, it is black. Then the image binaryzation is completed.

Table 1. Contrast of image before and after binaryzation.

Before binaryzation	After binaryzation
J5T6	J5T6
D36W	D36W
DT59	DT59
9KYA	9KYA
PN2R	PN2R

Table 2. Contrast of image before and after de-noising.

Before denoising	After denoising
J5T6	J5T6
D36W	D36W
AxBK	AxBK
9KYA	9KYA
LR3U	LR3U

3.2 *Image de-noising*

It is can observed that there are not many noises in the CAPTCHAs, noises are only some points. Therefore the connected domain based de-noising method is used, if the number of black dots within a connected domain is less than a threshold value, then this connected domain is treated as noise, and all the dots will be assigned white. Connected domain must select 8 connected domains. Threshold is obtained through many experiments, here it is 10.

4 CHARACTERS SEGMENTATION OF CAPTCHAS

It can be found, there are not always obvious boundaries between all the adjacent characters, the character is not twisted, so that to the adhesive characters, only one vertical dividing line need to be determined to separate them. For non-adhesive characters, a simple algorithm such as connected domain based algorithm or the x-axis projection algorithm can be used to split characters. This paper takes the algorithm based on the traditional projection method. The projection results were analyzed to determine a better division position. The process of the algorithm is as follows:

4.1 *Projection*

After the image de-noising, the results of x-axis projection are stored in the array p. The value of projected image is the number of black pixels in the column. And the results are displayed in the form of a line chart for analysis.

The points are labeled by circle in Figure 1 are referred as starting points, that is

$$\begin{cases} p(i) = 0 \\ p(i+1) > 0 \end{cases} \tag{1}$$

The number of starting points is the total number of characters. For example, if the number of starting points is 4, then the image has four separate characters; if the number of starting points is 3, there are two adhesive characters and two independent characters in the image; if the number of starting points is 2, then there are two sets of two adhesive characters or one three adhesive characters and one independent character in the image; if the number of starting points is 1, then all the characters are adhesive.

The end point that is corresponding to the starting point is defined as follows:

$$\begin{cases} p(i-1) > 0 \\ p(i) = 0 \end{cases} \tag{2}$$

The distance between the starting point and the end point is the character length, and it is labeled as len1, len2, len3 as shown Figure 1.

For better characters segmentation, we generally have to guess the length of a single character.

If the number of starting points is 3, it means there are two adhesive characters and two separate characters, we guess the character length as the mean of the two smaller length.

If the number of starting points is 2, then compare the two lengths, if one length is two times larger than the other length, it is considered that the image is composed of one independent character and three adhesive characters, the guessed character length is the smaller one. If the two

lengths are close to each other, it is considered that the image is composed of two sets of two adhesive characters, the guessed character length is half of the mean of the two lengths.

If the number of starting point is 1, the character length is guessed as fourth of the length.

4.3 *Characters segmentation of CAPTCHAs*

The trough point is defined as follows:

$$\begin{cases} p(i-1) > p(i) \\ p(i+1) > P(i) \end{cases} \tag{3}$$

The crest point is defined as follows:

$$\begin{cases} p(i-1) < p(i) \\ p(i+1) < P(i) \end{cases} \tag{4}$$

The extreme point is defined as follows:

$$\begin{cases} p(i-1) > p(i) \\ p(i+1) = P(i) \end{cases} \text{ or } \begin{cases} p(i-1) < p(i) \\ p(i+1) = P(i) \end{cases} \tag{5}$$

Or

$$\begin{cases} p(i-1) = p(i) \\ p(i+1) < P(i) \end{cases} \text{ or } \begin{cases} p(i-1) = p(i) \\ p(i+1) > P(i) \end{cases} \tag{6}$$

At the junction of the two characters, there is always hopping in the corresponding projection image. If two characters are adhesive, then let

$$\begin{cases} CUT = \dfrac{1}{2} BEGIN + \dfrac{1}{2} END \\ DELTA = \dfrac{1}{4} GUESSLEN \end{cases} \tag{7}$$

Here *BEGIN* is the starting point coordinates, *END* is the end point coordinates, *GUESSLEN* is the guessed length of a single character. Find the nearest point to the point *CUT* from the extreme points or crest and trough points from *p* (*CUT − DELTA*) to *p* (*CUT + DELTA*), the line which this point is located is chosen as a segmentation line.

If three characters are adhesive, then let

$$\begin{cases} CUT1 = \dfrac{2}{3} BEGIN + \dfrac{1}{3} END \\ CUT2 = \dfrac{1}{3} BEGIN + \dfrac{2}{3} END \\ DELTA = \dfrac{1}{4} GUESSLEN \end{cases} \tag{8}$$

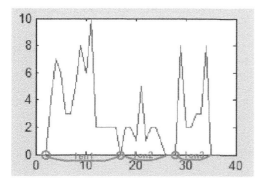

Figure 1.   The chart of projection.

Here *BEGIN* is the starting point coordinates, *END* is the end point coordinates, *GUESSLEN* is the guessed length of a single character. Find the nearest point to the point *CUT1* from the extreme points or crest and trough points from p (*CUT1* − *DELTA*) to p (*CUT1* + *DELTA*), the line which this point is located is chosen as one segmentation line. Find the nearest point to the point *CUT2* from the extreme points or crest and trough points from p (*CUT2* − *DELTA*) to p (*CUT2* + *DELTA*), the line which this point is located is chosen as another segmentation line.

If four characters are adhesive, then let

$$\begin{cases} CUT1 = \dfrac{3}{4}BEGIN + \dfrac{1}{4}END \\[2mm] CUT2 = \dfrac{1}{2}BEGIN + \dfrac{1}{2}END \\[2mm] CUT3 = \dfrac{1}{4}BEGIN + \dfrac{3}{4}END \\[2mm] DELTA = \dfrac{1}{4}GUESSLEN \end{cases} \quad (9)$$

Here *BEGIN* is the starting point coordinates, *END* is the end point coordinates, *GUESSLEN* is the guessed length of a single character. Find the nearest extreme points or crest and trough points to the points *CUT1*, *CUT2 and CUT3* in the *DELTA* range of these points, the lines which the points are located are chosen as the segmentation lines.

4.4 *The results of characters and analysis*

This paper selected 120 CAPTCHAs as samples, 97 of them can be segmented successfully. The segmentation results are compared with using basic projection segmentation method and minimum segmentation method proposed in the [2]. The segmentation rate of basic projection algorithm

Table 3. Results of character segmentation.

Before segmentation	After segmentation
9KYA	9 K Y A
JPVP	J P V P
D6B8	D 6 B 8
K43Y	K 4 3 Y

Table 4. Compare of segmentation results.

Segmentation method	Total samples	Success samples	Segmentation rate
Basic projection	120	15	12.5%
Minimum	120	45	37.5%
This method	120	97	80.8%

is 12.5%. The segmentation rate of minimum segmentation method is 37.5%. The minimum segmentation method selects the minimum from the trough points in the projection image within a certain range of widths.

There are adhesions in the most of the samples in this paper, basic projection algorithm cannot handle adhesive characters, so its segmentation rate is very low. There are difference between the samples in literature [2] and this paper, the characters in [2] are arranged in a more orderly level, so the projection location corresponding to the character adhesion tend to be small. The characters in this paper are not horizontally aligned, so segmentation rate of the minimum method is not high in this paper.

5 CONCLUSION

CAPTCHA is designed to distinguish whether the user is a human or a computer. Text-based scheme is the most widely used type. Attacking on the CAPTCHA can be used to find design flaws and weaknesses of CAPTCHA, and to improve the safety of CAPTCHA. The attacking of text-based CAPTCHA is divided into image segmentation and character recognition. Character adhesion mechanism increases the difficulty of characters segmentation, there isn't a very effective method to attack CAPTCHA with adhesion. In this paper we used CAPTCHAs from the educational administration site of Peking University as the samples, at first connected domain based de-noising is used, and then the improved projection algorithm based segmentation method is used which selects the extreme, trough or crest points within a certain range as a dividing line, segmentation rate reached to 80.8%. Because the use of projection, the dividing line is vertical, if there is some overlap between the characters, then the result may not be satisfactory. Combines with connected domains method, it may make better segmentation results.

ACKNOWLEDGMENT

Zhao Wang is the corresponding author of this paper. The authors would like to thank the anonymous reviewers for their helpful suggestions. This work was supported in part by the NSFC under grant number 61371131 and China Scholarship Council (CSC).

REFERENCES

[1] Huang S.Y., Lee Y.K., Bell G., et al. 2008. A projection-based segmentation algorithm for breaking MSN and YAHOO CAPTCHAs. *Proceedings of the World Congress on Engineering 2008 Vol 1 WCE 2008, July 2–4, 2008,* London, U.K. *Proceedings of the international Conference of Signal and Image Engineering.* 2008.

[2] Huang S.Y., Lee Y.K., Bell G., et al. 2010. An efficient segmentation algorithm for CAPTCHAs with line cluttering and character warping. *Multimedia Tools and Applications,* 2010, 48(2): 267–289.

[3] Yan J., El Ahmad A.S. 2007. Breaking visual CAPTCHAs with naive pattern recognition algorithms. *Proceedings—23rd Annual Computer Security Applications Conference, ACSAC 2007. Dec 10–14, 2007,* Miami Beach, FL, United states. 2007: 279–291.

[4] Yan J., El Ahmad A.S. A Low-cost attack on a Microsoft CAPTCHA. *Proceedings of the 15th ACM conference on Computer and communications security. ACM 2008, Oct. 27–31,* Alexandria, VA, United states. 2008: 543–554.

[5] Ling Chen. 2012. A recognition algorithm of colored image CAPTCHA based on rough set. *Information and network security.* 2012 (6): 26–28.

[6] Honggang Liu. 2012. The research on recognition technology of CAPTCHA based character distortion and adhesion. [D]. Xidian University, Xi'an. China.

Information Technology and Applications – Li (Ed.)
© 2015 Taylor & Francis Group, London, ISBN 978-1-138-02677-3

Predict drug-gene interactions based on collaborative filtering

Lingxiao Ma & Yi Li

College of Information Science and Technology, Beijing Normal University, Beijing, China

ABSTRACT: Predicting drug-targets interactions is useful to select effective drugs for identified genes. In recent years, machine learning technologies are used in predicting drug-targets interactions. In this paper, we converted predicting drug-gene interactions into a recommendation system that can be solved by collaborative filtering. Then, we described a collaborative filtering algorithm optimized by CHAMELEON clustering and Support Vector Machine (SVM). To build the predictors, we applied this collaborative filtering to build two models—drug-based and gene-based model. To achieve a higher accuracy rate, the stacked regression was introduced to combine these two models into one predictor. The experiments show that this algorithm is efficient and well-performed.

Keywords: predict drug-gene interactions; clustering; Support Vector Machine; collaborative filtering; stacked regression

1 INTRODUCTION

With the development of biological technologies and machine learning, it is possible to predict the drugs' effect based on known drugs.

Machine learning-based metrics proposed so far can be classified into three types: feature vector-based, similarity-based and other approaches (Ding, H. et al. 2013). For feature vector-based approach, the input of general machine learning is represented by feather vectors. The feature vector in drug-target interaction instances is generated by combining (structural) chemical descriptors of these given drugs and sequence of targets. With this feature vector as inputs, we can use any standard machine learning method, such as Support Vector Machines (SVMs). For similarity-based method, a similarity matrix for drugs must be generated, of which the (i, j) element is the similarity of drug i and drug j. Similarly, a target similarity matrix must be generated in some way. These two similarity matrices have been used in many methods, including recommendation system, kernel regression (Yamanishi, Y. et al. 2008), Laplacian regularized lest squares (LapRLS) (Xia, Z. et al. 2010) and so on. These similarity-based methods do not need feature extraction or selection that is a complex and difficult process. And the measures of getting the similarity matrices have already fully developed and widely used. These advantages make the approach promising than other approaches. For other approaches, other information, including pharmacological information of drugs (Kuhn, M.

et al. 2008) and biomedical documents and so on, can be used.

In this paper, we focus on similarity-based method and predicting drug-gene interactions is equivalent to a recommendation system. First, this paper describes how to convert the problem that predicting drug-gene interactions into recommendation system, and introduces collaborative filtering (Goldberg, D. et al. 1992) algorithm to solve this problem. Second, a collaborate filtering optimized by CHAMELEON clustering (Karypis, G. et al. 1999) and the Support Vector Machine (SVM) (Cortes, C. et al. 1995) is introduced to predict drug-gene interactions. Third, apply this collaborate filtering to build two models—drug-based and gene-based. Finally, use stacked regression (Breiman, L. et al. 1996) to combine two models to get a better predictor. The experiments show that the combined predictor performed well.

This paper is organized as follows: Section 2 describes the approach that converting predicting drug-gene interactions into a recommendation system. Section 3 describes a collaborative filtering algorithm optimized by CHAMELEON clustering and support vector machine. Section 4 applies this collaborative filtering to building two models—drug-based and gene-based. Then, stacked regression is introduced to combine the two models to train a predictor of more accuracy. Section 5 analyzes the time complexity of this predicting algorithm and validates the performance through experiments.

2 CONVERT PREDICTING PROBLEM INTO A RECOMMENDATION SYSTEM

2.1 Introduction to recommendation system

A recommendation system (Jannach, D. et al. 2010) can be described as follows: Let m be the number of the user set $U = \{U_1, U_2, ..., U_m\}$ and n be the number of the item set $I = \{I_1, I_2, ..., I_m\}$. The user-item score is described as a $m \times n$ matrix $R_{m \times n}$. The score R_{ij} of user U_i and item I_j ($U_i \in U, I_j \in I$) indicates user's preference of the item. How to predict the user-item score matrix $R_{m \times n}$ is the core problem of recommendation systems.

There are three main algorithms in recommendation systems (Jannach, D. et al. 2010): collaborative filtering, content-based filtering, and knowledge-based filtering. The collaborative filtering has the highest accuracy rate in most situations.

In this paper, we select the collaborative filtering as the recommendation system model. There are two key models to design the collaborative filtering: one model is user-based collaborative filtering (Konstan, J.A. et al. 1997), and the other model is item-based collaborative filtering (Sarwar, B. et al. 2001).

2.2 Equivalence between predicting problem and recommendation system

Recommendation system is described as follows:

– Input: the similarity of users and items, the known user-item scores.
– Output: the new user-item scores.

The problem that predicting drug-gene interactions can be described similar to the format of recommendation system:

– Input: the similarity of drugs and genes, the known drug-gene interactions.
– Output: the new drug-gene interactions.

According to the description of recommendation system and predicting problem, it is obviously that using drugs and genes instead of users and items, the problem that predicting drug-gene interactions can be convert into a recommendation system.

3 COLLABORATIVE FILTERING OPTIMIZED BY CHAMELEON CLUSTERING AND SVM

3.1 Introduction to CHAMELEON clustering

CHAMELEON (Karypis, G. et al. 1999), an effective way for clustering, can cluster data of different shapes and sizes.

The procedure of CHAMELEON is as follows:

1. Calculate the k-nearest neighbor graph based on the user similarity matrix $SU_{m \times m}$.
2. Partition the k-nearest neighbor graph into many initial clusters $C = C_1, C_2, ..., C_k$.
3. Combine each pair of clusters C_i and C_j according to Relative Interconnectivity (RI) and Relative Closeness (RC) (Karypis, G. et al. 1999) until there is no pair to be combined. Then, the user clusters are calculated. For each user U_t, the cluster that U_t in is recorded as C_{U_t}.

3.2 Introduction to Support Vector Machine (SVM)

The SVM (Cortes, C. et al. 1995) is an effective way for regression. It does well in finding a best compromise between the complexity of the model and study ability according to the limited sample information, so that it can help require good generalization ability (Vapnik, V. 2000).

We collected the sample set as follows:

$$(X_i, Z_i), X_i \in R^n, Z_i = score \in 0,1 \qquad (1)$$

where, X_i is the input and the Z_i is the objection, which are from the data as a training sample.

According to the literature (Cortes, C. et al. 1995), the final regression function is as follows:

$$Z = f(x) = \sum_{i=1}^{l} (-\alpha_i + \alpha_i^*) K(X_i, X_j) + b \qquad (2)$$

where, α_i and α_i^* are Lagrange multipliers and kernel function

$$K(X_i, X_j) = \exp\left(-|X_i - X_j|/2\sigma^2\right) \qquad (3)$$

3.3 Collaborate filtering

As described above, there are two kinds of collaborate filtering—user-based and item-based. Here, take the user-based collaborative filtering as the example to describe this collaborative filtering.

The main procedure of the new collaborative filtering is designed as follows:

1. Use CHAMELEON clustering algorithm to cluster users based on user similarity matrix $SU_{m \times m}$.
2. This step should be divided into two situations:
   i. The number of evaluated items of user U_i is greater than threshold α_{CF}:
   a. For item I_j, let the users in the U_i's cluster C_{U_i} be the training set.
   b. Let set of items with scores of U_i be $ISET_i$.

c. Input the vectors constructed by $ISET_i$ evaluated by users in C_{U_i} into the support vector machine to train the support vector machine.

d. Input vector constructed by scores of items in $ISET_i$ evaluated by user U_i to predict score of item I_j evaluated by user U_i that is recorded as R_{ij}.

e. Calculate each score by the above steps.

ii. The number of evaluated items of user U_i is less than threshold α_{CF}:

a. For item I_j, the user-item score RT_{ij} can be calculated by equation 4.

$$RT_{ij} = \sum_{U_p \in C_{U_i}} SU_{ip} R_{pj} \bigg/ \sum_{U_p \in C_{U_i}} |SU_{ip}| \qquad (4)$$

b. If RT_{ij} is greater than β_{CF}, let $R_{ij} = 1$. Contrarily, let $R_{ij} = 0$ when RT_{ij} is less than β_{CF}.

c. Calculate each score by step (a) and step (b).

3. Calculate the recommendation matrix $R_{m \times n}$.

At the same time, the item-based collaborative filtering is similar to the user-based collaborative filtering by exchanging the position of users and items. It is not necessary to describe it in detail.

3.4 Analysis of this collaborate filtering

Let the number of users be M, the number of items be N and the number of unknown items be T. It is clear that the time complexity of the CHAMELEON clustering is $O(M^2)$ (Karypis, G. et al. 1999), the time complexity of the support vector machine is $O(N_{sv}^3)$ (Cortes, C. et al. 1995), and the time complexity of calculating equation 4 is $O(|C_{U_i}| \| ISET_i |)$. The time complexity of this collaborative filtering is as follows:

$$O(M^2 + Tmax\{N_{sv}^3, |C_{U_i}| \| ISET_i |\}) \qquad (5)$$

4 PREDICT DRUG-GENE INTERACTION BASED ON COLLABORATIVE FILTERING

4.1 Apply collaborative filtering to predicting drug-gene interactions

According to theories of collaborative filtering, there are two types of drug-gene interactions predictors—drug-based and gene-based model.

4.1.1 Drug-based model
According to the collaborative filtering described in section 3.3, the drug-based model can be designed in brief as follows:

1. Let the number of drugs be m, and the number of genes be n. Use CHAMELEON clustering to cluster drugs based on drug similarity matrix $SD_{m \times m}$.

2. This step should be divided into two situations:

i. The number of evaluated genes of drug D_i is greater than threshold α_D: For each gene G_j, use the SVM to calculate the prediction PDG_{ij}.

ii. The number of evaluated genes of drug D_i is less than threshold α_D: For each gene G_j, use equation 4 to calculate prediction $PDGT_{ij}$. If $PDGT_{ij}$ is greater than threshold β_D, let $PDG_{ij} = 1$. Contrarily, let $PDG_{ij} = 0$ when $PDGT_{ij}$ is less than β_D.

3. Calculate the prediction matrix $PDG_{n \times m}$.

4.1.2 Gene-based model
Similar to relation between user-based and item-based models, the prediction matrix $PGD_{n \times m}$ of gene-based model can be calculated like drug-based model by exchange the position of drugs and genes.

4.2 Combined model based on stacked regression

4.2.1 Introduction to stacked regression
If we have a set of predictors $v_1(x), v_2(x), ..., v_K(x)$ and these predictor are based on the same training dataset $L = \{(y_n, x_n), n = 1, ..., N\}$, a more accurate predictor can be obtained by combining these k predictors instead of selecting a single the best one from this set (Breiman, L. 1996, Q. Liu. 2011).

Stacked regression is a method for forming linear combinations of different predictors to improve prediction accuracy (Breiman, L. 1996). The main idea of stacked regression is to use cross-validation data and least squares under non negativity constraints to determine the coefficients of models in combination.

4.2.2 Combine drug-based and gene-based model
(Q. Liu. 2011) has proposed a framework to combine user-based and item-based model based on stacked regression. Here, we use similar approaches to combine drug-based and gene-based model.

Let gene-based model be $v_1(x)$ and drug-based model be $v_2(x)$. According to theory of stacked regression, the procedure of combining gene-based and drug-based models is designed as follows:

1. Use the 10-fold cross-validation procedure to train the gene-based model and the drug-based model:

a. Let PR be initial interaction matrix. Divide data into 10 part. Let jth part be recorded as R_j and let the other data be training data R^j.

b. Let P_D be the drug-based predictor and P_G be the gene-based predictor. Use the training data R^j to train the jth predictors P_D^j and P_G^j.

c. Input the data R_j into the jth predictors P_D^j and P_G^j to get the stacked regression training data shown in equation 6. Let r_{dg} be the label, $P_D^j(r_{dg})$ be prediction of r_{dg}, $P_G^j(r_{dg})$ be prediction of r_{dg}

$$TR_{l1} = \{(P_D^j(r_{dg}), P_G^j(r_{dg}), r_{dg}) \mid r_{dg} \in R_j,$$
$$j = 1, 2, \ldots, 10\} \tag{6}$$

2. Combine the drug-based and the gene-based models based on the generated data. The final predictor is described in equation 7. The key problem of this procedure is how to define the value of $\alpha^{(1)}$ and $\alpha^{(2)}$ (Breiman, L. 1996, Q. Liu. 2011), which can be solved by equation 8.

$$P(r_{dg}) = \alpha^{(1)} P_D^j(r_{dg}) + \alpha^{(2)} P_G^j(r_{dg}) \tag{7}$$

$$min_{a_1, \ldots, a_K} \left\{ \sum_n \left(y_n - \sum_k a_k z_{kn} \right)^2 \right\};$$
$$s.t. \ a_k \geq 0, k = 1, 2. \tag{8}$$

5 ANALYSIS AND EXPERIMENTS

5.1 *Analysis of this predicting algorithm*

As the time complexity of the collaborative filtering in Section 3 is $O(M^2 + Tmax\{N_{sv}^3, |C_{U_l}| \|ISET_i|\})$, training the drug-based model and the gene-based model should cost $O(M^2 + Tmax\{N_{sv}^3, |C_{U_i}| \|ISET_i|\})$. The stacked regression should cost K times time. So the time complexity of the final model is as follows:

$$O(2K(M^2 + Tmax\{N_{sv}^3, |C_{U_i}| \|ISET_i|\})) \tag{9}$$

5.2 *Experiments*

To validate this predicting algorithm, we collected the dataset from (Bleakley K. 2009) and did the experiments.

According to CHAMELEON clustering, Drugs are partitioned into 13 clusters and Genes are partitioned into 18 clusters.

1. Let hsa55800 be the example: predicting the interaction between hsa55800 and D00332.
   It is clear that the predicted interaction is correct.
2. Let hsa55800 be the example: predicting the interactions between hsa55800 and all of the drugs.
   Most predictions of hsa55800 in Table 2 are right.

Table 1. Interaction of hsa55800 and D00332.

Name	D00294	...	D00303	D03826	D00332
hsa10008	1	...	0	0	0
hsa10369	0	...	0	0	1
hsa27133	1	...	0	0	0
hsa27345	1	...	0	0	0
hsa3708	0	...	0	0	1
hsa3710	0	...	0	0	1
hsa55515	0	...	1	0	1
hsa6330	0	...	1	0	1
...	...	...	...	...	...
hsa9992	1	...	0	0	0
hsa55800	0	...	1	0	1
hsa55800p	0	...	1	0	1

Table 2. Interaction of hsa55800 and other drugs.

Name	D00294	...	D00303	D03826	D00332
hsa10008	1	...	0	0	0
hsa10369	0	...	0	0	1
hsa27133	1	...	0	0	0
hsa27345	1	...	0	0	0
hsa3708	0	...	0	0	1
hsa3710	0	...	0	0	1
hsa55515	0	...	1	0	1
hsa6330	0	...	1	0	1
...	...	...	...	...	...
hsa9992	1	...	0	0	0
hsa55800	0	...	1	0	1
hsa55800p	0	...	0	0	1

6 CONCLUSIONS

Finally, this paper proposes a collaborative filtering algorithm for this predicting problem and applies this algorithm to predicting drug-gene interactions. Then, we studied the advantages and the disadvantages of the gene-based model and the drug-based model and proposed an approach to combine the two models to achieve a higher accuracy rate based on the stacked regression. The experiments validate the effects of the combined model.

REFERENCES

Bleakley, K. & Yamanishi, Y. 2009. Supervised prediction of drug–target interactions using bipartite local models. *Bioinformatics*, 25(18), 2397–2403.

Breiman, L. 1996. Stacked regressions. *Machine learning*, 24(1), 49–64.

Cortes, C. & Vapnik, V. 1995. Support-vector networks. *Machine learning*, 20(3), 273–297.

Ding, H. & Takigawa, I. & Mamitsuka, H. & Zhu, S. 2013. Similarity-based machine learning methods for predicting drug–target interactions: a brief review. Briefings in bioinformatics, bbt056.

Goldberg, D. & Nichols, D. & Oki, B.M. & Terry, D. 1992. Using collaborative filtering to weave an information tapestry. *Communications of the ACM*, *35*(12), 61–70.

Jannach, D. & Zanker, M. & Felfernig, A. & Friedrich, G. 2010. *Recommender systems: an introduction*. Cambridge University Press.

Karypis, G. & Han, E.H. & Kumar, V. 1999. Chameleon: Hierarchical clustering using dynamic modeling. *Computer*, *32*(8), 68–75.

Konstan, J.A. & Miller, B.N. & Maltz, D. & Herlocker, J.L. & Gordon, L.R. & Riedl, J. 1997. GroupLens: applying collaborative filtering to Usenet news. *Communications of the ACM*, *40*(3), 77–87.

Kuhn, M. & Campillos, M. & González, P. & Jensen, L.J. & Bork, P. 2008. Large-scale prediction of drug–target relationships. *FEBS letters*, *582*(8), 1283–1290.

Liu, Q. 2011. *Research on recommender systems based on collaborative filtering*, Master's thesis, University of Science and Technology of China.

Sarwar, B. & Karypis, G. & Konstan, J. & Riedl, J. 2001. Item-based collaborative filtering recommendation algorithms. In *Proceedings of the 10th international conference on World Wide Web* (pp. 285–295). ACM.

Vapnik, V. 2000. *The nature of statistical learning theory*. springer.

Xia, Z. & Wu, L.Y. & Zhou, X. & Wong, S.T. 2010. Semi-supervised drug-protein interaction prediction from heterogeneous biological spaces. *BMC systems biology*, *4*(Suppl 2), S6.

Yamanishi, Y. & Araki, M. & Gutteridge, A. & Honda, W. & Kanehisa, M. 2008. Prediction of drug–target interaction networks from the integration of chemical and genomic spaces. *Bioinformatics*, *24*(13), i232–i240.

Information Technology and Applications – Li (Ed.)
© 2015 Taylor & Francis Group, London, ISBN 978-1-138-02677-3

Predict NBA players' career performance based on SVM

Yi Li & Lingxiao Ma
College of Information Science and Technology, Beijing Normal University, Beijing, China

ABSTRACT: Predicting the National Basketball Association (NBA) players' career performance learning from history data can be amazing for the one's who paying attention to NBA. To predict NBA players' performance, this paper first adopts a new metric to evaluate players' performance based on factorial analysis through combining some popular existing metrics in NBA analysis. Then after defining feature vector according to the new metric to evaluate players' performance, we construct the training set for the player predicted after clustering players. Finally, learning algorithm is implemented based on Support Vector Machine (SVM) to get the prediction of player. The experiments show that this new way is efficient and well-performed.

Keywords: predict NBA players' career performance; factor analysis; clustering; Support Vector Machine (SVM)

1 INTRODUCTION

The National Basketball Association (NBA) is one of the most popular events conducted in United States annually. All kinds of analytic reports spread around the world try to predict the performance of one specific player in future career, and what level the player can reach. Predicting NBA players' performance in future career is vital, because team managers and coaches can control the states of teams in future, make right adjustment for teams, and make wise trade players and so on. Therefore, this paper mainly talks about predicting the performance of NBA players in future career learning from history data.

The history data here includes: the data of the whole entired and current players in NBA. To realize our prediction, we must solve two main problems: one is how to use a new metric to evaluate the performance of entired and current players in fair way, and the other is how to model the prediction to achieve higher accuracy. For the first problem, currently there are a number of "advance" metrics that are being employed by all kinds of NBA teams. Some examples are NBA Efficiency Formula, Player Efficiency Rating (PER), Win-Shares, and Plus-Minus/Adjusted. These statistic models can measure players in different aspects. (Piette, James et al. 2011) adapted a network-based algorithm to estimate centrality scores and players' corresponding statistical significances. For the second problem, some researches showed some similar prediction. (Hwang, Douglas. 2012) used Weibull-Gamma Statistical Timing Model to forecast NBA players'

performance, which quantified by players' scores. (Shi, Zifan et al. 2013) used machine learning techniques to predict NCAAB match outcomes. (Yang, Jackie B. et al. 2012) predicted NBA championship by learning from history data. (Bryan Cheng et al. 2013) predicted the Betting Line in NBA Games with SVM model. These studies inspired us to predict NBA players' future performance quantified by new metric proposed.

This paper proposes a new method to evaluate the performance of NBA players, putting together the accepted statistics—NBA Efficiency Formula, Player Efficiency Rating (PER), Win-Shares, and Plus-Minus/Adjusted based on principal component analysis. And then, we construct the training set with the whole data, adopt Support Machine Vector model to predict the performance.

This paper is organized as follows: Section 2 describes how to collect data and preprocess the collected data. Section 3 build a model to evaluate NBA players' performance based on accepted formulas and factor analysis. Section 4 describes the procedure that builds the predictor based on SVM and predicting NBA players' career performance from the trained model. Section 5 describes the experiments; the experiments show that this algorithm performed well in predicting NBA players' career performance.

2 DATA PREPROCESSING

2.1 *Data collection*

To predict NBA players' performance, we must get the actual NBA data. Then we collected the data from web www.basketball-reference.com/players. However, we can only get one player data from one web page respectively. Therefore, if we want to get the whole history data, web crawler must be established to get the complete data. Through web crawler, we collected average data per-season for every player's playing seasons. Of course, the whole players in NBA include entired and current players. The different kinds of stats collected for each player per-season can be described in Table 1.

2.2 Data preprocessing

After collecting data, what we must do next is filtering the data, because many data cannot contribute for predicting. The first step is filtering the players' data whose average playing time per-season is less than 5 minutes, because the playing time cannot reflect their actual skills and abilities. And then the players whose playing seasons are less than 2 should be cancelled from the whole data set, for their experience in NBA is so less that we cannot make accurate prediction from their history playing data. Similarly players

whose playing seasons is greater than 18 should be cancelled (because there are almost no values doing predicting to those so old players). Finally, we should divide the whole data after filtering into two parts: one is entired players, the other is current players, for the reason that entired players offer complete career data, and the current players are the ones whose performance which we'll going to predict.

3 EVALUATE PLAYERS' PERFORMANCE

3.1 Some accepted formulas

These statistics described above are the average of players in one particular season. But the problem with these statistics is that they are all raw numbers, which limit their expressiveness. We must search for a model to summarize the performance from all the features above. There are four main formulas to evaluate NBA players' performance: NBA Efficiency Formula (Hofler, Richard A. et al. 1997), Player Efficiency Rating (PER) (Hollinger, J. et al. 2009), Win-Shares (James, Bill. 2002) and Plus-Minus/Adjusted (Macdonald, Brian. 2011).

3.2 Combine formulas based on factor analysis

It is obviously that the NBA players' performance cannot be evaluated precisely based on only one formula. In order to evaluate NBA players' performance more precisely, factor analysis (Kim, Jae-On. et al. 1978) is introduced to combine these four formulas.

Factor analysis (Kim, Jae-On. et al. 1978) is a common statistical method, mainly dealing with the following issues:

– Correlation coefficients between the variables is large, and is required to be erased;
– Put the variables with strong correlation into a factor, which can reflect most information;
– Evaluate the importance of the factors.

The main procedure of factor analysis is as follows:

1. With dimension reduction, p variables can be expressed by m common factors, that is

$$V = AF + \varepsilon \qquad (1)$$

where, $A = \begin{pmatrix} a_{11} & a_{12} & ... & a_{1m} \\ a_{21} & a_{22} & ... & a_{2m} \\ & ... & ... & \\ a_{p1} & a_{p1} & ... & a_{pm} \end{pmatrix}$ is called factor

loading matrix;

Table 1. Statistics of NBA players.

Age	
Age of Player at the start of February 1st of that season.	
2PA	
2-point Field Goal Attempts	
MP	FG
Minutes Played	Field Goals
FTA	FT%
Free Throw Attempts	Free Throw Percentage
G	AST
Games	Assists
FT	STL
Free Throws	Steals
FG%	BLK
Field Goal Percentage	Blocks
DRB	TOV
Defensive Rebounds	Turnovers
3PA	
3-Point Field Goal Attempts	
2P%	
2-Point Field Goal Percentage	
FGA	PF
Field Goal Attempts	Personal Fouls
ORB	PTS
Offensive Rebounds	Points
2P	Pos
2-Point Field Goals	Position
TRB	3P
Total Rebounds	3-Point Field Goals
3P%	
3-Point Field Goal Percentage	

$F = (F_1, F_2, ..., F_m)^T$ is called common factors matrix;

$\varepsilon = (\varepsilon_1, \varepsilon_2, ..., \varepsilon_p)^T$ is called special factors matrix.

2. Factor Rotation: adopt the maximum variance cross rotation in order to make sense of the common factors, therefore we get a new factor loading matrix.

$$B = \begin{pmatrix} \beta_{11} & \beta_{12} & \cdots & \beta_{1m} \\ \beta_{21} & \beta_{22} & \cdots & \beta_{2m} \\ & \cdots & \cdots & \\ \beta_{p1} & \beta_{p1} & \cdots & \beta_{pm} \end{pmatrix} \qquad (2)$$

3. Calculate the scores: in the factor rotated matrix, β_{ij} is the loading of variable X_j on factor F_i. The weight of X_k called w_k can be calculated as follows:

$$b_{ik} = |\beta_{ik}| \Big/ \sum_{j=1}^{p} |\beta_{ij}| \qquad k = 1, 2, ..., p \quad i = 1, 2, ..., m$$

$$c_k = \sum_{i=1}^{m} b_{ik} \qquad k = 1, 2, ..., p$$

$$w_k = c_k \Big/ \sum_{j=1}^{p} c_j \qquad k = 1, 2, ..., p \qquad (3)$$

And the factor score is:

$$x = \sum_{i=1}^{p} w_i v_i \qquad (4)$$

After this step, we can combine the four formulas' score into one score so that the NBA players' performance can be evaluated more precisely.

4 PREDICTING PLAYERS' PERFORMANCE MODEL BASED ON SVM

To predict players' performance accurately, we should group the players. Therefore it's necessary to use clustering algorithm to group the players through their playing data. Given the player we'll make prediction for belongs to *Kth* cluster, obviously the players in the same cluster are the comparators. In other words, the training set consists of the data of the players in the same cluster as the player predicted.

4.1 Cluster players

4.1.1 Introduction to CHAMELEON
CHAMELEON is a well-performed clustering algorithm proposed by Karypis, George,

et al. (Karypis, G. et al. 1999). The key idea of CHAMELEON clustering is to create initial clusters from k-nearest neighbor graph and combine clusters according to relative interconnectivity and relative closeness.

4.1.2 Distance: similarity scores
Here we use similarity scores to measure the distance between two players in CHAMELEON algorithm. The similarity scores were derived using a method similar (no pun intended) to the one used by Doug Drinen over at Pro-Football-Reference.com. This method does not attempt to find players who were similar in style of play. Rather, it attempts to find players whose careers are similar in terms of quality and shape. Another important item to note is that players are only compared with other players who played a comparable position. Using the method, we can get the distance between two players.

4.1.3 Clustering
After getting the distance between two players and setting some thresholds for clustering algorithm, we can obtain multiple player clusters through CHAMELEON algorithm. Given that players are divided into m clusters after clustering, i_{th} cluster can be recorded as S_i, we can conclude that players' set $S = S_1 \cup S_2 \cup ... S_i ... \cup S_m$.

4.2 Using SVM to predict players' performance

Set the maximum career length in NBA among any players M, if one current player (recorded as N) have played i seasons (where $i < M$), then we expect to predict the performance of the following years. For similarity we take predicting the following two years' performance as an example.

4.2.1 Feature vector
For any j (where $j < i$) season, we can know his mark in this mark from X, recorded as X_{N_j}, and define feature vector $\overline{X_N}$, $\overline{X_N} = (X_{N_1}, X_{N_2}, X_{N_3} ..., X_{N_j} ..., X_{N_i})$.

4.2.2 Training set
After representing every player in every season with feature vector, we can easily conclude that the training set (recorded as T) results from putting feature vectors of the whole history data of every one in any season together. To be specific, to predict the performance of player N belonging to *kth* cluster in S, the training set of player N is the set of feature vectors of the whole players in the same cluster as player N (*kth* cluster).

4.2.3 Learning algorithm
One particular player N has played i years in NBA, belonging to *kth* cluster in S, and the feature is

$(X_{N_1}, X_{N_2}, X_{N_3}, ..., X_{N_j}, ..., X_{N_i})$. To get the value of $X_{N_{i+1}}$ $(i+1 < M)$, do as the following steps:

1. Make training set $T = \varnothing$.
2. For every player P (record P_{fv} as his feature) in kth cluster, if P's career length is greater than $i+1$, then $T = T \cup \{P_{fv}, X_{P_{i+1}}\}$ ($X_{P_{i+1}}$ represents the value of P's i+1 mark, training result).
3. Use SVM model to train T.
4. Let $(X_{N_1}, X_{N_2}, X_{N_3}, ..., X_{N_j}, ..., X_{N_i})$ as input of SVM model, resulting in $X_{N_{i+1}}$.

If we want to predict the value of $X_{N_{i+2}}$ $(i+2 < M)$, there exist two ways: one is using the value of $X_{N_{i+1}}$ obtained just now as one element of feature vector of N, meaning that we treat the $X_{N_{i+1}}$ as true value. By doing this way, the feature vector of N is $(X_{N_1}, X_{N_2}, X_{N_3}, ..., X_{N_j}, ..., X_{N_i}, X_{N_{i+1}})$, the training result is $X_{N_{i+2}}$. And use the same above strategy to get $X_{N_{i+2}}$ of N. The other is omitting the value of $X_{N_{i+1}}$, meaning that the feature vector of N is still $(X_{N_1}, X_{N_2}, X_{N_3}, ..., X_{N_j}, ..., X_{N_i})$, and then follow the same way to get $X_{N_{i+2}}$ of N. If we want to predict more seasons' performance of player, the same strategy can be used, but the accuracy can also decrease in some way.

5 EXPERIMENTS

To validate this predicting algorithm, we collected NBA history data from the website (www.basketball-reference.com/players) and use this algorithm to predict NBA player's career performance.

Take LeBron James, Dwayne Wade as examples:

1. Use the factor analysis to calculate players' scores:
   The KMO and Bartlett's Test shows that it is suitable to do factor analysis.
   According to Table 3 and Table 4, we can calculate the final score of each NBA player.
2. Train the SVM based on the NBA history data.
3. Predict LeBron James's career performance: the performance of LeBron James is shown in Figure 1.

Table 2. KMO and Bartlett's test.

Kaiser-Meyer-Olkin measure of sampling adequacy.		0.780
Bartlett's test of sphericity	Approx. chi-square	90982.689
	Df	6
	Sig.	0.000

Table 3. Component matrix*.

	Component	
	1	2
NBAEF	0.966	−0.071
PER	0.970	−0.059
WinShares	0.857	−0.383
PML	0.806	0.563

Extraction method: principal component analysis.
*. 2 components extracted.

Table 4. Rotated component matrix*.

	Component	
	1	2
NBAEF	0.817	0.522
PER	0.812	0.533
WinShares	0.915	0.206
PML	0.309	0.933

Rotation method: Varimax with Kaiser normalization.
*. Rotation converged in 3 iterations.

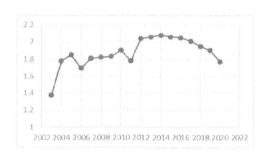

Figure 1. LeBron James's career performance.

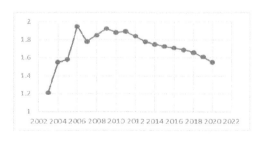

Figure 2. Dwayne Wade's career performance.

4. Predict Dwayne Wade's career performance: the performance of Dwayne Wade is shown in Figure 2.

6 CONCLUSION

This paper adopts a new metrics to evaluate NBA players' performance based on factorial analysis. All NBA players can get one mark per-season according to the whole NBA history data. A new way to predict players' performance is proposed based on SVM model. The experiments show that this new metric is efficient and well-performed. In the future work, we will focus on achieving higher predicting accuracy using larger data.

REFERENCES

Bryan Cheng & Kevin Dade & Michael Lipman & Cody Mills. 2013. Predicting the Betting Line in NBA Games. 2014-6-20. http://cs229.stanford.edu/proj2013/ChengDadeLipman Mills-PredictingThe-BettingLineInNBAGames.pdf.

Hofler, Richard A. & James E. Payne. 1997. Measuring efficiency in the national basketball association. *Economics Letters*, 55(2): 293–299.

Hollinger, J. 2009. The player efficiency rating. 2009-04-07. *http://insider.espn.go.com/nba/hollinger/statistics*.

Hwang, Douglas. 2012. Forecasting NBA Player Performance using a Weibull-Gamma Statistical Timing Model. *MIT Sloan Sports Analytics Conference*.

James, Bill & Jim Henzler. 2002. *Win shares.* STATS Pub..

Kim, Jae-On & Charles W. Mueller, eds. 1978. *Factor analysis: Statistical methods and practical issues.* Vol. 14. Sage.

Karypis, G. & Han, E.H. & Kumar, V. 1999. Chameleon: Hierarchical clustering using dynamic modeling. *Computer, 32*(8): 68–75.

Macdonald, Brian. 2011. A regression-based adjusted plus-minus statistic for NHL players. *Journal of Quantitative Analysis in Sports*, 7(3): Article 4.

Piette, James & Lisa Pham & Sathyanarayan Anand. 2011. Evaluating basketball player performance via statistical network modeling. *MIT Sloan Sports Analytics Conference*.

Shi, Zifan & Sruthi Moorthy & Albrecht Zimmermann. 2013. Predicting NCAAB match outcomes using ML techniques–some results and lessons learned. *Machine Learning and Data Mining for Sports Analytics ECML/PKDD 2013 workshop*.

Yang, Jackie B. & Ching-Heng Lu. 2012. Predicting NBA championship by learning from history data. *Proceedings of Artificial Intelligence and Machine Learning for Engineering Design*.

Information Technology and Applications – Li (Ed.)
© *2015 Taylor & Francis Group, London, ISBN 978-1-138-02677-3*

Fish-eye lens rectification based on equidistant model

Chaoqing Xu & Xiafu Peng
Department of Automation, Xiamen University, Xiamen, Fujian, China

ABSTRACT: Fish-eye lens with large field of view are widely used in the field of navigation applications and surveillance. However, the images captured by the fish-eye lens will introduce undesirable effects. By far, the most evident of these effects is radial distortion. Generally, the fish-eye images can be effectively utilized only after corrected. In this paper, we model the fish-eye lens using equidistant projection function which is the most widely used model. Local images are corrected by mapping the fish-eye images point onto a window rectification plane. In addition, several examples with real-time virtual roaming from real fish-eye images demonstrate the feasibility of the rectification procedure.

Keywords: fish-eye lens; equidistant model; images rectification; virtual roaming

1 INTRODUCTION

Fish-eye lens have a wide field of view and short focal length. As fish-eye images have serious barrel distortion, therefore we need to restore the fish-eye images into general perspective projection image. The most widely used model of fish-eye images correction method is hemispherical [1] and parabolic model [2]. But these models are not based on fish-eye lens model itself, so the corrected effect of images is not very good. Many other models, such as Spherical perspective projection model [3], Radial and Tangential Distortion Model [4], Logarithmic Fish-Eye Lenses Distortion Model [5], Polynomial Fish-Eye Distortion Mode [5] and Rational Function Distortion Mode [6] and so on, are used for image correction. However, these approaches require calibration of the corresponding parameters and the actual process is very complex.

In this paper, we apply equidistant projection model and window planes to correct fish-eye images. In fact, we are concerned about the whole images. At this point, we need to continuously change the position of the window planes to achieve virtual roaming effect.

2 FISH-EYE LENS MODEL

There are several different types of fish-eye projection function. The equidistant projection is the most widely used model, this paper is based on this model. Figure 1 gives the representation of fish-eye equidistant projection as quadric projection [7,8]. Quadric ε is a spherical surface with the lens focal length as the radius and ε_1 is the quadric of equidistant

model. Pinhole plane (window rectification plane) is the imaging plane of conventional camera.

An incident light from the space point P to origin point intersect pinhole plane, quadric ε_1 and ε at the points of P_0, P_1, and P_3, respectively. A vertical line P_1B through the point P_1 perpendicular to the Z-axis at the point B. In the similar way, A vertical line P_1C through the point P_1 perpendicular to the Y-axis at the point P_2 and intersect pinhole surface at the point C. The equidistant function is $r = f\delta$, where f = the fish-eye lens focal length; δ = the angle of incidence; and r = the ideal image height. We can know that the length of line P_1B and arc P_3A are equal to each other by simple geometrical computation. So P_1B is image height of the space point P on the fish-eye image and the point P_2 is the image of the point P. The set of imaging points of all space points form a circular field which is the intersectant region between the quadric ε_1 and plane *XOY*.

Figure 1 shows that $P_2O = P_1B = CA$ and the length of line P_0A is pinhole image height without

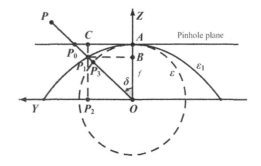

Figure 1. Equidistant projection model.

distortion of the point P. So the length of line P_0C is distorted fish-eye image height of the point P. So we can obtain the rectification of a distorted point after restoring fish-eye image height to pinhole image height [7].

Let $P(X, Y, Z)$, $P_0(x, y, z)$, $P_2(u, v)$, and $P_1(u, v, z_1)$. The following formulas can be easily obtained from Figure 1.

$$r = f\delta = f \cdot \arctan\left(\frac{\sqrt{x^2 + y^2}}{z}\right) \tag{1}$$

$$\frac{u}{r} = \frac{x}{\sqrt{x^2 + y^2}} \quad \frac{v}{r} = \frac{y}{\sqrt{x^2 + y^2}} \tag{2}$$

We solve equations (1) and (2) in terms of r and equate to obtain the following formulas [7,9]:

$$u = \frac{f \cdot x}{\sqrt{x^2 + y^2}} \arctan\left(\frac{\sqrt{x^2 + y^2}}{z}\right)$$
$$v = \frac{f \cdot y}{\sqrt{x^2 + y^2}} \arctan\left(\frac{\sqrt{x^2 + y^2}}{z}\right) \tag{3}$$

Let (u_0, v_0) be the coordinates of the principal point of the image plane and (x_0, y_0) be the center of the window correction plane. The fish-eye lens focal length is $f = R/\omega$, where R = the radius of the circular region of the fish-eye image; ω = the maximum angle of incidence. Therefore the formulas (3) have the form as follows:

$$\begin{cases} u = \dfrac{R \cdot (x - x_0)}{\omega \cdot \sqrt{(x-x_0)^2 + (y-y_0)^2}} \\ \qquad \arctan\left(\dfrac{\sqrt{(x-x_0)^2 + (y-y_0)^2}}{z}\right) + u_0 \\ v = \dfrac{R \cdot (y - y_0)}{\omega \cdot \sqrt{(x-x_0)^2 + (y-y_0)^2}} \\ \qquad \arctan\left(\dfrac{\sqrt{(x-x_0)^2 + (y-y_0)^2}}{z}\right) + v_0 \end{cases} \tag{4}$$

The window planes correction images can be magnified and shrunken by means of increasing and decreasing the value of z (the height of window plane).

3 CONTOUR EXTRACTION OF IMAGES

One of the most important steps before we rectify the fish-eye images is extracting the information of

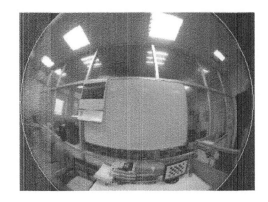

Figure 2. Fish-eye image.

the circular region of images, such as the coordinates of center and the value of radius. There are lots of approaches such as area method [12], scan line approach algorithm [10] and region growing algorithm [11] or the combination of these approaches. In this paper, we manually select a number of edge points of circular region. And then, the optimized circle can be obtained by the least squares fit these points, finally we can get the values of the center and the radius. Figure 2 is a fish-eye image, the size of it is 960×720 in pixels. The yellow circular curve is the extractive periphery of effective region of the image using this method. The coordinates of image center is (477,368) in pixels and the radius of it is 472 in pixels.

4 VIRTUAL ROAMING OF THE RECTIFICATION IMAGES

In fact, we concentrated not only on the area of the fish-eye image which near the center but also the edge of the image. In order to get any region of fish-eye perspective images, we just need to rotate the window planes at an angle. Moreover, in order to make the final Z-axis after rotation perpendicular to the center of the window plane, rotating plane or the coordinate axis can achieve the same effect.

Figure 3 is a schematic diagram of the coordinate axis rotation about the Y-axis. When the window plane is neither parallel to the Y-axis nor parallel to the X-axis, we need to rotate coordinate axes about the Y-axis and X-axis at the angle of θ and α, respectively. The coordinates of a space point $P(x, y, z)$ before and after rotation have the following relationship [12,13].

$$z = z'' \cos\theta \cos\alpha - y'' \sin\alpha \cos\theta - x'' \sin\theta$$
$$y = y'' \cos\alpha + z'' \sin\alpha$$
$$x = x'' \cos\theta + z'' \cos\alpha \sin\theta - y'' \sin\alpha \sin\theta \tag{5}$$

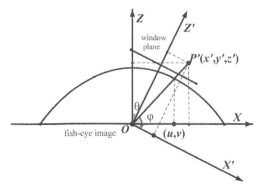

Figure 3. Rotation of coordinate axis.

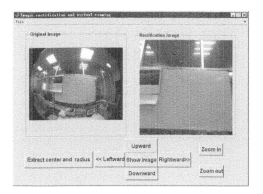

Figure 4. Graphical user interface.

The coordinate relationship between fish-eye image points and window rectification plane points can be obtained by combining (5) and (4). Then we can obtain the rectified fish-eye images.

5 REALIZATION OF RECTIFICATION AND VIRTUAL ROAMING

Anywhere of fish-eye image can be corrected by changing the values of the θ and α. In this paper, fish-eye lens gives about a 180° field of view. Rectification has been computed in a Pentium(R) Dual Core E6700 3.2 GHz with 2 Gbytes of RAM implemented in Matlab R2008a. Figure 4 is a graphical user interface which gives four arrow keys to control the directions of virtual roaming. Furthermore, we can also click zoom-in and zoom-out buttons to zoom correction images. Figures 5–8 show the effect of different region correction and virtual roaming.

Figure 5. Left side of image.

Figure 6. Right side of image.

Figure 7. Upside of image.

Figure 8.   Downside of image.

6   CONCLUSION

In this paper, we used equidistant model to rectify fish-eye images. Moreover, the virtual roaming of correction images is realized by rotating the window rectification planes. Experimental results indicate the feasibility of the rectification procedure. In order to obtain more accurate correction effect, parameters model should be introduced. High precision correction of fish-eye lens can be obtained by calibrating the fish-eye lens to obtain intrinsic and extrinsic parameters of it. Finally, this will be the direction of future research.

REFERENCES

[1] Huang Y., Su H. A Simple Transforming Model from Fisheye Image to Perspective Projection Image [J]. Journal of System Simulation, 2005, 17(1): 29–32.

[2] Wang J., Yang X., Zhang C. Environments of Full View Navigation Based on Picture Taken by Eye Fish Lens [J]. Journal of System Simulation, 2001, 13: 66–68.

[3] Ying X., Hu Z., Zha H. Fisheye lenses calibration using straight-line spherical perspective projection constraint [M] Computer Vision–ACCV2006. Springer Berlin Heidelberg, 2006: 61–70.

[4] Tsai R.Y. A versatile camera calibration technique for high-accuracy 3D machine vision metrology using off-the-shelf TV cameras and lenses [J]. Robotics and Automation, IEEE Journal of, 1987, 3(4): 323–344.

[5] Basu A., Licardie S. Alternative models for fish-eye lenses [J]. Pattern Recognition Letters, 1995, 16(4): 433–441.

[6] Ricolfe-Viala C., Sanchez-Salmeron A.J. Lens distortion models evaluation [J]. Applied optics, 2010, 49(30): 5914–5928.

[7] Li G., Fei Z., Yang S. Fish-eye image correction method and technique based on geometrical imaging model [J]. Journal of Mechanical & Electrical Engineering. 2013, 30(1): 0.

[8] Hughes C., Denny P., Jones E. et al. Accuracy of fish-eye lens models [J]. Applied optics, 2010, 49(17): 3338–3347.

[9] Abraham S., Förstner W. Fish-eye-stereo calibration and epipolar rectification [J]. ISPRS Journal of photogrammetry and remote sensing, 2005, 59(5): 278–288.

[10] Jiang F., Yuan Z. Contrast of Contour Extraction of Fish Eye Image [J], Computer Systems & Applications, 2011, 20(4): 214–218.

[11] Yang D., You L., Fisheye Image Contour Extraction Algorithm Based on Region Growing [J]. Computer Engineering, 2010, 36(8): 217–218.

[12] Zhang C., Algorithms research on virtual tour and moving object detect in fisheye image [D]. Northeastern University, 2011.

[13] Su X., Man J., Cheng Y. A Method Transforming Fisheye Image to Perspective Image and Simple Navigation [J]. Natural Science Journal of Xiangtan University, 2007, 29(2): 118–122.

Information Technology and Applications – Li (Ed.)
© 2015 Taylor & Francis Group, London, ISBN 978-1-138-02677-3

Emotional intelligence system for smart education in ubiquitous learning

Xuan Zhou
Section of Social Science, Sicuan Conservatory of Music, Chengdu, China

Shuang Huang
Overseas Training Center, Shanghai Foreign Language Studies, Shanghai, China

Ye-Sho Chen
E.J. Ourso College of Business, Louisiana State University, LA, USA

Weihui Dai
School of Management, Fudan University, Shanghai, China

ABSTRACT: Ubiquitous learning has become a new way of education in today's society. However, increasing the learners' initiative and their community cohesion is still the problem to be deeply thought. Emotional intelligence can help to find and respond to the learners' interest and emotions online, and therefore brings the new concept of smart education. This paper firstly analyzed the expression and detection of the learners' emotions in ubiquitous environment, and then discussed the key technologies on intelligent monitoring and automatic recognition of the learners' emotions by the voice analysis and behavior data mining, and finally designed an emotional intelligence system for the smart education in ubiquitous learning.

Keywords: emotional intelligence; smart education; ubiquitous learning

1 INTRODUCTION

With the development of the Internet, the Internet of things, the mobile communications and the increasing popularity of cloud computing, ubiquitous learning (U-Learning) has become a new way of education and the excellent complement to classroom teaching in today's society [13]. In the circumstances, the learners can access online educational resources anytime, anywhere and through any multimedia terminal. Exposed to all kinds of scenarios and cases enabled by the virtual reality technology, they can communicate and interact in an open learning community [15]. For example, in language education, it not only allows learners substantial freedom to choose their learning styles, environment, content and time but also "'immerses' them in the communicative context of a virtual target language culture and a suitable learning atmosphere." [9].

However, many researchers have also found the existing problems in ubiquitous learning. The major problem is how to increase learners'

initiative and their community cohesion, and help them search for and locate suitable learning resources among the massive amount available on the Internet [9, 13]. A possible solution is to use intelligent technology to analyze learners' behavioral data and explore their interest and needs so as to provide accurate, efficient resource organization model, retrieval technology and demand-based teaching strategies. Many empirical researches show that learners' emotional experiences play an important role in stimulating interest in learning and improving the efficiency as well as the effectiveness of teaching [14].

In 1990, Salovey and Mayer introduced the concept of "Emotional Intelligence" [10]. In 1997, Picard R.W. from Massachusetts Institute of Technology held that computer can capture, process and reproduce human emotions in his book "*Affective Computing*" [7]. Therefore in 2012, Dai defined the concept of "smart" as that machine can perceive and respond to human emotional needs and provide the full humanized services combining both rational and emotional intelligence [3].

This concept has been applied in wide areas, such as smart city, smart healthcare, smart service, etc. Emotional intelligence can help to detect the learner's emotional reactions online, and therefore stimulate their interest and the participation willingness by adjusting the teaching and improving the enjoyable experience, which brings the new concept of smart education.

2 EXPRESSION AND DETECTION OF EMOTIONS

2.1 Expression of emotions

Researches in cognitive neuroscience show that human emotion arises from the external signal stimulus, and is the result of a series of neural activities dominated by the brain mechanism [1, 2]. The signals of external scenario stimuli are transmitted to the brain via the sensory paths of human's body, where the emotion is produced and lead to the activated responses in the corresponding brain regions.

The brain reactions will afterwards trigger a series of physiological effects through the control paths of human's body adjusted by the neural regulation mechanism, and result in the variation of related peripheral physiological signals (e.g. EEG, ECG, EDR, respiration, skin temperature, etc.) as well as external performances (e.g. speeches, facial expressions, gestures, movements, etc.), and may cause subsequent behaviors. So the emotion expression may be reflected on the variations of the four aspects: brain reactions, peripheral physiological signals, external performances, and subsequent behaviors. In the meanwhile, those variations will be fed back to the brain and form the specific emotional experience which originates from the external scenario stimuli [12]. We concluded the generating mechanism and the expression of emotions as shown in Figure 1.

Under the ubiquitous learning environment, the learners log into the Internet via PC, smart phone, Pad or other multimedia terminal equipment, and participate in learning and teaching activities in an online virtual classroom. The learners' emotional information may be obtained mainly through the audio and visual signals as well as the behavior data produced by their online activities. We considered the audio signals and behavior data which are more easily and precisely processed by the computer.

2.2 Detection of emotions

Many researchers have found that human's voice contains rich emotion information, which has become the important "information textures" to identify different emotions [5, 11].

Vocal emotions are mainly reflected in such parameters as speech speed, voice intensity, pitch frequency, LPCC (Linear Prediction Cepstrum Coefficient) and MFCC (Mel Frequency Cepstrum Coefficient) [8, 16], among which, MFCC is based on the known variation of the human ear's bandwidths. It has the frequency characteristics linearly below 1000 Hz and logarithmically above 1000 Hz, which match well with the auditory characteristics of human speech signals. Therefore, we adopted MFCC as the feature parameters for the detection and recognition of the learners' emotions. The calculation process of MFCC coefficients is shown as in Figure 2.

The behavior data are mainly acquired from the server's log files or by some online tracking tools. The switching frequency and retention time of web pages, the locations and movements of the mouse, and the keyboard operations are all the important parameters as the features to reflect the learners' emotions. Through the behavior data mining of the learners, we can detect their emotional state and the interested points.

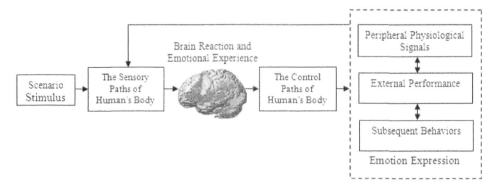

Figure 1.   Generating mechanism and the expression of emotions.

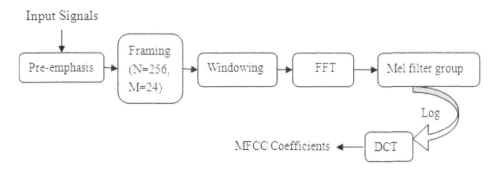

Figure 2. Calculation of MFCC for the detection and recognition of emotions.

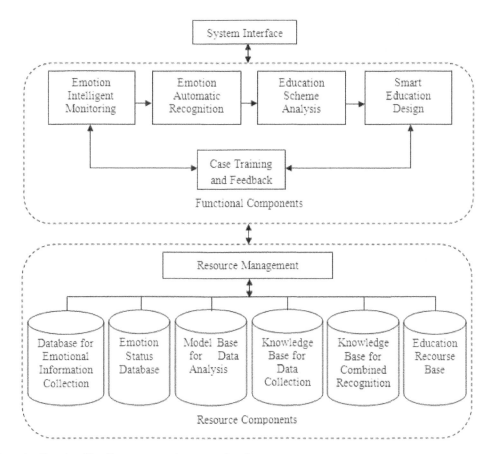

Figure 3. Emotional intelligence system for smart education.

3 INTELLIGENT MONITORING AND AUTOMATIC RECOGNITION

3.1 *Intelligent monitoring*

Intelligent monitoring of learner's emotions can be achieved through the webpage analysis and the data acquisition of mouse and keyboard.

Webpage analysis is realized by the web spider program based on Java technology to extract its layout structure, information type and the operation records for bookmarking. Every change of the elements on the webpage, such as the appearing of voice, moving, clicking and scrolling of the mouse or the operation of the keyboard, will all trigger

169

the corresponding JavaScript function. In order to meet the requirements of real-time data collection, we adopted the PHP language to program the above JavaScript function.

3.2 *Automatic recognition*

We classified the emotions of the learners into five categories: happiness, disgust, sadness, anxiety and anger. This classification was presented by Professor Ortony, et al in 1988 [6]. It is very suitable for the description of the learners' emotional status as well as their state of interest.

The recognition of emotions was fulfilled by the algorithm of Support Vector Machine (SVM) based on the sample training and machine learning in typical scenarios. The average recognition rate of the learners' emotions can come up to about 80% either in voice signals or behavior data, and reach more than 90% if both of them are available. In our solution the multi-agent technology was utilized to realize the monitoring and recognition of emotions.

4 EMOTIONAL INTELLIGENCE SYSTEM

Figure 3 shows the design of emotional intelligence system for smart education in ubiquitous learning.

It includes three parts: system interface, functional components and resource components. Among them, system interface offers the access and control for its users. Functional components are composed of five modules: Emotion Intelligent Monitoring, Emotion Automatic Recognition, Education Scheme Analysis, Smart Education Design, and Case Training and Feedback. Its resource components include two databases, one model base, two knowledge bases, and one hypermedia education resource base, which provide the resources for the operation of each functional component in this system.

Based on the multi agent technology, this system can realize the intelligent monitoring and automatic recognition of the learners' emotions. By analyzing the learners' emotion status and its varying characteristics in the learning process, we can obtain the statistical distribution in different periods and scenarios, and find the learners' ROI (Region of Interest), and therefore we can design the education scheme and teaching plans which are on-demand, appealing and enjoyable to the learners [4].

5 CONCLUSION AND DISCUSSION

In ubiquitous learning, the learners' emotional experiences have significant impacts on stimulating their interest and improving the teaching efficiency. Therefore, conducting the smart education in ubiquitous learning has been the issue worthy of our think. That means we should perceive and respond to the learner's emotional needs online and provide the full humanized services with not only the rational intelligence but also the emotional intelligence.

Emotion changes will not only result in the brain responses, but also trigger a series of physiological effects which may be reflected in the peripheral physiological signals, external performances, and subsequent behaviors. In ubiquitous learning environment, the voice signals and behavior data are more easily and precisely processed by computers, and based on which the intelligent monitoring and automatic recognition of the learners' emotions can be realized by the multi-agent technology.

Through the voice analysis and behavior data mining, we can reach the fairly good recognition rate of the learners' emotions, and design the emotional intelligence system for facilitating the smart education in ubiquitous learning. Research work of this paper provided valuable references for further research in this field.

ACKNOWLEDGMENTS

This research was supported by China Scholarship Council (No. 201308310364), the 2nd Regular Meeting Project of Science and Technology Cooperation between China and Serbia (No. 2-9/2013), Shanghai Philosophy and Social Sciences Plan, China (No. 2014BGL022) and Project of Sicuan Conservatory of Music (CY2014173). Miss Xuan Zhou and Dr. Shuang Huang are the joint first authors of this paper. Professor Weihui Dai is the corresponding author.

REFERENCES

[1] Damasio, A.R., *Descartes Error: Emotion, Reason and the Human Brain*. New York, USA: Gosset/Putnam Press, 1994.
[2] Davidson, R.J., Jackson, D.C. & Kalin, N.H., Emotion, plasticity, context, and regulations: perspectives from affective neuroscience, *Psychological Bulletin* 2000(126): 890–909, 2000.
[3] Dai, W.H. 2012, Context awareness and emotional intelligence: the gateway to smart city. *Urban Management* 2012(4): 29–32, 2012.
[4] Dai, W.H., Huang, S., Zhou, X., Yu, X.E., Ivanovi', M. & Xu, D.R., Emotional intelligence for ubiquitous learning: towards the smart foreign language education in the digital age, in *Proceedings of 12th International Conference on IT Applications and Management (ITAM-12)*: 87–93, Nairobi, Kenya, 8–9 July 2014.

[5] Jin, X.C., *Emotion Recognition Research Based on Speech Signals*. Hefei, China: University of Science and Technology of China, 2007.

[6] Ortony, A., Clore, G. & Collins, A., *The Cognitive Structure of Emotion*. Cambridge, England: Cambridge University Press, 1988.

[7] Picard, R.W., *Affective Computing*. Cambridge, USA: MIT Press, 1977.

[8] R. Doddipatla & Umesh, S., VTLN using analytically determined linear-transformation on conventional MFCC. *IEEE Transactions on Audio, Speech, and Language Processing* 20(5): 1573–1584, 2012.

[9] Sa, Z.Q. & Liu, D.D., A research on college English teaching reform based on ubiquitous learning. *Overseas English* 2008(8): 32–33, 55, 2008.

[10] Salovery, P. & Mayer, J.D., Emotional intelligence. *Imagination, Cognition, and Personality* 9(3): 185–211, 1990.

[11] Ververidis, D. & Kotrropoulos, C., Emotional speech recognition: resource, features, and methods, *Speech Communication*. 2006(48): 1162–1181, 2006.

[12] Wang, Y.H., Hu, X.H., Dai, W.H., Zhou, J. & Kuo, T.Z., Vocal emotion of humanoid robots: A study from brain mechanism, *Scientific World Journal*. 2014(Article ID 216341): 1–7, 2014.

[13] Wei, X.F., Zhang, Y.H. & Wei, Z.H., Transition from E-learning to U-learning: an interview with Prof. Kinshuk. *Education Research* 2012(2): 4–8, 2012.

[14] Xing, M.Y., The influences of affective factors on college English learning and teaching: theoretical and empirical research. *Foreign Languages and Their Teaching* 2003(3): 23–26, 2003.

[15] Yang, X.T., Ubiquitous learning: theory, pattern, and resources. *Distance Education in China* 2011(6): 69–73, 2011.

[16] Yuan, Y.J., Zhao, P.H. & Zhou, Q., Research of speaker recognition based on combination of LPCC and MFCC, in *Proceedings of IEEE International Conference on Intelligent Computing and Intelligent Systems (ICIS)*: 765–767, 2010.

Information Technology and Applications – Li (Ed.)
© 2015 Taylor & Francis Group, London, ISBN 978-1-138-02677-3

Emotion expression and affective computing on international cyber languages

Xuan Zhou
Section of Social Science, Sicuan Conservatory of Music, Chengdu, China

Shuang Huang
Overseas Training Center, Shanghai Foreign Language Studies, Shanghai, China

Xueer Yu
Department of Psychology, Washington University, St. Louis, MO, USA

Zhenyi Yang & Weihui Dai
School of Management, Fudan University, Shanghai, China

ABSTRACT: As the rapid development and wide application of the Internet, cyber space has become an important place where people share information and exchange their opinions, and therefore it gives rise to the booming of cyber languages. Characterized by the customary symbol system and vivid expression pattern, cyber languages act as not only the tools for convenient communication but also the carriers of abundant emotions. Based on the three languages of Chinese, English and Spanish which are used by the largest population in the world, this paper analyzed the expressive characteristics of emotions in international cyber languages, discussed the affective computing by the technology of knowledge base, and finally presented an effective computing method which has been successfully applied to the analysis of international public opinions about emergency events, customer comments and internet marketing.

Keywords: language emotions; cyber language; affective computing; knowledge base

1 INTRODUCTION

In today's society, cyber space has become an important place for people to share information and exchange their opinions. Due to its virtuality, autonomy, openness, inclusiveness as well as the high expressiveness by various technologies of new media, the language creativity of people has been inspired to the extreme and therefore it gives rise to the booming of cyber languages.

As to the cyber language (network language, Internet language or web language), there has not been a consistent definition so far. Professor Genyuan Yu at Communication University of China pointed out that cyber language is a "unique natural language" commonly used in cyber space [14]. According to the Ferdinand de Saussure's semiotic theory, Chinese scholars classified cyber languages into the two categories of readable symbols and non-readable symbols, and studied their symbol system, ideographic features and the formation rules [1, 5, 8].

With the rapid development of modern communication networks and new media technology, the concept of cyber space has been extended to ubiquitous environment. In the meantime, the expressions of cyber languages are full of variety by the hybrid of texts, images, audio or video signs, as well as their shapes, colors and brightness, so we define cyber language as "a symbol system that people have agreed on and widely used in communication under the ubiquitous environment".

Cyber language is showing the fast development trend in all the world major languages such as Chinese, English, Japanese, German, French, and Spanish. In the context of globalization, cyber language has brought new vitality to international communication with concise and vivid symbols, and the strong capacity for expressing emotions. Affective computing originally presented by Picard in 1997 [10] indicates that the emotional information are perceived, processed and computed by the machine [10], which has been applied to cyber language in recent years for the automatic analysis of public opinions and customer online comments [2, 4]. It has also become the focus of attention in the fields of Internet Translating Common Technology (ITCT) and intelligent recognition of natural languages.

As a symbol system in ubiquitous environment, cyber languages are very rich in the expression of emotions by either the simple assembly of readable and non-readable symbols, or the complexes of texts, images, audio or video signs, and their hybrid. Although great progress has been made in the research of natural languages in the past decades, there is still difficulty in the affective computing of cyber languages. By examining three languages, Chinese, English and Spanish, which are used by the largest number of people in the world, this paper will explore the effective method of affective computing of international cyber languages.

2 EXPRESSION OF EMOTIONS IN INTERNATIONAL CYBER LANGUAGES

2.1 *Description of emotions in cyber languages*

The description of the emotions in international cyber languages is usually from the typically cognitive perspective of the information receivers on the basis of statistical significance. Human's emotional experiences may be classified and described in different dimensional coordinates as is shown in Figure 1.

In the above graph, the one-dimensional coordinate describes only positive and negative emotions with their strengths. The two dimensional coordinates are based on the Hidenori H. and Fukuda T.'s unit circle of Emotional Space [6]. The two opposite coordinates, Peace vs. Excitation and Happiness vs. Sadness, are generally used and all emotions are represented in a circle with a radius of one vector.

There are a variety of 3-D coordinates used. Wilhelm M. Wundt [12], Schlosberg H. H. [11] and Izard C. E. [7] suggested different 3-D models. Of all the models, the one that has received most attention is the PAD model with the 3 dimensions of pleasure (positive and negative characteristics of emotional state), arousal (individual neural physiological activation levels) and dominance (individual control

over situations and others). Through the statistical experiments and studies on large populations, Fu, Liu & Tao designed the measurement scale system which could describe and position human's emotions in the PAD dimensions very well [9].

2.2 *Expression of emotions in international cyber languages*

Cyber languages boast of a wide range of emotional expression patterns such as text, images, audio and video signs and the hybrid of them. Any changes in the components, shapes, colors, layouts or presentation sequences may deliver different emotional messages. International cyber languages are characterized by widely-ranged vocabulary that is profuse in sentiments and updated rapidly. Some of the latest examples are shown in Table 1 [13].

The utilization of emotional vocabulary is the basic way of the expression of emotions in international cyber languages. The researches of cyber languages worldwide have built the corpuses of cyber languages by collecting and sorting out frequently used vocabulary and symbols.

The corpuses include a large quantity of emotional terms which are vital to the analysis of emotions. For example, China's HowNet has collected 52,000 Chinese terms and 57,000 English words. Among all the published ones, there are 219 words describing the intensity of emotions, 3,116 negative, 1,254 derogatory, 3,730 positive, 836 approbatory and 38 ones for making a proposition [13]. HowNet's Semantics Dictionary has also included a large collection of lexical semantic entries, each of which is composed of semantics and its description of a term. It offers guidance on how to analyze the above-mentioned emotional expressions in the specific context.

The emotional messages of cyber languages are decided not only by the emotional words used in the sentence, but also by the emotional expression patterns in the whole sentence. Therefore, the same emotional word can be completely opposite in meaning when expressed in a different pattern.

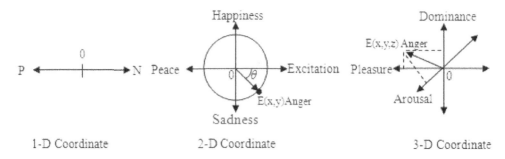

Figure 1.   Description of emotions in different dimensional coordinates.

Table 1. Latest vocabulary in international cyber languages.

Latest vocabulary	Definition
Cliché (French)	An idea or expression that has been used too often and is often considered a sign of bad writing or old-fashioned thinking
Logo (Portuguese)	Suddenly, all of a sudden
Fraquinho (Portuguese)	A joking way to say a dumb
Afeminado (Spanish)	A feminine man
Smuggled goods (English)	Goods taken from a place secretly and often illegally
Cross-dresser (English)	Usually a man who likes to wear clothing of the opposite sex

For example, "When I just heard the news, I was quite upset. But after having lurked for a while, I found hikers were right about that, so just want to show up today to share my joy." There are two emotional words in this sentence, the negative "upset" and the positive "joy" together with a concession word "but". Suppose P stands for the positive emotion word, N for the negative one and T for the concession, the emotion expression pattern of the above sentence can be generalized as the follows:

$$N + T + P \Rightarrow P$$

In cyber language, conjunctions are critical to the understanding of the emotional messages delivered in sentences and thereafter are important objects to be considered in the analysis of emotional structure and expression patters. Of course, a complete emotional expression pattern also involves the adverb of degree and negative words, especially, as well as punctuation marks and emoticons such as "!", "?", ":)", etc, which usually demonstrate very distinct sentiment orientations.

Without strict rules regulating cyber language expression which is ever changing, we will have to use computer to capture automatically new entries and modify them with subjective cognition in order to build open corpuses of emotional expression patterns in sentences and analyze the messages delivered by them.

3 AFFECTIVE COMPUTING ON INTERNATIONAL CYBER LANGUAGES

3.1 *Knowledge base for affective computing*

The recognition of the emotion will not only involve the mechanism of cognitive psychology, but also correlate with the cultural background, language context and social environment. Therefore, the most effective approach to affective computing is to establish an open and updated timely knowledge base as is shown in Figure 2.

In Figure 2, we build up the semantic dictionary based on the ontology of international cyber languages, which includes both emotional vocabulary and the non-readable symbols. The expression patterns of emotions in phrases and sentences are represented by the knowledge such as the templates and rules which descript the commonly used structures along with the conjunctions, adverbs and punctuation marks. Based on the knowledge base, affective computing can be carried out in a paragraph.

3.2 *Method of affective computing*

Based on the analysis of the characteristics in emotion expressions, we presented the method of affective computing on international cyber languages as shown in Figure 3. It includes three steps in the computing of a cyber language paragraph.

The first step is Sentence Segmentation and Keywords Retrieval. A paragraph of cyber language will be segmented into a series of sentences or phrases for further processing by the structure analysis.

Chinese, English and Spanish as well as the other international cyber languages have their different patterns in sentences or phrases, which can be mostly structured by the retrieved keywords such as conjunctions, adverbs and punctuation marks and stored in the knowledge base.

The second step is Analysis of Expression Pattern. This will be applied to each single segmented sentence or phrase based on the semantics dictionary which included the emotional vocabulary and non-readable symbols and the knowledge base which stored their commonly used combination with the conjunctions, adverbs and punctuation marks as templates and rules.

The third step is the Synthetic Affective Computing. Based on the semantics dictionary, the basic emotional polarity as well as its degree of the emotional vocabulary and non-readable symbols can be evaluated and calculated in a typical statistical significance, and will hereafter be adjusted by the analysis of emotion expression pattern according to the templates and rules from the knowledge base.

What remains a problem in affective computing is the variation and variety in emotion expression. The emotional vocabulary and non-readable symbols may be dynamically changed. In order to find out the value of emotional polarity and its degree of a new emotional word or non-readable symbol in cyber languages, we can use the PMI (Pointwise Mutual Information) to calculate the similarity

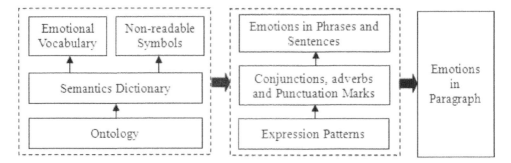

Figure 2. Knowledge for affective computing on international cyber languages.

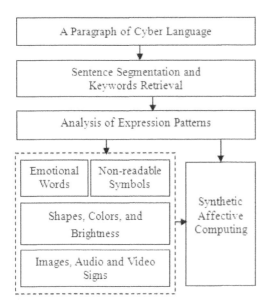

Figure 3. Affective computing on international cyber languages.

of which with the benchmarks in the semantic dictionary.

Any additional emotional information such as the meaningful characteristics in shapes, colors and brightness as well as the images, audio and video signs, will be specially processed by a further computing with the help of the knowledge base and the general cognitive psychological characteristics.

Finally, the computing results of all sentences or phrases may be adjusted and synthetized as the changing series of emotions by their appearing orders in a paragraph. This affective computing method has been successfully applied to the analysis of international public opinions in various areas as emergency events, customer comments and internet marketing [3, 13].

4 CONCLUSION AND DISCUSSION

Cyber languages not only provide the convenient symbol system for people's communication in ubiquitous environment, but also carry an abundance of emotion information with important significance on public opinion analysis.

Due to the open, virtual and dynamic language environment, the variation and variety in emotion expression of international cyber languages require the affective computing to be built based on knowledge base and synthetic analysis. After analyzed the expressive characteristics of emotions in the cyber languages of Chinese, English and Spanish, this paper presented an effective method for the affective computing on international cyber languages, and offered the reference for further research on this issue.

ACKNOWLEDGMENTS

This research was supported by Natural Science Foundation of China (No. 91324010), the 2nd Regular Meeting Project of Science and Technology Cooperation between China and Serbia (No. 2–9/2013), Shanghai Philosophy and Social Sciences Plan, China (No. 2014BGL022) and Project of Sicuan Conservatory of Music (CY2014173). Miss Xuan Zhou and Dr. Shuang Huang are the joint first authors of this paper. Professor Weihui Dai is the corresponding author.

REFERENCES

[1] Chen, W.T., *The Study on Symbols Formation and Development of Cyber Language*. Lanzhou, China: Northwest Normal University, 2010.
[2] Dai, W.H., Context awareness and emotional intelligence: the gateway to smart city. *Urban Management* 2012(4): 29–32, 2012.

[3] Dai, W.H., Public cognition morale mechanism of unexpected incidents in cities and the responding strategies. *Urban Management* 2014(1): 34–37, 2014.

[4] Dai, W.H., Wan, X.Q. & Liu, X.Y., Emergency event: Internet spread, psychological impacts and emergency management, *Journal of Computers* 6(8): 1748–1755, 2011.

[5] He, H.F., "Web Language" as viewed from perspective of symbolism. *Journal of Jianghan University (Humanities Sciences)* 22(2): 74–78, 2003.

[6] Hidenori, I. & Fukuda, T., Individuality of agent with emotional algorithm, in *Proceedings of 2001 IEEE/RSJ International Conference on Intelligent Robots and System* (2):1195–1200, 2001.

[7] Izard, C.E., *The Psychology of Emotions.* New York: Plenum press, 1991.

[8] Lu, Y.F., *A Study on the Representing Principles of Netspeak Symbols.* Harbin, China: Harbin Normal University, 2013.

[9] Liu, Y., Fu, X.L. & Tao, L.M., Emotion measurement based on PAD 3-D space, *Communications of The CCF* 6(5): 9–13, 2010.

[10] Picard, R.W., *Affective Computing.* Cambridge, USA: MIT Press, 1997.

[11] Schlosberg, H.H., Three dimensions of emotion. *Psychological Review* 61(2): 81–88, 1954.

[12] Wundt, W.M., *Outlines of Psychology.* Oxford, England: Engelmann, 1897.

[13] Yang, Z.Y., *The Method Of Emotional Calculation And Research Of Realizing Technology Of International Network Language,* Shanghai, China: Fudan University, 2014.

[14] Yu, Y.G., *The Generality of Network Language,* Beijing, China: China Economic Publishing House, 2001.

Information Technology and Applications – Li (Ed.)
© 2015 Taylor & Francis Group, London, ISBN 978-1-138-02677-3

Matching multi-sensor images based on Haar Binary Coding feature

Zhuang Li & Xiang Zhou
National University of Defense Technology, Changsha, China

Guoyong Tu & Weijian Li
Jiuquan Satellite Launch Center, Jiuquan, China

ABSTRACT: Multi-sensor images matching is a challenging problem in image process research field. As SAR (synthetic aperture radar) and optical images have significant difference, most existing methods can't achieve satisfied matching result. To give response to this issue, we present a new image descriptor, Haar Binary Coding feature (HBC). When calculating the HBC feature, we first project the input image to different Haar template and then sort the results and expressed them in binary form. To increase efficiency, we take integrated image to compute Haar template projection. Benefit from binary representation and comparison, the computation complexity is much lower than traditional method, while the common characters are better revealed. Synthetic Aperture Radar (SAR) and optical images are used to test the new algorithm. Experimental results show that our algorithm has a higher success rate than traditional methods and shorter computing time. In our method, the template image can be processed offline, then, the online processing time is further reduced.

Keywords: image matching; multi-sensor image; Haar template; binary coding; HBC feature

1 INTRODUCTION

Matching multi-sensor images is the key step in many image processing applications such as vision navigation for aircraft, remote sensing image fusion and medical image analysis.

The purpose of matching multi-sensor images is to geometrically align images of the same scene taken by different sensors or modalities. In multi-sensor imagery, the relationship between the intensity values of the corresponding pixels is complex and unknown. Contrast reversal may occur in some regions, and the contrasts of the images may differ from one another. The multiple-intensity values in one image may map to a single intensity value in another. Further, features present in one image may not appear in another, and vice versa. The matching of multi-sensor images thus constitutes a challenging problem.

Existing multi-sensor image matching techniques broadly fall into two categories: the feature-based methods, the area-based methods. The area-based methods are widely used to match multi-modal medical images, which usually have no clear regular features. The most common adopted area-based methods include Sum of Squared Difference (SSD), correlation and Normalized Cross Correlation (NCC), but they are seldom directly applied on multi-modal images in most time. A normalized

gradient fields based similarity measure is proposed in [1]. In [2], a registration technique based on the correlation of wavelet features is presented. A novel similarity measure strategy is invented in [3] by aligning the locations of gradient maxima of images. The alignment is achieved by iteratively maximizing the magnitudes of intensity gradients of a set of pixels in one image. Mutual Information (MI) is one of the most powerful tool for multi-modal medical image registration, and is well research in recent years [4,5,6]. Though some rapid MI algorithms are provided [7], the high computation complexity of MI is a main obstacle for its use in remote sensing image registration. An information theoretic measure called Cross-Cumulative Residual Entropy (CCRE) is proposed in [8]. Phase congruency is another promising method for medical image registration [9].

Our method is mainly inspired by the algorithm used for Content Based Image Retrieval (CBIR) and has significantly difference with all the above methods. In this method, compressed expression feature is achieved by Haar templates projection, feature pooling, binarization and coding. Images similarity is calculated by comparing corresponding features using bit manipulation while the maximum value indicates matching result. Benefit from binary representation and comparison, the computation complexity is much lower than traditional method, while the common characters are better revealed.

The rest of this paper is organized in the following manner. In section 2, we propose the Haar binary coding feature. In section 3, the new feature is used to matching multi-sensor images. In section 4, our method is validated in the application of multi-sensor image registration and we show its capability and compare performance against several popular methods. We conclude in section 5.

2 HAAR BINARY CODING FEATURE

The Haar binary coding feature computing process contains four step: projecting, pooling, binarizing and coding, as shown in Figure 1. In the projecting step, image is convoluted to a group of Harr images. In the pooling step, the convolution results are pooled by fixed scale (the scale parameter is 4 in Fig. 1). The pooling results are sorted and binarized in each position in the binarization step. Finally, in the coding step, the HBC feature is outputted.

2.1 Haar-like rectangle template

Wavelet is a strong tool that are widely used in image compression, feature extraction, object detection. When facing a large number of images windows, taking wavelet transform may take too much time. In some areas such as face detection, it only needs approximate wavelet feature representation, where precise wavelet transform result has no obvious difference. Papageorgiou [10] proposes a group of Haar-like rectangle templates in face recognition application. Some templates are shown in Figure 2.

Each template consists of some rectangle regions with white are gray color. The value in white regions is set 1 and in gray regions is set −1. The projection feature is defined by inner product of template and input image window. The projection feature value expresses bright variation information in specified orientation, usually corresponded to edges, lines, etc.

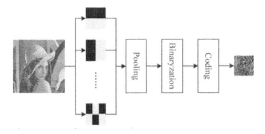

Figure 1. HBC feature computing process.

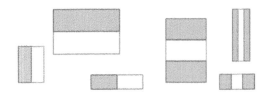

Figure 2. Papageorgiou's Haar-like rectangle templates.

The template's size and position are variable. Thus, each template is a function of type, size and position. By changing its parameter, many different templates can be generated. For example, there are more than 160000 different templates in a 24*24 image window. In most applications, the total number of templates is set by experience or by machine learning.

In HBC feature, we use only 8 different template, which are shown in Figure 3. There are 2 edge template (#1, #2), 1 diagonal template (#3), 2 stripe template (#4, #5), 1 speckle template (#6) and 2 corner template (#7, #8).

In object detection application, there are only a small number of candidates window, and the integral image method is widely used to save computing time [11].

In HBC feature, each template is convoluted with input image to generate a correspond feature value map. Convolution can be computed fast in frequency domain through FFT. We compared frequency domain convolution with integral image proposed by Viola. In HBC feature, frequency domain convolution is about 5 times faster than integral image.

2.2 Binary coding

Each Haar template is convoluted with input image. To adapt to gray level reversal, we take the absolute value of convolution results.

When the image size is $w \times h$, for arbitrary position (x, y) in image, the convolution results of all the templates can be expressed by a n-dimensional vector $v(x, y)$, where n is the number of templates. Thus, the entire convolution result includes $w \times h \times n$ integer numbers. It is obviously that the data size of convolution result is much larger than original image. For efficiency, we will compress these data.

Pooling is a common compress method. We implement pooling operation on convolution results. The input image is divided into $k \times k$ pools. For pool $p_{m,n}$, where m, n indicates the horizontal and vertical mark separately, the convolution vectors are summed.

$$s(m,n) = \sum_{y=nk+1}^{nk+k} \sum_{x=mk+1}^{mk+k} v(x,y) \qquad (1)$$

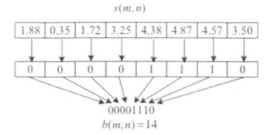

Figure 3. Haar templates used in HBC feature.

$s(m,n)$

1.88	0.35	1.72	3.25	4.38	4.87	4.57	3.50
0	0	0	0	1	1	1	0

00001110

$b(m,n) = 14$

Figure 4. Binary coding process.

The vector $s(m,n)$ is the pooling result of pool $p_{m,n}$. The pooling results are arranged in 2-dimentional order as input image. After pooling, the data size is reduced from $w \times h \times n$ to $w \times h \times n/(k \times k)$.

The pooling results are binarized. For each pool $s(m,n)$, it is expressed in vector form $[a_1(m,n),a_2(m,n),...,a_n(m,n)]$. The d largest value (in this paper, we set $d = 3$) are set 1, others are set 0. The new vector is converted into a binary number, where each dimension of the vector corresponds to one bit. When $s(m,n)$ is a 8-dimension vector, the new binary number $b(m,n)$ has 8 bits. Therefore, the binary coding feature can be reduced to $w \times h \times n/(8 \times k \times k)$ bytes. In this paper, we use 8 Haar template. So, $n = 8$, the Haar binary coding feature size is $w \times h/(k \times k)$. The binary coding process is shown in Figure 4.

The Haar binary coding feature is a kind of image description of projection length on different Haar template. The pooling operation has a smooth effect, which will increase the feature stability. The binarization operation improves feature's adaptability to image gray level difference. Binary number expression further saves the feature's storage space, and will contribute to reduce computation complexity in feature matching. A pair of multi-sensor images and their HBC feature are given in Figure 5. The HBC feature is smaller than original image. In next sector, we will show that the HBC feature is suit for multi-sensor image match.

Figure 5. Multi-sensor images and corresponding HBC features.

3 MULTI-SENSOR IMAGE MATCH

3.1 Image matching model

The purpose of image matching is to search a group of best geometric transformation parameters between two input images. According to their size, the two input images are recorded as small image I_s and large I_l. Image matching problem can be expressed as an optimization problem. In the parameter space U_T, the goal is to find the T_m that maximize image similarity S.

$$T_m = \arg\max_{T \subset U_T} \left(S(T(I_s),I_l) \right) \tag{2}$$

In the above equation, $T(I_a)$ is the transformation result of small image I_s, $S(T(I_s),I_l)$ is the similarity measure of $T(I_s)$ and I_l. For multi-sensor images, the image similarity is difficult to obtain directly from image intensity. Instead of intensity similarity, we take HBC feature similarity as image similarity measure. Then, the maximum value of HBC feature similarity indicates matching result.

3.2 HBC feature similarity

To reduce computation complexity, the similarity measure is computed by bit operation. Given a pair of image of same size (the small image and a window of the large image), their HBC features are B_1 and B_2, the feature similarity is defined as:

$$S(B_1,B_2) = \sum_{m,n} f_{bit}(B_1(m,n) \& B_2(m,n)) \tag{3}$$

& means bit and operation. $f_{bit}(\cdot)$ is a bit accumulation function, whose output is the number of bits with value 1 in the input binary number. Take binary number 00011100 as an example, its bit accumulation result $f_{bit}(00011100)$ is 3. We use bit shift and bit and operations to realize $f_{bit}(\cdot)$ efficiently. In addition, according to system bus width, the feature similarity can be computed in 32 bits or 64 bits, which can save running time.

4 EXPERIMENT

To compare different matching method, we first construct a dataset of multi-sensor images. Our test dataset consists of 13 group of images. Each group includes a optical large image (400*400 pixels) and several SAR small images (200*200 pixels). The geometric transformation is mainly translation and slight affine deformation.

The matching success rate is define as equation 4, where $N_{Success}$ counts the number of success times (the distance between matching result and truth position is less than 3 pixel), N_{Total} counts the total matching times.

$$R_{Success} = \frac{N_{Success}}{N_{Total}} \qquad (4)$$

We run Mutual Information (MI), Phase Congruency (PC) and our method (HBC) on test dataset. The MI source code is provided by Hosang Jin. The PC code is shared by Peter Kovesi. All three test methods are running in Matlab without multi-core parallel process.

The success rate of HBC is much higher than MI and PC. The running time of HBC is 0.36 s, in which, 0.29 s is used to compute HBC feature of large image. In some application, such as scene matching guidance, the HBC feature of large image can be calculated offline (before the mission start). Then, the online running time is reduced to 0.07 s. In MI matching, the 2-dimension histogram and combination entropy take up most running time. These operations need both images as input, thus can't running offline. In PC matching, the PC feature of large image can also be computed offline. Then, its online running time is 0.42 s, which is longer than HBC method.

Some matching results are given in Figure 6 and Figure 7. In each figure, (a) is large image, (b) is small image, (c), (d), (e) are similarity map of HBC, PC, MI respectively. The brightest positions indicate their matching results. In both test, only HBC finds the correct matching parameters.

Table 1. Comparison result.

Method	Success rate	Running time
MI	67.5%	40.89 s
PC	33.3%	2.11 s/0.42 s
HBC	93.3%	0.36 s/0.07 s

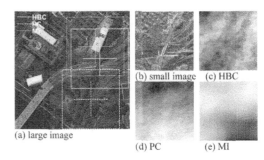

Figure 6.   Matching result of 3 methods in group 3.

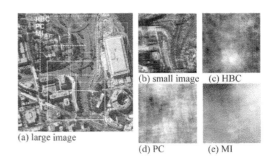

Figure 7.   Matching result of 3 methods in group 5.

5 CONCLUSION

The HBC feature and its corresponding multi-sensor image matching method are proposed in this paper. The HBC feature is easy to be computed. Both its running time and storage space are small, which makes it very suit to implement on computation platform such as embedded system. Experimental results show that, the HBC method has higher success rate and lower running time than traditional method. At present, the HBC method is mainly designed for translation transformation. In further research, we will expand its affine adaptability and make it able to match multi-sensor images with large affine transformation.

REFERENCES

[1] Haber E., Modersitzki J. Intensity gradient based registration and fusion of multi-modal images [M]// Medical Image Computing and Computer-Assisted Intervention–MICCAI 2006. Springer Berlin Heidelberg, 2006: 726–733.

[2] Le Moigne J., Campbell W.J., Cromp R.P. An automated parallel image registration technique based on the correlation of wavelet features [J]. Geoscience and Remote Sensing, IEEE Transactions on, 2002, 40(8): 1849–1864.

[3] Keller Y., Averbuch A. Multisensor image registration via implicit similarity. IEEE tansaction on pattern analysis and machine intelligence, 2006, 28(5): 794–801.

[4] Wells W.M. 3rd, Viola P., Atsumi H., Nakajima S., Kikinis R. Multi-modal volume registration by maximization of mutual information [J]. Medical Image Analysis. 1996, 1(1): 35–51.

[5] Maes F., Collignon A., Vandermeulen D., et al. Multimodality image registration by maximization of mutual information [J]. Medical Imaging, IEEE Transactions on, 1997, 16(2): 187–198.

[6] Pluim J., Maintz J., Viergever M. Mutual information based registration of medical images: a survey. IEEE Transactions on Medical Imaging. 2003, 22(8): 986–1004.

[7] Talbi H., Draa A., Batouche M. A novel quantum-inspired evolutionary algorithm for multi-sensor image registration [J]. International Arabic Journal on Information Technology, 2006, 3(1): 9–15.

[8] Pickering M.R., Xiao Y., Jia X. Registration of multi-sensor remote sensing imagery by gradient-based optimization of cross-cumulative residual entropy [C]//SPIE Defense and Security Symposium. International Society for Optics and Photonics, 2008: 69660U-69660U-10.

[9] Alexander W., Jeff Orchard. Robust Multimodal Registration Using Local Phase-Coherence Representations [J]. Journal of Signal Processing Systems. 2009, 54(1–3): 89–100.

[10] Papageorgiou C.P., Oren M., Poggio T. A general framework for object detection [C]//Computer vision, 1998. sixth international conference on. IEEE, 1998: 555–562.

[11] Viola P., Jones M. Rapid object detection using a boosted cascade of simple features [C]//Computer Vision and Pattern Recognition, 2001. CVPR 2001. Proceedings of the 2001 IEEE Computer Society Conference on. IEEE, 2001, 1: I-511-I-518 vol. 1.

Information Technology and Applications – Li (Ed.)
© 2015 Taylor & Francis Group, London, ISBN 978-1-138-02677-3

Applications of ICT for participatory sustainable rural development: A case study of Iran

Ahmad Khatoonabadi

Isfahan University of Technology, Iran

ABSTRACT: This paper aims to demonstrate the results and achievements of an action research project on the application of ICT in a participatory manner within four Iranian villages. The action research process involved: action planning, determination of stakeholders to participate in action program in an integrated manner, establishment and implementation of ICT centers, and evaluation of the behavioral changes among rural beneficiaries for further reflections. As a result four pilot ICT centers for rural communities were established through a participatory process. Consequently, the results of the survey study on the changes in the behavioral patterns of the local users in terms of socio-cultural, economical and technical was significant. One of the peripheral objectives of the project was to study the impacts of such ICT centers in order to be used in developing a comprehensive model for extending information technologies to all rural areas across Iran. The results from the studies conducted and the analyses thereupon indicate that the mere establishment of single ICT centers in scattered points in villages will not suffice to address the demand for ICT and that it is crucially necessary to develop local ICT networks. It was also concluded that sustainability of rural ICT centers greatly depends in their initial stages of support from local institutions and companies in neighboring urban centers. The logistic provisions for rural IT centers must be provided in an organized manner.

Keywords: ICT centers; rural; sustainable; human development; Iran

1   INTRODUCTION

Information and Communication Technologies (ICT) has shown the fastest growth among other industries today. The investment in IT has risen from US$ 2200 Billion in 1999 to US$ 3000 Billion in 2003, the greatest portion of this development taking place in industrialized countries. The number of internet users is impressively on the increase, rising from 20 million users in 1995 to 400 million in the year 2000 and it is estimated that the number will still increase to one billion users by 2005. Citizens in developed countries account for a major portion of internet hosts and users. This is while even less than 0.1% of the population in developing countries is connected regardless of the quality and bandwidth.

Most experts and theorists of development consider IT as the main or even the only solution to the realization of development for underdeveloped nations. More specifically, the key to development in the 21st century underdeveloped countries is unanimously regarded to be knowledge-based development facilitated by IT. Moreover, it is generally agreed that access to IT is one of the ways to realizing justice in the sense of providing equally opportunities to all.

The new age is generally associated with a new phenomenon called "the Digital Divide". Access to and skills in using information play important roles in enhancing the capacities of both individuals and societies alike in terms of education, employment, and economic or industrial activities. This access to information (or lack of it) has created two types of societies: one that possesses information and one that does not. This division is the one that has come to be called the "Digital Divide" and that entails the dividing line between the developed from the developing countries, the old from the new generations, the urban from the rural communities, and the educated from the uneducated individuals. Overcoming this divide is nowadays a major factor in the realization of development.

Rural areas are especially endangered by this digital divide. Such areas (particularly in developing countries) are characteristically deprived of most of the essential components of development (education, social amenities, communication facilities, and employment). Now with the advent of information technologies in urban areas, the threat is that this gap may widen more than ever. The lack of hardware, software, and communication facilities and the absence of skills in these areas make this deplorable situation even more probable to happen.

Over the past decade, international development agencies such as UNDP have focused their attention on rural internet centers and on training rural people on internet and the required skills such that the development of these centers in rural areas now accounts for a major part of the activities by these organizations. In their recommendations to countries, they have placed access to internet and connectivity on a par with such basic facilities such access to healthy drinking water and power. IT has also been recognized as one of the major components of poverty reduction activities by the UNDP and by the World Bank.

Establishment of internet connectivity centers in rural areas is the main approach to make sure of the development of ICT among rural communities. Despite the development of these technologies in urban areas in Iran, the rural areas are unfortunately still far behind any acceptable standards and no comprehensive plan yet exists for the extension to these facilities and services to rural communities.

The project titled "Application of Information Technologies for Sustainable Human Development (IT for SHD) is one implemented over the period from March 2003 to March 2005 by the Research Bureau of the Management & Planning Organization (MPO) with collaboration from UNDP within the framework of cooperation between the Government of the I.R. of Iran and UNDP Office in Tehran over the period. The basic scope of the project is ICT. Two types of results were expected to obtain from the project: establishment and operation of four pilots rural IT Centers and the holding of technical IT workshops. The main objective was geographical extension of information technologies across Iran through the establishment of pilot ICT centers and to draw upon international ICT expertise to enhance local skills.

The feasibility study of the project was completed in 2003 and the execution of the project took place in 2004–2007. All execution phases were conducted and monitored by a steering committee consisting of national IT experts and the experts of the UNDP Office in Tehran. In addition to the grant by UNDP used for the procurement of certain equipment and facilities, financial inputs from TAKFA (The National Agenda for ICT Development) were also secured to conduct studies and to organize workshops. Public funds are also secured to finance part of the logistics for rural IT centers under this project.

The project has proved to be a successful example of cooperation of international agencies with government organizations as well as with local communities. Through its mechanisms and structures, it has also provided a precedent for the cooperation of local people and organizations with government and international bodies such

that now and as a result of the implementation of this project, local organizations and communities utilize the facilities developed in the rural IT centers and are actively engaged not only in the management and planning of these centers but also in providing full support for their operation in cooperation with local companies.

The second part of this project involved the holding of technical workshops on IT. Drawing upon the international relations ensuing from the contributors of the project and investing on the experiences of UNDP consultants working in the field of IT, a number of workshops were organized and held to raise ICT knowledge among managers of public and private organizations in Iran. The topics of these workshops included Management Project in IT Projects, IT Strategies, and E-government. Each workshop was held in three different places and covered all three topics.

The present paper explains the methods developed/employed and the achievements from the project. It is a brief report on the studies and planning conducted as well as on the methods employed in the implementation of the plans. The results and impacts of the project will also be addressed. Finally, recommendations and practical plans for the continuation and/or replication of the project in establishing rural IT centers will be presented.

The objective followed by the present paper is to provide a summary of the achievements as, evidently, reporting on all aspects of the project goes well beyond its scope. The authors hope that this report can illustrate the cooperation and collaboration among managers and experts of Management & planning Organization and the UNDP Office in Tehran, on the one hand, and local people and institutions, on the other and that the achievements presented herein can be useful in replication and up-scaling of the project.

2 LITERATURE REVIEW

There is widespread attitude that while Information and Communication Technologies (ICTs) are a "new social and economic force in the world economy," their adoption and utilization in developing countries are constrained by, among other problems, inadequate infrastructure, limited human resource capacity, absence of national policies and low ICT literacy (Anderson, et al., 199 and Moyo, 1996). In the rural development processes, ICT centers create a context for social interaction and foster what Bawden (1996) considers experiential and action learning (1996), more than what could have been achieved earlier within rural radio forums in India and elsewhere. Forth et al., (2004) make a distinction between the adoption of ICTs and the intensity of their use in order to address key questions relating to the impact of skill constraints on firm-level performance. In particular, they investigate the extent to which reported ICT skill deficiencies at firm level: (i) restrict the adoption of ICTs; and (ii) limit the benefits which are actually gained from using ICTs once the required investments have been made.

Khalil-M and Khatoonabadi (2013) highlighted the issues of free accessibility and connectivity in the adoption process of ICT within the rural ICT centers of Iran. Mildorf and Charvat jr. (2012) described the status of current development and research in ICT for Agriculture, Food and Environment and analyzed the technological progress of the 10 years period since 2002 in terms of the results achieved in the research area, implementation in practice and success of the implementation as well as the current challenges having the Influence on ICT for agriculture and environment. The others have also emphasized on the crucial role of ICT in sustainable rural development, although recognizing the slow paces of ICT development in the rural sector (Anderson, 1999; Chapman et al., 2002; Malhotra et al., 2004; Puri, et al., 2007; Meera, 2010; Blake et al., 2010; Nayak et al., 2010; Chapman, and Slaymaker 2002; Kumar et al., 2012). In 2009, ICT firms in the US contributed about $1 trillion to U.S. GDP, or 7.1 percent of GDP, generating $2.5 billion in annual benefits (Shapiro and Mathur, 2011). ICT's contribution to Nigeria's GDP will shoot up to about $50 billion by 2019" (OLALEYE, 2014). The contribution of the Information Communication Technology (ICT) sector to the Kenyan economy increased to 12.1 per cent in 2013 from 8.9 per cent in 2006 (Mwenesi, 2014).

3 METHODOLOGY

A survey was carried out of the social and ethnic foundations of behavior and the variables affecting public behavior were identified. Also the norms in the community under study were identified in an attempt to define the following variables in our survey of the statistical population. Indexes were established for each variable in order to measure behavioral changes. The indexes and the variables are shown in Table 1.

Users of rural IT centers were classified as follows:

- According to jib: 1. School students; 2. University students; 3. Farmers; 4. Livestock breeders-Orchard growers—Shopkeepers-Housewives
- According to sex: 1. Men; 2. Women;
- According to village: Maranak; Mahabad; Tiss, Vorogh;
- According to stage of center development when users called to use services: 1. At early opening days of the center; 2. Months after opening.

4 DATA COLLECTION

Data collection was accomplished using three questionnaires. The IT center managers played a major role in administering the questionnaires. The questionnaires used were as follows:

Questionnaire No. 1: General characteristics and the level of development in the village. This questionnaire was used to gain a description of the natural and human geography of the village.

Questionnaire No. 2: This questionnaire was designed on the basis of the variables and the

Table 1. The research variables and indices.

No.	Variable	Index
1	Leisure time	No. of hours spent on listening to radio, watching TV, reading papers, reading books, playing local games, attending the village teahouse, doing sports, visiting friends, working with computer (and the comparison with the same before the introduction of the IT center)
2	Social	5-level responses (using Likert spectrum) on the following changes: Management skills, reduced No. of times referring to urban public service centers, voluntarily attending training, improved job and technical knowledge, increased participation in public decision-making process, improved general knowledge, improved health knowledge, improved women's participation, change in patterns of information exchange, better identification and knowledge of problems
3	Cultural	Reduction of the role of indigenous cultural values, youths' behavioral change and delinquency, deterioration in educational level of students, improved personal tolerance and ethics, improved social behavior (such as observing traffic regulations)
4	Economic	Easier access to production resources, easier and cheaper access to vocational and skills training, facilitated access to bank loans, improved marketing, increased income, increased production
5	Computer	Types of uses users make of computers, use of the internet and its resources, utilization of different applications (changes in use of hardware and software)

Table 2. Characteristics of the four villages served by the project.

Village	Maranak	Tiss	Mahabad	Vorogh
Province	Tehran	Sistan & Baluchistan	Isfahan	Ghazvin
Township	Damavand	Chabahar	Ardestan	Avaj
Population	1250	4000	5000	500
Distance from provincial capital (km)	120	300	100	150
Climate	Mountainous	Warm & humid	Desert	Mountainous
No. of school students	260	1100	1200	65
No. of schools	3	7	6	1
Roads	Asphalt	Asphalt	Asphalt	Paved earth road
Drinking water	Yes	Yes	Yes	Yes
Power supply	Yes	Yes	Yes	Yes
No. of fixed phone lines	100	50	100	2
Estimated No. of potential users	250	200	200	100

indexes as described in Table 1 to identify the behavioral indexes prior to the establishment of the IT center. Potential users were asked by center managers to sign for services when they were also given the questionnaire to fill out.

Questionnaire No. 3: After four to six months had passed since the establishment of the IT center, questionnaire No. 3 was handed over to IT center managers for administration. This questionnaire was almost similar in content to questionnaire No. 2, the only difference being that it was now being administered after the establishment of IT centers.

5 VILLAGE SERVED BY THE PROJECT

After the feasibility studies, four villages or rural areas were selected for the establishment of the IT centers. The basic criteria employed in the selection process were the level of initial support and participation by the local community and authorities as evaluated by the project consultants in their meetings with them. Another criterion was the population and infrastructure (connectivity). With regard to this factor, the villages were selected in such a way to allow for accurate impact assessment and evaluation.

The four villages selected were:

Maranak, a village in the Damavand Township County in Tehran Province; Tiss, a village in the Chabahar Township County in Sistan & Baluchistan Province; Mahabad; a village in the Ardestan Township County in Isfahan Province; and Vorogh, a village in the Avaj Township County in Ghazvin Province. The general characteristics of these four villages are presented in Table 2.

6 CONCLUSION

The present study was carried out to assess the feasibility of ICT development in rural areas. The impacts of such developments on target groups are commonly evaluated by comparing the states before and after the implementation of the project. Below are presented some of the more important criteria before and after the establishment of ICT centers in the study areas. The results have clarified the values for each criterion before and after the implementation.

One of the criteria used in the evaluation is the number of hours users worked with computers. It was observed that 74% of the people never used computers before the establishment of the ICT centers in their villages while 18.5% of these people have been working with computers for at least one hour per week. It must be mentioned that the variance between the two states is so great that no special statistical test is required to show the significant difference.

The Kruskal Wallis Test was used in the comparison of the opinions of the subjects from different ICT centers on the changes in each of the cultural, economic, and social behavioral changes. The results indicated no significant changes in social behavioral change and that all three centers have undergone almost similar changes. In the case of the economic and cultural indices, however, the results indicate differences between Tiss and Maranak centers. While Tiss has witnessed more of a cultural change, Maranak center has experienced more economic changes.

As a general conclusion, it may be said that the establishment of ICT centers will have a great contribution to learning and the use of computer among the rural community. A well defined program will be required if a cultural change and a proper backbone for development are required. This is borne out by the realization of some of the objectives such as public participation, model development, etc. and also by the general attitude of the people towards these centers. The project titled "applications of IT for Sustainable Human

Development" which aimed to create pilot IT centers for rural communities was accomplished. One of the peripheral objectives of the project was to study the impacts of such IT centers in order to be used in developing a comprehensive model for extending information technologies to all rural areas across Iran. The results from the studies conducted and the analyses thereupon indicate that the mere establishment of single IT centers in scattered points in villages will not suffice to address the demand for IT and that it is crucially necessary to develop local ICT networks. It was also concluded that sustainability of rural IT centers greatly depends in their initial stages on support from local institutions and companies in neighboring urban centers. This support for rural IT centers must be provided in an organized manner.

Motivating the local people to participate in the execution and in the management of the IT center was one of the most important achievements of the project. The studies conducted in the initial stages of the project had indicated great possibility for people's resistance against the project due to their ignorance of IT functions and nature and future role of the IT center in their village. Local authorities had even expressed their concern and worries about the cultural and administrative aspects of establishing such centers. In response to these worries, the following steps were taken to attract local people participation:

- Preliminary talks with local and provincial authorities;
- Holding meetings both in Tehran and in the village with local authorities;
- Introduction of the project and its functions to stakeholders through lectures and pamphlets;
- Creating intangible competition among the local people for making decision on the location of the center and utilization of its services and facilities;
- Participation of local authorities and using local facilities in establishing and furnishing the centers and delivery of equipment to the centers;
- Preliminary training for managers in each locality and temporary submission of the equipment to them;
- Request for monthly reports on the performance of each center and sight visits; and
- Holding workshops on SWOT for users as per needs in each case.

REFERENCES

Anderson, J.L. Van Crowder and D. Dion 1999. Special: The first mile of connectivity: Applying the lessons of participatory communication and training to rural telecentres; *FAO Forest Conservation Research and Education Service.*

Bawden. R. 1996. A learning approach to sustainable agriculture and rural development: reflections from Hawkesbury. *In Training for Agriculture and Rural Development (TARD).* Rome: FAO.

Chapman, R. & Slaymaker, T. 2002. ICTs and Rural Development: Review of the Literature, Current Interventions and Opportunities for Action. *Overseas Development Institute 111 Westminster Bridge Road.* London. SE1 7 JD.

Forth, J. & Mason, G. 2004. Information and Communication Technology (ICT) Adoption and Utilization, Skill Constraints and Firm-Level Performance: Evidence from UK Benchmarking Surveys, *National Institute of Economic and Social Research*, London NIESR Discussion Paper No. 234 March 2004.

International Centre for Science and High Technology. 2008. *Rural ICT Participatory Appraisal—RIPA.*

Joseph, Meera K. 2010. Using participation and participatory approaches to introduce ICTs into rural communities. *Univ. of Johannesburg.* Johannesburg. South Africa.

Khalil-M, B. & Khatoonabadi, A. 201. Factors affecting ICT adoption among rural users: A case study of ICT Center in Iran". *Tele-communications Policy*, 37 Pages:1083–1094.

Kumar, A. & Singh., M.S. & Krishna, M. 2012. Role of ICTs in Rural Development with Reference to Changing Climatic Conditions; ICT for Agricultural Development Under Changing Climate, Krishna M. Singh, M.S. Meena, eds., Narenda Publishing House, 2012. Available at SSRN: http://ssrn.com/abstract=2027782 or http://dx.doi.org/10.2139/ssrn.2027782

Malhotra, C. & Chariar, V.M. & Das, L.K. & P.V. Ilavarasan. 2004. ICT for Rural Development: An Inclusive Framework for e-Governance; *Computer Society of India.*

Mildorf, T. and Karel Charvat jr. Editors (2012) ICT for agriculture, rural development and environment; *Czech Centre for Science and Society*, Wirelessinfo

Moyo, L.M. (1996). Information technology strategies for Africa's survival. Information technology for development, 7: 17–27. *IOS Press.*

Mwenesi, S. 2014. "ICT contribution to Kenya's GDP now at 12.1%", Home to African Tech: Humanipo; http://www.humanipo.com/news/46203/ict-contribution-to-kenyas-gdp-now-at-12-1/.

Nayak, S.K & Thorat, S.B. & Kalyankar, N.V. 2010. Reaching the unreached: A Role of ICT in sustainable Rural development; (IJCSIS) International Journal of Computer Science and Information Security, Vol. 7, No. 1.

Olaleye, O. 2014. "ICT's contribution to Nigeria's GDP", The Sun News: Voice of the Nation. http://sunnewsonline.com/new/?p=74067

Puri, S.K. & Sahay, S. 2007. Role of ICTs in participatory development: An Indian experience. *Information Technology for Development.* Volume 13. Issue 2. Pages 133–160. Wiley Interscience.

Shapiro Robert, J. & Mathur, A. 2011. The Contributions of Information and Communication Technologies To American Growth, Productivity, Jobs and Prosperity. *SONECON.*

II *Computer science*

Information Technology and Applications – Li (Ed.)
© 2015 Taylor & Francis Group, London, ISBN 978-1-138-02677-3

An improved Apriori algorithm based on Hash table for borrowing data of library

Y. Feng, M. Kang, X.X. Liu, H.Y. Xu, L. Gao & H.M. Zhou
School of Information, Liaoning University, Shenyang, Liaoning, China

ABSTRACT: Borrowing data of library contains a huge application value, and mining association rules contained in the data is beneficial to improve the resources utilization and service level of library. Oriented to the actual situation of library borrowing management and the shortcomings of the classic Apriori algorithm, an improved Apriori algorithm based on Hash table (Apriori-Hash Algorithm) was proposed in this paper. Based on the Apriori algorithm, the proposed algorithm applied Hash function to generate frequent 2-item sets directly, and used Hash table to record different width of the transaction identifiers. So the location of transaction can be achieved quickly. Finally, the algorithm was applied to mine the borrowing data of library to verify that the given algorithm can not only improve mining efficiency, but also raise the utilization rate of resources and the personalized service level of library.

Keywords: association rules mining; Apriori algorithm; Hash table; books borrowing

1 INTRODUCTION

In recent years, with the continuous improvement of the library automation degree, each library accumulated a large amount of historical data. The borrowing information has the extremely important application value. By analyzing this information, we can find there are some association rules in the process of borrowing books, and these rules will help to produce reasonable borrowing recommendation, improve the utilization rate of library resources and the satisfaction degree of the readers. The association rules mining can find the potential links among the data through the analysis of massive borrowing data, and provide support for scientific decision-making [5]. Therefore, applying association rules mining in the library is beneficial to realize the value of the borrowing data.

Currently, association rules mining has got certain application in the library field. For example, the literature [7] introduced the reason why correlation analysis method is suitable for statistics and analysis library information, and analyzed the service standards of association mining technology application in library borrowing data. Some researchers [1] applied Apriori algorithm in historical borrowing data analysis directly, and dug out the potential rules among the books borrowed. Although it can improve the quality of the library service to a certain extent, it is inevitable to inherit the deficiency of this algorithm such as low mining efficiency, the lack of personalized service and so on. In order to improve the efficiency of mining, the literature [8]

used the Boolean matrix to indicate the transaction database, and adopted affairs compression and project compression to improve the Apriori algorithm. This method reduced the number of times of scanning database, but it needed a large number of comparative judgments when generating candidate item sets. The literature [4] proposed pruning frequent item sets, connection optimization and database structure optimization strategies and so on, which greatly enhanced the performance of the Apriori algorithm. But there were still some of the waste in the memory overhead, especially when there were many candidate items.

Against the above shortages, this paper puts forward an improved Apriori algorithm based on the Hash table, namely Apriori-Hash algorithm, in view of the Hash table's advantages of less storage cost and positioning things faster. The algorithm applied a Hash function to generate frequent 2-item sets directly during the process of scanning database of borrowing for the first time on the basis of original algorithm, and used the Hash table to store item sets to reduce the number of times of scanning database. Finally, via the relevant applications the proposed algorithm was verified to have good mining efficiency, and improve the utilization rate of library resources.

2 APRIORI-HASH ALGORITHM

Apriori algorithm is a classical association rule mining algorithm, which belongs to the database traversing algorithm. Its core is recursion algorithm based on two-stage mining frequent item sets [6]. The working principle of the algorithm is described as follows. The frequent tempests' subset is frequent, the non-frequent item sets' superset is non-frequent [10]. But this algorithm would produce the candidate item sets excessive each connection, need multiple scanning database, and need a lot of judgments and comparisons in every connection operation. So the algorithm is inefficient. This paper puts forward an association rule mining algorithm based on Hash table, namely Apriori-Hash algorithm, whose advantages include saving memory cost, fastening search speed, reducing the number of times of scanning database, reducing the number of candidate item sets as well as avoiding connection operation.

2.1 Basic thoughts

Apriori-Hash algorithm focuses on improving the efficiency of the Apriori algorithm whose the basic idea is as follows: (1) after get frequent 1-itemsets L_1, use Hash function [3] to produce frequent 2-itemsets quickly; (2) after get L_1 and L_2, use L_1 and L_2 to trim the original database, delete the independent transactions, further compress storage space, and establish Hash table of the trimmed database according to the width l of the transactions. The width of Hash table is M − 2 (frequent 1-item sets and frequent 2-item sets don't need to join the Hash table). We use Hash function h(l) = l mod (M − 2), where M is the maximum width of the transactions, to store the transactions logo in the corresponding Hash table node chain, so that we can directly find corresponding transactions in the database by using the transactions logo stored in Hash table when statistic support degree is counted. It avoids repeatedly scanning database, saves a lot of contrast time; (3) according to the reference [4], generate C_k and L_k through the iterative cycle, until $|L_k| < k + 1$ ($|L_k|$ means the elements number of L_k).

2.2 Algorithm steps

Apriori-Hash algorithm uses the Hash table to optimize database, and combines pruning frequent item sets strategies and connection optimization strategies. So it greatly improves the efficiency of association rule mining. The basic steps of the algorithm are described as follows:

1. Single scan borrowing database D, calculate support of each 1-item sets to get frequent 1-item sets, then by the Hash function showed as formula (1), produce frequent 2-item sets quickly.

$$h(x,y) = |L_1| * (order(x) - 1)$$
$$- \frac{order(x) * (order(x) - 1)}{2}$$
$$+ order(y) - order(x) \qquad (1)$$

where $|L_1|$ signifies the number of items in frequent 1-item sets, $order(x)$ signifies the index of x in L_1, Hash table's length is $|L_1|*(|L_1| − 1)/2$.

2. Utilize items to prune the database. Because there are a large number of infrequent transaction records in the database, so after scanning the database, utilize frequent 2-item sets produced in the first step to prune the database, delete the transaction record which does not contain the items in the database. After pruning, the transaction will be deleted if its length is less than 3. This can reduce the transaction record of database and compress the storage space when establishing Hash table.

3. Create a Hash table based on the width of the transaction in the database after pruning. In order to be able to quickly find out all frequent item sets, after pruning database utilize frequent 2-item sets to establish a Hash table according to the width of the transaction for trimmed database. Its practice is to determine the maximum width of the affairs M. Establish a Hash table header whose length is M − 2, and then use the function h(l) = l mod (M − 2) to store the transaction logo whose width is l in the corresponding node chain of Hash table.

4. Generate candidate item sets. Referring to the pruning and connection strategy proposed by the literature [4], candidate item sets will be generated through the iterative cycle.

5. Generate frequent item sets. When calculating the support degree of the candidate item sets whose length is k, only need use the function h(k) = k mod (M − 2) to position the corresponding location in the Hash table, find the transaction logo whose width is k to scan database quickly. If k < M, continue calculating h(k + i) = (k + i) mod (M − 2) (i = 1, 2, ……, and k + i ≤ M), find the transaction logos whose width is greater than k to quickly scan the database. Finally get the support of the candidate item sets, and compare them with the minimum support, then get the frequent k item sets L_k. When generating L_k, calculate $|L_k|$. If $|L_k| < k + 1$, the algorithm end. Otherwise, return to step (4).

2.3 Algorithm case description

This case adopts the database shown in Table 1. There are 20 transaction records in the table, and the minimum support threshold is 25%

Table 1.	Database D.
TID	Itemsets
1	ABC
2	C
3	ABDE
4	ACE
5	AC
6	ACE
7	C
8	BCE
9	ABCD
10	AD
11	ABC
12	ACE
13	BCD
14	C
15	BC
16	ADE
17	BD
18	E
19	ABC
20	ABCD

Table 2. L_1 frequent 1-item sets.

Item	A	B	C	D	E
Support count	11	10	14	8	7

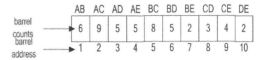

Figure 1. Candidate 2-item sets barrel distribution sample.

Table 3. Database D*.

TID	Itemsets
1	ABC
3	ABD
9	ABCD
11	ABC
13	BCD
19	ACD
20	ABCD

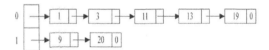

Figure 2. Hash table.

Table 4. Candidate 3-item sets.

Item sets	ABC	ABD	ACD	BCD
Support counts	4	3	3	3

(the corresponding support count is 5). The processing steps are as follows.

Step 1: Through single scanning the database D (shown in Table 1), calculate the support of 1-item sets, get frequent 1-item sets (shown in Table 2).

Step 2: Aim at content of Table 2, $L_1 = \{A,B,C,D,E\}$, $|L_1| = 5$, the Hash table's length is $|L_1|*(|L_1|-1)/2 = 10$, set $order(A) = 1$, $order(B) = 2$, $order(C) = 3$, $order(D) = 4$, $order(E) = 5$. Scan each transaction of the database, for example transaction ABC contains candidate item sets AB, AC, BC. Through the formula (1) calculate respectively that $h(1,2) = 1$, $h(1,3) = 2$, $h(2,3) = 5$. Thus the Hash bucket addresses of AB, AC, BC respectively are 1, 2, 5. Then make the count of barrel which addresses are 1, 2, 5 to add 1. Calculated barrel addresses of other candidates 2-itemsets which are shown in Figure 1. Finally compare the barrel counts with minimum support counts, get frequent 2-item sets $L_2 = \{AB, AC, AD, AE, BC, BD\}$.

Step 3: By the trimming frequent sets strategy calculate to get $|L_2(j)|$ of each element in L_2 and determine whether it is less than 2 or not. Because $|L_2(E)| = 1 < 2$, so remove elements AE, obtain $L_2' = \{AB, AC, AD, BC, BD\}$. Thus the item sets $I = \{A, B, C, D\}$ in L_2 are used to generate L_3, and delete the item of each transaction which does not belong to I in the database D. Moreover, if the width of transaction after trimming is less than 3, the entire transaction will be deleted from the database, because these transactions are useless to

generate frequent 3-item sets. The database D^* is shown in Table 3 after trimming.

Then according to the width of transactions build Hash table, and the Hash table from database D^* is shown in Figure 2.

Step 4: According to optimizing connection strategy, make L_2' after trimming connect with oneself to generate candidate 3-item sets which are shown in Table 4. Then through the Hash table scan database and calculate every item set support, get frequent 3-itemsets $L_3 = \Phi$. Due to the frequent 3 item sets are empty, namely $|L_3| < 4$, the algorithm end.

3 APPLICATION OF APRIORI-HASH ALGORITHM IN THE LIBRARY BORROWING DATA

This paper applied Apriori-Hash algorithm in the books management system to use the algorithm to mine association rules from books borrowing data, by which we can find readers' borrowing rules and predict the readers' information demand, lay a foundation for providing readers with personalized books recommendation. This paper selected a given university library's four years borrowing records from 2008 to 2011 as the basic data, which were in total of more than 7 million. Here we only considered id number and call number of records in order to reduce the amount of data. The book borrowing records after reducing are shown as Table 5. Then we transformed it into the readers' borrowing transaction database which was shown in Table 6 through combination and conversion pretreatment.

This paper had set minimum support to 0.20, set minimum confidence to 0.45, used Apriori-Hash algorithm to mine books association rule. We can get a lot of meaningful rules, for example, 52% of the students who borrowed books about web pages production also borrowed books about graphic design. The experimental results were consistent with the practical work of the library and the reader survey results. It is proved that the proposed algorithm in this paper is feasible.

In order to further verify the performance of the algorithm given here, this paper realized respectively AprioriMend algorithm [2] and TMMFI algorithm [9] which all were used widely in library based on the same computer environment for

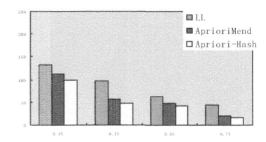

Figure 3. Contrast of execution time.

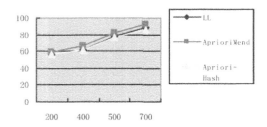

Figure 4. Comparison of memory utilization.

comparing with the algorithm proposed by this paper. Figure 3 showed the execution time contrast in the different minimum support. Figure 4 showed the memory utilization comparison in different sizes sample data sets. Through these contrasts, it is known that algorithm given by this paper has the better time complexity and space complexity.

4 CONCLUSIONS

Facing the actual demand of library borrowing management, this paper proposed an improved Apriori algorithm based on Hash table. The proposed algorithm generated frequent 2-item sets directly using Hash function for the first time process of scanning database on the basis of the original algorithm. And through the Hash table the algorithm rapidly scanned database in order to reduce the number of times of scanning database. Finally, through the application analysis it was verified that the proposed algorithm can enhance the performance of the library borrowing data mining, and improve the utilization of library resources and the level of personalized service.

ACKNOWLEDGMENT

This work was supported in part by Social Science Youth Foundation of Ministry of Education of China under Grant Nos. 12YJCZH048.

Table 5. Borrowing records.

Id number	Call number
A000022320	TP391.4/146
A000022320	TP313.C/123
20030018	TP393.4/170
20030018	TP393.09/121
... ...	... ...

Table 6. Readers borrowing transaction database.

Id number	Itemsets
A000022320	TP391.4/146, TP313.C/123, ...
20030018	TP393.4/170, TP393.09/121, ...
10044491	TP393.09/64, TP393.092/22, TP312.C/55, ...
... ...	... ...

REFERENCES

[1] Cai, H. Zhu, J. Cai, R. 2005. Application of data mining based association rule in the system of library. *Journal of Nanjing University of Technology* (1): 85–88.

[2] Gui, H. Meng, X. 2009. Research on Efficient Algorithm of Association Rules Mining Based on Apriori Algorithm. *Journal of Anhui University of Science and Technology (Natural Science)* 29(4): 55–58.

[3] Holt, J.D. Chung, S.M. 2007. Parallel mining of association rules from text databases. *The Journal of Supercomputing* (39): 273–299.

[4] Jiang, L. 2011. Application of Improved Apriori algorithm in the library lending data. *Information Research* (5): 16–19.

[5] Kantardzic, M. 2003. *Data mining: concept, model, methed and algorithm.* Beijing: Tsinghua University Publishing House.

[6] Li, J. Xu, Y. Wang, Y. et al. 2009. Strongest Association Rules Mining for Personalized Recommendation. *Systems Engineering—Theory & Practice* 29(8): 144–152.

[7] Ren, X.J. 2010. Study on Association Rule of Data Mining in Book-lending Service. *Information Science* 28(5): 720–731.

[8] Run, J. 2009. The Application of an Improved Apriori Algorithm in Library's Recommendation Service, *Sci-Tech Information Development & Economy* 9(10): 1–3.

[9] Salam, A. Khayal, M.S.H. 2012. Mining top-k frequent patterns without minimum support threshold. *Knowledge and information systems* 30: 57–86.

[10] Wang, X. Xu, T. Tang, L. 2008. *SQL Server2005 Data Dining Sample Analysis.* Beijing: China Water-Power Press.

Information Technology and Applications – Li (Ed.)
© *2015 Taylor & Francis Group, London, ISBN 978-1-138-02677-3*

Two-phase-precoding design for STBC MIMO systems

Meiyan Ju, Yueheng Li, Guoping Tan, Ping Huang, Jia Qian & Zhe Lu
College of Computer and Information, Hohai University, Nanjing, China

ABSTRACT: Mobile terminals tend to be less complicated and more portable due to limited size. We present a two-phase-precoding scheme combined with Space-Time Block Coding (STBC) to improve system performance and reduce the receiver's complexity in downlink Multiple-Input Multiple-Output (MIMO) systems. In the proposed scheme, there are two phases of precoding, one of which is implemented before the space-time block encoder, and the other of which is implemented just after the STBC module. We compared the proposed schemes with the existing schemes by simulations. The results showed that the proposed scheme can achieve better Symbol Error Rate (SER) performance with simpler processing at the receiver.

Keywords: Multiple-Input Multiple-Output (MIMO); precoding; Space-Time Block Coding (STBC); Symbol Error Rate (SER)

1 INTRODUCTION

Multiple-Input Multiple-Output (MIMO) techniques have attracted great attention because MIMO channels arising from the use of multiple antennas both at the transmitter and at the receiver provide an important increase in capacity over Single-Input Single-Output (SISO) channels under some uncorrelations [1]–[2]. The approaches to implement MIMO systems include Spatial Division Multiplication (SDM) and Space Diversity (SD). Space-Time Block Coding (STBC) [3]–[4] as a typical SD technique to combat the fading channel has attracted wide attention. Linear precoding [5]–[6] uses Channel State Information at the Transmitter (CSIT) to improve MIMO capacity and system performance, which has been a hot research topic. As STBC can exploit channel diversity without CSIT and precoding can exploit CSIT to enhance system performance, the combination of STBC and linear precoding is robust to the fading of wireless channels and can exploit the available CSIT at the same time.

Most of related references studied to design the precoding matrix for STBC codewords according to some criterion or under different system conditions. In these cases, STBC is performed before precoding. References [7]–[9] studied the design of linear precoding for a space-time coded system based on different criteria. A novel optimized non-unitary linear precoding design was proposed for Orthogonal Space-Time Block Codes (OSTBCs) considering the precoding codebook in [10].

There are also some exceptions. Reference [11] proposed a straightforward combination of conventional equalizer and space-time block decoder, which put precoding before STBC at the transmitter and decodes the received signal with Minimum Mean-Squared Error (MMSE) equalization at the receiver. References [12]–[13] mixed the STBC process and the precoding process aiming to improve diversity gain and coding gain.

To be simple, here we study a MIMO system with separate STBC module and precoding module, which means that STBC and precoding are not mixed. Practically, on the one hand, we want to design linear precoding to improve performance of a STBC system, and on the other hand, we do not want to make the system very complicated, especially the mobile terminal. To be inspired, we suggest a precoding-STBC-precoding scheme, which includes two phases of precoding. One is for improving OSTBC system performance, and the other is for pre-equalization which makes the receiver recover the original signal without the need of complex processing. Furthermore, we also explained that ZF criterion is equivalent to MMSE criterion for the pre-equalization process in this proposed scheme which uses OSTBCs.

The rest of the paper is organized as follows. The system model is introduced in section II. In section III, we described the precoding-STBC-precoding scheme. Simulation results and performance analysis are given in section IV. Section V concludes this paper.

2 SYSTEM MODEL

For the clarification, we standardize a few notations used. Vectors and matrices are denoted by lowercase bold and uppercase bold letters, respectively. We use T, H, and $*$ for transpose, conjugate transpose and conjugate respectively.

In this section, we consider a MIMO communication system with N_t transmit antennas and N_r receive antennas. The system model is shown in Figure 1. At the transmitter, QPSK mapped input signal s of size $M \times 1$ is first performed precoding with the $N \times M$ precoding matrix F_1 to get

$$x = F_1 s, \tag{1}$$

Here F_1 is for pre-equalization to reduce the complexity at the receiver.

The $N \times 1$ signal vector x continues to be input into space-time block encoder to get the codeword C of size $N_t \times T$ (T denotes time intervals spanned in one codeword). Here we adopt orthogonal space-time block encoders. Then the second precoding process with the $N_t \times N_t$ precoding matrix F_2 is followed. Finally, the signal is transmitted from transmit antennas.

At the receiver, the received signal is expressed as follows:

$$R = \frac{1}{\beta}(HF_2C + N), \tag{2}$$

where $N_r \times N_t$ H is the MIMO channel matrix between N_t transmit antennas and N_r receive antennas, each entry of which is independently identically distributed (i.i.d.) Rayleigh. H is assumed to be perfectly known both at the receiver and at the transmitter. R denotes the received signal with the size of $N_r \times T$, and the $N_r \times T$ N is the additive complex white Gaussian noise with zero mean and σ^2 being the noise power per spatial dimension. β is a factor to control the transmit power in designing F_1. The codeword C has been scaled according to the transmit power. Here F_2 is for improving the system performance.

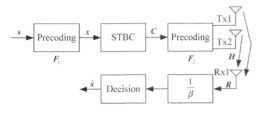

Figure 1. A STBC MIMO system with two-phase-precoding.

3 THE TWO-PHASE-PRECODING SCHEME WITH STBC

In this section, we will clarify how to design the two precoding matrices F_1 and F_2. Based on the orthogonal space-time block codewords, we first design F_2 according to some criteria and then design F_1. Here suppose Alamouti scheme [14] is used for STBC. The number of transmit antennas N_t is 2 and the number of receive antennas N_r is 1.

3.1 Design the precoding matrix F_2

As the codeword matrix C is orthogonal, here we can design precoding F_2 according to Pairwise Error Probability (PEP) or capacity criteria [15].

According to PEP criterion or ergodic capacity criterion, the design of precoding matrix F_2 can be unified into the following equation:

$$\max_{F_2} \quad E_H\left[\log\det\left(a\gamma HF_2QF_2^*H^* + I\right)\right]$$
$$s.t. \quad tr\left(F_2F_2^*\right) = 1 \tag{3}$$

where $\gamma = P/\sigma^2$ is the Signal-to-Noise Ratio (SNR), P is the average sum transmit power, and σ^2 is noise power. Q is the normalized correlation matrix of codeword C, which is denoted as $Q = 1/TP \ E[CC^*]$. a is a factor, which is 1/2 when PEP criterion is used and it (which) is 1 when ergodic capacity criterion is used.

Therefore the two criteria have the same precoding design for F_2. Now consider how to get F_2. If CSIT is known, the optimal solution has three parts: shaping, power allocation, and beamforming, which can be expressed as follows:

$$F_2 = U_{F_2} \Lambda_{F_2} V_{F_2}^*, \tag{4}$$

where V_{F_2} is shaping, Λ_{F_2} is power allocation, and U_{F_2} is beamforming.

Shaping part V_{F_2} is to decorrelate the input signals. The correlation matrix of input symbols can be decomposed through Singular Value Decomposition (SVD) to get

$$Q = U_Q \Lambda_Q V_Q^*. \tag{5}$$

The optimal V_{F_2} is given by $V_{F_2} = U_Q$. As for the orthogonal input structure, e.g., Alamouti scheme, the correlation of input symbols does not exist, that is, $U_Q = I$. So here V_{F_2} can be negligible.

The channel H is decomposed by SVD into

$$H = U_H \Lambda_H V_H^*. \tag{6}$$

When the transmitter knows Channel State Information (CSI) perfectly, MIMO channel can be decomposed into independent parallel additive white Gaussian noise channels. In order to achieve that, the beamforming part U_{F_2} is equal to V_H, that is, $U_{F_2} = V_H$.

Here we consider the power allocation, and the waterfilling method are (is) used. The waterfilling power allocation can be expressed by

$$p_i = \left(\mu - \sigma^2 \sigma_i^{-2} \right)^+, \qquad (7)$$

where p_i is signal power allocated to sub-channel i, μ is the water level, σ^2 is white noise power, σ_i is the i-th singular value of channel matrix H, and

$$(x)^+ = \begin{cases} x & x \geq 0 \\ 0 & x < 0 \end{cases}.$$

As (Since) the MIMO system here is configured with $N_t = 2$ and $N_r = 1$, the MIMO channel H has the rank as 1. $\mathbf{\Lambda}_{F_2}$ should be as $\mathbf{\Lambda}_{F_2} = \begin{pmatrix} \sqrt{P} & 0 \\ 0 & 0 \end{pmatrix}$ or $\mathbf{\Lambda}_{F_2} = \begin{pmatrix} 0 & 0 \\ 0 & \sqrt{P} \end{pmatrix}$.

3.2 Design the precoding matrix F_1

According to the Alamouti scheme, we suppose $x = [x_1 \quad x_2]^T$ and the codeword C is as follows:

$$C = \begin{bmatrix} x_1 & -x_2^* \\ x_2 & x_1^* \end{bmatrix}. \qquad (8)$$

Equation (2) can be transposed to get

$$R^T = \begin{bmatrix} r_1 \\ r_2 \end{bmatrix} = \frac{1}{\beta} \left(\begin{bmatrix} x_1 & x_2 \\ -x_2^* & x_1^* \end{bmatrix} (HF_2)^T + N^T \right), \qquad (9)$$

where $N^T = \begin{bmatrix} n_1 \\ n_2 \end{bmatrix}$. The matrix multiplication HF_2 is a vector with size $N_r \times N_t$ (where $N_r = 1$ and $N_t = 2$), which can be denoted as

$$HF_2 = [hequ_1 \quad hequ_2]. \qquad (10)$$

Conjugate the second entry of R^T in (9) and use (1), we can get

$$\begin{bmatrix} r_1 \\ r_2^* \end{bmatrix} = \frac{1}{\beta} \left(\begin{bmatrix} hequ_1 & hequ_2 \\ hequ_2^* & -hequ_1^* \end{bmatrix} \begin{bmatrix} x_1 \\ x_2 \end{bmatrix} + \begin{bmatrix} n_1 \\ n_2^* \end{bmatrix} \right)$$

$$= \frac{1}{\beta} \left(\begin{bmatrix} hequ_1 & hequ_2 \\ hequ_2^* & -hequ_1^* \end{bmatrix} F_1 s + \begin{bmatrix} n_1 \\ n_2^* \end{bmatrix} \right)$$

$$= \frac{1}{\beta} \left(\overline{Hequ} F_1 s + \begin{bmatrix} n_1 \\ n_2^* \end{bmatrix} \right), \qquad (11)$$

where $s = [s_1 \quad s_2]^T$, $\overline{Hequ} = \begin{bmatrix} hequ_1 & hequ_2 \\ hequ_2^* & -hequ_1^* \end{bmatrix}$, and the noise vector $n' = \begin{bmatrix} n_1 \\ n_2^* \end{bmatrix}$ has the noise power σ^2.

Now we design F_1 to make the following equation exist,

$$\begin{bmatrix} r_1 \\ r_2^* \end{bmatrix} = \frac{1}{\beta} \left(\overline{Hequ} F_1 s + \begin{bmatrix} n_1 \\ n_2^* \end{bmatrix} \right) = \tilde{s}, \qquad (12)$$

Here $\tilde{s}$ is estimated value of s.

As F_2 has been derived as mentioned above, we can get $\overline{Hequ}$ via (10) and then design the precoding matrix F_1 according to some criteria [16], e.g., Zero-Forcing (ZF) and MMSE.

According to ZF criterion, we can get

$$F_1 = \beta \overline{F_1}, \qquad (13)$$

where $\overline{F_1} = \overline{Hequ}^H (\overline{Hequ}\,\overline{Hequ}^H)^{-1}$, $\beta = \sqrt{P/tr(\overline{F_1}\,\overline{F_1}^H)}$, and P is the total transmit power.

According to MMSE criterion, we can get

$$F_1 = \beta \overline{F_1}, \qquad (14)$$

where $\overline{F_1} = \overline{Hequ}^H (\overline{Hequ}\,\overline{Hequ}^H + \sigma^2/\sigma_s^2 I)^{-1}$, $\beta = \sqrt{P/tr(\overline{F_1}\,\overline{F_1}^H)}$, and σ_s^2 is the symbol power.

It is obvious that $\overline{Hequ}$ is orthogonal, therefore we can easily get $\overline{Hequ}\,\overline{Hequ}^H = \lambda I$, where λ is a positive constant, and then (13) and (14) are different only in a constant value, which make them have the same effect, that is, ZF criterion is equivalent to MMSE criterion here. For the simplicity, we can use ZF criterion in place of MMSE criterion in the proposed scheme.

4 SIMULATION RESULTS AND PERFORMANCE ANALYSIS

In this section, considering a STBC MIMO system with 2 Tx antennas and 1 Rx antenna, two-phase-precoding schemes are verified here. QPSK mapping is used and the Alamouti scheme is adopted in the STBC module. The entries of H matrix are zero mean i.i.d. complex Gaussian random variables with variance 1. The CSI is supposed to be perfectly known at the receiver and also at the transmitter by free-error and zero-delay feedback links. Here the two power allocation methods are adopted for comparison including the average power allocation and the waterfilling power allocation. Their difference only lies in the power allocation part $\mathbf{\Lambda}_{F_2}$ of the precoding matrix F_2. We simulate Symbol Error Rate (SER) performances of the proposed schemes (two-precoding noted in Fig. 3), the STBC scheme with no precoding (no-precoding noted in Fig. 2 and Fig. 3) and the STBC schemes with one precoding (one-precoding noted in Fig. 2). In one-phase-precoding schemes only the precoding F_2 is designed, and the precoding F_1 being omitted, that is $F_1 = I$. For no-precoding scheme and one-phase-precoding schemes, the receivers eliminate the channel interference to recover the signal with ZF criterion.

Figure 2 is the SER performance comparison of different schemes including the STBC scheme with no precoding, the one-phase-precoding scheme with average power allocation, and the one-phase-precoding scheme with waterfilling power allocation. From Figure 2, we can observe that the STBC scheme with no precoding has the same performance with the one-phase-precoding scheme with average power allocation, and the

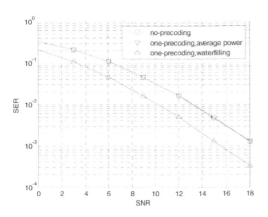

Figure 2. SER versus SNR for the STBC scheme with no precoding and one-phase-precoding schemes.

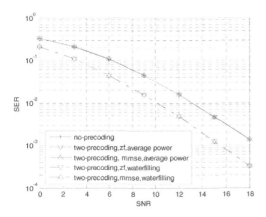

Figure 3. SER versus SNR for the STBC scheme with no precoding and two-phase-precoding schemes.

one-phase-precoding scheme with waterfilling has the best performance of them. For the one-phase-precoding scheme with average power allocation, the precoding design does not take full advantage of CSI and the STBC scheme with no precoding has full diversity gain and can achieve the same performance with the precoding scheme with average power allocation. The one-phase-precoding scheme with waterfilling power allocation makes full use of the CSI at the transmitter to achieve the best performance of the three schemes.

Figure 3 is the SER performance comparison of different schemes including the STBC scheme with no precoding, and several two-phase-precoding schemes with two power allocation methods and two criteria of designing F_1 including ZF and MMSE. From Figure 3, we can observe that the ZF criterion has the same performance with the MMSE criterion in designing the precoding F_1 when the other conditions are same, which is shown by the coinciding curves. The STBC scheme with no precoding has the same performance with the two-phase-precoding schemes with average power allocation. The reason is the latter schemes do not take full use of the CSI during designing the precoding F_2 and the precoding F_1 of the latter schemes is only used for pre-equalizing the transmitted signal without improving system performance. The two-phase-precoding schemes with waterfilling power allocation are the best of the different schemes because they use the CSI best via waterfilling power allocation. We can conclude that the one-phase-precoding scheme with waterfilling power allocation has the same performance with the two-phase-precoding schemes with waterfilling power allocation. It is notable that the two-phase-precoding schemes pre-equalize the signal at the transmitter and the receivers do not need to equalize the received signal, whereas the STBC scheme with no precoding and the one-phase-precoding schemes still need to remove the interference of channel to recover the original signal.

5 CONCLUSIONS

This paper presents a two-phase-precoding scheme for MIMO systems with STBC. In this proposed scheme, one precoding process is before STBC, and the other is after STBC. They are used for different aims, one of which is for pre-equalization and the other of which is for improving MIMO system performance. Furthermore, ZF criterion is equal to MMSE criterion for pre-equalization in the proposed scheme. Based on the proposed scheme, the receiver only needs to decide the received signal to recover the original one, which is very simple for receivers. The proposed scheme is easy to be extended to other orthogonal space-time block codes.

REFERENCES

[1] Telatar, I.E. 1999. Capacity of multi-antenna Gaussian channels. *Eur. Trans. Telecommun.* 10(6): 585–595.

[2] Foschini, G. & Gans, M. 1998. On limits of wireless communications in a fading environment when using multiple antennas. *Wireless Pers. Commun.* 6: 311–335.

[3] Tarokh, V., Jafarkhani, H., Calderbank, A.R. 1999. Space-time block codes from orthogonal designs. *IEEE Transactions on Information Theory* 45(5): 1456–1467.

[4] Tarokh, V., Jafarkhani, H., Calderbank, A.R. 1999. Space-time block coding for wireless communications: performance results. *IEEE Journal on Selected Areas in Communications* 17(3): 451–460.

[5] Liu Tong, Zhang Linbo, Zhang Shu. 2008. Study of preprocessor of multi-user linear-dispersion code and spatial multiplexing combined system. *Harbin Gongcheng DaxueXuebao/Journal of Harbin Engineering University* 29(12): 1330–1334.

[6] Vu Mai & Paulraj, A. 2007. MIMO wireless linear precoding. *IEEE Signal Processing Magazine* 24(5): 86–105.

[7] Sampath, H. & Paulraj, A. 2002. Linear precoding for space-time coded systems with known fading correlations. *IEEE Commun. Lett.* 6(6): 239–241.

[8] Alexiou, A. 2007. Error probability, diversity and coding gain considerations in space-time coded systems with linear precoding. *International Workshop on Antenna Technology: Small and Smart Antennas Metamaterials and Applications:* 215–218.

[9] Bhatnagar, M.R. & Hjorungnes, A. 2010. Linear precoding of STBC over correlated Ricean MIMO channels. *IEEE Transactions on Wireless Communications* 9(6): 1832–1836.

[10] Huang Haiyang, Wu Gang, Li Shaoqian. 2009. Optimized non-unitary linear precoding for orthogonal space-time block codes. *IEEE Communications Letters* 13(6): 414–416.

[11] Zhang Wei-dang, Lou Ming, Zheng Li-min. 2011. Research of an equalization algorithm based on STBC pre-coding. *Electronic Design Engineering* 19(8): 110–112.

[12] Hong Zhihong & Sayeed, A.M. 2003. Space-time block codes Based on precoding. *IEEE Global Telecommunications Conference* 2: 636–640.

[13] Yan Xin, Wang Zhengdao, Giannakis, G.B. 2003. Space-time diversity systems based on linear constellation precoding. *IEEE Transactions on Wireless Communications* 2(2): 294–309.

[14] Alamouti, S. 1998. A simple transmit diversity technique for wireless communications. *IEEE Journal on Selected Areas in Communications* 16(8): 1451–1458.

[15] Vu, M. 2006. Exploiting Transmit Channel Side Information in MIMO Wireless Systems. PhD Dissertation, Stanford University.

[16] Joham, M., Utschick, W., Nossek, J.A. 2005. Linear transmit processing in MIMO communications systems. *IEEE Trans. on Signal Processing* 53(8): 2700–2712.

Information Technology and Applications – Li (Ed.)
© *2015 Taylor & Francis Group, London, ISBN 978-1-138-02677-3*

Research on designing and developing the "Computer Platform" course website aiming at interactive learning

Xiaoying Wang, Po Li, Xiaojing Liu, Jianqiang Huang, Yu-an Zhang & Tengfei Cao
Department of Computer Technology and Applications, Qinghai University, Xining, China

ABSTRACT: "Computer Platform" is an important fundamental course in the cultivation plan of students in computer science major. To encourage the students to study by themselves outside class, it's necessary to design and develop a high-quality website which contains various staff about this course and supports interactive learning. This paper presents a course website designed for the "Computer Platform" course aiming at promoting the interactive activities of the students after class. The whole framework of the system is illustrated and the components are described in detail. By accessing this website, students can download documents and presentation slides, upload homework and experiments files, do exercise and judge themselves, and can also communicate with the teacher. The website plays an important role as the complementary part of the classroom teaching.

Keywords: Computer Platform; website design; interactive study

1 INTRODUCTION

1.1 *The "Computer Platform" course*

The "Computer Platform" Course is a professional fundamental course in the curriculum of the Computer Science major students, which is opened for sophomores. The basic concept of this course is to provide a holistic view of the computer platform, from the lowest level to the upper levels. The computer architecture is viewed as a whole platform, including both hardware principles and operating system implementation and working mechanisms. By studying the "Computer Platform" course, the students are expected to be familiar with both hardware and software systems of the computer, to set up a holistic concept of the entire computer system, and to gain the ability to use computer for practical problems in real-world.

In the end of year 2013, this course has been funded as the only First-class Elite Course in Qinghai University.

1.2 *The goal of course website design*

As the information technology develops rapidly, the construction of courses website for teaching plays an important role in the university (Mike et al. 2010). In order to make full use of the advantages of multimedia and network, the teachers need to provide an effective means to guide students to

learn by themselves, and to help students understand the focus and difficulty of the course (Tao et al. 2011). The course website should be designed to provide teachers with a platform for interactive teaching and learning so as to better serve the students, which provides students a self-learning platform as a classroom for student to supplement knowledge (Zhu-ying et al. 2010). It can also facilitate curriculum reform to open up a good channel to the domestic and international publicity (Xiaoying et al. 2008).

The course website introduced in this paper includes not only static webpages showing the teaching staffs, but also the interactive parts for students to upload homework and communicate with the teachers. JSP and ASP.NET are jointly employed in the system for different components. Both MySQL and SQL Server are adopted to provide the back-end database services.

2 FRAMEWORK DESIGN

2.1 *Three-tier architecture*

By analyzing the requirements, the course website should have the functions of post publishing, file uploading and sharing, online video streaming, teacher/student communication and so on. Hence, the website should be dynamically designed, so that it's contents could be updated in time. To facilitate the development, test and maintenance of the system, the website is designed hierarchically and separated into different modules. The popular three-tier architecture is employed here, including the presentation tier, the logic tier and the data tier. The presentation tier is in charge of showing webpages directly to the users. This tier communicates with other tiers by sending results to the browser and other tiers in the network. The logic tier is pulled from the presentation tier. It controls application functionality by performing detailed processing. The data tier houses database servers where information is stored and retrieved. Data in this tier is kept independent of the logic tier and the presentation tier.

Three-tier architecture allows any one of the three tiers to be upgraded or replaced independently. Because the programming for a tier can be changed or relocated without affecting the other tiers, the three-tier model makes it easier the software packager to continually evolve an application as new needs.

2.2 Main components

This subsection describes the main components of the course website system, including the presentation tier, the logic tier and the data tier, as shown in Figure 1.

1. The presentation tier is the front-end of the website system. It exhibits various materials such as course introduction, faculty and staff introduction, the syllabus of this course, and also provides all kinds of resources for the students to download, including slides, tutorials and videos.
2. The logic tier is in charge of dealing with kinds of management issues including user management, class management, homework management, knowledge management and exercise management, and so on. For instance, users could be added, deleted or modified under the functionality of this tier. Also, knowledge points could be maintained by teachers, which

remind the students which points are important and crucial. The homework submission and similarity comparison logic are also conducted at this tier. Besides, students can start up their own exercises or tests, or join in the published exams, which is under the control of the exercise management module.
3. The data tier is used to encapsulate the data entities and operations. It provides the interfaces of connecting and disconnecting with the database. It also provides basic operations such as inserting, deleting and updating data records. These operation interfaces could be called by the logic tier to facilitate its interaction with the database.

2.3 Database design

The kernel data model designed for the course website system is as shown in Figure 2, which illustrates several important entities of the database. They are as follows:

1. The user entity has attributes of ID, name, password, user role and its login time. A user might publish multiple posts or upload multiple copies of resources.
2. The column entity has attributes of ID, name, link, order, introduction and isVisible (to show it on the main page or not). Multiple posts or resources might belong to the same column.
3. The post entity has attributes of ID, author, title, contents, updateTime (what time it's last updated), and columnID (which column it belongs to).
4. The resource entity has attributes of ID, name, type, size, author, introduction, updateTime, columnID and URL (where to download it).

Due to the space constraints, the other entities which are less important are not shown in Figure 2.

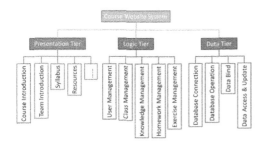

Figure 1. The holistic framework and main components of the course website system.

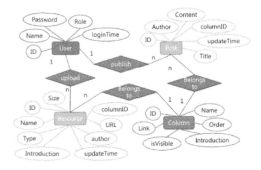

Figure 2. The entity-relationship diagram of several main entities in the database design.

3 DEVELOPMENT AND IMPLEMENTATION

3.1 *Database connection*

The database connection operation is encapsulated in the JSP pages. Part of the key codes is given as in Figure 3.

3.2 *Adding users in batch*

Usually, a class of students forms a unit for the teachers to manage. It's tedious for each students to register a new account and login. Thus, it's necessary to implement the function of adding users in batch.

Here in our system, the administrator can upload the Excel file including student's IDs and names, and the system will parse the file and obtain the info. Then, the user account of the students could be generated automatically, assigned with an initial password. Figure 4 shows the code of adding users

```
<%@ page import="java.sql.*"%>
<%@ page import="java.sql.DriverManager" %>
<%
String driver = "com.mysql.jdbc.Driver";
String url = "jdbc:mysql://127.0.0.1:3306/lipo";
String user = "lipo";
String password = "1234";
Class.forName(driver);
Connection conn = DriverManager.getConnection(url, user,
password);
%>
```

Figure 3. The encapsulation of data connection operations in JSP pages.

```
private int importExcelToSheetList(string FilePath)
    {
        string strConn;
        strConn = "Provider=Microsoft.Jet.OLEDB.4.0;" + "Data
Source=" + FilePath + ";Extended Properties=Excel 8.0;";
        OleDbConnection conn = new OleDbConnection(strConn);
        conn.Open();
        System.Data.DataTable dtSheetName =
conn.GetOleDbSchemaTable(OleDbSchemaGuid.Tables, new object[]
{ null, null, null, "Table" });
        string[] strTableNames = new string[dtSheetName.Rows.Count];
        for (int k = 0; k < dtSheetName.Rows.Count; k++)
        {
            strTableNames[k] = dtSheet-
Name.Rows[k]["TABLE_NAME"].ToString();
        }
        if (strTableNames.Length == 0)
        {
            Response.Write(MessageBox.ShowMessage("No available
Sheet!"));
            return -1;
        }
        for(int k=0;k<strTableNames.Length;k++)
            this.SheetList.Items.Add(strTableNames[k]);
        conn.Close();
        return 0;
    }
```

Figure 4. The codes used for parsing the Excel sheets containing user information.

in batch by using OLEDB to parse the Excel files and list key information of the Excel sheets.

3.3 *Post publishing*

When the administrator or teachers want to publish new posts, rich text is usually practical and useful. In the implementation of our system, we used an open-source system called xheditor.

The purpose of using xheditor is to change the exhibition effect of the traditional "textarea" element, so that the contents could be edited online and make sure that what you get is what you see. The editor saves both contents and its styles, which is easy and convenient to use.

3.4 *Embedded video streaming*

In order to publish the lecture videos in time for the students to watch online, we design en embedded video streaming method to embed the video in web pages. Users can watch the videos in stream while they access the web pages of the course website.

Here, we use CuPlayerMini V3.0 as the video player which is embedded in the video playing page. This is a free software release which supports .flv and .mp4 video files. Online self-test and exercise.

Aiming at self-learning and self-testing, we also incorporate an online self-test and exercise system into the course website. The teachers could upload the prepared exercise questions into the system in batch. The documents in a certain format could be parsed through java POI techniques, and split into single questions. Then, students can access the system to do exercise randomly or start an exam which is predefined by the teacher.

By doing exercises and simulating the exams, the students could check the knowledge points after class, and get feedback through the exercise scores given by the system. This greatly facilitates the self-learning and self-training demands of the students and also helps teachers to follow the study progress of each student.

4 WEBPAGE EXHIBITION

4.1 *Portal and main pages*

The portal page of the course website is as shown in Figure 5. The navigating banner consists of ten columns, including the course introduction, the syllabus, resource download center, message feedback, online exercise, homework review and so on. The columns could be dynamically added or moved in the back-end administration system. Besides, a video of brief introduction of this course will be played on a continuous loop in the main page. Important news keeps scrolling in the

Figure 5.   The portal webpage of the course site.

Figure 6.   The welcome page for teachers of the online self-test exercise system.

center, and latest notices are released and displayed at the right side.

In the back-end administration system, the administrator can upload new article, resource files, documents and videos on demand. These files could be classified into different categories, and will be shown in corresponding columns of the portal page.

4.2   Online self-test exercise system

One of the subsystems of the course website is the online self-test exercise system. The login page of a teacher is illustrated as in Figure 6. As it can be observed, the steps of the entire exam management process are shown in the welcome page. The user as a teacher role should strictly follow the steps to set up exam plans.

First, the information of students and classes should be inputted into the system. Then, the teacher can add some knowledge points for students to learn by themselves. A necessary step for the teacher is to upload exercise questions beforehand, including the description, the type,

Figure 7.   The page shown to teachers when they are reviewing students' homework online before similarity comparison.

and the answers of these questions altogether. Then, an exam paper could be automatically generated according to the restraints given by the teacher. Finally, the teacher can establish a specific exam plan at a certain date for the students to attend.

4.3   Homework submission and similarity detection system

Here, the homework submission and similarity detection system is another important subsystem of the course website. Figure 7 depicts the page shown to teachers when they are reviewing students' homework online before similarity comparison. We have also implemented the online review components, which read out the contents of homework documents and turn them into web pages. Then, on the browser side, the users can view the contents without installing any extra software or plugins. The teachers could give corresponding scores and comments after viewing the documents online, without download the files to their local hard disks.

After some students submit their homework, the teacher can choose to review and let the system help compare their homework. The comparison is done by using an improved version of the Smith-Waterman algorithm (Smith et al. 1981).

5   CONCLUSION

This paper presents the design and implementation of the "Computer Platform" course website. The design of the whole framework is first introduced, including the relationship of all the components. Then the database design is described in detail, with the illustration in entity-relationship diagrams. Based on the design, we presented some of the important and key codes in the implementation of the course website. At last, the portal page and two important subsystems are illustrated one by one, showing the convenience and practicality of our website. By using this website as an associated method in the teaching process, the self-learning opportunity could be enlarged, and the students could use it as a tool during the study of this course. In addition, teachers can also get benefit from this system, through the online resource sharing and homework reviewing functionalities.

ACKNOWLEDGEMENT

This paper is partly supported by the High Education Research and Reform Project in Qinghai Province ("Implementation and Application of Homework Similarity Detection Tools") and also the Course Construction Project in Qinghai University (number KC-13-1-1, KC-12-2-3).

REFERENCES

Mike Eisenberg, Doug John, Bob Berkowitz. 2010. Information, Communications, and Technology (ICT) Skills Curriculum Based on the Big6 Skills Approach to Information Problem-Solving. Library Media Connection, 28(6): 24–27.

Smith, Temple F.; and Waterman, Michael S. 1981. Identification of Common Molecular Subsequences. Journal of Molecular Biology, 147: 195–197.

Tao Wang, Guo-yong Qiu, Wei Song, Shi-hong Song. 2011. Research and Practice on Constructing Elaborate Course Website with CMS. Modern Educational Technology. 21(6): 120–122.

Xiaoying Wu, Shining Wu, Zhijian Li, Rude Yang. 2008. The Design and Development on the Website of the Exquisite Course of Biopharmaceutical Technology. Chemical High Education, (04): 22–25.

Zhu-ying Liu, Lei Shi, Hong-wei Lin, Qi Zhang. 2010. Design and Implement of Excellent Courses Website Based on Interactive Learning, Computer Education, No.21: 20–25.

Information Technology and Applications – Li (Ed.)
© 2015 Taylor & Francis Group, London, ISBN 978-1-138-02677-3

A secure multi-party protocol of association rules mining in vertically distributed environment

Ling Song

School of Computer, Electronics and Information, Guangxi University, Nanning, China

ABSTRACT: In the vertical distributed environment, the key is how to calculate the global support of item sets safely. The protocol to calculate the global support safely only applies to two parties involved and needs to generate a lot of random numbers and complex calculations, computational efficiency is not high. In this paper, a protocol called SMGSP is proposed to calculate the global support of item sets safely and can be applied to multiple parties. SMGSP calculates the sum easily based on the paillier's homomorphism encryption technology. SMGSP is secure and true and has good performances in theory, experiment results prove that SMGSP can reduce the cost of computation while the results of mining is accurate.

Keywords: association rules mining; privacy preserving; vertically distributed environment; paillier's homomorphism encryption

1 INTRODUCTION

The key of privacy preserving association rule mining algorithm in vertically distributed environment is how to compute global support safely, but not to reveal privacy data and local support of each site (participants) [1, 2]. Secure two-party computational protocol currently is seldom, and cannot be applied to multi-party (more than two parties participated), mainly represented by the dot product protocol, and need to produce a large of random numbers and complex computation [4, 7]. Therefore, we design a protocol called SMGSP that calculate the global support of item sets safely and can apply to multiple parties, it is based on the paillier's homomorphism encryption technology to calculate the sum easily.

SMGSP is analyzed to be safe and true and has good performances in theory. Finally, experiment results also proved that the SMGSP reduced the cost of computation while the results of mining are accurate.

The rest of the paper is organized as follows. In section 2, we presented the background and relate work. In section 3, we introduced a new secure multi-party protocol to get global support, named SMGSP. In section 4, we analysis the correctness, security, computation and communication cost of SMGSP. In section 5, experimental results are provided. Finally, concluding is presented in section 6.

2 BACKGROUND AND RELATE WORK

Considering n records in the heterogeneous database D are distributed over parties A and B, of which items $x_1, x_2, \ldots, x_a$ are distributed in party A, their corresponding values is $x_{i1}, x_{i2}, \ldots, x_{ia}$, the rest of items $y_1, y_2, \ldots, y_b$ are distributed in party B, their corresponding values is $y_{i1}, y_{i2}, \ldots, y_{ib}$. We can compute the support of items by computing the scalar product of $S = \sum_{i=1}^{n} x_i * y_i$. However, vector Y must be exposed to party A or vector X must be exposed to party B for computing the scalar product, it is against the principle of privacy preserving obviously.

Jaideep Vaidya [8] proposed a secure two-party computational protocol by computing the scalar product, which called dot product of privacy preserving association rules mining algorithm (PPVDR). Jaideep Vaidy [9] proposed a secure multi-party computational protocol based on secure set intersection on the basis of PPVDR. It converted the vector in order to protect the values of original data by producing a large number of random numbers and matrix combining with the hash technique. After that, it appeared a variety of dot product for the study of computing complexity and communication traffic.

Except that, it also proposed a secure two-party protocol based on ElGamal encryption, only the protocols given by Jaideep Vaidya [8, 9] and Saeed S. [6] were applied to multi-party. Saeed and Miri [6] used additive homomorphic encryption technique, the vector of one site are decomposed into n vectors plus (assumed n parties in cooperation), the rest $n-1$ vectors are sent to the other site after encrypted with additive homomorphic encryption algorithm. The sites calculated one by one and decrypted, then encrypted intermediate results by generating random sequence to calculate the global support. The results showed that it improved computing efficiency comparing with the algorithm of Jaideep Vaidya [8, 9].

Kumbhar, M. N. [3] combined homomorphic encryption with Shamir secret sharing technology to achieve privacy preserving in vertically distributed environment, and combined RSA encryption with homomorphic encryption to achieve privacy preserving in horizontal distributed environment.

Zhu, Yu-quan [10] proposed a secure two-party dot product protocol based on semi-honest model, the third party generated two random numbers, and then sent to site A. Site A used one of the random number multiplied by its vector, and sent to site B after adding another random number. Site B multiplied the received vector with its vector, and then sent the multiplication results and its vector to the third party. The third party used the two random numbers that are sent to the site A to calculate the vector product of the two sites, and got the global support.

Although this algorithm improved the computational efficiency without generating a large number of random numbers and matrix, but it reduced the safety, vectors of site B was fully revealed to the third party.

Boris Rozenberg et al. [5] added the false information to the original transaction database to achieve vertical distribution computing. Security and computation efficiency of the algorithm were improved for reducing the accuracy of the mining results.

3 A NEW SECURE MULTI-PARTY PROTOCOL TO GET GLOBAL SUPPORT—SMGSP

3.1 Relate definitions

Privacy preserving association rule in vertically distributed environment can be described as: Set up the database D has P items $(I_1, I_2,...,I_p)$, t records, distributed in sites $S_1, S_2,...,S_n$. P items are divided into $P_1, P_2,...,P$, and $P_1+P_2+...+P_n = P$, every item of each site stores all of the records.

Definition 1 if support of itemsets is greater than or equal to the minimum support min_sup, itemsets are frequent itemsets.

Definition 2 if confidence is greater than or equal to the minimum confidence min_conf, the rule is a strong association rule.

Definition 3 transaction vector $X_i^{S_i} = (x_{1i}, x_{2i}, ... x_{ti})$ represents every item in items $\{I_{p_{i-1}+1}, I_{p_{i-1}+2},...I_{p_i}\}$ of site S_i, and x_{ji} $(1 \leq i \leq n, 1 \leq j \leq t)$ represents the value of the jth column and ith row, $x_{ji} = 1$ represents item I_i is appeared in the transaction, and $x_{ji} = 0$ represents item Ii is not appeared.

Definition 4 E(x) is paillier function of additive homomorphic encryption, generating the key (e,d) to encrypt sequence $x_1, x_2,...,x_n$,

$$d(e(x1) * \cdots e(x_n)) = d(e(x_1 + x_2 + \cdots + x_n))$$
$$= x_1 + x_2 + \cdots + x_n$$

Definition 5 vector $X = (x_1, x_2,...,x_n)$, $Y = (y_1, y_2,...,y_n)$, the sum of two vectors $X+Y = (x_1+y_1, x_2+y_2,...,x_n+y_n)$.

Definition 6 local support of item Ii is lsup, it can be computed by transaction vector $X_i^{S_i} = (x_{1i}, x_{2i},...,x_{ti})$, $l\sup = \left(\sum_{j=0}^{j \leq t} x_{ji} = 1\right)$.

Definition 7 resuming that itemsets $I_1, I_2,...,I_k$ are distributed in sites $S_1, S_2,...,S_k$, transaction vector $X_k^{S_k} = (x_{1k}, x_{2k},...,x_{tk})$ represents every item I_k, therefore the number of $x_{j1} = x_{j2} = ... = x_{jk} (1 \leq j \leq t)$ is the global support of itemset.

3.2 SMGSP

We can find that association rule is mainly to get the support of itemsets from the definitions 1 and 2. Therefore the privacy preserving association rule mining algorithm in vertically distributed environment is mainly to solve safely getting global support of items without exposing local support and privacy information of raw data to other participants (site).

This section presents a new secure multi-party protocol called SMGSP to get global support, detailed descriptions are as follows:

1. Data center SP produces key (e,d), sends the public key e to each site, and sends −1 to site $S_2, S_4,...,S_k$ (k is multiple of 2, when $k = 2$, only send to the site S_2);
2. The site which has not received −1, in transaction vector $X_{p_i}^{S_i} = (x_{1i}, x_{2i},...x_{ti})$, the item of $x_{ji} = 1$ remains the same, the item of $x_{ji} = 0$ transforms into random number $u(u \geq t)$, and is encrypted to $e(X_{p_i}^{',S_i})$ by using the public key after the transformation; The site which has received −1, the item of $x_{ji} = 1$ transforms into $x_{ji} = -1$, the item of $x_{ji} = 0$ transforms into random number $u(u \geq t)$ in transaction vector $X_i^{S_i} = (x_{1i}, x_{2i},...x_{ti})$, and is encrypted to $e(X_{p_i}^{',S_i})$ by using the public key after the transformation; (if the items of item set distributed in one site, $x_{ji} = 1$ when all of the items is 1, $x_{ji} = u(u \geq t)$ otherwise);

3. Site S_1 sends $e(X'^{S_1}_{p_1})$ to the next site;
4. When $K = 2$, site S_2 multiplies the receiving encrypted data with its encrypted transaction vector (that is $e(X'^{S_1}_{p_1}) * e(X'^{S_2}_{p_2})$), then sends to data center SP, turn to step (12); Otherwise, execute the next step;
5. Site S_i ($2 \le i \le n-3$) multiplies the receiving encrypted data with its encrypted transaction vector (that is $e(X'^{S_i}_{p_i}) * \prod_{i=1}^{i} e(X'^{S_{i-1}}_{p_{i-1}})$), then sends to the next site. When candidate itemsets K are distributed in an odd number of sites, site S_{n-3} also sends to the site S_{n-1} except that sends the results of multiplying to the next site (site S_{n-2}), and then execute the next step; When candidate itemsets K are distributed in an even number of sites, turn to step (10);
6. Site S_{n-1} multiplies encrypted data of receiving with its encrypted transaction vector ($(e(X'^{S_{n-1}}_{p_{n-1}}) * \prod_{i=1}^{i} e(X'^{S_i}_{p_i})(1 \le i \le n-3))$ then sends to the site S_n. Site S_{n-2} multiply data of receiving with its encrypted transaction vector (that is $\prod_{i=1}^{i} e(X'^{S_i}_{p_i})(1 \le i \le n-2)$), then sends to the site S_{n-1};
7. Site S_{n-1} multiplies data of receiving ($\prod_{i=1}^{i} e(X'^{S_i}_{p_i})(1 \le i \le n-1)$) with its encrypted transaction vector ($e(X'^{S_{n-1}}_{p_{n-1}}) * \prod_{i=1}^{i} e(X'^{S_i}_{p_i})(1 \le i \le n-3)$), then sends to the data center SP, Site S_n multiplies encrypted data of receiving with its encrypted transaction vector ($e(X'^{S_{n-1}}_{p_{n-1}}) * e(X'^{S_n}_{p_n}) \prod_{i=1}^{i} e(X'^{S_i}_{p_i})(1 \le i \le n-3)$), then sends to the data center SP;
8. Data center decrypts the $\prod_{i=1}^{i} e(X'^{S_i}_{p_i})(1 \le i \le n-1)$ of receiving with the private key d, gets the vector $X = (x_1, x_2,...,x_n)$, data center decrypts the $e(X'^{S_{n-1}}_{p_{n-1}}) * e(X'^{S_n}_{p_n}) \prod_{i=1}^{i} e(X'^{S_i}_{p_i})(1 \le i \le n-3)$ of receiving with the private key d, then gets the vector $X_2 = (x_1, x_2,...,x_n)$;
9. Data center counts the number that both the values of vector X_1 and X_2 are 0. The number is global support H_sup of candidate itemsets K, ends the operation;
10. Site $S_i(n-2 \le i \le n-1)$ multiplies encrypted data of receiving with its encrypted transaction vector ($e(X'^{S_i}_{p_i}) * \prod_{i=1}^{i} e(X'^{S_{i-1}}_{p_{i-1}})$), then sends to the site S_n;
11. Site S_n multiplies encrypted data of receiving with its encrypted transaction vector (that is $\prod_{i=1}^{i} e(X'^{S_i}_{p_i})(1 \le i \le n)$), then sends to the data center SP;
12. Data center SP uses the private key d to decrypt the receiving data, then gets the vector $X = (x_1, x_2,...,x_n)$;
13. The number of 0 in vector $X = (x_1, x_2,...,x_n)$ is global support H_sup of candidate itemsets K, ends the operation.

4 ANALYSIS

4.1 Correctness of the algorithm

Denoting the candidate itemsets K are distributed in K sites, we prove the correctness of the algorithm as follow when value of K is even or odd.

(1) Value of K is even

We can observe from step (1) of SMGSP protocol, item of $x_{ji} = 1$ in transaction vector $X^{S_i}_{p_i} = (x_{1i}, x_{2i}, ... x_{ti})$ on sites $S_1, S_3,...,S_{n-1}$ remains the same, and item of $x_{ji} = 1$ in transaction vector $X^{S_i}_{p_i} = (x_{1i}, x_{2i}, ... x_{ti})$ on their adjacent to the next site takes the opposite value. We can conclude from definition 4.4 that the number of $\sum x_{ji} = 0$ is the global support of the candidate itemsets K when $1 \le j \le t$. For example, candidate itemsets 6 {A, B, C, D, E, F}, one of its row vector is $(1, 1, 1, 1, 1, 1)$, transformed vector is $(1, -1, 1, -1, 1, -1)$, $1 + (-1) + 1 + (-1) + 1 + (-1) = 0$ according to the step 1 of SMGSP protocol.

The data center receives encrypted vector $\prod_{i=1}^{i} e(X'^{S_i}_{p_i})(1 \le i \le k)$, therefore we can get the following conclusion according to the definition 4:

$$\prod_{i=1}^{i} e(X'^{S_i}_{p_i}) = e(X'^{S_1}_{p1}) * e(X'^{S_2}_{p2}) * ... * e(X'^{S_k}_{pk})$$
$$= e\left(X'^{S_1}_{p1} + X'^{S_2}_{p2} + ... + X'^{S_k}_{pk}\right)$$
$$= \begin{cases} e(1+(-1)+...+(-1)) = 0 \ all of \ X^{S_k}_{pk} = 1 \\ e(1+(-1)+...+(-1)) = 0 \ X^{S_k}_{pk} = 1 or 0 \\ e(1+(-1)+...+(-1)) = 0 \ all of \ X^{S_k}_{pk} = 0 \end{cases} \quad (1)$$

($u_k \ge t$, so the value of $1 + u_1 +...+ (-1)+...+ u_k$ impossibly equals to 0)

(2) Value of K is odd.

In this case, it needs to prove two points: the value of item on vector $X_1 = (x_1, x_2,...,x_n)$ or $X_2 = (x_1, x_2,...,x_n)$ equal to 0 Whether represents original value of each item on all sites equals to 1; the values of item on vector $X_1 = (x_1, x_2,...,x_n)$ and $X_2 = (x_1, x_2,...,x_n)$ equal to 0 Whether represents original value of candidate itemsets K on each site are all equal to 1.

$X_1 = (x_1, x_2,...,x_a) = d(\prod_{i=1}^{i} e(X'^{S_i}_{p_i})(1 \le i \le k-1))$, only the site S_k is not involved in operation, so it is equivalent to the K value is an even proved above. $X_2 = (x_1, x_2,...,x_a) = d(e(X'^{S_{k-1}}_{p_{k-1}}) * e(X'^{S_k}_{pk}) \prod_{i=1}^{i} e(X'^{S_i}_{p_i})(1 \le i \le k-3))$, only site S_{k-2} is not involved in operation, so it is also equivalent to the K value is an even proved above. Therefore the first point is proved.

When any item $x_i = 0$ of $X_1 = (x_1, x_2,...,x_n)$, the value of items on sites {$S_1, S_2,...,S_{k-1}$} are all equal to 1 according to formula (4-1), when any item $x_i^2 = 0$ of $X_2 = (x_1, x_2,...,x_n)$, the value of items on sites $\{S_1, S_2,...,S_{k-3}, S_{k-1}, S_k\}$ are all equal to 1 according formula (4–1), therefore, when item $x_i = x_i^2 = 0$ of $\{S_1, S_2,...,S_{k-1}\} \cup \{S_1, S_2,...,S_{k-3}, S_{k-1}, S_k\} = \{S_1, S_2,...,S_k\}$, each item on every site is equal to 1, hence the above second point is also proved.

We can conclude from (1) and (2) that our algorithm is correct.

4.2 Security analysis

1. Except the second site, other sites only get the transaction vector addition result of front site even if it can decrypt, they do not know the specific transaction vector value of each site.
2. According to the properties of the additive homomorphic encryption, though the data center knows the private key, it only gets the transaction vector addition result of every front site through decryption, it does not know specific value of the vector of each site.
3. u is randomly generated and $u \geq t$, if data center SP finds any item's value of the sum of transaction vector is greater than 1 after decryption, it also only knows the value of item is 0 on one or more sites, but it cannot know on which site, let alone know the number of sites.
4. A site does not need to communicate with other sites except when getting global support, therefore there is no security issue in other steps.

In summary, the proposed protocol can securely calculate global support of K itemsets namely securely get association rules.

4.3 Computation and communication cost analysis

Obtaining global support of itemsets securely is the key for finding association rules of vertical distributed data. Computation and communication cost of SMGSP is analyzed as follows.

The number of participant site is n, the number of items in transaction vector is t (or records in the database), the maximum of number of values of 0 in per vector is a. Table 1 shows the complexity of getting the global support when the number of sites involved in the operation is odd. (Because when the number of sites involved in the operation is even, calculation is simple relatively, and we should show the maximal complexity). It is mainly the step (1), (2), (3), (5), (6), (7), (8) and step (9), a total of eight steps to get global support.

Saeed S. gave ulti-security getting global support protocol CSI (Cardinality of Set Intersection), which used additive homomorphic encryption and random number. Its' complexity is shown in Table 2.

Table 1. Complexity for calculating global support of SMGSP.

Step	Communication cost	Computation cost
(1)	n keys and $\lfloor n/2 \rfloor$ numbers	Public and private keys generation
(2)	–	A random numbers generation n vectors encryption
(3)	One vector	–
(5)	n–3 vectors	A group of vectors multiplication
(6)	Two vectors	A group of vectors multiplication
(7)	Two vectors	A group of vectors multiplication
(8)	–	Two vector decryption
(9)	–	One vector comparison

Table 2. Complexity of calculating global support of CSI.

Step	Communication cost	Computation cost
(1)	–	n random vectors generations
(2)	n–1 keys	Public and private keys generation
(3)	n–1 vectors	n–1 vectors encryptions
(4)	One vector	One vector encryption a group of vectors multiplication
(5)	n–3 vectors	n–3 vector encryptions a group of vectors multiplication
(6)	One vector	One random vector generation two vector encryptions a group of vectors multiplication
(7)	–	One vector decryption two vectors addition
(8)	n–2 vectors	One random permutation generation two vector permutations
(9)	n numbers	n–2 vectors subtraction and comparison one number addition

From the Table 1 and Table 2 we can observe that the communication cost of SMGSP is: $1 + (n-3) + 2 + 2 = n+2$ vectors, n keys and $\lfloor n/2 \rfloor$ numbers. CSI is: $(n-1) + 1 + (n-3) + 1 + (n-2) = 3n-4$ vectors, n–1 keys and n numbers. The deference of communication cost between two algorithm is $(3n-4) - (n+2) = 2n-6$, when $n = 3$, $2n-6 = 0$, the communication cost between two algorithm is

Table 3. Comparison of computation cost in algorithm of SMGSP and CSI.

	SMGSP	CSI
Vector of encryption and decryption	n+2	$(2n-1)+1$
Number of generation	A random numbers	$n+1$ random vectors 1 random permutation
Number of compared vector	1	$n-2$
Number of permutation	–	2
Number of subtracted vector	–	$n-2$
Number of added vector	–	1

Table 4. Global support of K itemsets in different environments.

		Global support of K itemsets	
K	N	Centralize	SMGSP
2	2	258	258
3	3	246	246
5	5	66	66
7	7	29	29
10	10	4	4

almost the same, but when the number of sites increase, the communication cost of CSI increases double than SMGSP.

Both of the computation cost of SMGSP and CSI need to generate public key and private key, and they both have three steps that need 1 group of vectors multiplication, SMGSP have n vectors to encrypt, two vectors to decrypt, a random number to generate, one vector to compare. CSI has $(n-1)+1+(n-3)+2 = 2n-1$ vectors to encrypt, one vector to decrypt, $n+1$ random vectors and one random permutation to generate, $n-2$ vectors to compare, two vectors to permutated randomly, $n-2$ vectors to subtract, and one number to add.

The differences between two algorithms are shown in Table 3, obviously, the computation cost of SMGSP is less than CSI.

5 EXPERIMENT RESULTS

We simulated multiple sites on one computer, use RMI (Remote Method Invocation) to realize communication between sites. Experiment running environment for the computer: Intel (R) Core (TM) i5-2430-m @ 2.4 GHz CPU 2.0 GB/Windows 7, experimental data is 0–1 matrix of transactional data after pretreatment.

Experiment 1: verifying the correctness of the proposed SMGSP.

SMGSP is proved to be correct, as long as verify the global support of K item sets in centralized mining to be equal to that in distributed mining. Table 4 shows the comparison of global support between distributed in N sites using SMGSP protocol and centralized environment, 1 K of

transactions in this experiment. From Table 4 we can observe that the global support calculated by using SMGSP is correct.

Experiment 2: execution time contrast of SMGSP when the item sets are distributed in different number sites.

The contrast execution time of SMGSP protocol between different number of transactions and item sets distribution in 2, 3, 4, 5, 7, 10 is shown in Figure 1.

It can be observed from the experimental results, execution time is basically the same when the item sets distributed in an even number of sites, also in an odd number of sites. N vectors need to encrypt for an odd number of sites from the analysis of the complexity of the previous section, these n vectors are encrypted one time in each of n sites respectively. Encryption process is executed in parallel, so no matter how many sites, just take a vector of the encryption time in fact; only a vector needs to decrypt when number of sites is even, two vectors need to decrypt when number of sites is odd, therefore execution time of even number of sites is smaller than odd number of sites. And due to the number of vector associated with the transaction number, the execution time of each site is linear increased as the number of transaction is increased.

Experiment 3: the contrast of execution time of getting global support protocol between two protocols.

The contrast execution time of SMGSP protocol and CSI protocol which proposed by Saeed S. is shown in Figure 2, transaction number is 3 K. X-axis represents the number of different sites, the Y-axis represents the execution time, and the unit is second.

From Figure 2 it can be observed, the more site numbers, the better performance of SMGSP. CSI has $2n-1$ vectors to encrypt, the $n-3$ vectors in step (5) are parallel encrypted, the other $n+2$ vectors cannot be executed in parallel, its encrypting vector number is influenced by the number of sites, but SMGSP always has only one vector to encrypt.

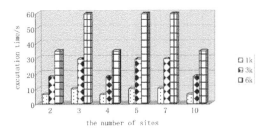

Figure 1. Execution time of the SMGSP.

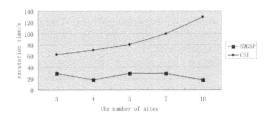

Figure 2. Execution time comparison of SMGSP and CSI in different number of sites.

6 CONCLUSIONS

In this paper, we proposed a new secure multi-party protocol named SMGSP to get global support of item sets. Experiment results show that the performance of SMGSP is better than CSI protocol.

ACKNOWLEDGEMENTS

Project supported by the National Natural Science Foundation of China (61103245) and the Guangxi Natural Science Foundation (2013GXNSFAA253003).

REFERENCES

[1] Cheung S. & Thinh N. 2007. Secure multiparty computation between distributed networks terminals [J]. *EURASIP Journal on Information Security*: 1–11.

[2] Dragos T. & Sanguthevar R. 2007. Fast cryptographic multiparty protocols for computing Boolean scalar products with applications to privacy preserving association rule mining in vertically partitioned data[C]//*LNCS 4654: Da Wak: Industrial Conference on Data Mining Posters and Workshops*: 418–427.

[3] Kumbhar, M.N. 2012. Privacy preserving mining of Association Rules on horizontally and vertically partitioned data: A review paper[C]// *Proc of 12th International Conference on Hybrid Intelligent Systems(HIS)*: 231–235.

[4] Luong, T.D., Ho, T.B. 2010. Privacy Preserving Frequency Mining in 2-Part Fully Distributed Setting[J]. *IEICE Transactions on Information and Sysytems E93 D*(10): 2702–2708.

[5] Rozenberg B. & Gudes E. 2006. Association rules mining in vertically partitioned databases [J]. *Data & Knowledge Engineering* 59(2): 378–39.

[6] Saeed S. & Ali M. 2009. Secure two and multi-party association rule mining[C]//*Computational Intelligence for Security and Defense Applications*: 1–6.

[7] Shen C. & Zhan, J. 2007. Information theoretically secure number product protocol[C]//Hong Kong: *Proceedings of the Sixth International Conference on Machine Learning and Cybernetics*: 19–22.

[8] Vaidya J. & Clifton C. 2002. Privacy Preserving Association Rules Mining in Vertically Partitioned Data[C]//*In the 8th ACM SIGMOD International Conference on Knowledge Discovery and Data Mining*: 639–644.

[9] Vaidya J. & Clifton C. 2004. Secure Set Intersection Cardinality with Application to Association Rule Mining[J]. *Journal of Computer Security* 4(13):593–622.

[10] Zhu, Yu-quan et al. 2011. A Privacy Preserving Algorithm for Mining Distributed Association Rules[C]// *Proc of 2011 International Conference on Computer and Management(CAMAN)*: 1–4.

Information Technology and Applications – Li (Ed.)
© 2015 Taylor & Francis Group, London, ISBN 978-1-138-02677-3

Cryptanalysis of a one-time two-factor authentication and key agreement scheme using smart card

Jie Ling & Guangqiang Zhao
Faculty of Computer, Guangdong University of Technology, Guangzhou, China

ABSTRACT: The one-time two-factor authentication scheme that combined password with smart card is one of the most widely used methods. In 2012, Xie showed that Holbl's scheme is vulnerable to the smart card lost attack, besides, the scheme suffers from impersonation attack and parallel session attack, and proposed a new one-time two-factor mutual authentication and key agreement scheme to eliminate their weakness and claimed that it is secure. However, we find that Xie's scheme is still insecure; it cannot resist stolen smart card attack and password guessing attacks, and impersonation attack. Besides, it cannot detect the wrong password timely. Meanwhile, it also suffers from user anonymity violation and clock synchronization problem.

Keywords: smart card; impersonation attack; anonymity; clock synchronization

1 INTRODUCTION

In 2000, Hwang-Li introduced the concept of smart card in password based on user authentication systems. Since then a considerable amount of research [1–10] has been carried out in this field. In 2009, Hsiang and Shih [1] showed that Liao-Wang [2] scheme cannot resist insider attack, masquerade attack, and server spoofing attack, meanwhile, the scheme cannot achieve mutual authentication. And they proposed an enhanced scheme on Liao-Wang scheme, However, Sood et al. [3] pointed out that Hsiang et al.'s scheme is still insecure; it fails to resist replay attack, impersonation attack and stolen smart attack. And they proposed a secure dynamic identity based authentication scheme. Unfortunately, Xiong Li et al. [4] found that Sood et al.'s scheme still suffers from some attacks such as leak-of-verifier attack, stolen smart card attack and impersonation attack. In 2010, Holbl et al. [5] showed that Shieh et al.'s [6] scheme is vulnerable to the smart card lost attack and does not achieve perfect forward secrecy. Further, they proposed a security enhanced scheme. However, in 2012 Xie [7] showed that Holbl et al.'s security enhanced scheme is still vulnerable to the smart card lost attack, and the scheme cannot resist impersonation attack and parallel session attack. Also, he proposed a new one-time two-factor mutual authentication and key agreement scheme. In 2012, He et al. [8] proposed an efficient ID-based scheme for mobile client-server environment on ECC without the MapToPoint function. The scheme attempts to cope with many of the well know security and efficiency problems of previous schemes. Despite of its claim of provable security, unfortunately, in 2013, Ding Wang et al. [9] pointed out that He et al.'s scheme cannot resist a reflection attack and also a parallel attack; it suffers from clock synchronization problem and user anonymity violation. And they proposed an enhanced scheme.

In this paper, we showed that Xie's scheme is in fact still insecure in the face of an active attacker. We demonstrated it by presenting stolen smart card attack and impersonation attack that breach the essential goal of mutual authentication. Besides, we pointed out that it also suffers from user anonymity violation and clock synchronization problem and cannot detect the wrong password quickly.

2 REVIEW OF XIE'S SCHEME

Before the review, we first notify the whole notations that will be used throughout the paper.

P, q: two large prime numbers, such as $q|p$-1;
g: a generator with order q of the group $GF(p)$;
U: the user;
ID: U's identify;
PW: U's password;
S: the server;
$(X, Y = g^X \bmod p)$: S's private-public key pair;
$h()$: a secure cryptographic hash function;
$\|$: string concatenation operation;
$\oplus$: bitwise XOR operation;

Xie's scheme consists of two phases: Registration phase and Login and key agreement phase. The authentication scheme performs as follows, and it is also shown in Figure 1.

2.1 Registration phase

U and S carry out the following steps during the user registration phase.

Step1: U chooses a password PW, and his identity ID, then submits ID to S.

Step2: S computes $d = h (ID \oplus X)$, and sends a smart card to U via a secure channel. The smart card contains ID, d, and h ().

Step3: U computes $R = d \oplus h (PW)$, and replaces d with that is, the smart card contains ID, $h()$ and $R = h(ID \oplus X) \oplus h(PW)$.

2.2 Login and key agreement phase

If when U is about to login to the remote server S, U completes the following operations.

Step1: U inserts his smart card into the card reader of a **terminal**, and keys PW.

Step2: The smart card generates a random number $c < p-1$, and computes: $d = R \oplus h(PW) = h(ID \oplus X)$, $C_0 = Y^c \bmod p$, $C_1 = g^c \bmod p$, $C_2 = h(d \| C_0 \| T_u)$ where T_u is the current timestamp. Then, the smart card sends $\{ID, C_1, C_2, T_u\}$ to S.

Upon receiving the message $\{ID, C_1, C_2, T_u\}$, S completes the following operations:

Step3: S checks if $(T_S - T_u) \leq \Delta T$, where T_S is S's current timestamp and ΔT is the expected valid time interval for transmission. If not, S rejects U's login request. Otherwise, S computes $C_2^* = h(h(ID \oplus X) \| C_1^X \| T_u)$, and checks if $C_2^* = C_2$, if not S rejects U's login request. Otherwise, U is authenticated by S.

Step4: S chooses a large random integer, $s < p-1$. Computes $C_3 = g^s \bmod p$, C_1^s, $C_4 = h(h(ID \oplus X) \| T_S \| C_1^s \| T_U)$, and the session key, $K = h(C_1^s)$, and returns $\{ C_3, C_4, T_S \}$ to U.

After receiving the message, $\{C_3, C_4, T_S\}$, the smart card checks the validity of T_S by $(T_u^* - T_S) \leq \Delta T$, where T_u^* is the current time. If not, it is terminated, otherwise, the smart card computes $C_4^* = h (d \| Ts \| C_3^c \| T_U)$, and compares it with the received C_4. If they are equal. S is authenticated by U. Then, the smart card computes the session key $K = h(C_3^c)$.

Thus, S and U share the session key K, for subsequent private communications.

3 CRYPTANALYSIS OF XIE'S SCHEME

3.1 Stolen smart card and password guessing attack

If a malicious adversary stolen U_i's smart card. And then he can easily extract the identity ID_i of U_i from the memory of smart card. He entered the smart card into a terminal. And inputs the true identity of U_i and a password PW^*. The smart card computes and compares $d = R \oplus h(PW) = h(ID \oplus X) \oplus h(PW) \oplus h(PW^*)$, $C_0 = Y^c \bmod p$, $C_1 = g^c \bmod p$, $C_2 = h(d \| C_0 \| T_u)$ and sends $\{ID_i, C_1, C_2, T_u\}$ to S. then S computes $C_2^* = h(h(ID \oplus X) \| C_1^X \| T_u)$, and checks if $C_2^* = C_2$. If they are not equal, it means the attacker enter a wrong password. However, the smart card and S do not limit the login times. Therefore, the attacker can make a limited number login attempts with a guessed password. Due to the small space and low entropy of password, users are likely to choose the password which is easily remembered. The possibility of get the correct password is very high. Once the adversary guesses the right password, we can observe the attack can login on the server S from the scheme.

3.2 Impersonation attack

The security of Xie's scheme depended on the value d, if the attacker gets d, he can carry out user impersonation attack and server impersonation attack. During login and key agreement phase, U sends login request $\{ID_u, C_1, C_2, T_u\}$ to S, a malicious adversary U_a can easily obtain the identity ID_u of U by intercepting any login request of U. Then U_a can impersonate U to cheat S as explained as follows:

1. U_a sends the registration request ID_u, where ID_u is the identity of U.
2. S sends U_a the smart card contains $\{ID_u, d, h ()\}$.
3. U_a extracts values d from the smart card.

It is easy to observe that, after the attacker get the value of d, he can pass the server authentication

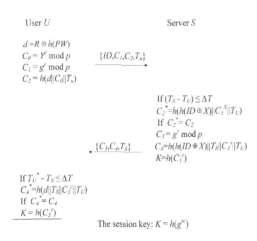

User U Server S

$d = R \oplus h(PW)$
$C_0 = Y^c \bmod p$ $\{ID, C_1, C_2, T_u\}$
$C_1 = g^c \bmod p$
$C_2 = h(d \| C_0 \| T_U)$

If $(T_S - T_U) \leq \Delta T$
$C_2^* = h(h(ID \oplus X) \| C_1^X \| T_U)$
If $C_2^* = C_2$
$C_3 = g^s \bmod p$
$\{C_3, C_4, T_S\}$ $C_4 = h(h(ID \oplus X) \| T_S \| C_1^s \| T_U)$
$K = h(C_1^s)$

If $T_U^* - T_S \leq \Delta T$
$C_4^* = h(d \| T_S \| C_3^c \| T_U)$
If $C_4^* = C_4$
$K = h(C_3^c)$

The session key: $K = h(g^{sc})$

Figure 1. Xie's authentication scheme.

218

verify and can compute the session key between U and S from the scheme.

Here we move one step forward from the previous demonstration. The attacker possessing identity ID_u and d of U and he can impersonate S to cheat U as explained as follows:

1. Suppose U sends the login message $\{ID_u, C_1, C_2, T_u\}$ to S.
2. U_a intercepts and blocks $\{ID_u, C_1, C_2, T_u\}$ from reaching to S.
3. U_a chooses a random integer $s<p-1$, computes $C_3 = g^s \bmod p$, C_1^s, $C_4 = h\ (d\|T_S\|C_1^s\|T_S)$, and returns $\{C_3, C_3, T_S\}$ to U.

It is easy to observe that, upon receiving the S (actually, U_a's) request, the user U will of course accept it as legal S because of the following reasons:

1. It contains valid identity ID_u of U and the fresh timestamp T_S.
2. The equivalence $C_4^* = C_4$ holds as $C_4^* = h\ (d\|T_S\|C_3^c\|T_S)$ where $C_3^c = C_1^s = g^{sc} \bmod p$.

3.3 Not preserving user anonymity and intractability

User anonymity requires that only the server knows the identity of the user with whom he is interacting, while any third party is unable to do this. While, user intractability requires that any adversary should be prevented from linking one unknown user interacting with the server to another transcript, that is to say, the adversary is not capable of telling whether he has observed the same user twice. In Xie's scheme, the user's identity ID is transmitted in plain-text, which may leak the identity of the logging user once the login messages were eavesdropped. That is to say, without employing any effort an adversary can distinguish and recognize the particular transactions performed by the specific user U. Moreover, the user's identity ID is static in all the login phases, which may facilitate the attacker to trace out the different login request messages belonging to the same user and to derive some information related to the user U. In summary, neither initiator anonymity nor initiator un-traceability can be preserved in their scheme [9].

3.4 The clock synchronization problem

It is well known that, remote user authentication schemes employing timestamps to provide message freshness may still suffer from replay attacks as the transmission delay is unpredictable in existing networks. Besides, clock synchronization is difficult and expensive in existing network environments,

especially in wireless and mobile networks [10] and distributed networks [11]. Hence, these schemes employing the timestamp mechanism to resist replay attacks are not suitable for mobile applications [12, 13]. In Xie's scheme, obviously, this principle is violated.

3.5 Detect wrong password slowly

In the login and key agreement phase of Xie's scheme, if the user puts the wrong password PW^* (which is not equal to PW), then the smart card does not detect the flaw and continues to compute the login message using the incorrect password as follows:

1. The smart card computes $d = R \oplus h(PW) = h(ID \oplus X) \oplus h(PW) \oplus h(PW^*)$, $C_0 = Y^c \bmod p$, $C_1 = g^c \bmod p$, $C_2 = h(d\|C_0\|T_u)$.
2. The smart card sends $\{ID_u, C_1, C_2, T_u\}$ to S.
3. S receives $\{ID_u, C_1, C_2, T_u\}$, computes and checks if $C_2^* = h(h(ID \oplus X)\|C_1^X\|T_u)$? $= C_2$. However, if PW^* is not a correct password. The value $d = R \oplus h(PW) = h(ID \oplus X) \oplus h(PW) \oplus h(PW^*)$ will not be equal to $h(ID \oplus X)$, of course $C_2^* = h(h(ID \oplus X)\|C_1^X\|T_u) \neq C_2$, therefore, S rejects the login, Unfortunately, it wasted unnecessary extra communication and computational overheads of S during this phase. If the validity of the password can quickly be checked at the beginning, this situation will not be happen.

4 CONCLUSIONS

In this paper we have reviewed Xie's one-time two-factor authentication and key agreement scheme. The improved scheme is equipped with a claimed proof of provable security. We have shown that besides the problems of clock synchronization and user anonymity violation. It suffers from other weakness. We described how an attacker can break the security walls of the scheme by merely obtaining user's identity from intercepted login request. We have enlightened that the presence of user's plaintext identity in login request is the main reason behind various vulnerabilities such as impersonation attack; also the improved scheme cannot resist stolen smart attack. In future, we plan to come up with more viable approach for user authentication with user anonymity.

ACKNOWLEDGEMENT

This research work is supported by the Key Program of the Natural Science Foundation of Guangdong Province (No. S2012020011071), and the Industry-Education-Research Cooperation Project of Guangdong Province and the Ministry of Education (number 2012B091000037 and 2012B091000041) and the Project of Guangzhou Science and Technology (No.2013J4300058).

REFERENCES

[1] Hsiang H-C., Shih W. Improvement of the secure dynamic ID based remote user authentication scheme for multi-server environment. Computer Standards & Interfaces 31(2009):1118–1123.

[2] Liao Y-P., Wang S-S. A secure dynamic ID based remote user authentication scheme for multi-server environment. Computer Standards & Interfaces 31(2009):24–29.

[3] Sood S-K., Sarje A-K., Singh K. A secure dynamic identity based authentication protocol for multi-server architecture. Journal of Network and Computer Applications 34(2011):609–18.

[4] Xiong Li, Yongping Xiong, Jian Ma et al. An efficient and security dynamic identity based authentication protocol for multi-server architecture using smart cards Journal of Network and Computer Applications. 35(2012):763–769.

[5] Holbl M., Welzer T., Brumen B. Attacks and improvement of an efficient remote mutual authentication and key agreement scheme. Cryptologia 34(2009):723–728.

[6] Shieh W.G., Wang F.M. Efficient remote mutual authentication and key agreement Computer & Security 25(2006):72–77.

[7] Qi Xie. Improvement of a security enhanced one-time two-factor authentication and key agreement scheme. Scientia Iranica 19(2012):1856–1860.

[8] D. He, J. Chen, J. Hu, An id-based client authentication with key agreement protocol for mobile client-server environment on ECC with provable security, Information Fusion 13(2012):223–230.

[9] Ding Wang, Chun-guang Ma, Cryptanalysis of a remote user authentication scheme for mobile client-server environment based on ECC. Information Fusion 14(2013):498–503.

[10] A. Giridhar, P. Kumar, Distributed clock synchronization over wireless networks: algorithms and analysis, in: Proceedings of the 45th IEEE conference on Decision and Control, IEEE, (2006):4915–4920.

[11] J. Han, D. Jeong, A practical implementation of IEEE 1588-2008 transparent clock for distributed measurement and control systems. IEEE Transactions on instrumentation and Measurement 59(2010):433–439.

[12] S. Islam, G. Biswas. A more efficient and secure id-based remote mutual authentication with key agreement scheme for mobile devices on elliptic curve cryptosystem, Journal of Systems and Software 84(2011):1892–1898.

[13] C. Chang. C. Lee, A secure single sign-on mechanism for distributed computer networks IEEE Transactions on Industrial Electronics 59(2012):629–637.

Information Technology and Applications – Li (Ed.)
© *2015 Taylor & Francis Group, London, ISBN 978-1-138-02677-3*

Low complexity adaptive robust algorithm for nonlinear Active Noise Control

Bin Wang & Zizhong Tan
Department of Electrical Engineering, Hebei University of Science and Technology, Shijiazhuang, China

ABSTRACT: To circumvent nonlinear distortions which occur in nonlinear Active Noise Control (ANC) system used for practical applications. A simplified robust filtered-s LMS (FsLMS) algorithm is proposed in this paper. The performance of the new model is evaluated in terms of computational convergence characteristics. We also propose the algorithm based on decomposing the long adaptive filter into smaller sub-filters. This leads to a significant improvement in the convergence rate of the algorithm with low computational overhead. However, the algorithm has a high final Mean-Square Error (MSE) at steady-state that increases as number of sub-filters increases. Computer simulations have been carried out to verify that the nonlinear adaptive filter with the new algorithm is more effective in eliminating nonlinear distortions in ANC systems than the linear FXLMS, FuLMS and FsLMS algorithm, the new algorithm requires less computations without suffering from any performance degradation of nonlinear ANC systems.

Keywords: robust; convergence; decomposing; sub-filter

1 INTRODUCTION

In recent years, active control of nonlinear noise processes for a single channel system has been the topic of much research under the following situations: (1) the noise signal received by a reference microphone is nonlinear and predictable (chaotic), while the secondary path transfer function between the speaker and the error microphone has a non-minimum phase; (2) the primary path exhibits nonlinearity [1]. In both cases, linear adaptive filter for ANC systems performs poorly, and even fails to work. Researches have reliably shown that nonlinear ANC based on the non-linear adaptive filter outperforms ANC based on linear adaptive filters, and shows good performance for attenuating nonlinear low frequency noises [1–4]. However, till today, there is no unique theory for modeling and characterizing nonlinear ANC systems. Adaptive nonlinear controllers for ANC systems can be mainly divided into two categories: Neural Networks (NNs) and adaptive polynomial filters. Nonlinear ANC controllers, based on neural networks (such as multilayer neural networks [1–4], Radial Basis Function Networks (RBFNS) [4], Recurrent Neural Networks (RNNs) [5], Functional Link Artificial Neural Networks (FLANNs) [6] and so on) using a training algorithm developed from a back propagation scheme, were reported in applications where the actuators exhibit nonlinear characteristics.

In [13], a stereophonic LMS algorithm based on the decomposition of the adaptive filter was introduced that leads to considerable improvements in convergence rate over the Stereo LMS (SLMS) and with a complexity that is almost the same as that of the SLMS. A popular approach to lowering the computational complexity of the adaptive algorithm is to update only a subset of the adaptive filter coefficients per iteration. The resulting algorithm needs fewer arithmetic operations compared with its full-update counterpart. In [14], the decomposition technique was applied to the Transform-Domain (TD) algorithm and algorithm complexity was reduced by using the Selective Coefficient Update (SCU) approach [15].

In this paper, a simplified and computationally efficient nonlinear ANC structure based on decomposing the long adaptive filter into smaller sub-filters is proposed which based on Filter-s LMS (FSLMS) algorithm as reported in [12]. Section 1 provides the whole model, and provides derivation of the adaptive algorithm based on the decomposition for nonlinear ANC systems in detail. Then, theoretical study of the FsLMS algorithm based on the model introduce in section 1 is analyzed in section 2. Computer simulations have been carried out to validate the proposed algorithm for nonlinear ANC systems. Some of the significant results are shown in section 3. Finally, concluding remarks follow in section 4.

2 FULL-UPDATE FXLMS ALGORITHM BASED ON THE DECOMPOSITION OF THE ADAPTIVE FILTER AND THE NEW ANC SYSTEM

In ANC, the signal-channel input signal vectors and filter coefficient vectors can be combined as $X(n) = [x(n)x(n-1) \ldots x(n-L+1)]$, $W(n) = [\omega_1(n) \omega_2(n) \ldots \omega_L(n)]$ respectively, and L is the adaptive filter length, the new model of the ANC system is shown in Figure 1.

In [11], the input vector $X(n)$ and the weight vector $W(n)$ are partitioned into M sub-vectors such that

$$X(n) = [X_1^T(n)X_2^T(n) \ldots X_K^T(n)]^T \tag{1}$$

$$W(n) = [W_1^T(n)W_2^T(n) \ldots W_K^T(n)]^T \tag{2}$$

where $X_k(n)$ and $W_k(n)$ have each P_k elements, and $2L = \sum_{k=1}^{M} P_k$, $k=1,2 \ldots K$ need not to be equal for implementation purposes. However, we will assume throughout this paper that all sub-filters have equal lengths, that is, $P = P_1 = P_2 = \ldots = P_k$. We introduce M error signals as [13]

$$e_k(n) = d(n) - \sum_{i=1}^{k} W_i^T(n)X_i(n)S(n), \quad k = 1,2,\ldots,K \tag{3}$$

And let each sub-filter $W_k(n)$ seeks the minimum of the corresponding cost function $J_k(ek(n)) = E\{e_k^2(n)\}$. Here, $d(n)$ is the output of the unknown system. Minimization of the instantaneous estimate of this cost function with respect to $W_k(n)$ results in the FXLMS equation for each sub-filter as [11],

$$W_k(n+1) = W_k(n) + \mu_k X_k(n)e_k(n) \tag{4}$$

where μ_k is the adaptation step size, and $e_k(n)$ has the order update equation

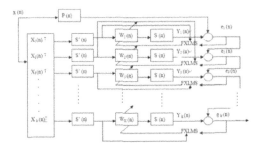

Figure 1. The adaptive noise control setup with filter decomposition.

$$e_k(n) = e_{k-1}(n) + W_k^T \mu_k X_k(n) \tag{5}$$

With $e1(n) = d(n) - W_1^T X_1(n)$, and the system output error is given as follows:

$$e_K(n) = d(n) - \sum_{m=1}^{K} W_m^T(n)X_m(n)$$

$$= d(n) - W^T(n)X(n) \tag{6}$$

The adaptive noise cancellation setup with filter decomposition is shown in Figure 1. The arithmetic complexity of the new algorithm, in terms of number of multiplications and additions, is almost the same as that of the traditional FXLMS algorithm. The algorithm adds only M multiplications for the calculation of $\mu_k e_k(n)$; $k=1, 2; \ldots; K$, which is a negligible amount.

3 LAYOUT OF TEXT

The controller proposed in (11) consists of a FLANN, the weights of which are updated using a novel robust algorithm. The input to a FLANN is non-linearly expanded using a suitable basis function. The basis function can be trigonometric, Chebyshev or Legendre. The use of a Chebyshev or Legendre basis function requires input normalization, which is not practically possible when the ANC scheme deals with impulsive noises. This paper uses a trigonometric expansion as it is simple to implement and does not require amplitude normalization. This trigonometric expansion not only increases the dimensionality, but also maps the linear inputs to nonlinear ones. Each of the expanded inputs are then linearly weighted and summed to produce the output. Compared with the other artificial neural structure such as multilayer artificial neural network or radial basis function network, the proposed structure involves low computational complexity.

Let $x_k(n)$ be the output for the reference signal of each part and the corresponding input signal vector is $x_k(n) = [x_k(n), x_k(n-1), \ldots, x_k(n-p+1)]^T$ the P-element input signal vector is trigonometrically expanded as follows:

$$x_{ks}(n) = \{x_k^T(n), \sin[\pi x_k^T(n)], \cos[\pi x_k^T(n)], \ldots,$$

$$\sin[\beta\pi x_k^T(n)], \cos[\boldsymbol{\beta\pi} x_k^T(n)]\}^T \tag{7}$$

With P as the number of tap-delayed input samples. The total length of the input vector, as well as the weight vector, is denoted by $N_k = P(2\beta+1)$, where β is the order of the FLANN filter. If $Akf(n) = [akf1(n), akf2(n), \ldots akfp(n)]T$, represents the adaptive weight vector. The output of the controller is given by

$$y_k(n) = A_{kf}^T(n)X_k(n) \qquad (8)$$

The residual noise measured by the error is given by

$$e_k(n) = e_{k-1}(n) - s_k(n) * A_{kf}^T(n)X_k(n), (d(n) = e^1(n)) \qquad (9)$$

The conventional FsLMS algorithm has been developed with the objective of minimizing the cost function $E(e^2(n))$. Which is the expectation operator. The FsLMS algorithm [5] is given by

$$A_{kf}(n+1) = A_{kf}(n) - \mu_k e_k(n)X_k'(n) \qquad (10)$$

It is clear that for higher values of $e(n)$, as in the case of impulsive noise, the FsLMS algorithm may diverge. To alleviate this problem, a robust cost function [14] is defined as

$$\xi = E\left[\log(1 + e^2(n)/2\sigma^2)\right] \qquad (11)$$

Is used in the development of a Robust FsLMS (RFsLMS) algorithm for tuning the parameters of an ANC system. In (11) σ^2 is computed as an estimated variance of $e(n)$ using a sliding window approach with a window length N [8]. The weight vector of an RFsLMS algorithm based ANC scheme is updated using a gradient descent approach [1] which minimizes the cost function in (11). The update equation for the weight vector is given by

$$A_{kf}(n+1) = A_{kf}(n) - \mu_k\left[e(n)/e^2(n) + 2\sigma^2\right]X_k'(n) \qquad (12)$$

However, as in the filter, sometimes, the non-linear ANC system based RFsLMS becomes unstable. In addition, as the error function of the output-error filter is the nonlinear function of the coefficient values. It may converge to local minima while not to the global minimum. According to the suggestions [9] these problems can be avoided by choosing the step size carefully and like the leaky FXLMS algorithm. Therefore, the weight update EQ.(12) is written by applying the leaky algorithm

$$A_{kf}(n+1) = (1 - \mu_k\gamma)A_{kf}(n)$$
$$-\mu_k\left[e(n)/e^2(n) + 2\sigma^2\right]X_k'(n) \qquad (13)$$

where the parameter γ is a leaky factor, which is appositive constant very close to 1. We call this algorithm as LRFsLMS.

4 PHOTOGRAPHS AND FIGURES

To evaluate the effectiveness and robustness of the Leaky robust FsLMS proposed in this paper over the linear FXLMS, and FSLMS algorithms, computer simulations are carried out on various non-linear situations in the nonlinear ANC systems. In these entire simulations, the value of the memory size p is set to be 10. In addition, the Ensemble Average of Square Error (EASE) defined by [7]

$$EASE = 10\log10\left[\sum_{k=1}^{K} e_k^2(n)/K\right] \qquad (14)$$

Is obtained after averaging over 50 independent runs each consisting of 1000 iterations.

Assuming that the primary path exhibits non-linear behavior, the performance of different ANC controllers is further evaluated in the second experiment. The primary noise at the canceling point is generated based on the following polynomial model:

$$d(n) = t(n-1) + gt^3(n-2) \qquad (15)$$

where g denotes the varying of the third-order nonlinearity in the system, and $t(n)$ is obtained from the following linear convolution:

$$t(n) = x(n) * f(n) \qquad (16)$$

Note that the transfer functions with impulse response are $F(z) = z^{-3} - 0.3z^{-4} + 0.2z^{-5}$, [9]. And the reference noise $x(n)$ is a sinusoidal wave of 500 Hz sampled at the rate of 800 samples/s, that is

$$x(n) = \sqrt{2}\sin(2\pi * 100 * n/600) + v(n) \qquad (17)$$

And $v(n)$ is a whiter noise process with the Gaussian distribution. The signal power-to-noise power ratio (SNR) is chosen to be 40 dB. The length of each adaptive filter is 150. $K = 8$ and the SCU-BLFXLMS is used with $N = 20$ and $N = 40$. The non-minimum phase secondary path transfer functions are considered to following (41). The step sizes are used in the following: (a) FXLMS $u=0.00005$; (b) FuLMS $u1=0.00001$, $u2=0.00005$; (c) LRFsLMS $u1=u2=, \ldots uk=0.0001$. Figure 2 illustrates the comparison for the low nonlinearity case where $g = 0.04$. While the same is potted for the high nonlinearity case taking $g = 0.4$ in Figure 3. Various conclusions drawn from these figures

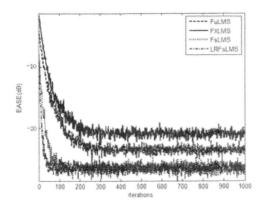

Figure 2. Comparison of convergence characteristics with periodic input signal, low nonlinear primary paths ($g = 0.04$), and non-minimum phase secondary path transfer function.

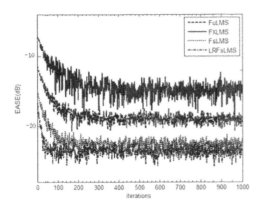

Figure 3. Comparison of convergence characteristics with periodic input signal, low nonlinear primary paths ($g = 0.1$), and non-minimum phase secondary path transfer function.

show that the LRFsLMS algorithms exhibit better convergence characteristics as compared with FXLMS and FuLMS algorithms in both cases, the LRFsLMS with the new model and traditional FsLMS algorithms exhibit same performance. The LRFsLMS still offers significant improvement in convergence rate when for the high nonlinearity case. On the other hand, the FU-BLFXLMS requiring more multiplications and additions.

As a consequence, the performance of the LRFsLMS with less computational burden is significantly superior to that of the FXLMS, FsLMS and FuLMS algorithms for nonlinear ANC systems in all cases. By applying the leaky algorithm rule, the stable performance of the LRFsLMS algorithm for nonlinear ANC systems is guaranteed.

5 CONCLUSIONS

A simplified robust nonlinear leaky algorithm has been developed in this paper. A new learning algorithm has been proposed for the FLANN base the nonlinear adaptive controller which based on the model of decomposing long adaptive filter into small sub-filters. This nonlinear adaptive filter which is a nonlinear extension of the FLANN filter. Can model nonlinear system accurately. Moreover, contrasted to the traditional FsLMS, it can reduce the computations and simplify the structure of the adaptive nonlinear filters during updating while does not degrade any performance of the system. Furthermore, due to applying the leaky algorithm, the stability of nonlinear ANC systems is also guaranteed by the LRFsLMS algorithm. Computer simulations illustrate that the new algorithm with less computational complexity, exhibits better performance than the linear FXLMS, FuLMS and FSLMS algorithms for nonlinear ANC systems. Consequently, the simplified FLANN filter for nonlinear ANC systems is more effective and stable.

ACKNOWLEDGEMENT

This work was supported by Hebei Province Nature Fund Grand (F2014208119) and HUST Doctor Fund Grant (010015).

REFERENCES

Bouchard M., Pailard B., Dinh C.T.L. Improved training of neural networks for nonlinear active control of sound and vibration. IEEE Trans Neural Networks 1999;10(2):391–401.

Bouchard M. New recursive-least-squares algorithms for non-linear active control of sound and vibrations using neural networks. IEEE Trans Neural Networks 2001;12(1):135–47.

Chang C.Y. Neural filtered-U algorithm for the application of active noise control system with correction terms momentum. Digital Signal Process 2010;20:1019–26.

Strauch P., Mulgrew B. Active control of nonlinear noise processes in a linear duct. IEEE Trans Signal Process 1998;46(9):2404–12.

Zhang Q.Z., Gan W.S., Zhou Y.L. Adaptive recurrent fuzzy neural networks for active noise control. J Sound Vibr 2006;296:935–48.

Das D.P., Panda G. Active mitigation of nonlinear noise processes using a novel filtered-s LMS algorithm. IEEE Trans Audio Speech Lang Process 2004;12(3):313–22.

Das D.P., Mohapatra S.R., Routray A., Basu T.K. Filtered-s LMS algorithm for multichannel active control of nonlinear noise processes. IEEE Trans Audio Speech Lang Process 2006;14(5):1875–80.

Mathews V.J., Sicuranza G.L. Polynomial signal processing. John Wiley & Sons, Inc.; 2000.

Sicuranza G.L., Carini A. Filtered-X affine projection algorithm for multichannel active noise control using second-order Volterra filters. IEEE Signal Process Lett 2004;11(11):853–7.

Napoli R., Piroddi L. Nonlinear active noise control with NARX models. IEEE Trans Audio Speech Lang Process 2010;18(2):286–95.

Kuo S.M., Wu H.T. Nonlinear adaptive bilinear filters for active noise control systems. IEEE Trans Circ Syst—I Regul Pap 2005;52(3):617–24.

Park D.C., Lee Y. Equalization of 16 QAM signals with reduced bilinear recurrent neural network. LNAI 2007;4570:601–10.

Mayyas K. Stereophonic low complexity algorithms based on the decomposition of the adaptive filter. In: 1st IEEE international symposium on signal processing and information technology ISSPIT'2001, Cairo, Egypt, December 28–30; 2001. p. 196–199.

Information Technology and Applications – Li (Ed.)
© 2015 Taylor & Francis Group, London, ISBN 978-1-138-02677-3

Stochastic Network Calculus based delay analysis for Window Flow Controller

Baoliang Li & Wenhua Dou
National University of Defense Technology, Changsha, P.R. China

Jie Zhao
State Key Laboratory of Mathematical Engineering and Advanced Computing, Zhengzhou, P.R. China

ABSTRACT: In the network with flow control, the output usually feedbacks some information to the input, controlling the data transmission and preventing buffer overflow. Flow control transforms the traditional forward networks into feedback networks, and the correlation between arrival and departure processes makes the performance modeling and analysis rather complex and difficult. In this paper, we take up the challenge by applying the theory of Stochastic Network Calculus (SNC) to characterize the feedback nature of static Window Flow Controller (WFC), and provide a probability end-to-end delay bound for network traffic regulated by WFC. We derive the delay bound based on the virtual-backlog-centric stochastic arrival curve and virtual-backlog stochastic strict service curve. As far as we know, it is the first time to model and analyze the WFC with the theory of SNC, which is of great value to both the guarantee of network Quality-of-Service (QoS) and the development of SNC theory. Finally, the influence of offered load on the end-to-end delay bounds is also discussed.

Keywords: delay bound; Window Flow Controller; Stochastic Network Calculus

1 INTRODUCTION

Window flow control is an effective solution to prevent a fast sender from overwhelming a slow receiver, and has been widely used in computer networks and interconnection networks. Window Flow Controller (WFC) is a simple and effective way to avoid buffer overflow and reduce the retransmission cost. The introduction of WFC transforms a feed-forward network into a feedback network, because in a system with flow control, the output should feedback some control information to the input to ensure that, the amount of data backlogged in the system never exceeds the window size. Although it is a very important and useful mechanism for a realistic network with finite buffer size, the correlation between actual input traffic and output traffic makes the Quality-of-Service (QoS) modeling and analysis very complex, as it is difficult to break the feedback control loop and directly characterize the admitted traffic for QoS analysis purpose. However, in the stage of network planning and design, a general mathematical theory is crucial to reveal what kind of service guarantee a network with WFC can support and how the performance varies with the different arrival and service processes.

To meet these great demands, the modeling and analysis of WFC has become a hot research topic, drawing lots of attention and much progress has been achieved. Because WFC can be viewed as a special Discrete Event Dynamic System (DEDS), the conventional theories used in DEDS, such as petri net, automata, queuing theory, max-plus algebra, Deterministic Network Calculus (DNC), etc. can also be utilized to model and evaluate the performance of WFC. As a commonly used WFC, the performance and dynamic behavior of Transmission Control Protocol (TCP) has been deeply investigated with max-plus algebra (Baccelli & Hong 2000), petri nets (Gaeta et al. 2003), automata (Billington & Han 2007) and queuing theory (Reiser 1979), etc. In addition, the so-called rate-control throttle with finite-capacity token bank and job buffer was analyzed with queuing theory in (Berger & Whitt 1992), which can also be treated as a variation of static window flow control. In (Agrawal et al. 1999, Chang 1998), DNC is applied to capture the dynamic behavior of WFC, and the equivalent deterministic service curve of end-to-end and hop-by-hop WFC are derived. Similar effort was made in the field of credit-based flow control (Qian et al. 2009), and the worst-case buffer requirement is considered.

In this paper, we try to tackle the same flow control problem with the newly developed Stochastic Network Calculus (SNC) (Jiang 2006) theory. Stochastic network calculus is the combination of DNC and probability theory, and characterizing the performance metrics in the form of probability distribution. For the analysis and provision of service guarantees in multimedia networks and wireless networks, SNC is more convenient and meaningful than DNC, and has been widely used in the field of cognitive radio (Gao & Jiang 2010) and retransmission channel (Wang et al. 2013), etc. However, to the best of our knowledge, current research on SNC is mainly focusing on the feed-forward networks with infinitive buffer size, few results on feedback networks can be found. The most related works to our research is the SNC-based retransmission model proposed in (Wang et al. 2013, Gao & Jiang 2010) but they still characterize the retransmission behavior in the scope of feed-forward networks. While analyzing the stochastic performance bounds of WFC, the key problem is that, the WFC regulates the actual input traffic injected into the network depending on the input traffic, output traffic, feedback delay and window size, making it difficult the characterize the actual input traffic with stochastic arrival curve directly, which is the basis of the delay analysis with SNC.

To take up the challenge, we first relax the flow control constraint by leveraging the causality of arrival process and actual input process. Then, we adopt the *v.b.c.* stochastic arrival curve and *v.b.* stochastic service curve to characterize the traffic process and service process and derive the delay bound. We will explain this procedure in details in the following sections. Our main contribution is the derivation of probabilistic delay bound with SNC for the admitted traffic regulated by WFC. These results are of great value to the performance evaluation and optimization of a network with WFC, which reveals the impact of offered load on the delay bound, and enriches the theoretical research on performance analysis. As far as we know, it is the first time to model and analyze the WFC with SNC theory, which is of great value to both the guarantee of network QoS and the development of SNC theory.

The rest of this paper is organized as follows. We introduce the notations and network model used in this paper in Section II. The probabilistic delay bound of WFC were derived in Section III. In Section IV, we present the experimental results to demonstrate the correctness and tightness of our results, the influence of the offered load on the delay bound is also investigated and explained. Finally, we summarize the paper and point out some future work in Section V.

2 PRELIMINARIES

2.1 Notation

In this paper, we denote by $\mathcal{F}$ the set of nonnegative and wide-sense increasing functions, and for any functions $a(x)$ in $\mathcal{F}$, we set $a(x) = 0$ for $\forall x < 0$. We denote by $\bar{\mathcal{F}}$ the set of non-negative wide-sense decreasing functions, and for any functions $a(x) \in \bar{\mathcal{F}}$, we set $a(x) = 1$ for $\forall x < 0$. We also denote by $\bar{\mathcal{G}}$ the subset of $\bar{\mathcal{F}}$, whose nth-fold integration still belongs to $\bar{\mathcal{G}}$. Let $A(t)$, $I(t)$, $S(t)$ and $D(t)$ be the accumulative arrival process, accumulative actual input process, accumulative service process and accumulative departure process, respectively. To simplify our expressions, we define four bivariate functions $A(s, t) \equiv A(t) - A(s)$, $S(s, t) \equiv S(t) - S(s)$, $D(s, t) \equiv D(t) - D(s)$ and $[x]_1 \equiv \min\{x, 1\}$. In addition, we use $\alpha(t)$, $\beta(t) \in \mathcal{F}$ to denote a service curve an arrival curve, and use $f(x)$, $g(x) \in \bar{\mathcal{F}}$ to denote the bounding functions, respectively. The min-plus convolution max-plus deconvolution and Stieltjes integration are denoted as $\otimes$, $\bar{\oslash}$ and *, respectively.

2.2 Network model

Typical implementations of WFC are TCP in packet switched networks and credit based flow control in Asynchronous Transfer Mode (ATM) (Kun & Morris 1995). Both TCP and credit based flow control feedback some control information to the input to prevent buffer overflow. As in (Agrawal et al. 1999, Chang 1998), one can view the entire network as a single node, shown in Figure 1. We assume that, in the flow control system, the amount of control information feedback from the output is in the same volume as the cumulative departures of the system, so that we can describe the cumulative admitted process $I(t)$ as

$$I(t) = \min\{A(t), D(t - \tau) + B\} \qquad (1)$$

where B is the window size and τ is the feedback delay. This assumption is reasonable and does not affect the generality of the results, as explained in (Agrawal et al. 1999). We also assume that the system is lossless and the controller has sufficient capacity to accommodate the unadmitted traffic.

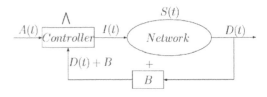

Figure 1. End-to-end flow control model.

The virtual delay of carried traffic and and backlog in the network are defined as:

$$Delay(t) = \inf \{d \geq 0 : I(t) - D(t + d) \leq 0\}.$$

From Eq. (1), we know that $I(t)$ depends on the accumulative arrival process $A(t)$ and accumulative departure process $D(t)$. In addition, $D(t)$ depends on the carried process $I(t)$ and the service process $S(t)$. The correlation among $A(t)$, $D(t)$, $I(t)$ and $S(t)$ makes it difficult to use $I(t)$ and $D(t)$ to derive the performance bound directly, because even with $A(t)$, $S(t)$ and B given, $I(t)$ is yet to be unknown. Fortunately, based on the causality between $A(t)$ and $I(t)$, i.e. $I(t) \leq A(t)$, we relax the constraint of flow control constraint, and utilize the traffic model of $A(t)$ and the service model of $S(t)$ to characterize the end-to-end delay bound indirectly, which will be explained in details in Section III. In this paper, what we will derive is the upper bound of the end-to-end probabilistic bound on the virtual delay of the admitted traffic, denoted by $P\{Delay (t) > d\}$, i.e. $P\{I(t) - D(t + d) > 0\}$.

2.3 Traffic and service models

Two fundamental concepts in SNC are stochastic arrival curve and stochastic service curve, which are of general concepts, and can be concreted with specific functions suitable for each particular application. We consider that the traffic is described with virtual-backlog-centric (v.b.c.) stochastic arrival curve (Jiang 2006), denoted by $< f(x), \alpha(t) >$, and the service is characterized by virtual-backlog (v.b.) stochastic strict service curve (Wu et al. 2010), denoted by $< g(x), \beta(t) >$. The promising property of v.b. stochastic strict service curve is that, its definition does not depend on the stochastic arrival process, which is convenient for independent case performance analysis. For more detailed description about these two curves, please refer to (Jiang 2006) and (Wu et al. 2010). Because the system is lossless, the arrival rate should not exceed the service rate provided by the system in long term, otherwise the system is unstable. Thus, unless explicitly stated, we assume the follow inequality holds

$$\lim_{t \to \infty} \frac{1}{t} [\alpha(t) - \beta(t)] \leq 0. \tag{2}$$

3 DELAY BOUND

In this section, we will derive end-to-end stochastic delay bound for WFC. The basic idea to get the worst-case performance bound of feedback

networks with DNC is intuitive. First derive an equivalent deterministic service curve for the controller (so-called throttle in (Agrawal et al. 1999, Chang 1998)) to break the control loop and transform the feedback network to feed-forward network. Then, concatenation theorem is applied to get the equivalent service curve of the transformed forward network. Finally, the backlog and delay bound can be easily obtained with the theory of DNC. However, it is difficult to imitate this approach in SNC, because the correlation between the traffic process and service process makes it hard to characterize the controller with stochastic service curve. To bypass this, we derive the probability delay bound directly, by skipping the step of characterizing the controller with equivalent stochastic service curve. From Eq. (1), we know that, for any time t, the carried traffic $I(t)$ is always equal to either $A(t)$ (corresponding to low offered load scenario) or $D(t - \tau) + B$ (corresponding to high offered load scenario). In addition, $I(t) \leq A(t)$ always holds for causality, which can be used to derive the upper delay bound, as presented in Theorem 1.

(Theorem 1) Consider a flow $A(t)$ passing through a network with WFC, and the network provides service process $S(t)$. Suppose the flow $A(t)$ has a v.b.c. stochastic arrival curve $< f, \alpha >$, $S(t)$ has a v.b. stochastic strict service curve $< g, \beta >$, the window size and feedback delay of the WFC are B and τ, respectively. Then, for a sufficient small feedback delay τ, the end-to-end delay bound of admitted traffic is

$$P \{Delay(t) > d\} \leq [f \otimes g(y'')]_1 \tag{3}$$

where $y'' = \beta \overline{\oslash} \alpha(d)$.

If $A(t)$ and $S(t)$ are independent, we can get a much tighter probability bound with Stieltjes convolution, which is

$$P\{Delay(t) > d\} \leq 1 - \overline{f} * \overline{g}(y'') \tag{4}$$

where $\overline{f} = 1 - [f]_1$, $\overline{g} = 1 - [g]_1$.

Before proving this theorem, we first introduce the concept of the beginning of last backlogged period, which will be used in our proof to establish the relationship between departure process $D(t)$, arrival process $A(t)$ and service process $S(t)$. Consider a work-conserving server $S(t)$ with arrival process $\mathcal{A}(t)$ and departure process $\mathcal{D}(t)$, the beginning of last backlogged period before time instance t is defined as

$$\Phi(t) = \sup \{s \in [0, t] : \mathcal{D}(s) = \mathcal{A}(s)\}.$$

From the definition of backlog, we know that the backlog keeps positive in period $(\Phi(t), t]$ if t

229

is in a busy period, else $t = \Phi(t)$. Let $\Phi(t)$ be the beginning of last backlogged period before t. Because the backlog at $\Phi(t)$ is zero, we have

$$\mathcal{D}(\Phi(t)) = \mathcal{A}(\Phi(t)). \tag{5}$$

For the work-conserving server, the amount of departures within the backlogged period $(\Phi(t), t]$ is the same as the service provided by the server. Thus,

$$\mathcal{D}(\Phi(t), t) = \mathcal{S}(\Phi(t), t). \tag{6}$$

Combining Eq. (5) and Eq. (6), we have

$$\mathcal{D}(t) = \mathcal{A}(\Phi(t)) + \mathcal{S}(\Phi(t), t). \tag{7}$$

These properties of the beginning of last back-logged period listed above provide a convenient way for us to derive the delay bound of WFC, as shown in the following proof.

Proof: Let t_{01} be the beginning of last backlog period of the network before $t + d$, then $0 \leq t_{01} \leq t + d$. According to Eq. (7), the departure process at time instance $t + d$ can be expressed as

$$D(t + d) = S(t_{01}, t + d) + I(t_{01}).$$

Actually, for a sufficient small feedback delay τ, we have $I(t_{01}) = A(t_{01})$. Or else, the buffered data in the flow controller will be fed into WFC and make the backlog positive, which conflicts with the fact that t_{01} is in an idle period. Thus

$$D(t + d) = S(t_{01}, t + d) + A(t_{01}). \tag{8}$$

From Eq. (1), we get

$$I(t) = \{A(t), D(t - \tau) + B\}$$
$$\leq A(t) \tag{9}$$

Thus, combining Eq. (8) and Eq. (9), we have

$$I(t) - D(t + d) \leq A(t_{01}, t) - S(t_{01}, t + d)$$
$$\leq \sup_{0 \leq s \leq t+d} \{\beta(t + d - s) - S(s, t + d)\}$$
$$+ \sup_{0 \leq s \leq t} \{A(s, t) - \alpha(t - s)\} - \beta \overline{\oslash} \alpha(d).$$

Thus, based on the definition of $v.b.$ stochastic strict service curve and $v.b.c$ stochastic arrival curve, we have

$$P\{Delay(t) > d\} = P\{I(t) - D(t + d) > 0\}$$
$$\leq [f \otimes g(y'')]_1$$

where $y'' = \beta \overline{\oslash} \alpha(d)$.

If $A(t)$ and $S(t)$ are independent, we also have

$$P\{Delay(t) > d\} \leq 1 - \overline{f} * \overline{g}(y'')$$

which ends the proof. We want to emphasize that, the WFC limits the amount of entering traffic $I(t)$ and enforces it to satisfy the flow control constraint of Eq. (1), which makes the end-to-end delay analysis very difficult. In this section, we first relax the strict flow control constraint in Eq. (9) and the establish a relationship between the end-to-end delay bound and the $v.b.c$ stochastic arrival curve of accumulative arrival process $A(t)$ together with the $v.b.$ stochastic strict service curve of service process $S(t)$.

4 NUMERICAL RESULTS

In this section, we take a simple feedback networks as example to verify the correctness of our results and demonstrate the tightness of our results. The feedback network discussed here has Poisson arrival with arrival rate λ and exponentially distributed service time with service rate μ. The reason of choosing this model is that, there already exists analytical results for the queueing system with Poisson arrivals and exponentially distributed service time, which ease the characterization of arrivals and services with $v.b.c$ stochastic arrival curve and $v.b.$ stochastic strict service curve. The influence of offered load $\rho = \lambda/\mu$ on the delay bound are also investigated.

4.1 Service and traffic model

In order to apply Theorem 1 to compute the delay upper bound, we have to find the $v.b.c$ stochastic arrival curve and the $v.b.$ stochastic strict service curve first.

1. The Poisson arrival process has independent and stationary increments. Suppose all the packets have the same length L, we can get the $v.b.c$ stochastic arrival curve $< e^{-\theta_1 \theta} e^{-\theta_1 x}, \lambda/\theta_1 (e^{\theta_1 L} - 1) + \theta t >$ for the arrival process (Jiang 2010), where $\theta \geq 0$ and $\theta_1 > 0$ are free parameters. To ease our analysis, let $L = 1\ kb$ and $\theta = 0$, we get a $v.b.c$ stochastic arrival curve $< e^{-\theta_1 x}, \lambda t/\theta_1 (e^{\theta_1} - 1) >$.

2. The service process of first model has independent and stationary increments. We can get a $v.b.$ stochastic strict service curve $< e^{-\theta \theta_2} e^{-\theta_2 x}, [\mu'(\theta_2) - \theta] \cdot t >$, where $\theta \geq 0$ and $\theta_2 > 0$ are free parameters, $\mu'(\theta_2) \leq (1/-\theta_2) \log E[e^{-\theta_2 S(1)}]$. The moment generating function $E[e^{-\theta_2 S(1)}] = e^{\mu}(e^{-\theta_2 L} - 1)$. Thus, let $\mu'(\theta_2) = (1/-\theta_2) \log E[e^{-\theta_2 S(1)}] = (\mu t/-\theta_2)(e^{-\theta_2 L} - 1)$, $\theta = 0$ and $L = 1\ kb$, we get

a simplified $v.b.$ stochastic strict service curve $\beta(t) = (\mu t / -\theta_2)(e^{-\theta_2} -1)$ with bounding function $g(x) = e^{-\theta_2 x}$.

4.2 *Results discussion*

The arrival process and service process of this feedback network are independent, hence we can applying Eq. (4) to get the delay bound. We first compute the theoretical results, and then verify it by simulation. The simulation is carried out with OMNeT++ Network Simulation Framework. The parameters are configured as $\lambda = 20\ s^{-1}$, $\mu = 25\ kbps$, $\tau = 0.01\ s$, B = 10 kb, L = 1 kb. Related results are shown in Figure 2. It can be observed that, the delay bound computed with Theorem 1 is very conservative, similar observation is also

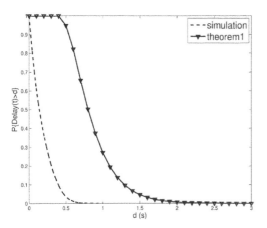

Figure 2. Delay comparison with simulation: $\lambda = 20\ s^{-1}$, $\mu = 25\ s^{-1}$, $\tau = 0.01\ s$, B = 10 kb, L = 1 kb.

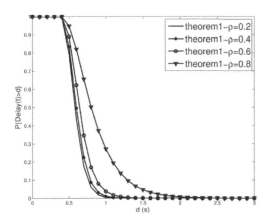

Figure 3. Influence of offered load on the delay bound: $\lambda = 5\ s^{-1}$, 10 s^{-1}, 15 s^{-1}, 20 s^{-1}, $\mu = 25\ s^{-1}$, $\tau = 0.01\ s$, B = 10 kb, L = 1 kb.

found while applying SNC to other fields, e.g. (Wang & Wang 2011), which indicates that, there is still need much improvement and some advanced probability theory techniques might be required when applying SNC. In addition, the influence of offered load $\rho = \lambda/\mu$ on the delay bound of feedback network is also investigated, as shown in Figure 3. It is found that smaller bound will be provided when decreasing the offered load ρ, due to the fact that higher offered load leads to larger average waiting time.

5 CONCLUSION

Flow control is an effective solution to prevent buffer overflow and network congestion, and has been widely used in computer networks and interconnection networks. However, the correlation between input and output in such system makes the QoS modeling and analysis complex and difficult. Our main contribution is that, by employing SNC and the concept of beginning of last backlogged period, we derived probability upper bounds of delay for static WFC. We also studied the influence of offered load and window size on the delay bound. To the best of our knowledge, this is the first work applying SNC to analyze the performance of feedback networks. Our work hence contribute to the development and application of SNC. In the future work, we expect to extend the analysis to support Gaussian process or Fractional Brown Motion (FBM) process, and to characterize the performance of WFC network fed with self-similar traffic.

ACKNOWLEDGMENT

The authors would thank the reviewers for their suggestions and comments. This research is supported by High Technology Research and Development Program of China (Grant No. 2012AA012201) and National Program on Key Basic Research Project of China (Grant No. 2012CB933504).

REFERENCES

Agrawal R. & Cruz R.L. Cruz et al. 1999. Performance bounds for flow control protocols. *Networking, IEEE/ACM Transactions on*, vol. 7, no. 3, pp. 310–323, 1999.

Baccelli, F. & Hong D. 2000. TCP is max-plus linear and what it tells us on its throughput. *ACM SIGCOMM Computer Communication Review*, vol. 30, no.4, pp. 219–230, Oct. 2000.

Billington J. & Han B. 2007. Formalising tcp's data transfer service language: A symbolic automaton and its properties. *Fundam. Inf.*, vol. 80, no. 1–3, pp. 49–74, Jan. 2007.

Berger A.W. & Whitt W. 1992. "The impact of a job buffer in a token-bank rate-control throttle," *Stochastic Models*, vol. 8, no. 4, pp. 685–717, 1992.

Chang C.-S. 1998. On deterministic traffic regulation and service guarantee: A systematic approach by filtering. *IEEE Transactions on Information Theory,* vol. 44, no. 3, pp. 1097–1110, May 1998.

Gaeta R. & Sereno M. et al. 2003. Stochastic petri nets models for the performance analysis of tcp connections supporting finite data transfer. *Quality of Service in Multiservice IP Networks,* ser. Lecture Notes in Computer Science, vol. 2601. Springer Berlin Heidelberg, 2003, pp. 372–391.

Gao Y. & Jing 2010. Performance analysis of a cognitive radio network with imperfect spectrum sensing. *INFOCOM IEEE Conference on Computer Communications Workshops,* 2010, pp. 1–6.

Jiang 2006. A basic stochastic network calculus. *Proceedings of the 2006 conference on Applications, technologies, architectures, and protocols for computer communications,* ser. SIGCOMM '06. New York, NY, USA: ACM, 2006, pp. 123–134.

Jiang Y. 2010. A note on applying stochastic network calculus. Department of Telematics, Norwegian University of Science and Technology, Technical Report, May 2010.

Kung N. & Morris R. 1995. Credit-based flow control for atm networks. *Network, IEEE*, vol. 9, no. 2, pp. 40–48, 1995.

Le Boudec J.-Y. & Thiran P. 2001. Network Calculus: A Theory of Deterministic Queueing System for the Internet. Berlin, Heidelberg: Springer-Verlag, 2001.

Qian Y. & Lu Z. et al. 2009. Analyzing credit-based router-to-router flow control for on-chip networks. *IEICE Transactions on Electronics*, vol. E92-C, no. 10, pp. 1276–1283, Oct. 2009.

Reiser M. 1979. A queueing network analysis of computer communication networks with window flow control. *Communications, IEEE Transactions on*, vol. 27, no. 8, pp. 1199–1209, 1979.

Wang H. & Schmitt et al. 2013. "Performance modelling and analysis of unreliable links with retransmissions using network calculus," in *Proceedings of the 25th International Teletraffic Congress (ITC 25)*. Shanghai, China: IEEE, Sept. 2013.

Wu K. & Jiang Y. et al. 2010. *On the model transform in stochastic network calculus.* IEEE, 2010, pp. 1–9.

Wang Y. & Wang T. 2011. Applying stochastic network calculus to 802.11 backlog and delay analysis. *Quality of Service (IWQoS)*, 2011 IEEE 19th International Workshop on, 2011, pp. 1–3.

Information Technology and Applications – Li (Ed.)
© 2015 Taylor & Francis Group, London, ISBN 978-1-138-02677-3

Design and simulation of wearable RFID tag antenna

Shudao Zhou & Haotian Chang
Institute of Meteorology and Oceanography, PLA University of Science and Technology,
Nanjing, Jiangsu, China

ABSTRACT: Due to the modest read distance and narrow bandwidth, design of effective wearable tag antenna in UHF band has drawn large attention. Based on the theory of wide bandwidth antenna, the geometry for RFID wearable tag antenna which adopts the bowtie shape is presented here through simulation. The design and simulation of the bowtie antenna are carried on with the software, HFSS, in which the S parameter, VSWR and radiation pattern are analyzed. The bandwidth of the optimized antenna that the S11 parameter is less than −10 dB achieves 144 MHz, ranging from 858 MHz to 1002 MHz. The tag antenna has a good flexibility which means that it fits human figures well and it is also easy to realize because of its simple structure.

Keywords: RFID; tag antenna; wide bandwidth; wearable

The possibility to identify and track targets automatically by means of low-power devices is one of the most interesting and attractive advantages of Radio Frequency Identification (RFID) techniques. Nowadays, it is feasible to deliver real-time human monitoring, logistic managing, applications of internet of things and animal tracking with RFID devices integrated with sensors and signal processing parts, which benefits a lot from advances in low-power electronics (Ochiuzzi, Cippitelli & Marrocco 2010).

The UHF (860–960 MHz) standard is especially attractive because of its high data rate and large read distances, while the typical applications in ID cards in HF and LF bands have much smaller read distances.

Wearable antennas have to meet the requirements of small dimensions, wide bandwidth and flexibility to human figures. These issues are common considerations of design of tag antennas in human applications (Manzari 2012). In this paper, we mainly focus on the design and simulation of a wide bandwidth wearable antenna. A brief introduction to the simulation software is presented in section 1. The modeling and configuration of the antenna are then discussed in section 2. Then the performances of the antenna due to changes of the antenna size are discussed in section 3, which contributes guidance to the optimization of the antenna in section 4. Finally, the paper draws a conclusion of the performances of the optimized tag antenna in section 5.

1 AN INTRODUCTION TO HFSS

High Frequency Structure Simulator (HFSS) is a 3D simulation software delivered by Ansoft and is the first commercial 3D structure electromagnetic simulation software in the world. HFSS has become the standard of industry of its kind, which provides the users with convenient user interface, accurate adaptive field solver and powerful ability of analysis and processing. The S parameters and full wave electromagnetic fields of any 3D passive structure can be solved by HFSS.

1.1 *Basic functions of HFSS*

Using HFSS, the following issues can be solved and simulated: 1) near field and far field radiation and the numerical value of electromagnetic field; 2) characterized impedance of ports and transmission constants; 3) S parameters and normalized S parameters of ports; 4) resonance solutions, and so on.

In addition, the solution to high frequency design, consisting of Ansoft HFSS and Ansoft designer solutions, is the only solution based on physical prototypes, which covers all the procedures of high frequency design, ranging from circuits to systems.

HFSS has equipped with powerful ability of antenna design, which is able to figure out various antenna parameters including gain, directivity, radiation pattern and 3D polar graph, and so on. Therefore, HFSS is adopted for our design simulation.

1.2 Simulation setup

The simulation setup can be described as the following procedures: 1) signing the solution type, normally interpolating; 2) creating a new project and insert new designs into the project; 3) modeling the antenna and signing the boundaries, materials and excitations (Wang et al. 2005); 4) analysis set up, including inserting infinite sphere for far field analysis, start frequency, end frequency step frequency and sweep mode, and so on.

2 MODELING OF THE TAG ANTENNA

2.1 Substrate properties

In order to optimize and sweep particular parameter of the antenna, all the parts of size are signed as parameters.

The substrate has a rectangular plane with a thickness of SubH, and has a size of SubX in length and a size of SubY in width. The material of the substrate is defined as a new material in HFSS, whose relative permittivity is 1.2.

The substrate's size remains constant because of the practical realization of the prototype. Meanwhile, SubH is only 1 mm, which ensures the good flexibility of the antenna.

2.2 Design of the patch pattern

The impedance of antenna needs to remain steady in order to achieve a wide bandwidth. Once the length of antenna is longer than the wavelength, there will be only a small proportion of the wave energy being reflected, which means most of the wave energy will be radiated to the free space. The bowtie antenna can sustain a constant impedance in a wide bandwidth and achieves good match with the free space, which prevents most of the wave energy being reflected from the boundary of the antenna and free space. Therefore, we adopt bowtie antenna to realize our design of the antenna with a wide bandwidth.

2.3 Modeling the patch pattern

As discussed earlier, according to the wide bandwidth antenna theory, the bowtie antenna is adopted. The geometrical center locates in the center of the substrate plane. Both parts of the bowtie lie along the Y-dimension, each with a length named arm_length. The distance between the two arms of the bowtie is signed as port_gap_width. In addition, the upper and lower edges of the bowtie are named as inner_width and outer_width, respectively.

It can be observed in Figure 1, the modeling of the antenna geometry is presented in Figure 1.

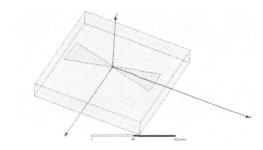

Figure 1. Modeling of the patch pattern on a substrate.

2.4 Analysis setup

As the performance in far field is our main concern, we insert Infinite Sphere in Radiation setup. The solution frequency is set up as 915 MHz, which is the standard of UHF applications in the USA. The sweeping frequency starts at 750 MHz and ends at 1050 MHz, with a step frequency of 1 MHz. Interpolating is chosen as the sweeping mode.

3 INFLUENCES OF GEOMETRY PARAMETERS TO ANTENNA PERFORMANCES

The origin size of the antenna is set up as 2 mm inner_width, 40 mm of outer_width and 55 mm of arm_length. Sweeping the parameters respectively gives us understanding of what influence each parameter brings to the antenna performance.

The results of S11 parameter according to 5 inner_width values, from 1.7 mm to 2.5 mm with a step of 0.2 mm, are shown in Figure 2. Later, sweeping of outer_width and arm_length are shown in Figure 3 and Figure 4, respectively.

It can be observed from the three following figures, inner_width and outer_width contribute to a much less influence to the S11 parameter than arm_length does. Specifically, in comparison of the 5 inner_width values, we can find out that S11 parameter is not sensitive to the change of inner_width, because the offset of central frequency between those 5 results reaches only 13 MHz at most. Similar results are found in the sweeping of outer_width as well. In contrast, influence to S11 parameter that changes of arm_length from 53 mm to 57 mm bring about is much more obvious. The largest offset of the S11 results corresponding to the sweeping of arm_length attains to 32 MHz. The S11 parameter of the antenna moves to higher frequencies as arm_length decreases, while it moves to lower frequencies as arm_length increases. Therefore, when designing and optimizing the antenna, arm_length is the most vital aspect of

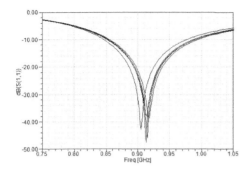

Figure 2. Influence that inner_width brings to S11 parameters.

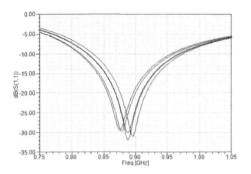

Figure 3. Influence that outer_width brings to S11 parameters.

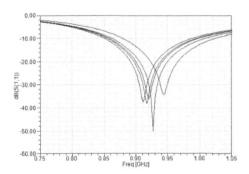

Figure 4. Influence that arm_length brings to S11 parameters.

consideration, which may lead to improvement and optimization of the antenna performances.

4 OPTIMIZATION OF THE ANTENNA

4.1 Optimization goal setup

Because the S11 parameters are sensitive to arm_length as discussed above, we choose arm_

length as the optimizing parameter in the design properties. The optimization goal is set up as signing the VSWR less than 1, which means the reflecting wave energy accounts for a little proportion of the radiated wave energy.

When optimized arm_length is acquired, inner_width and outer_width are then optimized based on the acquired arm_length so as to achieve the best design properties.

4.2 Optimization results

The optimized antenna has an inner_width of 2.37 mm, an outer_width of 32.78 mm and an arm_length of 55.23 mm. The results of the antenna performances are then shown below, S11 parameter in Figure 5 and VSWR in Figure 6.

It can be observed in Figure 5 and Figure 6, VSWR reaches 1.0571 and the S11 parameter hits the bottom of −23 dB at 915 MHz. Although there is a small offset of the central frequency of the S11 parameter result, only 5 MHz, but the band that S11 is less than −10 dB reaches 140 MHz, achieving a satisfactory bandwidth.

Because non-frequency-variant antennas should be infinitely long and truncations of antenna cause

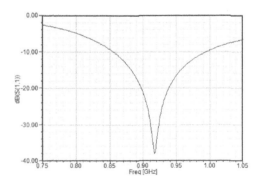

Figure 5. S11 parameter of the optimized antenna.

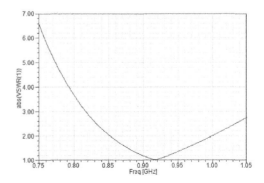

Figure 6. VSWR of the optimized antenna.

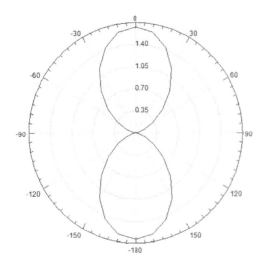

Figure 7. Radiation pattern of the XOY plane.

impedance change and reflecting radiation if without any impedance match at the truncation end, the current along the bowtie arm should be managed to attenuate to a very low level along with radiation.

Simulation results show that the surface current distributes mainly near the port and the boundary of the bowtie patch and the substrate. At the truncation ends hardly distributes any surface current, which satisfies the requirements of non-frequency-variant antennas, thus achieving the goal of wide bandwidth.

The far field performance is shown in Figure 7. The radiation pattern on the XOY plane is a spindle-like shape, while the radiations pattern on XOZ plane is a circle. The radiation pattern is similar to that of dipole antennas', because the bowtie antenna is a variant form of dipole antennas. But the dipole antennas are resonance-like devices which have high Q values. That is

to say, dipole antennas store a large amount of energy before radiating, which leads to intense variation of impedance and narrow bandwidth. In contrast, bowtie antennas have the advantage of self-complementary antennas, which helps to achieve a wide bandwidth.

5 CONCLUSION

With the developments of electromagnetic field science and microwave techniques, HFSS has become an indispensable tool for antenna designers. The influences that geometry sizes bring to antenna performances can be easily simulated and analyzed through HFSS, by which the optimization of antennas can also be easily carried on. The inner_width and outer_width contribute to a much less influences to performances than arm_length does. Hence, optimization mainly on the aspect of arm_length helps to acquire the most satisfactory design properties of the bowtie antenna. The results of the performances of optimized antenna show a good radiation pattern and a wide bandwidth of 140 MHz that the S11 is below −10 dB. Besides, the designed antenna has a thin thickness and a simple structure, which is easy to realize in practice.

REFERENCES

Manzari, S. et al. 2012. Feasibility of Body-Centric Systems Using Passive Textile RFID Tags. *IEEE Antennas and Propagation Magazine:* 54(4):49–62.
Ochiuzzi, C. et al. 2010. Modeling, Design and Experimentation of Wearable RFID Sensor Tag, *IEEE Transactions on Antennas and Propagation:* 58: 2490–2498.
Wang J.J. et al. 2005. Circuit model of microstrip patch antenna on ceramic land grid array package for antenna—chip codesign of highly integrated RF transceivers. *IEEE Transactions on Antennas and Propagation:* 53(12):3877–3883.

Information Technology and Applications – Li (Ed.)
© *2015 Taylor & Francis Group, London, ISBN 978-1-138-02677-3*

Attribute weighted fuzzy clustering algorithm for incomplete datasets

D. Li, C.Q. Zhong & Z.J. Chen
School of Control Science and Engineering, Dalian University of Technology, China

ABSTRACT: An attribute weighted fuzzy *c*-means algorithm for incomplete datasets is presented in this paper. The nearest-neighbor interval representation proposed in our previous researches is used here to describe the missing attribute values, and the widely used ReliefF algorithm is involved to determine the attribute weights. The proposed algorithm combines the attribute weights and fuzzy clustering by weighted Euclidean distance, and is successfully justified based on benchmark datasets.

Keywords: fuzzy clustering; fuzzy *c*-means; incomplete data; attribute weighted

1 INTRODUCTION

Clustering techniques, especially fuzzy clustering, have been widely applied to image segmentation, information retrieval and computer vision (Ren et al. 2012). However, most of the fuzzy clustering algorithms, including the widely used fuzzy *c*-means algorithm (FCM) can't deal with incomplete datasets directly. In many areas of applications, lots of datasets plagued by the problem of missing values for several reasons, including incorrect measurements, limited time, and equipment errors (Farhangfar et al. 2007). Thus, it is meaningful to investigate the problem of clustering incomplete datasets.

There are two general approaches to deal with the problem of missing values in clustering: ignorance and imputation (Farhangfar et al. 2007). Hathaway & Bezdek (2001) proposed four strategies to continue the FCM clustering of incomplete data. The Whole Data Strategy (WDS-FCM) and the Partial Distance Strategy (PDS-FCM) belong to the ignorance-based methods, and the Optimal Completion Strategy (OCS-FCM) and the Nearest Prototype Strategy (NPS-FCM) belong to the imputation-based methods. Then, based on kernel metric, Zhang & Chen (2003) proposed a fuzzy *c*-means algorithm to cluster incomplete datasets, Timm & Kruse (2004) developed a fuzzy clustering algorithm extended from the Gath and Geva algorithm, and Honda & Ichihashi (2004) partitioned the incomplete datasets into several linear fuzzy clusters. In view of the uncertainty of missing attribute values, interval representation of missing attribute values was proposed and combined into fuzzy *c*-means algorithms in our previous researches (Li et al., 2010; Li et al., 2013). Besides, Nuovo (2011), Li & Zhong (2012) and Azadeh

et al. (2013) applied the missing data analysis with fuzzy clustering into psychological scenario, image segmentation and randomized complete block design respectively.

In this paper, we continue to focus on the nearest-neighbor intervals of missing attribute values (Li et al., 2010), and attribute weights obtained by the ReliefF algorithm are introduced to emphasis the contribution of important attributes. The proposed algorithm can guide the clustering process to partition the incomplete datasets into more meaningful clusters, and get more satisfying results.

2 NEAREST-NEIGHBOR INTERVALS AND THE ReliefF ALGORITHM

2.1 *The determination of nearest-neighbor intervals*

Nearest-neighbor techniques have been often applied for missing values imputation. And in our previous researches (Li et al., 2010; Li et al., 2013), the Nearest-Neighbor Interval (NNI) representation of missing attribute values was proposed.

Let $\tilde{X} = \{\tilde{x}_1, \tilde{x}_2, ..., \tilde{x}_n\}$ be a set of *s*-dimensional incomplete dataset, which contains at least one incomplete datum with some (but not all) missing values. For an incomplete datum $\tilde{x}_b$, the partial distance calculation of $\tilde{x}_b$ and an instance $\tilde{x}_p$ (incomplete or complete) is given by

$$D_{pb} = \left[\sum_{j=1}^{s} \left(\tilde{x}_{jb} - \tilde{x}_{jp} \right)^2 I_j \right] \Bigg/ \sum_{j=1}^{s} I_j \qquad (1)$$

where $\tilde{x}_{jb}$ and $\tilde{x}_{jp}$ are the *j*th attribute of $\tilde{x}_b$ and $\tilde{x}_p$ respectively, and

$$I_j = \begin{cases} 1, & \text{if both } \tilde{x}_{jb} \text{ and } \tilde{x}_{jp} \text{ are nonmissing} \\ 0, & \textit{otherwise} \end{cases}, \quad (2)$$

For $p,b = 1,2,...,n$, and $j = 1,2,...,s$.

For the missing attribute value $\tilde{x}_{jb}$ of $\tilde{x}_b$, the method searches for the q nearest neighbors of $\tilde{x}_b$ based on (1) in the incomplete dataset $\tilde{X}$. Let x_{jb}^- and x_{jb}^+ be the minimum and maximum of the jth attribute values of the q nearest neighbors, thus the nearest-neighbor interval of $\tilde{x}_{jb}$ can be determined as $[x_{jb}^-, x_{jb}^+]$.

2.2 The ReliefF algorithm

For a complete dataset $X = \{x_1, x_2, ..., x_n\} \subset R^s$, the key idea of the ReliefF algorithm is to evaluate the contribution of each attribute to inter-class difference and intra-class similarity (Kira & Rendell, 1992). Firstly, an instance x_k should be selected randomly, then the ReliefF algorithm searches for its two nearest neighbors, one from the same class (termed the nearest hit) and the other from a different class (termed the nearest miss). The algorithm's estimation w_j for attribute j is an approximation of the following difference of probabilities

$$w_j = P(\text{different value of } jth \text{ attribute} \mid \text{nearest miss}) \\ - P(\text{different value of } jth \text{ attribute} \mid \text{nearest hit}).$$
$$(3)$$

The rational is that a useful attribute should different between instances from different classes and have the same value for instances from the same class. The ReliefF algorithm can only deal with the two-class problems, and then the ReliefF algorithm was proposed to handle noise and multi-class datasets (Kononenko, 1994). By searching for z nearest neighbors of x_k from the same class, called nearest hits $\mathbf{H}_r$, and also z nearest neighbors from each of the different classes, called nearest misses $\mathbf{M}_r(L)$, the ReliefF algorithm averages the contribution of all the hits and misses. The pseudocode of ReliefF is presented in Figure 1, where $\text{diff}(j, x_k, x_b)$ calculates the difference between the jth attribute value of two instances x_k and x_b. For continuous attributes

$$\text{diff}(j, x_k, x_b) = \frac{|\text{value}(j, x_k) - \text{value}(j, x_b)|}{\max(j) - \min(j)}. \quad (4)$$

The above-mentioned process is repeated for num times, and the weight vector w increases when the attribute value of the selected instance is different from that of the nearest miss $\mathbf{M}_r(L)$, and decreases when there is a difference between the attribute values of the selected instance and the

Algorithm ReliefF
Input: for each training instance a vector of attribute values and the class value
Output: the vector $\mathbf{w} = [w_1, w_2, ..., w_s]^T$ of estimations of the qualities of attributes
Process:
  set all weights $w_j := 0$;
  for $i := 1$ to num do{
  randomly select an instance $\mathbf{x}_k$;
  find z nearest hits $\mathbf{H}_r$;
  for each class $L \neq class(\mathbf{x}_k)$ do{
  from class L find z nearest misses $\mathbf{M}_r(L)$; }
  for $j := 1$ to s do{
$$w_j := w_j - \sum_{q=1}^{z} \text{diff}(j, \mathbf{x}_k, \mathbf{H}_r)/(num \cdot z)$$
$$+ \sum_{L \neq class(\mathbf{x}_k)} \left[\frac{P(L)}{1 - P(class(\mathbf{x}_k))} \right.$$
$$\left. \cdot \sum_{q=1}^{z} \text{diff}(j, \mathbf{x}_k, \mathbf{M}_r(L))/(num \cdot z) \right]$$
  }
  }

Figure 1. Pseudocode of ReliefF.

nearest hit $\mathbf{H}_r$. Compared with other filter methods, ReliefF usually perform better due to the performance feedback of the nonlinear classifier in searching for informative attributes.

3 THE PROPOSED ATTRIBUTE WEIGHTED FUZZY CLUSTERING ALGORITHM FOR INCOMPLETE DATASETS

In this paper, attribute weighting is incorporated into incomplete data clustering to emphasize the contribution of important attributes. And as the ReliefF algorithm can only determine attribute weights of complete datasets, so firstly, comparatively accurate imputations of missing attribute values and classification labels can be obtained by an existing algorithm, such as the OCS-FCM approach proposed by Hathaway & Bezdek (2001). Then, each attribute of the "completed" dataset can be evaluated by the ReliefF algorithm.

Based on the nearest-neighbor intervals of missing attribute values (Li et al., 2010; Li et al., 2013), firstly, the proposed attribute weighted fuzzy c-means algorithm for incomplete datasets (WI-FCM) transforms the incomplete dataset to an interval-valued one. In this transformed dataset, missing attribute value $\tilde{x}_{jb}$ can be represented by its corresponding nearest-neighbor interval

$[x_{jb}^-, x_{jb}^+]$, and non-missing attribute value $\tilde{x}_{jw}$ can be rewritten into interval form $[x_{jw}^-, x_{jw}^+]$. Note that $x_{jw}^- = x_{jw}^+ = \tilde{x}_{jw}$, that is, the original values of the non-missing attribute values are unchanged.

Then, the proposed WI-FCM algorithm combines the attribute weights and the fuzzy c-means algorithm by weighted Euclidean distance. Let $\bar{X} = \{\bar{x}_1, \bar{x}_2, ..., \bar{x}_n\}$ be the s-dimensional interval-valued dataset transformed from the incomplete dataset $\tilde{X}$, where $\bar{x}_k = [\bar{x}_{1k}, \bar{x}_{2k}, ..., \bar{x}_{sk}]^T$, $\forall j, k : \bar{x}_{jk} = [x_{jk}^-, x_{jk}^+]$. The WI-FCM algorithm minimizes the objective function:

$$J(U, \bar{V}) = \sum_{i=1}^{c} \sum_{k=1}^{n} u_{ik}^m \left\| \bar{x}_k - \bar{v}_i \right\|_w^2 \tag{5}$$

with the constraint of

$$\sum_{i=1}^{c} u_{ik} = 1, \quad for \ k = 1, 2, ..., n, \tag{6}$$

where c is the number of clusters; u_{ik} is the membership that represents the degree to which $\bar{x}_k$ belongs to the ith cluster, $\forall i, k : u_{ik} \in [0, 1]$, and let the partition matrix $U = [u_{ik}] \in R^{c \times n}$; m is a fuzzification parameter, $m \in (1, \infty)$; $\bar{v}_i$ is the ith interval cluster prototype, and let the matrix of interval cluster prototypes $\bar{V} = [\bar{v}_{ji}] = [\bar{v}_1, \bar{v}_2, ..., \bar{v}_c]$, where $\bar{v}_{ji} = [v_{ji}^-, v_{ji}^+]$, $\forall i = 1, 2, ..., c, j = 1, 2, ..., s$. The weighted Euclidean distance between $\bar{x}_k$ and $\bar{v}_i$ is defined as

$$
\left\| \bar{x}_k - \bar{v}_i \right\|_w
$$
$$
= \left[(x_k^- - v_i^-)^T W (x_k^- - v_i^-) + (x_k^+ - v_i^+)^T W (x_k^+ - v_i^+) \right]^{\frac{1}{2}}, \tag{7}
$$

where $W = diag(w_1, w_2, ..., w_s)$ is a diagonal matrix determined by the ReliefF algorithm; and $x_k^- = \left[x_{1k}^-, x_{2k}^-, ..., x_{sk}^- \right]^T$, $x_k^+ = \left[x_{1k}^+, x_{2k}^+, ..., x_{sk}^+ \right]^T$, $v_i^- = \left[v_{1i}^-, v_{2i}^-, ..., v_{si}^- \right]^T$, $v_i^+ = \left[v_{1i}^+, v_{2i}^+, ..., v_{si}^+ \right]^T$.

Similar with the method that used in our previous work (Yu & Fan, 2004; Li et al., 2010), the necessary conditions for minimizing (5) with the constraint of (6) are the update equations as follows:

$$v_i^- = \frac{\sum_{k=1}^{n} u_{ik}^m x_k^-}{\sum_{k=1}^{n} u_{ik}^m}, \quad for \ i = 1, 2, ..., c, \tag{8}$$

and

$$v_i^+ = \frac{\sum_{k=1}^{n} u_{ik}^m x_k^+}{\sum_{k=1}^{n} u_{ik}^m}, \quad for \ i = 1, 2, ..., c. \tag{9}$$

And if $\exists k, h, 1 \le k \le n, 1 \le h \le c, \forall j : \bar{x}_{jk} \subseteq \bar{v}_{jh}$, that is, $\bar{x}_k$ is within the convex hyper-polyhedron formed by $\bar{v}_h$, then $\bar{x}_k$ can be considered to belong fully to the hth cluster with membership 1, and belong to the other clusters with membership 0. Thus

$$u_{ik} = \begin{cases} 1, i = h \\ 0, i \ne h \end{cases}, \quad for \ i = 1, 2, ..., c, \tag{10}$$

else

$$u_{ik} = \left[\sum_{t=1}^{c} \left(\frac{\left\| \bar{x}_k - \bar{v}_i \right\|_w^2}{\left\| \bar{x}_k - \bar{v}_t \right\|_w^2} \right)^{\frac{1}{m-1}} \right]^{-1}, \quad for \ i = 1, 2, ..., c. \tag{11}$$

The procedure of WI-FCM is iteratively calculate the interval cluster prototypes $\bar{V}$ and the partition matrix U by using (8)-(11) until the stopping criterion is satisfied.

4 NUMERICAL EXPERIMENTS

In the experiments presented below, we tested the performance of the proposed WI-FCM for two well-known datasets: IRIS and Crude Oil. These databases are often used as standard databases to test the performance of clustering algorithms.

The IRIS data contains 150 four-dimensional attribute vectors, depicting four attributes of iris flowers: Petal Length, Petal Width, Sepal Length and Sepal Width. The three IRIS classes involved are Setosa, Versicolor and Virginica, each containing 50 vectors. Setosa is well separated from the others, while Versicolor and Virginica are not easily separable due to the overlapping of their vectors. Hathaway and Bezdek (1995) presented the actual cluster prototypes of the IRIS data:

$$
V^* = \begin{bmatrix} 5.00 & 5.93 & 6.58 \\ 3.42 & 2.77 & 2.97 \\ 1.46 & 4.26 & 5.55 \\ 0.24 & 1.32 & 2.02 \end{bmatrix}. \tag{12}
$$

The Crude Oil data contains 56 samples from three zones of sandstone. And the five attributes (chemical constitution) include vanadium, iron, beryllium saturated hydrocarbons and aromatic hydrocarbons.

In this paper, we consider the clustering problems where attribute values are Missing Completely At Random (MCAR). The scheme for artificially generating an incomplete dataset $\tilde{X}$ is to randomly select a specified percentage of components and designate them as missing. The random selection of missing attribute values is constrained so that:

1. each original attribute vector $\tilde{x}_k$ retains at least one component;
2. each attribute has at least one value present in the incomplete dataset $\tilde{X}$.

To test the clustering performance, the clustering results of WI-FCM and those of WDS, PDS, OCS, and the NPS versions of FCM proposed by Hathaway and Bezdek (2001) are compared. The initialization of the five approaches is partition matrix $U^{(0)}$ that satisfies (6) here, and the corresponding stopping criterion is $\|U^{(l)} - U^{(l-1)}\| < \varepsilon$. Besides, the missing attribute values are randomly initialized in OCS-FCM and NPS-FCM.

Choose fuzzification parameter $m = 2$, the number of clusters $c_{\text{IRIS}} = 3$, $c_{\text{Crude-Oil}} = 3$, convergence threshold $\varepsilon = 10^{-6}$, the number of nearest neighbors $q = 5$. To eliminate the variation in the results from trial to trial, Table 1 and Table 2 present the average number of misclassifications obtained over 10 trials on incomplete IRIS and Crude Oil

datasets. The same incomplete dataset is used in each trail for each of the five approaches, so that the results can be correctly compared. The optimal solutions in each row are highlighted in bold, and the suboptimal solutions are underlined.

For the actual cluster prototypes of the IRIS data are already known, Figure 2 shows the mean prototype error calculated by

$$\left\| V - V^* \right\|_F^2 = \sum_{j=1}^{s} \sum_{i=1}^{c} \left(v_{ji} - v_{ji}^* \right)^2 \qquad (13)$$

where V^* is the actual cluster prototypes of the IRIS data as shown in (12).

To illustrate the attribute weights determined by the ReliefF algorithm, as an example, Figure 3 shows the attribute weights obtained when the IRIS data miss 5% of attribute values.

From Table 1 and Table 2, as well as Figure 2, it can be seen that when the missing percentage of

Table 1. Average results of 10 trials using incomplete IRIS dataset.

| % missing | Mean number of misclassification | | | | |
	WDS	PDS	OCS	NPS	WI
0	16	16	16	16	6
5	17.2	16.8	16.7	<u>16.4</u>	**6.3**
10	<u>16.3</u>	16.9	17.1	16.9	**7**
15	<u>16.4</u>	16.9	17	17.1	**11.7**
20	16.9	16.9	16.8	<u>16.5</u>	**15.2**

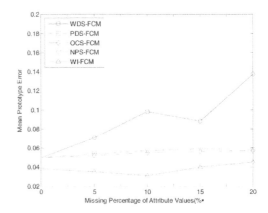

Figure 2. The mean prototype error of five algorithms on incomplete IRIS data.

Table 2. Average results of 10 trials using incomplete Crude Oil dataset.

| % missing | Mean number of misclassification | | | | |
	WDS	PDS	OCS	NPS	WI
0	23	23	23	23	**21**
5	<u>21.2</u>	22	21.9	22.2	**19.7**
10	<u>22.2</u>	22.3	22.3	22.4	**19.5**
15	<u>21.6</u>	21.9	21.8	22.3	**19.9**
20	<u>21.3</u>	21.9	21.9	21.7	**20.2**

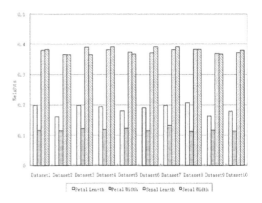

Figure 3. The attribute weights on 10 incomplete IRIS datasets with 5% attribute values missing.

attribute values is less than 20%, by emphasizing the contribution of important attributes, the proposed WI-FCM algorithm can achieve better clustering performance in terms of misclassification error, and the cluster prototypes obtained by the WI-FCM algorithm are closer to the actual ones. And from Figure 3, it can be seen that when the IRIS data miss 5% of attribute values, the weights of attribute "Sepal Length" and "Sepal Width" are much larger than those of attribute "Petal Length" and "Petal Width". Thus, through weighted Euclidean distance (7), the contribution of attribute "Sepal Length" and "Sepal Width" can be emphasized and involved in the clustering analysis, which is helpful to obtain more satisfying clustering results.

5 CONCLUSION

In this paper, an attribute weighted fuzzy clustering algorithm for incomplete datasets has been proposed, in which the missing attribute values are represented by nearest-neighbor intervals, and the ReliefF algorithm is used to determine the attribute weights. Compared with the compared methods, the proposed WI-FCM incorporates attribute weighting into incomplete data clustering by introducing weighted Euclidean distance, and emphasize the contribution of important attributes. The experiments performed on several datasets show that the proposed algorithm provides better clustering results than the compared methods.

ACKNOWLEDGEMENTS

This work is supported by the Natural Science Foundation of China under Grant 61305034, and the Fundamental Research Funds for the Central Universities DUT13JS03.

REFERENCES

Azadeh A., Asadzadeh S.M., Marandi R.J., Shirkouhi S.N., Khoshkhou G.B., Talebi S. & Naghavi A. 2013. Optimum estimation of missing values in randomized complete block design by genetic algorithm. Knowledge-Based Systems 37: 37–47.

Farhangfar A., Kurgan L.A. & Pedrycz W. 2007. A novel framework for imputation of missing values in data-based. IEEE Transactions on Systems, Man, and Cybernetics—Part A: Systems and Humans 37(5): 692–709.

Hathaway R.J. & Bezdek J.C. 1995. Optimization of Clustering Criteria by Reformulation. IEEE Transactions on Fuzzy Systems 3: 241–245.

Hathaway R.J. & Bezdek J.C. 2001. Fuzzy c-means clustering of incomplete data. IEEE Transactions on Systems, Man, and Cybernetics—Part B: Cybernetics 31(5): 735–744.

Honda K. & Ichihashi H. 2004. Linear fuzzy clustering techniques with missing values and their application to local principle component analysis. IEEE Transactions on Fuzzy Systems 12(2): 183–193.

Kira K. & Rendell L.A. 1992. A practical approach to feature selection in UK: The 9th Workshop on machine learning; Proc. intern. symp., Aberdeen, pp. 249–256.

Kononenko I. 1994. Estimating attributes: Analysis and extensions of Relief in Germany: The 7th European Machine Learning, Proc. intern. symp., Berlin, pp. 171–182.

Li D., Gu H. & Zhang L.Y. 2010. A fuzzy c-means clustering algorithm based on nearest-neighbor intervals for incomplete data. Expert Systems with Applications 37(10): 6942–6947.

Li D., Gu H. & Zhang L.Y. 2013. A hybrid genetic algorithm-fuzzy c-means approach for incomplete data clustering based on nearest-neighbor intervals. Soft Computing 17(10): 1787–1796.

Li D. & Zhong C.Q. 2012. Segmentation of images with damaged blocks based on fuzzy clustering. ICIC Express Letters 6(10): 2679–2684.

Nuovo A.G.D. 2011. Missing data analysis with fuzzy c-means: a study of its application in a psychological scenario. Expert Systems with Applications 38(6): 6793–6797.

Ren Y., Liu X.D. & Liu W.Q. 2012. DBCAMM: A novel density based clustering algorithm via using the Mahalanobis metric. Applied Soft Computing 12: 1542–1554.

Timm H., Doring C. & Kruse R. 2004. Different approaches to fuzzy clustering of incomplete datasets. International Journal of Approximate Reasoning 35(3): 239–249.

Yu C.H. & Fan Z.P. 2004. A FCM clustering algorithm for multiple attribute information with interval numbers. Journal of system engineering, 19(4): 387–393. (in Chinese)

Zhang D.Q. & Chen S.C. 2003. Clustering incomplete data using kernel-based fuzzy c-means algorithm. Neural Processing Letters 18: 155–162.

Information Technology and Applications – Li (Ed.)
© 2015 Taylor & Francis Group, London, ISBN 978-1-138-02677-3

Semantic automatic image annotation based on Non-Abstract Visual Keywords

Xiao Ke

College of Mathematics and Computer Science, Fuzhou University, Fuzhou, China

Guolong Chen

Fujian Provincial Key Laboratory of Networking Computing and Intelligent Information Processing, Fuzhou University, Fuzhou, China

ABSTRACT: Automatic image annotation is a significant and challenging problem in pattern recognition and computer vision. Existing models did not describe the visual representations of corresponding keywords, which would lead to appearing plenty of irrelevant annotations in final annotation results. These annotations did not relate to any part of images considering visual contents. We propose a new automatic image annotation model (NAVK) based on relevant visual keywords to overcome above problems. Our model focuses on non-abstract words. First, we establish visual keyword seeds of each non-abstract word, and then a new method is proposed to extract visual keyword collections by using corresponding seeds. Second, we propose adaptive parameter method and fast solution algorithm to determine similarity thresholds of each keyword. Finally, the combinations of above methods are used to improve annotation performance. Experimental results verify the effectiveness of proposed image annotation model.

Keywords: automatic image annotation; Non-Abstract Visual Keywords; adaptive threshold

1 INTRODUCTION

Automatic Image Annotation refers to automatically generating text labels according to images' visual contents. Automatic Image Annotation [1, 2] assigns tags to images by considering visual contents, which can be called content-based automatic image annotation. This kind of methods can build good links between visual contents and annotations.

Unfortunately, almost all the content-based automatic image annotation models had same problems. That is, existing models did not consider the visual representations of corresponding keywords, which would lead to appearing a lot of irrelevant annotations in final annotation results. We propose a new automatic image annotation model (NAVK) based on relevant visual keywords to overcome above problems. Firstly, we establish visual seeds of each non-abstract word, and then a new method is proposed to extract visual keyword collections by using corresponding seeds. Secondly, we propose adaptive parameter method and fast solution algorithm to determine similarity thresholds of each keyword. Finally, the combinations of above methods are used to improve annotation performance.

2 RELATED WORK

Duygulu et al. proposed Machine Translation Model [3], they considered image annotation as a translation problem between two languages: one language is visual vocabulary of image contents; the other is real text vocabulary. They use Normalized Cut to segment images, and then K-Means algorithm is used to cluster these regions. Image annotation can be regarded as translation processes from visual vocabulary *blobs* to the semantic keywords. Jeon et al. proposed Cross Media Relevance Model (CMRM) [4] which used joint probabilities of semantic labels and visual words to annotate images. They use same discrete features as translation model [3], so it will inevitably lose some helpful visual information. Lavrenko et al. put forward Continuous Relevance Model (CRM) [5], which directly make use of continuous features of image regions and use non-parametric Gaussian kernel to continuously estimate generation probability of visual contents. Feng et al. proposed Multiple Bernoulli Relevance Model (MBRM) [6] which used rectangular grids instead of complicated segmentation algorithms to partition images, and they applied Bernoulli distribution instead of multinomial distribution to describe generation distribution of vocabulary. Kang et al. proposed Correlated Label Propagation model (CLP) [7], which took into account diffusing multi labels by means of labels' correlations between adjacent images at the same time. Liu et al. proposed AGAnn model [8] to improve the annotation results by using adaptive graphs and words correlations.

3 AUTOMATIC IMAGE ANNOTATION BASED ON RELEVANT VISUAL KEYWORDS

At present, the dataset of automatic image annotation only gives several labels to certain entire images, that is, there are no labels corresponding to image regions. There are two reasons about above phenomenon. Firstly, although image segmentation algorithms have been developed, segmentation results are still hard to achieve a satisfactory degree of practicality. There are many blurry regions after image segmentation, and these regions can not correspond to any labels. Secondly, it is time-consuming and hard to annotate each image region if image library are relatively large. In fact, if we can build model of each label's visual expression, it will absolutely improve the performance of automatic image annotation. This is a key content in this paper.

3.1 Extraction of non-abstract labels' visual keyword collection

In this paper, we focus on the non-abstract words, such as "horse", "sky", "boat". Existing image library do not give visual descriptions of corresponding annotation words, and it is nearly impossible to directly separate visual regions of each label from original images by using certain algorithm. Therefore, visual seeds of each non-abstract word are manually selected. When visual seeds of each non-abstract word are confirmed, visual keyword collections will be constructed by using these visual seeds. We first form N_l collections that all images in one collection should share certain label, where N_l is the number of labels. Visual keywords of each label would be selected from corresponding collection. Visual keywords is selected by using Equation (1).

$$Col(vw_i^l) = \left\{ vw_{ij}^l \mid Sim(R_{ij}, R_l) >= \eta_{thres}^l, \right\}$$
$$i = 1, ..., |List(l)|, j = 1, ..., |\Lambda| \quad (1)$$

where $Col(vw_i^l)$ are visual keywords in i-th image of l-th label. vw_{ij}^l is j-th region in i-th image of l-th label's image collection. η_{thres}^l is the visual keyword threshold of l-th label, how to determine η_{thres}^l will

be given in next subsection. $|List(l)|$ is the size of l-th label's image collection. $|\Lambda|$ is the number of regions in one image. R_{ij} are visual features of j-th region in i-th image. R_l are visual features of l-th visual seeds. $Sim(R_{ij}, R_l)$ is similarity between R_{ij} and R_l, which can be calculated by Equation (2):

$$Sim(R_{ij}, R_l) = \exp\left\{ -\sqrt{\sum_{m=1}^{d} \left(R_{ij}^m - R_l^m \Big/ \sigma_m \right)^2} \right\} \quad (2)$$

where σ_m is standard deviation of m-th dimension's visual features, d is dimension of visual features. $Col(VW^l)$ is visual keyword collection of l-th label, which can be calculated by Equation (3):

$$Col(VW^l) = \left\{ Col(vw_1^l), Col(vw_2^l), ..., Col(vw_{|List(l)|}^l) \right\} \quad (3)$$

Above method can be used to extract visual keywords. However, there are still some irrelevant visual keywords in the collections. These irrelevant visual keywords will absolutely affect the quality of visual keyword collections. So it is necessary to remove them from visual keyword collections. The number of visual keywords is rather various among different labels. For each label, we select top (t%) visual keywords in visual keyword collection which are most similar to corresponding visual seed. The specific value of t will be given later in the experimental section.

3.2 Similarity threshold determination

Similarity threshold is very important to select visual keywords, which directly determine if selected visual keywords can be used to effectively describe visual features of corresponding object. Threshold should not be too big or too small.

In addition, different labels have different visual descriptions, so different labels should not use the same similarity threshold. For example, "sky" and "sea" are very consistent in visual descriptions, so similarity threshold should be set a little higher. On the contrary, other labels like "bear", "flower" and "car" should use a little smaller threshold. Because selected visual keywords may be only some parts of these objects, and descriptions of selected visual keywords have many diversities. Therefore, if all labels share the same similarity threshold, it will lead to selected visual keywords which could not describe all labels well. So it is necessary to set different similarity thresholds for each label. If an image contains label w, we know there are at least one region could be used to describe this label w. From this idea, we propose an adaptive method to determine similarity threshold. Threshold can be determined by Equation (4).

$$\eta_{thres}^l = \max\{\eta_{thres}^* \mid \forall vw_{ij}^l \in Col(vw^l),$$
$$\exists vw_{ij}^l \in Col(vw_i^l), s.t. \ Sim(R_{ij}, R_l) \geq \eta_{thres}^*\} \qquad (4)$$

From Equation (4), we can determine the similarity thresholds of each label. In addition, we also propose a fast algorithm to solve similarity thresholds of each label. Our algorithm can find each label's similarity threshold under time complexity $\log_2(n)$. The specific algorithm is as follows:

Alogrithm 1:
Input: R_l is visual seed of *l-th* label. R_{ij} is *j-th* region in *i-th* image that contains *l-th* label. $Sim(R_{ij}, R_l)$ is similarity between R_{ij} and R_l.
Output: η_{thres}^l is similarity threshold of *l-th* label.
Initialize: $min\_idx = 0$, $max\_idx = |R^l|-1$;

1. For each region R_{ij}, sort all $Sim(R_{ij}, R_l)$ by ascending order, the results are recorded in $AS[|R^l|]$. Initialize $min(R_l) = AS[0]$, $max(R_l) = AS[|R^l|-1]$;
2. While $(min(R^l) <= max(R^l))$
   2.1 $mid(R_l) = (AS[min\_idx] + AS[max\_idx])/2$, $mid\_idx = (min\_idx + max\_idx)/2$;
   2.2 if $(min\_idx = max\_idx \| min\_idx)1 = max\_idx)$
       if $(AS[max\_idx]$ meet Equation (4))
           return $\eta_{thres}^* = AS[max\_idx]$;
       else
           return $\eta_{thres}^* = AS[min\_idx]$;
   2.3 if $(\eta_{thres}^* = AS[mid\_idx]$ meet Equation (4))
       $min(R^l) = AS[mid\_idx]$,
       $min\_idx = mid\_idx$;
   2.4 else
       $max(R^l) = AS[mid\_idx]$,
       $max\_idx = mid\_idx$;
3. Output η_{thres}^* is similarity threshold of *l-th* label;

3.3 *Generation of candidate annotation words*

This subsection will discuss how to determine candidate annotation labels for each unlabeled image. Divide each unlabeled image into several regions, and then the joint probabilities of each region and each label are calculated by using Equation (5).

$$\ell(U_{ij}, w) = \sum_{p=1}^{|Col(VW^w)|} Sim(U_{ij}, VW_p^w) \bigg/ |Col(VW^w)| \qquad (5)$$

where U_{ij} is *j-th* region in *i-th* unlabeled image. VW_p^w is *p-th* visual keywords in visual keyword collection of label w. $|Col(VW^w)|$ is the number of label w's visual keywords. $Sim(U_{ij}, VW_p^w)$ is the similarity of U_{ij} and VW_p^w. $\ell(U_{ij}, w)$ is probability

that U_{ij} is annotated by label w. $lbl(U_{ij})$ is candidate label collection of region U_{ij}, which can be calculated by using Equation (6).

$$lbl(U_{ij}) = \{(w_1, w_2, ..., w_r) \mid Sort \ \ell(U_{ij}, w_k) \\ in \ descending \ order\} \qquad (6)$$

$lbl(U_i)$ is candidate label collection of unlabeled image U_i, which can be calculated by using Equation (7).

$$lbl(U_i) = \{W^* \mid \forall w \in [lbl(U_{i1}), lbl(U_{i2}), ..., lbl(U_{i|\Lambda|})], \\ s.t. \ \exists 1 \leq i, j \leq |W^*|, \forall i, j, i \neq j, w_i \neq w_j\} \qquad (7)$$

4 NAVK AUTOMATIC IMAGE ANNOTATION MODEL

4.1 *NAVK model*

Each image I is divided into a series of non-overlapping regions, record as $R_I = \{r_1, ..., r_{|\Lambda|}\}$. We use same partition method as MBRM, $|\Lambda|$ is the number of image regions. Extract *d*-dimensional feature vector F^i from each image region r_i, define $P_F(\sim | I)$ as visual generation probability of image regions. Generation probability of vocabulary is multiple Bernoulli distribution, which is a more reasonable way to describe vocabulary distribution comparing to polynomial distribution. Assuming that annotation sets W_I is independently sampled from $|V|$ Bernoulli distribution $P_V(\sim | I)$, where $|V|$ is the number of annotation words. So an image I can be seen that be make up of two independent distributions: one is generation probability distribution of region features, the other is generation probability distribution of vocabulary.

Assuming image U is an unlabeled image, and $F_T = \{F_T^1, ..., F_T^{|\Lambda|}\}$ are visual feature vectors of U, where F_T^i is feature vector of *i-th* region in image U. W_L is a subset of all annotation labels. We model U's joint probability of visual descriptions and vocabulary's descriptions, denoted as $P(F_T, W_L)$. In joint probability of $P(F_T, W_L)$, implied correlations between F_T and W_L are assuming similar to certain correlations between image's visual and vocabulary's descriptions in training set, however, we do not know the specific implied correlations. So we calculate joint probability's expectation of visual features and vocabulary between unlabeled image U and each image in training set. The processes of jointly generating F_T and W_L are as follows:

1. Obtain each label's visual seed and corresponding visual keyword collection by using methods in Section 3.1 and Section 3.2;

2. Select an image I from training set Γ with probability $P_\Gamma(I)$;
3. For each training image I, $i = 1, ..., |\Lambda|$ ($|\Lambda|$ is the number of image regions):

Generate visual descriptions from i-th image region by using conditional probability $P_F(\sim|I)$;
4. For each word v in annotation label set:

Generate annotation set W_I by using multiple Bernoulli distribution $P_V(\sim|I)$;

We are looking for several most probable labels of each unlabeled image. According to above generating processes, Equation (8) can be used to calculate joint probability of image visual descriptions and annotation labels in our NAVK model.

$$P(F_T, W_G) = \sum_{I \in \Gamma} P_\Gamma(I) \times \prod_{i=1}^{|\Lambda|} P_F(F_T^i | I) \times \xi_{W_G}$$
$$\times \prod_{v \in W_G} P_V(v | I) \times \prod_{v \notin W_G} (1 - P_V(v | I)) \quad (8)$$

4.2 Parameter estimation

In this section, we will discuss parameter estimation of Equation (8). $P_\Gamma(I)$ is the probability of selecting an image I from training set. Since there is not any priori knowledge, $P_\Gamma(I)$ can be assumed to obey uniform distribution, that is,

$$P_\Gamma(I) = 1/|\Gamma| \quad (9)$$

where $|\Gamma|$ is the size of training image set.

Conditional probability $P_F(\sim|I)$ is used to estimate visual generation probability of image regions. Non-parameter kernel density function is used to estimate distribution of $P_F(\sim|I)$. Assuming $F_I = \{F_I^1, ..., F_I^{|\Lambda|}\}$ are regional features of image I, $P_F(\sim|I)$ can be estimated as follows:

$$P_F(F_T^i | I)$$
$$= \frac{1}{|\Lambda|} \sum_{j=1}^{|\Lambda|} \frac{\exp\left\{-(F_G^i - F_I^j)^T \sum^{-1} (F_G^i - F_I^j)\right\}}{\sqrt{2^d \pi^d |\Sigma|}}$$
$$(10)$$

where $|\Lambda|$ is number of image regions, d is dimension of features. Equation (10) uses Gaussian kernel function to estimate visual description F_I^j of each region in image I. Gaussian kernel is determined by covariance matrix Σ, where $\Sigma = \mu \cdot I$, and μ is the width of Gaussian kernel which is used to determine the smooth level of P_F around F_I^i, I is an identity matrix.

$P_V(v | I)$ is v-th component of multiple Bernoulli distribution. It means probability of annotation set W_L which is generated by training image I. Bayesian estimation is used for each annotation label:

$$P_V(v | I) = \frac{\varepsilon \cdot \phi_{v,I} + N_v}{\varepsilon + |\Gamma|} \quad (11)$$

where N_v is the number of label v that appears in training set, $|\Gamma|$ is the size of training image set. $\phi_{v,I}$ is a binary function, if training image I contains label v, $\phi_{v,I} = 1$, else $\phi_{v,I} = 0$. ε is an empirical smooth parameter, which also could be seen as weight of $\phi_{v,I}$.

ξ_{W_G} is a binary function which can be calculated by Equation (12). We assign higher weights to W_G if W_G appear in candidate label collection of unlabeled image U_i, that is, $\xi_{W_G} = \varpi > 1.0$. Specific value of ϖ will be given in experimental section.

$$\xi_{W_G} = \begin{cases} \varpi, & \text{if } W_G \in lbl(U_i) \\ 1.0, & \text{else} \end{cases} \quad (12)$$

4.3 Framework of NAVK model

The framework of NAVK model is in Figure 1. Given a training library, methods in Section 3.1 are used to extract visual seeds of each non-abstract label, and then our adaptive threshold method in Section 3.2 can be used to select corresponding visual keywords. Given an unlabeled image, this unlabeled image should be segmented at first, candidate labels of each region should be generated, and then candidate labels of each unlabeled image should also be generated by using method in Section 3.3. Relevance model is used to annotate unlabeled image. Generation probability of each image region and each label should be calculated. If a label is in candidate label collection of the unlabeled image, we give a higher weight to this label. Finally, we output annotation results of each unlabeled image.

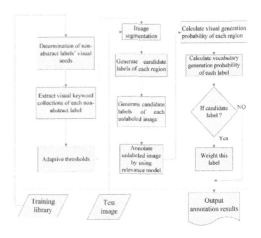

Figure 1. Framework of NAVK model.

246

5 EXPERIMENTS

5.1 Experiments set-up

In order to verify the validity of our NAVK model, we compare our model with other models using the same dataset on Corel dataset [3]. This paper mainly focuses on how to construct an effective automatic image annotation model. For the convenience of comparing with other models, we do not use some new features. We use same 30-dimensional features, including: 9-dimensional RGB color moments, 9-dimensional Lab color moments, 12-dimensional Gabor texture features which consist of 3 scales and 4 directions.

As other automatic image annotation model, we use precision, recall and F-measure to evaluate annotation results. Assume w is a label, $\#(s)$ is number of images that annotated by w, $\#(c)$ is number of images that correctly annotated by w, $\#(t)$ is number of images that contained w in ground-truth annotation.

$$Pr(w) = \frac{\#(c)}{\#(s)} \quad Re(w) = \frac{\#(c)}{\#(t)}$$

$$F(w) = \frac{2 \times Pr(w) \times Re(w)}{Pr(w) + Re(w)} \quad (13)$$

In addition, we also count the labels that are correctly annotated at least once as some other models, denote as "NZR".

5.2 Experimental comparisons

5.2.1 Parameters determination
We need to determine one parameter value in our NAVK model by experiments, that is, parameter ϖ in Section 4.2. Figure 2 are automatic image annotation results when we set different values of ϖ under $t = 100$. $\varpi = 1.0$ means that we do not weigh

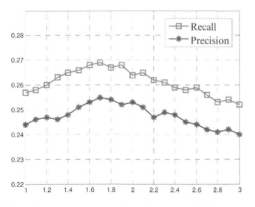

Figure 2. Annotation results when we used different ϖ.

on candidate labels. From Figure 2, we can find that it obtains best annotation results when $\varpi = 1.7$. Precisions and recalls of annotation results would not be improved if ϖ increases. This is because higher ϖ will impact labels' generation probability distribution, in further, the whole annotation performance will be affected.

5.2.2 Experimental results
We will verify the performance of our NAVK model in this subsection. We compare NAVK model with some state-of-the-art models, CMRM [4], MBRM [6], CLP [7]. The experimental results are shown in Table 1.

From Table 1, we can find that our NAVK model is very effective. Annotation results are better than those state-of-the-art models. NAVK obtains the highest precision 0.25 which is at least 4% higher than the other four models, and recall achieves 0.26 which is at least 4% higher than the other four models. F-measure of NAVK achieves 0.26, and it is about 8% higher than the other four models. In addition, in the evaluation criterion of "NZR" which reflects the coverage of annotation words, our NAVK model reaches 134, and it is also the highest in all models.

5.2.3 Analysis of results
Table 2 is the annotation results of several test images. We compare our NAVK model with the results of ground truth, and the rankings of annotation labels are sorted in descending order by annotation probability. If labels are in ground-truths, we use bold type. In addition, we also

Table 1. Experimental results.

Models	Precision	Recall	F-measure	NZR
CMRM	0.10	0.09	0.09	66
MBRM	0.24	0.25	0.24	122
CLP	0.21	0.25	0.22	125
NAVK	0.25	0.26	0.26	134

Table 2. Comparisons of annotation results.

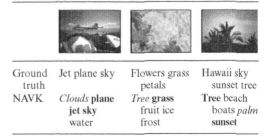

Ground truth	Jet plane sky	Flowers grass petals	Hawaii sky sunset tree
NAVK	*Clouds* **plane jet sky** water	*Tree* **grass** fruit ice frost	**Tree** beach boats *palm* **sunset**

247

find that some annotation words do not appear in ground truth annotations of the dataset, but some of these words can also describe the contents of image. That is, some of right annotations are ignored by some people who help to annotate the dataset. These labels are in italic type. For example, "*clouds*" dose not belong to the ground truth annotations in first image, but the label can be used to describe the contents of first image without question.

6 CONCLUSIONS

In this paper, we propose a new automatic image annotation model (NAVK) based on relevant visual keywords to overcome above problems. Firstly, we establish visual keyword seeds of each non-abstract word, and then a new method is proposed to extract visual keywords by using corresponding seeds. Secondly, we propose adaptive parameter method and fast solution algorithm to determine similarity thresholds of each label. In future work, we will consider how to reduce time complexity of automatic image annotation models.

ACKNOWLEDGMENT

This work is supported by National Natural Science Foundation of China (Grant No. 61103175), Natural Science Foundation of Fujian Province, China (Grant No. 2013J5088).

REFERENCES

[1] Yahong H., Fei W., Qi T., et al. Image Annotation by Input–Output Structural Grouping Sparsity [J]. IEEE Transactions on Image Processing, 21(6), 2012: 3066–3079.

[2] Zechao L., Jing L., Changsheng X., et al. MLRank Multi-correlation Learning to Rank for image annotation [J]. Pattern Recognition, 46(10), 2013: 2700–2710.

[3] Duygulu P., Barnard K., Freitas J., Forsyth D. Object Recognition as Machine Translation: Learning a Lexicon for a Fixed Image Vocabulary. Proceedings of the 7th European Conference on Computer Vision, Copenhagen, Denmark, 2002: 97–112.

[4] Jeon J., Lavrenko V., Manmatha R. Automatic image annotation and retrieval using cross-media relevance models. Proceedings of the 26th Annual International ACM SIGIR, Toronto, Canada, 2003: 119–126.

[5] Lavrenko V., Manmatha R., Jeon J. A model for learning the semantics of pictures. Proceedings of Advance in Neutral Information Processing, Vancouver, Canada, 2003.

[6] Feng S.L., Manmatha R., Lavrenko V. Multiple Bernoulli relevance models for image and video annotation. Proceedings of the IEEE Computer Society Conference on Computer Vision and Pattern Recognition, Washington, D.C., USA, 2004:1002–1009.

[7] Kang F., Jin R., Sukthankar R. Correlated Label Propagation with application to Multi-label Learning. Proceedings of the 2006 IEEE Computer Society conference on Computer Vision and Pattern Recognition, New York, USA, 2006: 1719–1726.

[8] Liu J., Li M.J., Ma W.Y., Liu Q.S., Lu H.Q. An adaptive graph model for automatic image annotation. Proceedings of the ACM SIGMM Workshop on Multimedia Information Retrieval, Santa Barbara, USA, 2006: 61–69.

Information Technology and Applications – Li (Ed.)
© *2015 Taylor & Francis Group, London, ISBN 978-1-138-02677-3*

JVM based traffic simulation oriented computer language framework

Y. Bin, Z. Xiaoqing, W. Weiyang & W. Yanqing
School of Transportation Engineering, Tongji University, Shanghai, China

ABSTRACT: Currently the mainstream traffic simulation platforms embed logic of concrete traffic models into platforms themselves in the form of compiled binary code. The platforms will achieve benefit of faster execution speed by doing like this. But, in the meantime, as there is a close coupling between concrete traffic models and the platforms, the platforms suffer a shortcoming of less flexibility in terms of model simulation. This shortcoming can be easily spotted by realizing the fact that none of the platforms can easily simulate different traffic models targeting similar behaviors such as car following behavior. For this purpose, the authors propose a concept of building a strong static data type computer language oriented to traffic simulation upon Java byte code and Java virtual machine. The core of concept is to decouple close relationship between simulation platform and traffic model, thus improve platform's flexibility in terms of model simulation. Theoretically, a platform built upon an idea similar to the one proposed only needs less coding efforts in order to simulate any microscopic traffic model.

Keywords: JVM; traffic simulation; computer language; advanced emulator

1 INTRODUCTION

Traditional traffic simulation platforms [1, 2] usually embed logic of concrete traffic models into the system itself in the binary code form. However, this will lead to less inflexibility in terms of traffic simulation on these platforms. These defects were observed at the same time learn from success stories of other industries such as Mathworks Inc. Matlab, Yu Bin extract the following key ideas: simulation platforms' own code should not rely on any specific traffic models. Stripping out the logic of a particular model of traffic simulation platform brings many benefits to the platform itself, such as it can make the simulation platform mode independent from the model. That can be simulated on the platform macro, meso, micro, and even cellular automata. You can also theoretically arbitrary traffic simulation model. Also due to the model simulation platform independence, on the platform for secondary development, relative to conventional simulation systems, traditional traffic simulation platform will also be greatly simplified.

In order to verify the feasibility of the idea, Yu Bin [3] developed a simulation system Advanced Emulator, referred to as AE. In order to reduce the dependence on particular traffic models, AE is a specially developed computer simulation system based on traffic script language which is the core for traffic simulation. This language allows the user to customize the logic of specific traffic model. AE's core consists of three parts, namely the Language Compiler, Runtime Virtual Machine, and Just-In-Time Compiler. These three parts is completed with the combination of ANSI C language, and compilation-like language DynASM. The process of AE system is generally described as follows. First, the user use specialized language which is implemented by AE to run the specific traffic model (taking into account that the users are often non-computer major, or no computer programming experience, AE reduced the difficulty of user programming that the language has been designed and implemented for a dynamic data types interpreted scripting language). Second, a script file that contains user-written logic specific traffic models will be input to the compiler, which is responsible for the use of AE interpreted scripting language to translate traffic simulation model into bytecode. Furthermore, these bytecode will be the input to the virtual machine, which will run these byte codes one by one to complete the traffic simulation. Dynamic compiler is specifically being introduced to enhance the operational efficiency of the system. Taking into account the low efficiency interpretive bytecode execution, dynamic compiler module translates bytecode which is dynamically running frequently and commonly known as hotspot into an equivalent machine code in order to improve the overall operational efficiency of the AE.

Figure 1 shows the flowchart of AE traffic simulation system, where each module is completely designed and developed by the author.

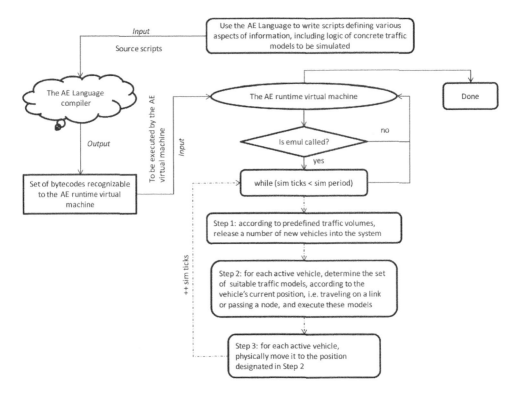

Figure 1. The flowchart of AE traffic simulation.

Although Yu Bin, through the development of AE, demonstrated the feasibility and effectiveness of the simulation thought, it is worth noting that the traffic studies developed from scratch in terms of units and individuals such simulation system similar AE is a very time-consuming and labor-intensive research work (in fact, it takes eight years for Yu Bin). So for many reasons, it is necessary to explore these ideas in the field of traffic simulation specifically about its implementations.

2 JVM BASED TRAFFIC SIMULATION ORIENTED COMPUTER LANGUAGE

It is not difficult to find that the key of achieving the simulation ideas above is a traffic simulation-oriented computer language. The language should contain specific language mechanism or structure in the semantic level, allowing researchers to define logical rules for specifically traffic model in the code level. As for the issues how to translate these logical rules into machine instructions that can run on a physical machine should be considered by system developers and be irrelevant and transparent for the final user such as transportation researchers. In other words, on the premise of the consistency of

the semantics and syntax, internal implementation of the language is varied.

Java language [4] is created by the U.S. SUN Microsystems (America Oracle company received Java language IPR in 2010) in 1995, which is a computer language using for writing network applications. One of the features of the Java language is compiled once, run everywhere. For this feature, Java language is built on Java Virtual Machine (JVM). In the earlier age, Java language and the Java Virtual Machine are the two concepts which always closely linked together, and there is no obvious boundary between them, that is the Java language equivalent to the Java Virtual Machine. However, with the continuous developing of Java, these two concepts began to differ from each other. Currently the popular view in IT circles is that the Java language is a kind of computer language runs on the Java virtual machine.

The presence of these two views is essentially different. Compared to the first point of view, the second point of view is much wider. According to this view, Java virtual machine is the infrastructure to run computer language, including such as Hotspot based Just-In-Time compiler, multiple garbage collector and so on. In particular, a byte code collection implemented of Java virtual machine is language-independent, so developers can run a variety of different computer languages on the Java Virtual Machine. The Just-In-Time compiler, garbage collection, etc. not only put forward high demands to master all aspects of computer science and technology knowledge, but also requires a huge work of computer programming, which made a very high technical requirements for system developers. However, the design and development of appropriate computer languages in the Java virtual machine can effectively circumvent these technical difficulties, which greatly simplifies the difficulty of the system developing and speeds up the process. It is also aware of the advantages of the Java virtual machine, there are now new computer languages such as Clojure [5], Scala [6]. Although these new languages have their own scope and objects such as Scala mainly for parallel programming, they both compile language source code to Java byte code so that run on the Java virtual machine to research and development.

Authors believe that this method can be used to develop a traffic simulation-oriented computer language, and a framework is proposed in this article. In this framework, firstly, design a specialized computer language based on the own characteristics of traffic simulation (Note: only related to the level of semantic and syntactic structures). On this aspect, developers can get reference from Yu Bin [3]. Secondly, design and write Java byte code compiler which compile the source code of

this language into Java byte code. Then the Java Virtual Machine is responsible for running the Java byte code to complete the traffic simulation. Figure 2 shows a flow chart of the simulation framework, in which the dashed box partial division is responsible by the system development staff. Comparing with Figure 1, it is not difficult to find traffic simulation-oriented computer language which will greatly simplify the overall traffic simulation system development difficulty and workload on a Java virtual machine.

3 CONCLUSIONS

The authors through the development of AE system demonstrated the value of this simulation thought that traffic model is away from the system itself. Given the difficulty of the simulation system developed from scratch, based on this idea, the authors combined current research achievements in the fields of computer science and traffic simulation, on this basis, proposed development traffic simulation computer language based on the Java virtual machine and concept of traffic simulation based on this. The simulation system is based on this idea, while simulation ideas are presented before the organic integration can effectively reduce the difficulty of the simulation system development, and for building the next generation of traffic simulation system has a certain significance and scientific value.

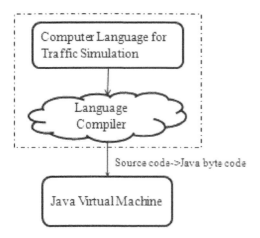

Figure 2. The flowchart of JVM-based traffic simulation system.

REFERENCES

[1] Laufer, J. (2007). "Freeway Capacity, Saturation Flow and the Car Following Behavioral Algorithm of the VISSIM Microsimulation Software" 30th Australasian Transport Research Forum.
[2] Yang, Q., Koutsopoulos, H.N. (1996). "A Microscopic Traffic Simulator for Evaluation of Dynamic Traffic Management Systems" Transportation Research Part C, Vol 4.
[3] Bin, Y., Chiu, L., Zhongren, W., (2014). "Development of a Computer System for Simulation of Traffic Models", Journal of Computing in Civil Engineering.
[4] Gosling, J., Joy, B., Steele, G., Bracha, G., (1999) "The Java Language Specification, 2nd Edition".
[5] Cbas, E., Brian, C., Cbristopbe, G., (2013) "Clojure Programming", O'Reilly.
[6] Martin, O., (2014) "The Scala Language Specification Version 2.9".

Information Technology and Applications – Li (Ed.)
© 2015 Taylor & Francis Group, London, ISBN 978-1-138-02677-3

A novel pattern recognition algorithm

Xiaoqing Guan & Baoling Han
School of Mechanical Engineering, Beijing Institute of Technology, Beijing, China

Zhuo Ge & Kai Niu
School of Mechatronics, Beijing Institute of Technology, Beijing, China

ABSTRACT: In view of shortages of the traditional target recognition algorithms, this paper proposed a kind of improved mutual information combined with particle swarm optimization algorithm of target recognition algorithm. This novel method can easily compute transformation matrix and has smaller computational complexity than the traditional target recognition algorithm based on mutual information, and also can realize effective data dimension reduction. The experimental results show that compared with the traditional LDA and PCA algorithm, the new algorithm has greater advantage in recognition performance, complexity of algorithm and noise resistance. This proposed algorithm can effectively improve the efficiency of image recognition, convenient for real-time processing.

Keywords: target recognition; mutual information; particle swarm optimization algorithm

1   INTRODUCTION

The main purpose of automatic target recognition system is to reduce human workload, and to eliminate the danger of unmanned control system. This is a challenging task, because the system must be able to adapt to complex scenes, to ensure a low false alarm rate and to guarantee real-time performance.

Traditional feature extraction methods include Principal Components Analysis (PCA), Linear Discrimination Analysis (LDA), Independent Component Analysis (ICA) and etc. PCA has fast computation speed. However, since noise information has great influence on the results, PCA is not suitable for low image quality infrared images. In a manner of speaking, ICA is an extension of PCA method. While the eigen vectors of ICA method is statistical independent between each dimension, but for a specific classified recognition problem, the feature vector set is not necessarily optimal.

LDA is considered as a kind of optimal linear probability density function, but during transformation, the feature dimension has to be less than or equal to number of classes minus 1. Aiming at many types of problems, Devijver and Kittler proposed mutual information theory based on Information theory. The mutual information theory has been introduced into pattern recognition. Due to excellent performance, mutual information has become another important analysis method of feature extraction, but its development was limited by great computational complexity.

Aiming at disadvantages of the algebraic feature transform method, this paper proposes a new feature extraction method based on mutual information theory retained the advantages of the original methods, and solved the problem of computational complexity.

2   THE BASIC PRINCIPLES OF MUTUAL INFORMATION

Mutual Information (MI), commonly represented by entropy, is a measure of Information theory which is used to describe the statistical correlation between two variables or the amount of information of one variable contained by another variable, and is a measure of the mutual dependence of the two random variables. Entropy represents complexity and uncertainty of a system. Shannon information theory proposed a measure of uncertainty using entropy. The uncertainty is defined as

$$H(x) = -\sum P(x)\log[P(x)] \tag{1}$$

If distribution of different classes is $P(c)$, $c = 1, 2 \ldots N_c$, then the entropy of output class is

$$H(C) = -\sum_{c}^{N_c} P(c) \log P(c) \qquad (2)$$

If feature vector f is given, conditional entropy is defined as

$$H(C|F) = -\sum_{c=1}^{N_f} P(f) \sum_{c=1}^{N_c} P(c|f) \log P(c|f) \qquad (3)$$

$P(c|f)$ is the conditional probability of vector f be classified as class C.

Generally, the conditional entropy is bigger than or equal to the original entropy, only when the output class and feature vector completely independent, the conditional entropy is equal to the original entropy. The decrement of uncertainty is defined as mutual information:

$$I(C;F) = H(C) - H(C|F) \qquad (4)$$

This function is symmetric for C and F, and it is ordinary algebraic operation.

$$I(C;F) = I(F;C) = \sum_{c,f} P(c,f) \log \frac{P(c,f)}{P(c)p(f)} \qquad (5)$$

Mutual information method has strong robusticity, high identification accuracy, and especially when one image's data partly missing, MI method can get good recognition results. Therefore, in recent years, mutual information method has obtained the widespread attention.

3 IMPROVED TARGET RECOGNITION ALGORITHM AND ITS IMPLEMENTATION

It is well known that, as a class separability criterion, the traditional mutual information needs to be discretized in actual applications. The process of transformation leads to high data redundancy, great time consumption, large computational complexity and low approximation accuracy. It is the enormous computational complexity seriously hindered development and application of the traditional mutual information. In order to overcome these shortcomings, this paper improves the mutual information start with reducing computational complexity.

In mutual information criterion, $H(C)$ causes computational complexity. Therefore $H(C)$ is replaced by normal entropy $H_g(C)$ in this paper

$$H_g(C) = \frac{1}{2} \log \left((2\pi e)^n |\sigma| \right) \qquad (6)$$

Substituting the above equation into the mutual information criterion formula, we can get a new function

$$I' = H_g(C) - H(C|F) \qquad (7)$$

This function has smaller computational complexity than the original mutual information criterion formula, and orthogonal properties, linear invariance, monotonicity of the function could be proved. Thus, this function can be defined as new classifiable evaluation function of feature extraction methods. With the three properties, the new function has orthogonal decomposition property and could be feature extracted.

To calculate the new evaluation function, $H(C|F)$ has to be calculated first. Parameter estimation and non-parametric estimation are two different ways of computing entropy value. Devijver and Kittler pointed out that, compared to parameter estimation, the recognition results of non-parametric estimation have large calculation error because of large computational complexity and low accuracy. Therefore, normal parameter estimation is adopted

$$H(C|F) = \frac{1}{2} \log \left((2\pi e)^n |\sigma_i| \right) \qquad (8)$$

Thus, the new evaluation function is

$$I'(C;F) = \frac{1}{2} \left[\log(|\sigma|) - \sum_{i=1}^{c} \log(|\sigma_i|) p_i \right] \qquad (9)$$

The fundamental purpose of feature extraction is to reduce dimension and to reduce redundancy among the feature vectors. Convert the original space to feature space by transformational matrix T. The transformation will obtain maximum class separability criterion. Dimension reduction means calculate the transformational matrix T,

$$T = \mathbf{argmax} \left\{ I'(\bar{F};C) : \bar{F} = TF \right\} \qquad (10)$$

In the equation,

$$TT^T = I_m \qquad (11)$$

Make use of derivative to calculate T.

The result of first derivative is a divergent function which could not calculate extremum.

In this paper, the transformational matrix T has zero second derivative value.

$$\frac{\partial^2 I'}{\partial T^2} = \frac{\partial \left(T\sigma T^T\right)^{-1}T\sigma}{\partial T} - \sum_{i=1}^{c}\left(T\sigma_i T^T\right)^{-1}T\sigma_i p_i$$

$$= \left(T\sigma T^T\right)^{-1}\sigma - \left(\sigma T^T\right)\left(T\sigma T^T\right)^{-T}\left(T\sigma T^T\right)^{-1}T\sigma$$

$$- \sum_{i=1}^{c}\left(\left(T\sigma_i T^T\right)^{-1}\sigma_i - \left(\sigma_i T^T\right)\right.$$

$$\left.\times\left(T\sigma_i T^T\right)^{-T}\left(T\sigma_i T^T\right)^{-1}T\sigma_i\right) \qquad (12)$$

Set

$$\frac{\partial^2 I'}{\partial T^2} = 0 \qquad (13)$$

Make equation (12) equal to zero, the matrix T calculated is the transformational matrix.

In order to reduce the computation we use particle swarm optimization algorithm. The flow chart of this particle swarm optimization algorithm is shown in Figure 1.

The process of target recognition algorithm based on improved mutual information and particle swarm optimization algorithm as follows:

Step 1: Determine the original training sample set;

Step 2: Determine the feature space dimension m, then calculate transformational matrix T based on equation (12) by particle swarm optimization algorithm;

Step 3: Input the sample under test, then map the original sample space (n) to feature space (m) using the transformation matrix;

Step 4: Input the classifier and calculate recognition error rate.

4 EXPERIMENTAL RESULTS AND ANALYSIS

4.1 Recognition accuracy experiment

Choose four types of ground vehicles as object classification as shown in Figure 2. Comparison of recognition error rate of three recognition algorithms is as shown in Tables 1 and 2.

Tables 1 and 2 show that the new method has minimum recognition error rates with different classifiers. Overall, PCA cannot obtain satisfying recognition results. Recognition error rates of the new method are amount to the recognition error rates of LDA. However, LDA is restricted by number of classes, while PCA and the new method can search classified feature information in higher dimensional space.

4.2 Anti-interference experiment

In order to verify anti-interference performance of the new method, we recognize and identify the target image with inputted noise. First, input

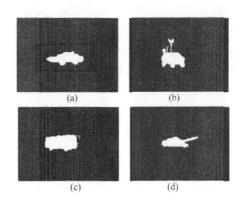

Figure 2. Images of four typical vehicles.

Table 1. Comparison of recognition error rate (1).

Methods	Classifier	Feature space (m)		
		3	10	14
LDA	L	23.25	–	–
	Q	23.09	–	–
PCA	L	40.85	33.26	32.17
	Q	34.79	27.60	23.68
The new	L	29.05	23.26	22.51
method	Q	23.26	15.59	13.59

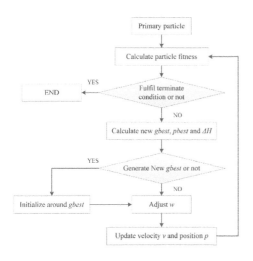

Figure 1. Flow chart of particle swarm optimization algorithm.

Table 2. Comparison of recognition error rate (2).

Methods	Classifier	Feature space (m)		
		15	16	18
LDA	L	–	–	–
	Q	–	–	–
PCA	L	35.55	34.75	35.15
	Q	23.90	24.05	26.34
The new	L	22.46	21.23	22.08
method	Q	13.82	14.00	14.36

Table 3. Comparison of recognition rate (%) added noise.

Methods	Variance of noise				
	0.01	0.02	0.03	0.04	0.05
PCA	68.23	58.61	60.29	55.67	55.98
The new method	86.19	85.51	87.26	82.98	80.37

Table 4. Comparison of recognition time.

Time (s)	(a)	(b)	(c)	(d)
PCA	4	0.8	2	6
The new method	0.6	0.2	0.4	0.8

Figure 3. Sample image of anti-interference experiment.

multiplicative noise which has zero mean value and variable variance into sample image as it shown in Figure 3, and conduct filtering with median filtering method. Then accomplish classification and recognition process in classifier. Table 3 shows results of classification and recognition using nearest neighbor classifier.

Table 3 shows that the new method remains high recognition rates under different variance noises and has great anti-interference performance, obviously better than PCA.

4.3 Computation speed experiment

In order to examine computational complexity of different methods, average time of each target classification from Figure 2 has been compared. Table 4 shows results applying nearest neighbor classifier.

According to Table 4, computational complexity of the new method is much better than PCA

and the recognition time is less than one second. Therefore the new method could be applied to automatic target recognition system, and satisfy the real-time requirements of the system.

5 CONCLUSION

This paper proposed a kind of target recognition algorithm based on improved mutual information and particle swarm optimization algorithm. This new algorithm can easily calculate transformation matrix and has smaller computational complexity than the traditional target recognition algorithm based on mutual information, and also could realize effective data dimension reduction. The experimental results show that compared with the traditional LDA and PCA algorithm, the new algorithm has greater advantage in recognition performance, complexity of algorithm and noise resistance. This presented algorithm will effectively improve the efficiency of image recognition, convenient for real-time processing.

REFERENCES

Bi Ying-wei, Qiu Tian-shuang. An Adaptive Image Segmentation Method Based on a Simplified PCNNt [J]. Acta Electronica Sinica, 2005, 33(4): 647–650.

Cheng Gong, Zhao Wei Mao Shi-yi. Study on fast improving KLDA criterion for MSTAR SAR feature extraction and recognition [J]. Acta Aeronautica et Astronautica Sinica, 2007, 28(3): 672–678.

Cui X., Hardin T., Ragade R.K. A Swarm Approach for Emission Sources Localization [C]. The 16th IEEE International Conference on Tools with Artificial Intelligence, Boca Raton, Florida, USA, 2004:424–430.

Ho S.L., Yang S, Ni G. A Particle Swarm optimization-Based Method for Multiobjective Design Optimizations [J]. IEEE Transactions on Magnetics, 2005, 41(5): 1756–1759.

Iona Cristian, Trelea. The particle swarm optimization algorithm: convergence analysis and parameter selection [J]. Information Processing Letters, 2003, 85: 317–325.

Krusienski D.J., Jenkins W.K. Design and performance of Adaptive Systems Based on Structured Stochastic Optimization Strategies [J]. IEEE Circuits and Systems Magazine, 2005, 5(1): 8–20.

Liu Yi-tong, Fu Ming-yin. Muti-threshold infrared image segmentation based on the modified particle swarm optimization algorithm [C]. Proceedings of the Sixth International Conference on Machine Learning and Cybernetic.

Ma K., Jannorone R.J., Gorman J.W. FAST: parallel airplane pattern recognition [C]. Twenty-Second Southeastern Symposium on System Theory, Cookeville, TN, USA: [s. n.]1990: 7–11.

Mishra A.K., Mulgrew B. Radar signal classification using PCA-based features [C]. IEEEICASSP, Toulouse, France: [s. n.], 2006: 1104–1106.

Pan Feng, Chen Jie, Gan Ming-gang, et al. Model analysis of particle swarm optimizer [J]. Acta Automatica Sinica, 2006, 32(3): 148–153.

Schumacher R., Schiller J. Non-cooperative target identification of battle field targets—classification results based on SAR images [C]. IEEE International Radar Conference, 2005: 167–172.

Yu Jing, You Zhi-sheng. Survey of Automatic Target Recognition and Tracking Method [J]. Application Research of Computers, 2005 (1): 12–15.

Zeng Peng-xin, Chen Peng, Yang Chen-hui, et al. Integration method for the detection of moving multi-targets in dynamic scenes [J]. Control and Decision, 2006, 21(3): 331–335.

Zhang Yan, Meng Yu, Li Wen-hui. Particle Swarm Optimization-Based Texture Synthesis and Texture Transfer [J]. Proceedings of 2004 International Conference on Machine Learning and Cybernetics, 2004, 7:4037–4042.

Zhang Yi-guang, Yang Jun, Yin Zhi-xiang, et al. Study on the Infrared Imaging Target Recognition Algorithm of the Homing System [J]. Laser & Infrared, 2007, 37(9): 381–384.

Zhou Li-wei. Target detection and recognition [M]. Beijing: Beijing Institute of Technology Press, 2002.

Information Technology and Applications – Li (Ed.)
© *2015 Taylor & Francis Group, London, ISBN 978-1-138-02677-3*

Cancellation of loop-interference in full-duplex MIMO relays

Mengmeng Wang & Xianyi Rui
School of Electronic Information, Soochow University, Suzhou, Jiangsu, China

ABSTRACT: Loop-interference in full-duplex Multiple-Input Multiple-Output (MIMO) relays has a negative effect on the relay performance. In this paper, the relay employs transmit and receive weight vectors for removing the loop-interference signal. Maximizing the ratio between the power of the useful signal to the loop-interference power at the relay reception and transmission and forming a criterion function for jointly calculating the optimal receive and transmit suppression vectors, respectively. Simulation results confirm that the proposed scheme improves the channel capacity and mitigates the self-interference significantly.

Keywords: full-duplex relays; loop-interference cancellation; transmit and receive weight vectors

1 INTRODUCTION

The use of relays in wireless networks has been identified as a promising technique for combating the shadowing effect, extending the network coverage and increasing the network throughput [1] [2]. Conventionally, relays are assumed to operate in a half-duplex mode. Half-duplex relays require two orthogonal channels for transmission and reception. Hence, the half-duplex mode incurs a spectral efficiency penalty since it requires two time slots to send one symbol. To recover the throughput loss owing to the additional time resource, full-duplex relays can receive and transmit simultaneously [3]. However, the main limiting factor is the resulting loop-interference signal caused by coupling from the relay's transmission to its own reception, which deteriorates the relay performance [4]. It is crucial to minimize the loop-interference to render full-duplex relays feasible. Typically, the loop-interference is suppressed by estimating the interference signal and subtracting it from the relay input [5]. The main drawback of the solution is that it does not exploit the spatial domain in the case of MIMO relays. In [6], the authors propose null space projection and minimum mean square error filters for self-interference suppression at the relaying node. In [7], a solution exploits the Singular Value Decomposition (SVD) in order to find null space. However, the SVD-based scheme does not take account of the desired signal, which degrades the performance significantly.

In this paper, an alternative technique for suppressing the loop-interference signal at the relay is proposed. Similarly to null space projection, the relay employs transmit and receive weight vectors

that cancel loop-interference channel. However, the proposed technique is based on the orthogonal complement, which considers the desired signal as well as the loop-interference. Our simulation results show that the proposed scheme improves the full-duplex MIMO relay system throughput and nearly achieves the performance of the upper bound system as if there were no loop-interference.

2 SYSTEM MODEL

Consider the MIMO relay channel shown in Figure 1. It consists of the source, the destination, and a full-duplex MIMO relay. The source and the destination are equipped with N_S and M_D antennas, respectively. The relay has M_R receive and N_R transmit antennas. $H_1 \in C^{M_R \times N_S}$ and $H_2 \in C^{M_D \times N_R}$

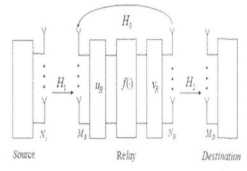

Figure 1. System model for loop-interference suppression.

represent the source-relay, and relay-destination channel matrices, respectively. For simplicity reasons, we assume that the direct source to destination link is blocked. The relay loop-interference matrix is denoted by $H_0 \in C^{M_R \times N_R}$.

Let $x_S \in C^{N_S \times 1}$ and $x_R \in C^{N_R \times 1}$ be the signal vectors transmitted by the source and the relay. The received signals of the relay and the destination can be modeled as

$$y_R = H_1 x_S + H_0 x_R + n_R \tag{1}$$

$$y_D = H_2^H x_R + n_D \tag{2}$$

where $n_R \in C^{N_R \times 1}$ and $n_D \in C^{M_D \times 1}$ are complex Gaussian noises. Since the term $H_0 x_R$ is generated by the relay itself, it is called loop-interference. As shown in Figure 1, the relay applies the receive weight vector $u_R \in C^{M_R \times N_S}$ and the transmit weight vector $v_R \in C^{N_R \times M_D}$ for cancelling the loop-interference channel H_0.

$$u_R^H H_0 v_R = 0 \tag{3}$$

Hence, the transmit signal at the relay, x_R, is generated as $x_R = v_R f(u_R^H y_R)$, where $f(\cdot)$ represents the demodulation and re-modulation operation. Then, the measure of the relay system performance in this paper is the channel capacity expressed as:

$$C = \min\left(\log_2(1 + SNR \left\| u_R^H H_1 \right\|_F^2), \right.$$
$$\left. \log_2\left(1 + SNR \left\| v_R^H H_2 \right\|_F^2 \right) \right) \tag{4}$$

The remaining work is the choice of the weight vectors, u_R and v_R that maximize the capacity while cancelling the loop-interference.

3 LOOP-INTERFERENCE CANCELLATION

In this section, we briefly discuss the main idea behind the SVD-based scheme proposed in [10], and then we proceed by presenting our proposed loop-interference cancellation algorithm.

3.1 SVD-based scheme

The SVD-based scheme has been identified as an efficient technique for cancelling loop-interference. It exploits the SVD in order to find the null space. The SVD of H_0 can be written as

$$H_0 = U_R \Sigma_R V_R^H \tag{5}$$

where Σ_R is a diagonal matrix with non-negative singular values in non-increasing order, and U_R and V_R are unitary matrices consisting of left and right singular vectors, respectively. Let $U_R^{[k]}$ and $V_R^{[k]}$ denote the kth column vectors of U_R and V_R, respectively. To satisfy (3), we may set the receive and transmit vectors to be a left singular vector and a right singular vector corresponding to different singular values:

$$u_R = U_R^{[i]}, v_R = V_R^{[j]} \quad i \neq j \tag{6}$$

However, one limitation of this approach is that it does not take account of the channel gains H_1 or H_2 at all. In some extreme case, if u_R is nearly orthogonal to H_1, then the receiver vector cancels the desired signal as well as the loop-interference, which degrades the performance significantly. Therefore, we need to design a new pair of transmit and receive vectors.

3.2 Proposed self-interference suppression scheme

In this section, we present our proposed algorithm that calculates the optimal suppression filters by taking into account not only the loop-interference but also the useful signal. We first rearrange the capacity equation given in (4) to get a constrained optimisation problem:

maximise $\min\left(\left\| u_R^H H_1 \right\|_F^2 , \left\| v_R^H H_2 \right\|_F^2 \right)$

subject to $u_R^H H_0 v_R = 0$ (7)

$$\left\| u_R \right\|_F^2 = \left\| v_R \right\|_F^2 = 1$$

Ideally, we would like to form a criterion function for jointly calculating the optimal receive and transmit suppression vectors, u_R and v_R, respectively. For instance, we could either choose to maximize the Signal to Interference Ratio (SIR) at the relay input or maximize the SIR at the relay output. However, since it is not possible to arrive at a tractable, closed form solution for the optimal vector, we propose a simpler two step approach.

First, we begin by designing the receive weight vector, u_R and neglecting for the moment the transmit weight vector v_R. Our goal is to reduce the power of the loop-interference signal and at the same time improve the useful signal power received by the relay. Hence, we aim at maximizing the SIR at the relay input

maximise $\left\| u_R^H H_1 \right\|_F^2 \Big/ \left\| u_R^H H_0 \right\|_F^2$ (8)

or equivalently,

maximise $Tr\{u_R^H H_1 H_1^H u_R\}/Tr\{u_R^H H_0 H_0^H u_R\}$ (9)

If we assume that the matrix $H_0 H_0^H$ is invertible, then the solution to (9) is obtained by solving the generalized eigenvalue problem. Therefore, the optimal receive suppression vector can be expressed as

$$u_{R,opt} = U^H (H_0 H_0^H)^{-1/2} \qquad (10)$$

The columns of U are the corresponding eigenvectors of the matrix $(H_0 H_0^H)^{-1/2} H_1 H_1^H (H_0 H_0^H)^{-1/2}$.

Then we proceed with the design of the transmit weight vector v_R. In this second step, we aim at maximizing the ratio between the power of the useful signal at the relay output to the remaining loop-interference power. In other words, we maximize the SIR at the relay side. The remaining loop-interference can be modeled as

$$H = u_{R,opt} H_0 \qquad (11)$$

where $u_{R,opt}$ is given by (10).

We can now formulate the SIR maximization problem at the relay output

$$\text{maximise} \left\| v_R^H H_2 \right\|_F^2 / \left\| v_R^H H \right\|_F^2 \qquad (12)$$

or equivalently,

$$\text{maximise} \ Tr\{v_R^H H_2 H_2^H v_R\}/Tr\{v_R^H H H^H v_R\} \qquad (13)$$

Similarly to (10), the optimal transmit weight vector is expressed as

$$v_{R,opt} = (H^H H)^{-1/2} V \qquad (14)$$

The matrix V is obtained by the eigenvalue decomposition of the matrix $(H^H H)^{-1/2} H_2^H H_2 (H^H H)^{-1/2}$ and its columns are the corresponding eigenvectors.

4 SIMULATION RESULTS

In this section, the theories presented in the previous sections are verified through computer simulations. We assume the Rayleigh fading channel and the average received SNRs at the source node, the destination node, and the relay node are all identical. The simulation conditions are as follows: $P_S = P_R = P$, $\sigma_R^2 = \sigma_B^2 = \sigma^2$ and $M_R = N_R = N$.

Figure 2 illustrates the capacity of the full-duplex relay system, in which the relay node is

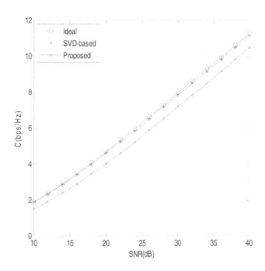

Figure 2. Comparison of capacities for different schemes.

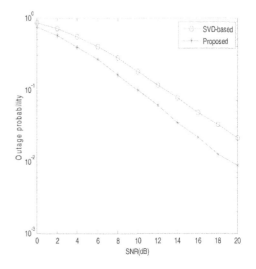

Figure 3. Comparison of outage probabilities for different schemes.

equipped with two antennas, $N = 2$. For a benchmark, we consider a full-duplex relay system with no loop-interference as an upper bound to the proposed scheme. The capacity of this upper bound corresponds to twice the capacity of the half-duplex relay systems because the required time slots are reduced by half. The proposed scheme is superior to the SVD-based scheme and performs very close to the upper bound. Figure 3 shows the outage probalities for the SVD-based scheme and the proposed scheme. We can visually verify that the proposed scheme outperforms the SVD-based scheme.

5 CONCLUSION

In this paper, we propose a loop-interference cancellation algorithm for full-duplex relays. The proposed scheme cancels the loop-interference signal by applying receive and transmit weight vectors at the relay. The optimal vectors are designed by taking into account not only the directions of the interference signal, but also the useful signal. Simulation results found that the proposed scheme outperforms the SVD-based scheme and significantly improves the system performance.

ACKNOWLEDGMENT

This work is supported by Natural Science Found of China (No. 61201213).

REFERENCES

[1] Kim T.M., Paulraj A. Outage probability of amplify-and-forward cooperation with full duplex relay [C]. IEEE Wireless Communications and Networking Conference, 2012:75–79.

[2] Khafagy M., Ismail A., Alouini M.-S., and Aissa S. On the outage performance of full-duplex selective decode-and-forward relaying [J]. IEEE Communication Letters, 2013, 17(6): 1180–1183.

[3] Rankov B., Wittneben A. Spectral efficient protocols for half-duplex fading relay channels [J]. IEEE Selected Areas in Communications, 2007, 25(2): 379–389.

[4] Haneda K., Kahra E., Wyne S., et al. Measurement of loop-back interference channels for outdoor-to-indoor full-duplex radio relays [C]. European Conference on Antennas and Propagation, 2010: 1–5.

[5] Nasr K.M., Cosmas J.P., Bard M., and Gledhill J. Performance of an echo canceller and channel estimator for on-channel repeaters in DVB-T/H networks [J]. IEEE Transactions on Broadcasting, 2007:609–618.

[6] Riihonen T., Werner S., Wichman R. Spatial loop interference suppression in full-duplex MIMO relays [C]. IEEE Asilomar Conference on Signals, Systems and Computers, 2009: 1508–1512.

[7] Riihonen T., Werner S., Wichman R. Mitigation of loopback self-interference in full-duplex MIMO relays [J]. IEEE T-Signal Processing, 2011, 59(12): 5983–5993.

Information Technology and Applications – Li (Ed.)
© *2015 Taylor & Francis Group, London, ISBN 978-1-138-02677-3*

Improved interleaving PAPR reduction scheme of OFDM signals with BPSK inputs

Lingyin Wang & Xinghai Yang

School of Information Science and Engineering, University of Jinan, Jinan, China

ABSTRACT: As one of the PAPR reduction schemes, interleaving scheme is viewed as an efficient solution for obtaining good PAPR reduction performance in OFDM systems. But in original interleaving scheme, many IFFT operations must be needed for generating OFDM candidate signals, which induces large computational complexity. In this paper, an improved interleaving scheme with low computational complexity for PAPR reduction of OFDM signals with BPSK inputs is proposed. In the proposed scheme, two different signal constellations of BPSK are adopted, and some additional OFDM candidate signals are generated without IFFT operations. Theoretical analysis and simulation results show that, compared with original interleaving scheme, the proposed interleaving scheme can achieve low computational complexity with keeping similar performance for PAPR reduction.

Keywords: OFDM; PAPR; interleaving; computational complexity

1 INTRODUCTION

Orthogonal Frequency Division Multiplexing (OFDM) is viewed as an attractive technique for high data rate communications because of its immunity to selective fading and high spectral efficiency [1]. But the main shortcoming of OFDM signals is the large Peak-to-Average Power Ratio (PAPR), which may cause the in-band distortion and the out-of-band radiation in the nonlinear region of high power amplifier and bring on Bit Error Rate (BER). For this reason, the peak power reduction of OFDM signals has received many researchers' interests.

In recent years, some PAPR reduction schemes of OFDM signals have been proposed [2,3], where multiple signal representation techniques [4] are viewed as one of the most promising PAPR reduction schemes due to their good PAPR reduction performance and no signal distortion, such as interleaving scheme [5–7], selected mapping [8,9] and partial transmit sequence [10,11]. Among these multiple signal representation techniques, interleaving scheme is the simplest one. However, with the purpose of gaining good PAPR reduction performance, many Inverse Fast Fourier Transform (IFFT) operations must be needed for generating a lot of OFDM candidate signals, which results in the large computational complexity.

In this paper, an improved interleaving PAPR reduction scheme of OFDM signals with BPSK inputs is proposed, which can reduce computational complexity clearly and obtain similar PAPR reduction performance compared with original interleaving scheme. In the proposed interleaving scheme, two different signal constellations of BPSK are adopted, and the corresponding OFDM candidate signals are achieved. Finally, the candidate signals employing different signal constellations are combined for generating additional OFDM candidate signals instead of performing IFFT operations. Theoretical analysis and simulation results show that the proposed interleaving scheme can obtain dramatic computational complexity reduction with keeping similar PAPR reduction performance compared with original interleaving scheme.

The rest of this paper is organized as follows. In Section 2 and Section 3, the system model of OFDM and the original interleaving scheme are described. In Section 4, an improved interleaving scheme is proposed in detail, and its computational complexity is discussed. In Section 5, the simulation results to illustrate the performance of proposed interleaving scheme are shown. Finally, a brief conclusion is given in Section 6.

2 OFDM SYSTEM DESCRIPTION

In an OFDM system with N subcarriers, an OFDM signal in the discrete time domain can be expressed by

$$x_n = \frac{1}{\sqrt{N}} \sum_{k=0}^{N-1} X_k e^{j2\pi kn/N}, \ 0 \le n \le N-1 \qquad (1)$$

where $X_k, k = 0, 1, ..., N-1$ denote the input modulated symbols in the frequency domain.

The PAPR of an OFDM signal in the discrete time domain is defined as the ratio of the peak power to the average power of this signal, which can be expressed by

$$\text{PAPR}(\boldsymbol{x}) = 10\log_{10} \frac{\max\limits_{0 \le n \le N-1} \{|x_n|^2\}}{E\{|x_n|^2\}} \text{dB} \qquad (2)$$

where $\boldsymbol{x}$, $\max\limits_{0 \le n \le N-1} \{|x_n|^2\}$ and $E\{|x_n|^2\}$ denote an OFDM signal, the peak power of this OFDM signal and the average power of this OFDM signal, respectively.

Normally, an oversampling factor L is employed for capturing all the signal peaks. It can be obtained by LN-point IFFT of the sequence with $(L-1)N$ zero-padding, and the real PAPR results can be approached when the oversampling factor L is four [12].

As for the PAPR reduction performance of an OFDM system, Complementary Cumulative Distribution Function (CCDF) [13] can be utilized for evaluating and comparing this performance of any PAPR reduction schemes, which can be expressed by

$$\text{CCDF}(N, \text{PAPR}_0) = \Pr\{\text{PAPR} > \text{PAPR}_0\}$$
$$= 1 - (1 - e^{-\text{PAPR}_0})^{LN} \qquad (3)$$

where N, PAPR_0 and L denote the number of subcarriers in an OFDM system, a certain value of PAPR and the oversampling factor, respectively.

3 ORIGINAL INTERLEAVING SCHEME

In original interleaving scheme, the OFDM candidate signals can be generated by the block interleaver. In an OFDM system with N subcarriers, the input modulated sequence X is firstly divided into several disjoint sub-block sequences $X_i, i = 1, 2, ..., U$. Then, by using the pseudo-random order or the periodic order, all the sub-block sequences can be permuted for generating different OFDM candidate signals. For instance, after the permutation of all the sub-block sequences being performed, the original sequence $[X_1, X_2, ..., X_i, ..., X_U]^T$ becomes $[X_{\pi(1)}, X_{\pi(2)}, ..., X_{\pi(i)}, ..., X_{\pi(U)}]^T$, where $\{i\} \rightarrow \{\pi(i)\}$ is the one-to-one mapping and $\pi(i) \in \{1, 2, ..., U\}$. Finally, the OFDM candidate signal with the lowest PAPR is selected for transmitting. The block

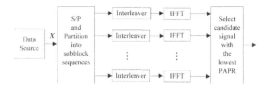

Figure 1. The block diagram of original interleaving scheme.

diagram of original interleaving scheme is shown in Figure 1.

As for interleaving scheme, it is a probabilistic PAPR reduction scheme, and many OFDM candidate signals are generated for representing the same input sequence. Therefore, in terms of Eq. (3), the CCDF about the interleaving scheme can be given by

$$\text{CCDF}_{\text{interleaving}} = \left[\Pr\{\text{PAPR} > \text{PAPR}_0\} \right]^C$$
$$= \left[1 - (1 - e^{-\text{PAPR}_0})^{LN} \right]^C \qquad (4)$$

where C is the number of OFDM candidate signals generated in the original interleaving scheme. It can be seen from Eq. (4) that the PAPR reduction performance of interleaving scheme is mainly decided by the number of OFDM candidate signals. That is to say, the more OFDM candidate signals are generated, the better PAPR reduction performance can be achieved. However, because each OFDM candidate signal is obtained by employing an IFFT operation, the computational complexity of original interleaving scheme is very large.

Moreover, the side information should be required for recovering the input sequence successfully at the receiver. When the binary symbols are used, $\lceil \log_2 C \rceil$ bits should be allocated to represent this side information, where C and $\lceil \cdot \rceil$ denote the number of OFDM candidate signals and the elements to the nearest integers toward infinity, respectively.

4 ORIGINAL INTERLEAVING SCHEME

In this section, an improved interleaving scheme for PAPR reduction of OFDM signals with BPSK inputs is proposed, which aims to obtain dramatic computational complexity reduction with similar PAPR reduction performance compared with original interleaving scheme.

4.1 Basic ideas of proposed scheme

In the original interleaving scheme, with the purpose of gaining good PAPR reduction performance, a sufficient number of OFDM candidate signals need to be generated, which results in the large number of IFFT operations. Thus, the computational complexity of original interleaving scheme is very large. In addition, as for interleaving scheme, because its PAPR reduction performance is mainly decided by the number of OFDM candidate signals, it is desirable if the number of IFFT operations could be reduced without decreasing the number of OFDM candidate signals (i.e. the PAPR reduction performance).

In the proposed interleaving scheme, the input binary sequence is firstly modulated by employing two different signal constellations of BPSK, and the constellations of BPSK are shown in Figure 2.

Let X_A and X_B be the modulated sequences with constellation A and constellation B, respectively. Then, the two kinds of modulated sequences X_A and X_B are partitioned into several disjoint sub-block sequences. After the sub-block sequence partition being performed, X_A and X_B can be expressed by $X_A = [X_{A,1}, X_{A,2}, ..., X_{A,K}]^T$ and $X_B = [X_{B,1}, X_{B,2}, ..., X_{B,V}]^T$, where K and V denote the number of sub-block sequences in X_A and X_B respectively. By employing the permutation of all the sub-block sequences and IFFT operations, two sets of OFDM candidate signals can be obtained based on X_A and X_B.

Let m_i, $i = 1, 2, ..., G$ and n_j, $j = 1, 2, ... H$ be the OFDM candidate signals from the above two sets of OFDM candidate signals generated by performing the block interleaving on the modulated sequences X_A and X_B, where G and H denote the number of OFDM candidate signals in the two sets of OFDM candidate signals respectively. Then, in terms of the linear property of IFFT, these two OFDM candidate signals m_i and n_j could be combined to generate an additional OFDM candidate signal instead of performing an IFFT operation, which can be expressed by

$$y_{i,j} = \frac{\sqrt{2}}{2}\left(m_i + n_j\right) \quad (i = 1, 2, ..., G; j = 1, 2, ... H)$$

$$= \frac{\sqrt{2}}{2}\left[\text{IFFT}\left(M_i\right) + \text{IFFT}\left(N_j\right)\right]$$

$$= \text{IFFT}\left[\frac{\sqrt{2}}{2}\left(M_i + N_j\right)\right] \tag{5}$$

where $M_i, i = 1, 2, ..., G$ and $N_j, j = 1, 2, ..., H$ are the permutations of $X_A = [X_{A,1}, X_{A,2}, ..., X_{A,K}]^T$ and $X_B = [X_{B,1}, X_{B,2}, ..., X_{B,V}]^T$, respectively.

Because the elements of M_i and N_j take the values in $\{1, -1\}$ and $\{j, -j\}$, the elements of $M_i + N_j$ belong to $\{1+j, 1-j, -1+j, -1-j\}$. Compared with the sequences M_i and N_j, each element of the new one

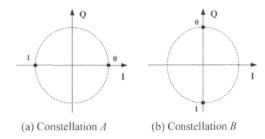

(a) Constellation A (b) Constellation B

Figure 2. BPSK constellations.

$M_i + N_j$ is equivalent to be multiplied by the phase weighting factor $\exp(j\pi/4)$ or $\exp(-j\pi/4)$. In addition, because each element of the new sequence $M_i + N_j$ has non-unit magnitude, $\sqrt{2}/2$ must be the adjust magnitude factor to make each element of the sequence $M_i + N_j$ has unit magnitude. Based on the above discussion, $y_{i,j} = \sqrt{2}\left(m_i + n_j\right)/2$ can be considered as the new OFDM candidate signal.

In this way, all the OFDM candidate signals in the proposed interleaving scheme mainly consist of two parts: (i) By performing IFFT operations on the permutations of $X_A = [X_{A,1}, X_{A,2}, ..., X_{A,K}]^T$ and $X_B = [X_{B,1}, X_{B,2}, ..., X_{B,V}]^T$, the OFDM candidate signals m_i, $i = 1, 2, ..., G$ and n_j, $j = 1, 2, ..., H$ can be obtained; (ii) According to Eq. (5), additional OFDM candidate signals can be achieved by combining m_i, $i = 1, 2, ..., G$ and n_j, $j = 1, 2, \cdots, H$ instead of performing IFFT operations directly.

Finally, among all the OFDM candidate signals, the one with the lowest PAPR is selected for transmitting.

Since G OFDM candidate signals and H OFDM candidate signals are generated in the first part of OFDM candidate signals, $G \times H$ additional OFDM candidate signals could be obtained. Therefore, the total number of OFDM candidate signals in the proposed interleaving scheme is $C' = G + H + G \times H$, such as

$$\left\{m_1, m_2, ..., m_G, n_1, n_2, ..., n_H, \frac{\sqrt{2}}{2}\left(m_1 + n_1\right),\right.$$
$$\left.\frac{\sqrt{2}}{2}\left(m_1 + n_2\right), ..., \frac{\sqrt{2}}{2}\left(m_G + n_H\right)\right\} \tag{6}$$

For instance, when $G = H = 2$, two sets of OFDM candidate signals in the first part of OFDM candidate signals are $\{m_1, m_2\}$ and $\{n_1, n_2\}$. Thereupon, as for the proposed interleaving scheme, the total number of OFDM candidate signals C' is eight, such as $\{m_1, m_2, n_1, n_2, \sqrt{2}\left(m_1 + n_1\right)/2, \sqrt{2}\left(m_1 + n_2\right)/2, \sqrt{2}\left(m_2 + n_1\right)/2, \sqrt{2}\left(m_2 + n_2\right)\}$.

Moreover, in order to recover the received signals successfully, similar to original interleaving scheme, the side information still needs to be required in the proposed interleaving scheme.

4.2 *Computational complexity analysis*

In the proposed interleaving scheme, the computational complexity reduction mainly comes from the generation of $G \times H$ additional OFDM candidate signals, which can be obtained by combining the OFDM candidate signals $m_i, i = 1, 2, ..., G$ and $n_j, j = 1, 2, ..., H$ instead of performing IFFT operations directly.

It is known that a LN-point IFFT needs $LN/2 \log_2(LN)$ complex multiplications and $LN \log_2(LN)$ complex additions, where N and L denote the number of subcarriers in an OFDM system and the oversampling factor.

As for the original interleaving scheme, its computational complexity comes from the IFFT operations. Suppose there are C OFDM candidate signals generated in the original interleaving scheme. Thereupon, $CLN/2\log_2(LN)$ complex multiplications and $CLN \log_2(LN)$ complex additions are needed for performing C IFFT operations. Hence, the computational complexity of original interleaving scheme can be given as follows.

The number of complex multiplications in the original interleaving scheme:

$$\frac{CLN}{2}\log_2(LN) \tag{7}$$

The number of complex additions in the original interleaving scheme:

$$CLN \log_2(LN) \tag{8}$$

As for the proposed interleaving scheme, all the OFDM candidate signals consist of two parts, and the IFFT operations are only needed for generating the first part of OFDM candidate signals. Since there are $G + H$ OFDM candidate signals generated in the first part of OFDM candidate signals, $G + H$ IFFT operations should be needed, which results in $(G+H)LN/2\log_2(LN)$ complex multiplications and $(G+H)LN\log_2(LN)$ complex additions. On the other hand, as for the second part of OFDM candidate signals, because each additional candidate signal needs LN complex multiplications and LN complex additions, the computational complexity of this part needs $GHLN$ complex multiplications and $GHLN$ complex additions. Therefore, the computational complexity of the proposed interleaving scheme can be expressed as follows.

The number of complex multiplications in the proposed interleaving scheme:

$$\frac{(G+H)LN}{2}\log_2(LN) + GHLN \tag{9}$$

The number of complex additions in the proposed interleaving scheme:

$$(G+H)LN\log_2(LN) + GHLN \tag{10}$$

Here, in order to illustrate the advantage of proposed interleaving scheme in computational complexity over original interleaving scheme, Computational Complexity Reduction Ratio (CCRR) [14] is defined as

$$\text{CCRR} = \left(1 - \frac{\text{Complexity of Proposed Scheme}}{\text{Complexity of Original Interleaving}}\right) \times 100\% \tag{11}$$

In terms of Eq. (11), multiplicative CCRRs and additive CCRRs can be easily achieved, named as $\text{CCRR}^\times$ and CCRR^+. For comparison purposes, suppose the total number of OFDM candidate signals generated in the proposed interleaving scheme is the same as that in original interleaving scheme, i.e. $C' = C$. Thereupon, the CCRR of the proposed interleaving scheme over the original interleaving scheme with the typical values of the total number of OFDM candidate signals is shown in Table 1, where G and H denote the number of OFDM candidate signals from two different sets of OFDM candidate signals in the first part of OFDM candidate signals of proposed interleaving scheme.

Table 1. Multiplicative CCRRs and additive CCRRs of proposed interleaving scheme over original interleaving scheme.

	$\text{CCRR}^\times$	CCRR^+
The total number of OFDM candidate signals		
$C' = C = 8$ ($G = H = 2$)	38.9	44.4
$C' = C = 15$ ($G = H = 3$)	46.7	53.3
$C' = C = 24$ ($G = H = 4$)	51.9	59.3
$C' = C = 35$ ($G = H = 5$)	55.6	63.5

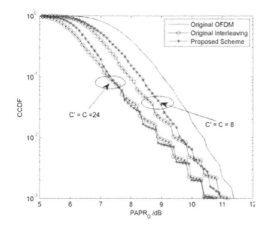

Figure 3. PAPR reduction performances of the proposed interleaving scheme and the original interleaving scheme.

5 SIMULATION RESULTS

In this section, the PAPR reduction performance of the proposed interleaving scheme and the original interleaving scheme are investigated by simulations. The correlative parameters of OFDM system for simulations are $N = 128$ subcarriers, BPSK and the oversampling factor $L = 4$.

Figure 3 shows a PAPR reduction performance comparison of the proposed interleaving scheme and the original interleaving scheme. For comparison purposes, the total number of OFDM candidate signals in proposed interleaving scheme is the same as that in original interleaving scheme.

It can be seen from Figure 3 that, compared with the original interleaving scheme, the proposed interleaving scheme can obtain similar PAPR reduction performance. That is to say, the proposed interleaving scheme and the original interleaving scheme have almost the same PAPR reduction performance under the condition that the total number of OFDM candidate signals is same.

6 CONCLUSION

In this paper, an improved interleaving PAPR reduction scheme with low computational complexity of OFDM signals with BPSK inputs is proposed. By using two different signal constellations of BPSK, additional OFDM candidate signals can be generated without IFFT operations.

Theoretical analysis and simulation results show that compared with original interleaving scheme, the proposed interleaving scheme can obtain dramatic computational complexity reduction with similar PAPR reduction performance.

ACKNOWLEDGEMENTS

This work was supported in part by Science Research Award Fund for the Outstanding Young and Middle-aged Scientists of Shandong Province of China (No. BS2013DX014), Doctor Fund of University of Jinan (No. XBS1309), a Project of Shandong Province Higher Educational Science and Technology Program (No. J11 LF02).

REFERENCES

[1] Prasad, R. 2004. *OFDM for Wireless Communications Systems*. Boston: Artech House Publishers.
[2] Jiang, T. & Wu, Y. 2008. An overview: peak-to-average power ratio reduction techniques for OFDM signals. *IEEE Trans. Broadcast.* 54(2): 257–268.
[3] Han, S.H. & Lee, J.H. 2005. An overview of peak-to-average power ratio reduction for multicarrier transmission. *IEEE Wireless Commun.* 12(2): 56–65.
[4] Jayalath, A.D.S. & Athaudage, C.R.N. 2004. On the PAR reduction of OFDM signals using multiple signal representation. *IEEE Commun. Lett.* 7(8): 425–427.
[5] Jayalath, A.D.S. & Tellambura, C. 2000. The use of interleaving to reduce the peak-to-average power ratio of an OFDM signal. *IEEE Global Telecommunications Conference (GLOBECOM)*, USA: San Francisco, 1: 82–86.
[6] Ryu, H.-G., Kim, S.-K. & Ryu, S.-B. 2007. Interleaving method without side information for the PAPR reduction of OFDM system. *International Symposium on Communications and Information Technologies*, Australia:Sydney, 1: 72–76.
[7] Malathi, P. & Vanathi, P.T. 2008. Improved interleaving technique for PAPR reduction in OFDM-MIMO system. *Second Asia International Conference on Modeling and Simulation*, Malaysia: Kuala Lumpur, 1: 253–258.
[8] Bäuml, R.W., Fisher, R.F.H. & Huber, J.B. 1996. Reducing the peak-to-average power ratio of multicarrier modulation by selected mapping. *IET Electron. Lett.* 32(22): 2056–2057.
[9] Park, J., Hong, E. & Har, D.S. 2011. Low complexity data decoding for SLM-based OFDM systems without side information. *IEEE Commun. Lett.* 15(6): 611–613.
[10] Müller, S.H. & Huber, J.B. 1997. OFDM with reduced peak-to-average power ratio by optimum combination of partial transmit sequences. *IET Electron. Lett.* 33(5): 368–369.

[11] Hou, H., Ge, J. & Li, J. 2011. Peak-to-average power ratio reduction of OFDM signals using PTS scheme with low computational complexity. *IEEE Trans. Broadcast.* 57(1): 143–148.

[12] Tellado, J. 1999. Peak to average power reduction for multicarrier modulation. Ph.D. Thesis, Stanford University.

[13] Jiang, T., Guizani, M., Chen, H.-H., Xiang, W. & Wu, Y. 2008. Derivation of PAPR distribution for OFDM wireless systems based on extreme value theory. *IEEE Trans. Wireless Commun.* 7(4): 1298–1305.

[14] Lim, D.-W., Heo, S.-J. & No, J.-S. 2006. A new PTS OFDM scheme with low complexity for PAPR reduction. *IEEE Trans. Broadcast.* 52(1): 77–82.

Information Technology and Applications – Li (Ed.)
© 2015 Taylor & Francis Group, London, ISBN 978-1-138-02677-3

CssQoS: A load balancing mechanism for cloud serving systems

W.K. Hsieh
General Education Center, De Lin Institute of Technology, New Taipei, Taiwan

W.H. Hsieh
Department of Electrical Engineering, National Taiwan University of Science and Technology, Taipei, Taiwan
Department of Computer and Communication, De Lin Institute of Technology, New Taipei, Taiwan

J.L. Chen & P.J. Yang
Department of Electrical Engineering, National Taiwan University of Science and Technology, Taipei, Taiwan

ABSTRACT: This study presents a novel QoS-aware load balancing mechanism called CssQoS system for scheduling Virtual Machines (VMs) to leverage the capability of the Eucalyptus platform and improves the overall performance of cloud computing systems. Additionally, the Hadoop platform is integrated into the CssQoS system to improve its ability to accommodate real-time services. The system performance is assessed in terms of CPU load, memory load and system throughput. Analytical results indicate that the proposed CssQoS system can reduce both CPU load by 12.92% and memory load by 5.99%, eventually improving system throughput by 6.94% over those of the original Eucalyptus platform with Hadoop mechanisms.

Keywords: Virtual Machine; Eucalyptus open cloud; Cloud Serving System Quality of Service (CssQoS); load balancing system

1 INTRODUCTION

Cloud computing enables developers to deploy applications automatically during task allocation and storage distribution by using distributed computing technologies in numerous servers (Ahlgren et al. 2011). The cloud computing architecture is divided into three layers. The bottom layer, Infrastructure as a Service (IaaS), has a service-oriented architecture (i.g. Amazon EC2/S3). The middle layer, Platform as a Service (PaaS), is a service platform that allows developers to deploy their own applications (i.g. Amazon Web Services and Google App Engine). The top layer, Software as a Service (SaaS), enables users to access services based on their requirements (i.g. Microsoft's online update service).

By using virtual devices, cloud computing reduces server power costs and minimizes hardware costs (Xiao et al. 2013). Also, the enormous data scale enables users to access multiple data storage. Developers must design mechanisms that optimize the use of architectural and deployment paradigms to gain the maximum benefit from cloud computing (Jin et al. 2011).

Eucalyptus (Elastic Utility Computing Architecture for Linking Your Programs to Useful Systems), an IaaS platform, is an open-source framework. The operations of VMs across heterogeneous physical resources can be executed and coordinated using this platform. However, the data transfer rate obviously decreases when VMs calculate large amounts of data on the Eucalyptus cloud computing platform, resulting in an unsatisfactory network QoS. Consequently, VMs encounter difficulty in exchanging data, causing overloading (Monhamed et al. 2013 & Wang et al. 2013). To solve the above problem, this study develops a novel load balancing system, capable of improving the performance of VMs. This Cloud Serving System, called CssQoS, is devised by improving the Eucalyptus platform. In the CssQoS system, agents are created to monitor the load balancing mechanism. Figure 1 schematically depicts the CssQoS system architecture.

According to Figure 1, the lower level of the Eucalyptus platform comprises numerous VMs that contain dispersive data (Sempolinski et al. 2010). The computing capability of VM decreases when a large number of users access these services. To handle this problem, the proposed load balancing mechanism leverages the features of Hadoop frameworks. A Hadoop framework consists mainly of: the Hadoop Distributed File System (HDFS),

Figure 1. CssQoS system architecture.

Figure 2. Eucalyptus architecture.

Map Reduce and Hadoop Database (HBase) (Fagui et al. 2010). HDFS consists of a master system and slave systems. Map Reduce is a simple distributed system for parallel-oriented computing. Both HDFS and Map Reduce can handle large amounts of data.

2 BACKGROUND KNOWLEDGE

2.1 Cloud computing systems

The proposed CssQoS system is implemented using two cloud open platforms, Eucalyptus and Hadoop. These platforms are described as follows.

2.1.1 Eucalyptus platform
The Eucalyptus platform architecture supports VMs that operate on the top of Xen hypervisor layer, as shown in Figure 2.

- Cloud Controller (CLC): CLC is responsible for managing underlying virtualized resources such as servers, storages and networks.
- Cluster Controller (CC): CC runs on the cluster front-end machine to collect VMs information, scheduling VMs execution on a specific Node Controller and managing the configuration of VMs' network.
- Node Controller (NC): NC is a component executed on the bottom of cloud computing to perform the start-up, shutdown, cleanup, inspection and management of VMs.
- Storage Controller (Walrus): Walrus is responsible for putting, getting, deleting services, as well as setting access control policies.

2.1.2 Hadoop framework
Hadoop is characterized by a high scalability, efficiency and reliability. Hadoop consists of HDFS and Map Reduce components. HDFS is a master-slave file system. HDFS requires only a major name node and numerous data nodes (Warneke et al. 2011).

2.2 Load balancing

As a load balancing solution for Linux systems (Kansal et al. 2012), Linux Virtual Server (LVS) attempts to provide a basic framework to construct a high-performance, scalability and reliability server. Additionally, LVS offers advanced IP load balancing software (IPVS), application-level load balancing software (KTCPVS) and cluster management components.

2.2.1 Linux Virtual Server via Network Address Translation (LVS/NAT)
NAT function maps one group IP addresses to another set of local IP addresses. On the LVS/NAT workflow, when a user accesses the service in the server cluster, the request packet is destined for the external virtual IP address for the load balancer.

2.2.2 Linux Virtual Server via IP Tunneling (LVS/TUN)
While encapsulating IP packets within an IP datagram, IP tunneling can be built on the virtual server where the load balancer can tunnel the request packets to the different servers.

2.2.3 Linux Virtual Server via Direct Routing (LVS/DR)
The load balancer has an interface for configuration along with a virtual IP address, where the interface can route the packets to the chosen server directly.

3 CssQoS SYSTEM

To improve the VMs services in Eucalyptus cloud computing, Figure 3 illustrates the proposed CssQoS system architecture, including Hadoop, a load balancer and an agent-based monitor. Hadoop Map Reduce technology processes vast amounts of data in hardware clusters in a fault-tolerant, simple and reliable manner.

Figure 4 illustrates the proposed CssQoS system framework, including Map Reduce Module, Distributed File System (DFS) Module, HBase Module and Load Balancing Module. This framework combines Hadoop technology to support large amounts of data operations and a load balancer to schedule VM resources.

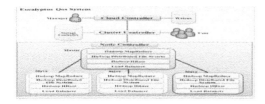

Figure 3.   Proposed CssQoS system architecture.

Figure 4.   CssQoS system framework.

3.1   MapReduce module

Figure 5 illustrates the MapReduce operation procedure. The client no de provides MapReduce actions to the JobTracker node. The JobTracker node is responsible for reducing the action to optimize the scheduling tasks. The TaskTracker node performs the map/reduce operation, while the HDFS stores the events of map/reduce operation. By including JobTracker and TaskTracker functions that can analyze a large amount of data, the MapReduce Module provides optimal services to users.

3.2   DFS module

The proposed DFS Module is derived from Hadoop HDFS with Eucalyptus cloud computing. DFS consists of a NameNode and many DataNodes. NameNode manages namespace operations like closing, opening and renaming files, as well as determines the mapping of blocks to DataNodes. DataNode manages the storages attached to the node and provides read/write requests from users of the file system. DataNode also performs block creation and replication from NameNode.

3.3   HBase module

This module combines the Hadoop HBase with Eucalyptus cloud computing, which consists mainly of HMaster and HRegion functions.

3.3.1   HMaster server

In addition to assign regions to HRegion, HMastermanages HRegion loading. If a HRegion server fails, HMaster reassigns the work to another HRegion.

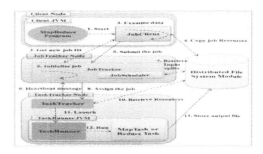

Figure 5.   MapReduce operation procedure.

3.3.2   HRegion server

HRegion handles users' read/write requests, forms a list of regions from HMaster and transmits heartbeat messages to HMaster, which checks that HRegion is still alive. The HRegion server includes many HRegions, which consist of HStores. HStores consist of MemStore and Store File, which includes HFile. MemStore is a sorted memory buffer that stores the client data. When full, MemStore flushes out a Store File.

3.4   Load balancing module

This study proposes a novel load balancing module to distribute the system workload of VMs in Eucalyptus cloud computing. Figure 6 illustrates the functions of the proposed load balancing module. The module consists of two components: a load balancer and agent-based monitor. The load balancer supports three capabilities that are incorporated into three mechanisms: balance triggering, CssQoS scheduling and VM controlling. The balance triggering mechanism receives data of VMs from agent-based monitor. The CssQoS scheduling mechanism performs a load balancing method by analyzing the data of VMs. The VM control mechanism provides the balancing control of the VMs' load based on the actions taken by the CssQoS scheduling.

3.4.1   Load balancer

The CssQoS scheduling capability captures the parameters and computes the parameters where the results are stored in the weighted table. The VM control module then distributes the loading of VMs, according to the value stored in the weighted table.

1. Balance Trigger: State parameters of VMs are obtained by the agent-based monitor. If unable to capture the data, the balance trigger module interprets that VM is dead. The balance trigger module then sends messages about the VM state to the CssQoS scheduling module. Parameter W is defined as the VM state. A situation in which

Figure 6.   Functions of load balancing module.

W = 0, implies that VM is running and its state is currently alive. A situation in which W = 1 implies that the VM state is dead. Meanwhile, a situation in which W = 2 implies that the VM state remains the same as the previous state.

2. CssQoS Scheduling: The CssQoS scheduling module uses parameters from the balance trigger module to schedule the VM loading. The load balancer captures VM data when the service begins. The CssQoS scheduling capability then generates the weighted table according to the VMs' state that consists of six parameters. Table 1 lists the notations of the parameters. Figure 7 shows the flow chart of the CssQoS scheduling. Operations of the CssQoS scheduling are listed as follows.

  1. *Step 1:* The balance trigger module receives the agent-based monitor message containing W value. A situation in which W = 1 is established implies that the VM current state as dead. Therefore, the load balancer does not need to retrieve data from the agent-based monitor.
  2. *Step 2:* The balance trigger capability receives data from the agent-based monitor. The data is classified as *Mcpu_idle, Mmem used, Mth_rx, Mth_tx, Mpl_rx and Mpl_tx*.
  3. *Step 3:* The CssQoS scheduling capability receives the parameters from the balance trigger and builds the VM table. The attributes of VM table are {*VM—Hostname, Mcpu_idle, Mmem_used, Mth_rx, Mth_tx, Mpl_rx, Mpl_tx*}. The VMs hostname is allocated from the Eucalyptus Cluster Controller.
  4. *Step 4:* The CssQoS scheduling capability sorts the weighted VMs according to the performance of the VMs state.
  5. *Step 5:* The CssQoS scheduling capability generates the weighted table. The attributes of weighted table are {*VM—Hostname, icpu_idle, imem_used, ith_rx, ith_tx, ipl_rx, ipl_tx*}.
  6. *Step 6*: The VM control capability receives the weighted table from the CssQoS scheduling capability and, then distributes the task to the VMs according to the weighted value.
3. VM Control: The tasks are distributed using weighted round-robin method. Figure 7 describes the scheduling of the VM control capability.

Table 1.   Attributes of weighted table.

Symbol	Unit	Definition
M_{cpu_idle}	%	The rate of the physical CPU time that the VM not used
M_{mem_used}	%	The rate of memory that VM used
M_{th_rx}	KB/s	The average rate of successful message delivery about receiving action
M_{th_tx}	KB/s	The average rate of successful message delivery about transmission
M_{pl_rx}	%	The packet loss ratio about receiving action
M_{pl_tx}	%	The packet ratio about transmission

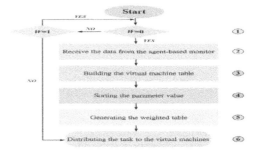

Figure 7.   Flowchart of CssQoS scheduling.

3.4.2   *Agent-based monitor*

The agent-based monitor module consists of data collection, data analysis and data transmission modules. The operation of the module; its capabilities are described as follows.

1. Data Collection: The information of VMs states is collected, including their CPU, memory, disk space and bandwidth.
2. Data Analysis: The data analysis mechanism analyzes the collected data and updates the value of W, indicating whether the data differs from the previous data.
3. Data Transmission: The data transmission mechanism is responsible for transmitting the data of VMs to the load balancer.

The operation of the load balancing module is assigned to the load balancer when users want to access services from the Eucalyptus platform. The balance trigger mechanism of the load balancer then sends messages to the agent-based monitor to obtain the data of VMs. The data collection mechanism gathers the parameters and passes the data to the data analysis mechanism. The data analysis mechanism analyzes the data of VMs that has been transmitted to the data transmission mechanism of the agent-based monitor.

Next, the load balancer receives the data of VMs. If W does not equal 1, the CssQoS scheduling mechanism obtains the VM data and generates the weighted table to the VM control mechanism, which balances the VM's loading, according to the values of the weighted table. The HDFS and HBasemodules store the data created when the VMs perform tasks requested by users. Finally, the MapReduce module handles the tasks based on the weighted table and transmits the results of the tasks to the users.

4 PERFORMANCE ANALYSIS

The proposed Eucalyptus QoS Mechanism (CssQoS system) is compared with Eucalyptus with Hadoop and, then, with Eucalyptus with Hadoop and the balancer module.

In this study, the performance of cloud computing services was evaluated using data sizes of 3.78 MB (1 Chunk), 7.56 MB (2 Chunks) and 11.34 MB (3 Chunks) starting at 20 seconds. The performance of Eucalyptus was then compared with that of Hadoop built, Eucalyptus with Hadoop and the original balancer built in, and the CssQoS system by using CPU, memory and throughput. All experiments were repeated 100 times. Table 2 lists the average execution times of the three cases.

Figure 8 shows the CPU loading. With an increasing amount of data, the CPU loading on CssQoS system resembles that performed on the other two systems. Table 3 shows the average CPU loading in these three cases. The CssQoS system increases the CPU ratios in that the loading gradually diminishes as the data size increases.

Figure 9 illustrates the memory loading performance, while Table 4 shows the average memory loading in all three cases. With an increasing amount of data, memory loading with the CssQoS system is nearly the same as that with the other systems. This phenomenon may be caused

Table 2. Average execution time.

Case	Average execution time (sec)		
	Original Eucalyptus and Hadoop	With original Hadoop balancer (improved rate)	CssQoS system (improved rate)
1 Chunk	118.23	117.56 (+0.57%)	116.32 (+1.62%)
2 Chunks	159.72	158.89 (+0.47%)	157.93 (+1.00%)
3 Chunks	236.56	235.78 (+0.32%)	234.67 (+0.79%)

Table 3. Average CPU load.

Case	Average CPU loading (%)		
	Original Eucalyptus and Hadoop	With original Hadoop balancer (improved rate)	CssQoS system (improved rate)
1 Chunk	36.86	33.33 (+9.57%)	32.1 (+12.92%)
2 Chunks	67.37	62.96 (+6.55%)	61.66 (+8.47%)
3 Chunks	76.55	72.76 (+4.93%)	71.48 (+6.61%)

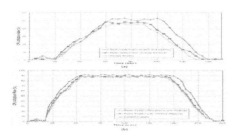

Figure 8. CPU load: (a) 3.78 MB; (b) 11.34 MB.

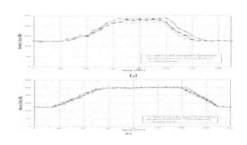

Figure 9. Memory load: (a) 3.78 MB; (b) 11.34 MB.

Table 4. Average memory load.

	Average memory loading (MB)		
Case	Original Eucalyptus and Hadoop	With original Hadoop balancer (improved rate)	CssQoS system (improved rate)
1 Chunk	195.04	186.85 (+4.19%)	183.35 (+5.99%)
2 Chunks	210.13	205.35 (+2.27%)	198.20 (+3.48%)
3 Chunks	223.12	218.22 (+2.19%)	215.67 (+3.33%)

Table 5. Average system throughput.

	Average system throughput (Kbps)		
Case	Original Eucalyptus and Hadoop	With original Hadoop balancer (improved rate)	CssQoS system (improved rate)
1 Chunk	65.4	67.23 (+2.79%)	69.95 (+6.94%)
2 Chunks	115.33	117.83 (+2.16%)	120.26 (+4.27%)
3 Chunks	142.25	144.02 (+1.24%)	145.47 (+2.26%)

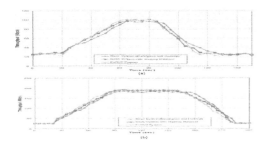

Figure 10. System throughput: (a) 3.78 MB; (b) 11.34 MB.

by the memory size of VMs, which is limited to 256 MB.

Figure 10 illustrates the system throughput performance, while Table 5 shows the average system throughput in all three cases. With an increasing data size, the master node should allocate many data blocks to slave nodes. If the loading of slave nodes are too large, the load balancer uses the round-robin method, causing the CssQoS system throughput performance to become nearly the same as that of the other two systems.

5 CONCLUSIONS

This study has developed the CssQoS system, which incorporates a Eucalyptus cloud computing infrastructure, Hadoop HDFS, Hadoop MapReduce, Hadoop HBase, load balancer and agent-based monitor technologies. In the proposed system, load balancer computing parameters are retrieved, and the loads are distributed to Eucalyptus cloud computing nodes using the IPVS toolkit. Emulation results indicate that the CPU loading, memory loading and system throughput performance comparable to that of the original Eucalyptus system with Hadoop technology and the original Eucalyptus system with the original built-in Hadoop load balancer. Analytical results demonstrate that the proposed CssQoS system can reduce the loading of virtual machines and improve the performance of cloud computing systems.

REFERENCES

Ahlgren, B.; Aranda, P.A.; Chemouil, P.; Oueslati, S.; Correia, L.M.; Karl, H.; Sollner, M. & Welin, A. 2011. Content, Connectivity and Cloud: Ingredients for the Network of the Future. *IEEE Communications Magazine* 49(7): 62–70.

Fagui, L.; Hao, Z. & Haiyan, Z. 2010. A Xen-Based Secure Virtual Disk Access-Control Method. *Proceedings of the International Conference on Multimedia Information Networking and Security* 375–378.

Jin, H.; Gao, W.; Wu, S.; Sli, X.; Wu, X. & Zhou, F. 2011. Optimizing the live migration of virtual machine by CPU scheduling. Journal of Network and Computer Applications 34(4): 1088–1096.

Kansal, N.J. & Chana, I. 2012. Existing Load Balancing Techniques in Cloud Computing: A Systematic Review. Journal of Information Systems and Communication 3(1): 87–91.

Monhamed, N.; AI-Jarood J. & Eid, A. 2013. A dual-direction technique for fast file downloads with dynamic load balancing in the cloud. Journal of Network and Computer Applications 36(4): 1116–1130.

Sempolinski, P. & Thain, D. 2010. A Comparison and Critique of Eucalyptus, Open Nebula and Nimbus. Proceedings of the 2nd IEEE International Conference on Cloud Computing Technology and Science 417–426.

Wang, Y.; Zhou, Z.; Liu L. & Wu, W. 2013. Replica-aided load balancing in overlay networks. Journal of Network and Computer Applications 36(1): 388–401.

Warneke, D. & Kao, O. 2011. Exploiting Dynamic Resource Allocation for Efficient Parallel Data Processing in the Cloud. IEEE Trans. on Parallel and Distributed Systems 22(6): 985–997.

Xiao, P.; Hu, Z.; Liu, D. & Yan, G. 2013. Virtual machine power measuring technique with bounded error in cloud environments. Journal of Network and Computer Applications 36(2): 818–828.

Information Technology and Applications – Li (Ed.)
© *2015 Taylor & Francis Group, London, ISBN 978-1-138-02677-3*

Dynamically Adjusted Detection Node Algorithm for Wireless Sensor Networks

Qing Yu & Yuguang Liu
Tianjin Key Laboratory of Intelligence Computing and Novel Software Technology,
Tianjin University of Technology, Tianjin, China

Jinlin Wang
Tianjin Zhonghuan Electronic Computer Corporation, Tianjin, China

ABSTRACT: Compared with traditional network, wireless sensor networks have many differences in storage, computing power, energy resources. And it also suffers intrusion from hacker. Maintaining continuous normal intrusion detection with limited energy storage is very difficulty. This paper proposes a Dynamic Adjustment Detection Node Algorithm (DADNA) for wireless sensor networks. It dynamically calculates the number of detection nodes, which make the wireless sensor network to not only achieve the best detection results, but also can balance the energy consumption. Due to the discard of invalid node, the number of detection nodes is decreased according to system balance. In order to ensure the detection accuracy, the number of detection node is correspondingly increased according to the number of malicious nodes lastly detected.

Keywords: Wireless Sensor Networks (WSNs); energy balance; dynamical adjust; detection node

1 INTRODUCTION

Wireless Sensor Networks (WSNs) [1,2] consist of a large number of sensor nodes and are usually distributed in the unattended natural environment (such as forest, grassland, ocean, etc.), which is used to monitor environmental changes or local action in the battlefield. Since the WSNs are distributed in the dangerous environment, they are vulnerable to damage. As with traditional networks, WSNs also suffer invasion of hackers or enemy [3,4]. But WSNs have many differences with the traditional network. Its own limitations: Limited storage space, computing capacity, bandwidth, transmission communication capacity and energy resources, etc. [5,6]. Environmental restrictions: physical security cannot be guaranteed in the layout area and the layout of latter node lacks of prior knowledge [7]. Because of the limited storage space, computing power and energy of the sensor, maintaining continuous normal intrusion detection work of the entire network is very difficult. In this paper, an algorithm is presented to dynamically calculate the number of detection nodes, which makes the whole (WSNs) can achieve the best effect of detection and can also balance the energy consumption. The structure of this paper is: the second part introduces the present situation and development of research at home and abroad, the third part proposes the algorithm, the fourth part is the simulation validation and the fifth part is the summary and future work.

2 RELATED RESEARCH

According to the limited resources of WSNs, Heinzelman et al. put forward the Low-Energy Adaptive Clustering Hierarchy (LEACH) [8] algorithm, the main idea of LEACH algorithm is to let each node in turn act as the cluster head, so as to balance the energy consumption of network and prolong the network life cycle. However, LEACH algorithm also has many shortcomings, such as lacking of consideration of the remaining energy, uneven distribution of cluster head, unstable number of cluster head, etc. XUY et al. proposed the GAF [9] (geographical adaptive fidelity) algorithm, it is a kind of clustering algorithm based on geographic location, which divides the monitor area into virtual cell and assigns node to the corresponding cell according to the geographical location information. GAF algorithm is a kind of plane model without considering the direct communication of node in the actual network, and it also needs to know the accurate location information. But the division idea of GAF algorithm is of great reference significance. Paper [10] proposed EEUC (Energy-Efficient Uneven Clustering) algorithm. In EEUC algorithm, each node randomly determines whether it becomes pseudo cluster head according to a prescribed threshold, and the pseudo cluster head becomes a cluster head through competition. Multiple hops are adopted between clusters; the cluster head of closer base station takes more tasks, the faster of energy it consumes. In order to balance the energy consumption of cluster head, cluster closed to the base station has smaller competition radius, which makes the number of members in the cluster relatively small, and it saves energy for the use of data forwarding of cluster head. Compared with LEACH algorithm, the energy consumption of this algorithm is more balanced, the lifetime of network is longer, but the control is more complicated. Paper [11] proposed a clustering algorithm BREND (based on residual energy and node degree), which takes the residual energy as the primary factor of the selection of cluster head. The algorithm firstly compares residual energy of each node, 6 percent of nodes is chosen as alternative temporary cluster head, then nodes not conform the requirements are eliminated according to the standard that there is only one temporary cluster head within one hip range, then the selection is continued from the rest regular node until the number of temporary cluster head is 6%. Ordinary nodes are added into cluster according to the distance, and then the final cluster head node is determined in accordance with the average residual energy within the cluster.

3 ALGORITHM

3.1 *The idea of algorithm*

This paper proposes a kind of algorithm based on energy balance, which can dynamically adjust the node in the WSNs, and the idea of the algorithm is that some sensor nodes in the WSN will be lapsed dynamically or discarded by system owing to the energy consumption of node itself or the attack of hackers, which makes the number of nodes in the entire network decrease, the persistent existence of the whole system will be affected if the number of detection nodes don't reduce. But when large scale of network intrusion happens, in order to ensure the accuracy of detection, the number of detection nodes should be appropriately increased under the condition that the lifetime of the whole system is not affected. In short, invalid nodes should be eliminated and the number of detection nodes should be decreased according to the balance of system. According to the number of invaded nodes last detected, the number of detection node is correspondingly increased to ensure the accuracy of detection, which can make the whole WSN achieve best detection results and also balance energy consumption.

We assume that there is a homogeneous WSN with a uniform distribution, and there are N sensor nodes distributes in the network, which distributes in L*L two-dimensional plane. All sensors have the same storage, computing power, energy and approximate load, so the adjacent sensors have similar behaviors under normal circumstances. Cluster is divided according to the residual energy and node degrees algorithm, after the division, the algorithm is called again in the cluster to determine the number of detection node. Figure 1 is the system structure.

3.2 *Description of algorithm*

Assuming that the total number of nodes distribute in WSNs is N; The number of detection nodes we initialized is M; The number of invaded nodes in each detection round is I; The number of failed nodes in each round is D; The number of reduced detection nodes in each round is S; The number of increased detection nodes is A. We define the standards of the system balance as

$$\Phi = M/N,$$

Formula (1)

According to the number of detected intrusion node in each round, we define the risk level of WSNs as

$$R = \frac{I}{(N - M)} \quad R \in \{0, 1\}$$

Formula (2)
Based on the system energy balance

$$\frac{M - S}{N - D} = \Phi$$

Formula (3)
From Formula (1) (3), we can get

$$S = M - \Phi(N - D)$$

Formula (4)

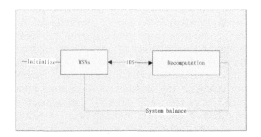

Figure 1.

278

A can be determined by Formula (2)

$$A = \begin{cases} 0 & R = 0 \\ \left\lceil \dfrac{1}{2} \right\rceil & 0 < R < 0.1 \\ \dfrac{1}{N}(N-M) & R \geq 0.1 \end{cases}$$

Formula (5)

So according to the principle of dynamic adjustment $M_{next} = M - S + A$, $N_{next} = N - D$

The pseudo code of algorithm

```
int M,N;
int D = I = 0, S = 0, A = 0;
for(t = 0;;t++){
    S = M-(N-D)*Φ;
    R = I/(N-M);
if(R = = 0)
    A = 0;
elseif(0 < R < 0.1)
    A = ⌈I/2⌉;
else
    A = I/N*(N-M);
    M = M-S+A;
    N = N-D;
}
```

4 EXPERIMENTAL ANALYSIS

Network model assumption is as follows:

1. The location of all sensor nodes and sink nodes in the entire WSNs are fixed;
2. The sensor nodes are randomly distributed in the network;
3. The sensor node is unique, no new sensor nodes will be added and energy is limited, which cannot be supplemented;
4. The sensor nodes in the entire network have the same structure including the same initial energy and the same function;
5. Algorithm in paper [12] is used as intrusion detect algorithm, detection nodes collect data information of packet loss rate, contract rate, forwarding time of sensor nodes around. The distance between nodes and mean value is calculated according to Markov Distance and then compared with threshold, if it is greater than the threshold, it is considered as the intrusion node.

In this paper, the lifetime of WSNs is from the network beginning to work to all sensor nodes run out of energy. The wireless communication energy consumption model is similar to paper [11]. When node 'a' sends x bit packet to node 'b', the distance is 'd', the energy consumption of 'a' is composed of transmission circuit loss and power amplifier loss,

Namely

$$E_{Tx,d\Upsilon} = \begin{cases} xE_{elec} + x\varepsilon_{fs}d^2, & d < d_0 \\ xE_{elec} + x\varepsilon_{mp}d^4, & d \geq d_0 \end{cases}$$

Among them: E_{elec} represents the transmission circuit loss of node, if the transmission distance is less than the threshold value d_0, then the power amplifier loss uses free space model; Otherwise it uses multipath attenuation model. ε_{fs}, ε_{mp} represent the energy for the power amplifier of the two models, respectively. When 'b' receives the information sent by 'a', the energy loss of wireless receivers is:

$$E_R(x) = xE_{elec}$$

In terms of energy consumption, this paper adopts the same set with BREND algorithm, and the parameters of simulation experiment are as shown in Table 1.

Figure 2 compares the energy consumption of DADNA algorithm with BREND algorithm, this paper is dynamically adjusts the number of detection nodes, and reduces the number of unnecessary detection nodes, which reduces the energy of the entire network and prolongs the network working time. Figure 3 compares DADNA algorithm with dynamic adjustment of detection node and method using fixed detection node under the condition that they all use the intrusion detection algorithm in literature [12]. As it can be seen from Figure 3, with the growth of time, the number of failed detection nodes are increased, which affects the detection rate. But DADNA algorithm dynamically adjusts the number of monitor node depending on the energy balance of the whole network, which enables the normal intrusion detection of entire network under the condition of losing a large number of nodes. From these we can know that DADNA

Table 1. Simulation parameter set.

Parameter	Value
Network size	$(100 * 100)$ m^2
Node number	100
Radio range	25 m
BS position	(50 m, 150 m)
Initial energy	2 J
E_{elec}	50nJ/bit
E_{BF}	5nJ/bit
ε_{fs}	10pJ/(bit·m^2)
ε_{mp}	0.0013pJ/(bit·m^4)
d_0	86.3 m
L	4 000 bit

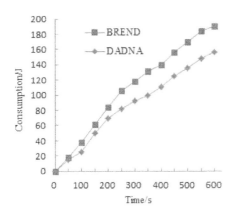

Figure 2.

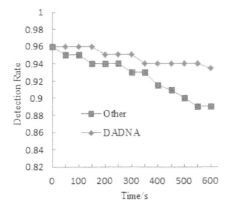

Figure 3.

algorithm can effectively prolong the whole network and achieve good detection results.

5 CONCLUSION

This paper proposes a kind of algorithm based on energy balance, which can dynamically adjust the number of detection node of the entire WSNs, the WSNs and the number of detection nodes are initialized by BREND algorithm, later in the operation of the network, the number of detection nodes are dynamically adjusted according to the balance of whole network. Compared with the method with fixed number of detection node, it not only prolongs the service life of the whole network, but also improves the detection accuracy. Future work will apply DADNA algorithm to the actual WSNs to test the algorithm.

ACKNOWLEDGMENT

This work was supported by the Science and Technology Support Project of Tianjin under Grant No. 11ZCKFGX00500 and the Science and Technology Support Project of Tianjin under Grant No. 10ZCKFGX03700.

REFERENCES

[1] R.A. Maronna, R.D. Martin & V.J. Yohai. 2006. Robust Statistics: Theory and Methods. Wiley Publisher chapter 6.9.1 pp. 205–208.
[2] Limin Sun, Jianzhong Li, Yu Chen. 2005: 4–20. Wireless Sensor Network [M]. Beijing: Tsinghua University Press.
[3] Ahmed, M. & Naser, A. 2013. A novel approach for outlier detection and clustering improvement. Industrial Electronics and Applications (ICIEA), 2013 8th IEEE Conference. pp. 577–582.
[4] P. Albert & M. Kuhn. 2002. Security in ad hoc networks: A general intrusion detection architecture enhancing trust based approaches. In First International Workshop on wireless Information Systems, 4th International Conference on Enterprise Information Systems, pp. 1–12.
[5] A.P. daSilva. 2005. Decentralized intrusion detection in wireless sensor networks. In ACM international Workshop on Quality of Service and Security in Wireless and Mobile Networks, pp. 16–23.
[6] F. Liu, X. Cheng, F. An, On the Performance of In-Situ Key Establishment Schemes for Wireless Sensor Networks, in IEEE GLOBECOM2006, San Francisco, CA, November 27–December 1, 2006.
[7] A. Perrig, J. Stankovic, D. Wagner, Security in wireless sensor networks, in CACM, June 2004, vol. 47, pp. 53–57.
[8] Heinzelman W.R., Chandrakasan A., Balakrishnan H. An application-specific protocol architecture for wireless microsensor networks [J]. IEEE Transactions on Wireless Communications, 2002, 1(4):pp. 660–670.
[9] Xu Y., Heidemann J., Estrin D. 2001. Geography-informed energy conservation for ad hoc routing [C]. Proceedings of the 7th ACM/IEEE International Conference on Mobile Computing and Networking (MobiCOM), Los Angeles: Lenter for Embedded Network Sensing: pp. 70–84.
[10] Chengfa Li, MaoYe, Guihai Chen, et al. 2005. An energy-efficient unequal clustering mechanism for wireless sensor networks [C]. Proceedings of IEEE International Conference on Mobile Ad hoc and Sensor System. Washington DC P: 8 pp. -606.
[11] Juping Fu, Xiaogang Qi. Clustering algorithm based on residual energy and node degree for WSNs. 2011, 28 (1):pp. 250–252.
[12] F. Liu, X. Cheng and D. Chen. 2007. Insider Attacker Detection in Wireless Sensor Networks. IEEE INFOCOM 2007 proceedings. pp. 1937–1945.

Information Technology and Applications – Li (Ed.)
© 2015 Taylor & Francis Group, London, ISBN 978-1-138-02677-3

A cross-layer energy saving routing finding method for Underwater Wireless Sensor Network

Changbing Huang, Xinwang Zheng, Guangsong Yang, Chaoyang Chen & Ling Gao
School of Information Engineering, Jimei University, Xiamen, Fujian, China

ABSTRACT: In order to improve the energy efficiency in Underwater Wireless Sensor Network (UWSN), the routing finding problem from source nodes to sink is studied. It was proved that the optimal energy hop distance must be existed when the total distance between source node and sink is given. A distributed energy saving routing finding method is proposed, which can get neighbor nodes' information by exchanging control packets without have full knowledge of the whole network topology. At the physical layer, different power levels were set according to the position of optimal hop nodes and feedback of neighbor nodes. At the network layer, the node closest to the idea node will be selected as the next hop. Simulation results show that the proposed protocol consumed less energy than the FBR in sparse networks, than VBF both in sparse and in dense networks.

Keywords: Underwater Wireless Sensor Network; energy efficient; cross layer; routing protocol

1 INTRODUCTION

Along with the increasingly human activities on exploration, development and utilization of marine resources, the idea of Underwater Wireless Sensor Network (UWSN) has received increasing interests which applies wireless sensor networks technology into underwater environments [1]. These sensor nodes are deemed to enable applications for oceanographic data collection, pollution monitoring, offshore exploration, disaster prevention, assisted navigation and tactical surveillance applications. Because the sensor nodes work in the underwater environment and usually powered by batteries, it is impossible and difficult to replace and recharge them in most application scenarios, energy constraint is a crucial factor to the lifetime of whole network. Energy consumption of UWSN is not only related to the physical characteristics of sensing circuit, but also related to the protocol design. Therefore, enormous efforts have been made for designing efficient protocols while considering the characteristics of underwater communication.

How to relay data to the sink effectively with limited energy resource, energy saving routing protocols need to be designed. Several energy saving routing protocols of UWSN were proposed in recently study, such as DFR (Directional Flooding-based Routing), VBF (Vector-Based Forwarding), and FBR (Focused Beam Routing) [2] etc. In DFR protocol [3], routes were found by using local flooding mechanic. The flooding zone is mainly determined by the position of source node, sink node and current forwarding node. DFR limits the forwarding number based on link quality, but the redundant forwarding packets still cannot be avoided. VBF [4] is vector based routing protocol which forms a virtual pipe between a source and a destination. The pipe width and the density of nodes in the pipe determine the successful delivery of data. VBF assumes that nodes have full knowledge of their locations and nodes have fixed transmission range. FBR [5] is a scalable routing technique for multi-hop communication based on location information. It employs various transmission power levels in order to minimize the energy consumption. Is has some disadvantages, for example, it uses a single transmitting cone that covers only a fraction of communication area. It needs to rebroadcast RTS every time it cannot find a candidate node within its transmitting cone. It lacks of a collision avoidance mechanism. CTS may collide easily in dense networks.

In this paper, we proposed an energy efficient cross-layer routing protocol, which gets neighbors information by broadcasting an RTS and CTS packet, finds energy efficient path and utilizes different transmission power levels during the selection of next relay node.

2 THE OPTIMAL ENERGY HOP NUMBERS WHEN TOTAL DISTANCE IS GIVEN

2.1 Underwater acoustic channel model

2.1.1 Passive sonar equation

To satisfy the requirement of receiver, the signal needs to overcome the influence of channel fading and noise in underwater acoustic channel, when the transmitted signal strength is SL, the received signal noise ratio must be greater than detection threshold DT, according to the passive sonar equation, need to meet the following conditions [6]:

$$SNR = SL - TL - NL + DI >= DT \qquad (1)$$

where, TL is the transmission loss caused by the underwater environment, NL is the ambient noise, DI is the directivity index of the hydrophone being utilized. All units in equation (1) are dB re μPa, 1 μP can be equivalent to 0.67×10^{-22} Watts/cm^2, i.e. 1 $\mu Pa = 0.67 * 10^{-18} watts/m^2$.

2.1.2 Transmission loss and source level

If we consider I_0 to be the intensity at a reference point located approximately 1 yard (1 yard = 0.9144 m) from the acoustic center of the source and I_1 as the intensity at a distant point, then we can define the transmission loss as:

$$TL = 10\log(I_0/I_1) = 10\log I_0 - 10\log I_1 \qquad (2)$$

The source level SL can be defined as [6]:

$$SL = 10\log(I_0/I_{ref}) = 10\log(I_0/1\mu Pa) \qquad (3)$$

So we can get the relationship between TL and SL as:

$$TL = 10\log\left(\frac{I_0}{I_1}\right) = 10\log\left(\frac{10^{SL/10}}{I_1}\right) = SL - 10\log I_1 \qquad (4)$$

2.2 Energy consumption

Transmission loss in deep water is a combination of the spherical spreading loss, attenuation and anomaly of transmission. As such, transmission loss distant r from source point can be expressed as [6]

$$TL = 20\log r + \alpha(f)r * 10^{-3} + A \qquad (5)$$

where α is the absorption coefficient in dB/km which can be obtained from the Thorp Equation [7], r is the range and A is transmission loss anomaly that accounts for multipath propagation, refraction and other phenomenon.

$$\alpha(f) = \frac{0.11f^2}{1+f^2} + \frac{44f^2}{4100+f^2} + 2.75 * 10^{-4} f^2 + 0.003 \qquad (6)$$

As such, using the information obtained from Equations 4 and 5 we can deduce the intensity located at a distance r from the source node I_1, which is:

$$I_1 = 10I^{(SL-NL)/10} = 10^{(SL-20\log r - \alpha r 10^{-3} - A)/10} \qquad (7)$$

So the transmit power level with spherical spreading loss model is

$$P(r) = 4\pi r^2 I_1 = 4\pi r^2 10^{(SL-20\log r - \alpha r 10^{-3} - A)/10} \qquad (8)$$

The energy consumption of each node is:

$$E(r) = E_t + E_r = kPT_{tx} + E_r \qquad (9)$$

where E_t is energy consumed for sending, which is the amount of energy to transmit k packets of data over a distance in the interval of T_{tx}, E_r is consumption energy for receiving, which is 50 nJ according to [8].

2.3 The energy-optimal number of hops

Place We consider a linear topology multi-hop system, which consists of a source node S and a destination node D separated by $N-2$ equidistantly placed intermediate relay nodes, the total distance between S and D is d_{total}, so the distance and consumed energy of each hop is d_{total}/N and $P_i = 4\pi(d_{total}/N)^2 I_i$, the total consumed energy of sending data from S to D is

$$E_{total} = NkE(d_{total}/N) \qquad (10)$$

where, k is transmitted packets of data, in order to minimize the total energy consumption, when the d_{total} is given, it should be satisfy

$$\partial(E(N))/\partial N = 0 \qquad (11)$$

So we can get the optimal hop numbers $Nopt$, which can use the least energy to send data from source to destination. We set $k = 1$, $f = 10$ KHZ, $T_{tx} = 40$ ms, and get the relationship between hop numbers and total consumed energy with different d_{total} in Figure 1. We can see there always an optimal solution exist according to equation (11).

2.4 An energy saving routing protocol in UWSN

The proposed scheme is a cross-layer based routing protocol in which different power levels were

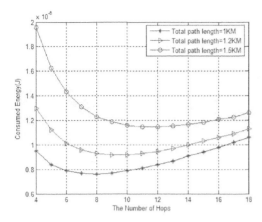

Figure 1. The relationship between energy consumption and hop numbers.

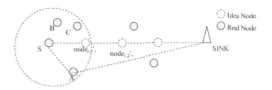

Figure 2. Explaining of route finding.

set according to the position and the feedback information of neighbor nodes in physics layer. In network layer, the node which closed to the idea energy-optimal node was selected as next hop. The purpose is minimizing the consumed energy.

We assume sensor nodes were uniformly randomly distributed in a circle area with its center at a SINK, each node knows itself position, and can adjust its power level, it also can estimate the distance between its neighbor and itself through control packets. The work process includes the following two phase.

2.4.1 *Phase 1. The initial phase*

A. Sink broadcast initial packet (include the position of sink itself) with the highest power level in the coverage area, node i who received the broadcast, can estimate the distance between the sink and itself d_{i-sink}, and know the energy-optimal path to sink by the method mentioned in section 2.2. The sensing data were sent to sink hop by hop delayed by nodei-1', nodei-2' ...

2.4.2 *Phase 2. Route finding phase*

If a source node S wants to send a message to sink, it first finds the idea energy-optimal path according phase 1. In the practical scenario, randomly distributed nodes do not necessarily located at the idea energy-optimal path, but we can select the node which closest to the idea node as our next hop, and get the energy-suboptimal path.

Firstly, node S broadcasts a packet to neighbor with fixed power level, which includes the idea position information of next hop node and itself. For example, in Figure 2, if node A receives this packet, it will compute the distance between sink and itself, if satisfy $d_{A-sink} < d_{S-sink}$, it will send ACK include itself position information. Similarly, node B, C in

the coverage area, will also send ACK as node A. After received these information, node S computes the distance between nodei-1' and these nodes, and select the closest node as next hop. Then repeats the process until the next hop is sink node. Once selected a path to sink, the data were sent along the path.

If nodes become sparse, it is possible that no node will lie within that forwarding cone of angle. If S has never received any reply information, it means there is no node in the coverage area, so the S has to increase its power level and repeats the same phase.

3 SIMULATION AND EVALUATION

In this section, we run simulations to evaluate the proposed scheme and compare it with VBF [4] and FBR.

Sensor nodes are deployed in acoustic scenario 1000 m*1000 m area. We use the channel model describes in section 1.1, sensor nodes were distributed in grid scenario and random scenario. The detection threshold of receiver is 20 dB, average ambient noise level NL set as 50 dB by considering the factors of wind turbulence, shipping, wind and thermal, directivity index DI = 1, we fixed the transmission frequency at 10 HKz, transmit time of a packet at 40 ms. Total energy consumption is the sum of all energy dissipated by all nodes. Each simulation will be run more than 10 times, and the results were an average of all runs.

In Figure 3 show the route find result in grid scenario, 100 nodes are deployed at square grid, distance between any two nodes are 100 m both in the X and Y directions, we set No. 31 as source node, No. 100 as destination node, the solid line represents the finding path by using our proposed routing protocols for the communication path, the dotted line represents the finding path by using FBR.

During real application, nodes are randomly assigned to a designated area. We change the node density in uniform and no-uniform way. Figure 4 and Figure 5 show the relationship between energy consumption and node density in the uniform distribution and no-uniform scenarios respectively.

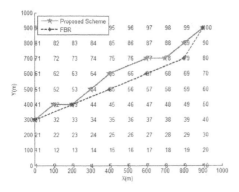

Figure 3. Route finding process of different protocol in grid scenario.

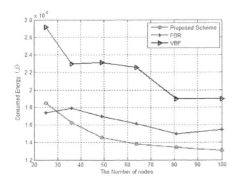

Figure 4. Energy consumption with uniformly distributed nodes.

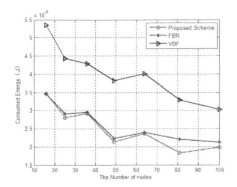

Figure 5. Energy consumption with non-uniformly distributed nodes.

In the above figure, we observed the impact of node density on the performance. We can see VBF consumed most energy among the three protocols in different node densities, and our proposed protocol has best energy efficiency.

Both in the uniformly or non-uniform distributed scenario, as the node density increases, the total energy consumption of the three protocol have declined. In VBF only those nodes close to the routing vector are involved in data forwarding, it is very sensitive about the routing pipe radius threshold which can affect the routing performance significantly with the increase of numbers of node. For FBR, the next relay node was found within the transmitting cone and the power level can be adjusted, the more nodes exist in this cone, the energy efficient node can be more likely to be found. So it has better energy efficiency than VBF. For our proposed protocol, with the increase of node numbers, we can easily found the nodes closest to idea energy optimal node, and has best energy performance among the three protocols. The simulation and operation results confirm that above-mentioned routing strategy was effective.

4 CONCLUSIONS

The above material should be with the editor before the deadline for submission. Any material received too late will not be published. Send the material by airmail or by courier well packed and in time. Be sure that all pages are included in the parcel. Energy efficiency is the most important issue to be considered in the UWSN. In this paper, the optimal energy path in linear topology is analyzed and proved. We proposed an energy saving protocols for UWSN based on the idea of cross-layer idea. Our proposed protocol has the following advantages. Firstly, Nodes just need know their own position and the position of the final destination of the packet, the location of other intermediate nodes is not required, so unnecessary flooding can be reduced. Secondly, the protocol is a distributed algorithm, in which a route is dynamically established as the data packet traverses the network towards its final destination. An energy sub-optimal route can be selected in the candidate selection process. Thirdly, the protocol is a cross-layer approach, in which the routing protocol, the medium access control and the physical layer functionalities are tightly coupled by power controlling, routes can be established on demand with minimal energy consumption.

We run the simulation and compare the performance of other two approaches employed in UWSN. The overall system performance is improved when use our proposed protocol both in sparse and dense scenarios. Future works involve several refinements and extensions, and combine the protocol with clustering schemes.

ACKNOWLEDGMENT

Project supported by Fujian Science Foundation (2007J0036, 2013J01203), Xiamen Science and Technology Project (3502Z20130005), and Scientific Research Foundation of Jimei University, China (ZC2014002, ZQ2014008).

REFERENCES

[1] Ian F. Akyildiz, Dario Pompili, Tommaso Melodia, "Underwater acoustic sensor networks: research challenges", Ad Hoc Networks, Volume 3, Issue 3, pp. 257–279, 2005.

[2] Ahmed, S., U. Khan, M.B. Rasheed, M. Ilahi1, R.D. Khan, S.H. Bouk, N. Javaid. Comparative Analysis of Routing Protocols for Under Water Wireless Sensor Networks. J. Basic. Appl. Sci. Res., 3(6)130–147, 2013.

[3] Hu T.S. and Y.S. Fei. QELAR: A Machine-Learning-Based Adaptive Routing Protocol for Energy-Efficient and Lifetime-Extended Underwater Sensor Networks [J], IEEE Transactions on Mobile Computing, Vol. 9, No. 6, June 2010.

[4] Xie, P., J. Cui H., L. Lao. VBF: Vector-Based Forwarding Protocol for Underwater Sensor Networks. Proceedings of Networking.2006, 5: 1216–1221.

[5] Jornet, J.M., M. Stojanovic, and M. Zorzi. Focused beam routing protocol for underwater acoustic networks, in Proc. ACM WUWNet, San Francisco, California, USA, Sept. 2008:75–82.

[6] Urick, R.J. Principles of Underwater Sound, 3rd ed, McGraw-Hill, 1983.

[7] L.M. Brekhovskikh, Y.P. Lysanov, Fundamentals of Ocean Acoustics, 3rd ed., Springer, New York, 2003.

[8] Arupa K. Mohapatra, Natarajan Gautam, and Richard L. Gibson, Jr. Combined Routing and Node Replacement in Energy-Efficient Underwater Sensor Networks for Seismic Monitoring, IEEE Journal of Oceanic Engineering, 2013.1, 38(1).

Information Technology and Applications – Li (Ed.)
© 2015 Taylor & Francis Group, London, ISBN 978-1-138-02677-3

A fast algorithm of double-scalar multiplication on Koblitz curves

Qing Yu & Chong Lin
Tianjin Key Laboratory of Intelligence Computing and Novel Software Technology,
Tianjin University of Technology, Tianjin, China

Jinlin Wang
Tianjin Zhonghuan Electronic Computer Corporation, Tianjin, China

ABSTRACT: Koblitz curves are a special class of elliptic curves. Using the Frobenius map on Koblitz curves can be significantly improved calculation of scalar multiplication kP. Three-dimensional Frobenius extensions have been shown to improve the speed of a single scalar multiplication. Inspired from this approach, we propose a new method to calculate the two-scalar multiplication in this paper. Comparing to traditional computing scalar τJSF method of double-scalar multiplication, the calculation speed has been obviously improved.

Keywords: Koblitz curve; double-scalar multiplication; multi-base

1 INTRODUCTION

Koblitz curves, also known as anomalous binary curves, are elliptic curves defined over F_2. The primary advantage of these curves is that point multiplication algorithms can be devised that do not use any point doublings [6]. Koblitz curves can be defined as:

$$E_a : y^2 + xy = x^3 + ax + 1, \tag{1}$$

where $a \in \{0,1\}$, $E(F_2m)$ is a group of points on E_a for some extension field F_2m and n is the group order of $E_a(F_2m)$. Any point $P \equiv (x,y) \in E_a(F_2m)$ has following properties:

$$(\tau)P = (x^2, y^2) \text{ and } -P = (x, x+y), \tag{2}$$

where τ is called the Frobenius map over $E_a(F_2m)$. Further, there exists a point at infinity denoted by O, The point at infinity satisfies the properties:

$$(\tau)O = O \text{ and } -O = O. \tag{3}$$

The Frobenius mapping of a point can be computed by squaring its coordinates. The spending of calculating the square is very cheap. In the digital signature verification of the elliptic cryptosystem, double scalar multiplication $kP + lQ(P, Q \in E_a(F_2m)$ and $k,l \in [1, n-1]$ cost most computational power. The scalar k,l is represented in τ – adic expansion

to obtain the advantage of Frobenius map by replacing point doubling to improve computing speed. We represent $Z[\tau]$ as the polynomial ring shaped like $\sum_{i=0}^{l-1} u_i \tau^i$, where 1 is the length of the polynomial, $u_i \in \{0, \pm 1\}$ for all $0 < i < 1 - 1$ and $u_{1-1} = \pm 1$. First, both scalars are represented and reduced in $Z[\tau]$ to complex numbers such that 1 is minimal. The reduction in $Z[\tau]$ is defined as $\rho \equiv k$ mod δ, where k is an integer in $[1, n-1]$ and $\delta = (\tau^m - 1)/(\tau - 1)$. The average Hamming weight of τ – adic non-adjacent form $\sum_{i=0}^{l-1} u_i \tau^i$ is computed and is 1/3 [4]. Then we used τNAFs as inputs to generate τ – adic joint sparse form (τJSF), the average Hamming weight of τJSF is 1/2 [4].

The multi-base representation of scalar has been researched by many people. Dimitrov et al has introduced the three dimensional Frobenius expansion $\sum_{i=1}^{d} S_i \tau^{a_i} (\tau - \mu)^{b_i} (\tau^2 - \mu\tau - 1)^{c_i}$, where $S_i = \pm 1$, $a_i, b_i, c_i \in Z \geq 0$ and this method has its advantage in computing single scalar multiplication. Inspired from this approach, we propose a new method to calculate the two-scalar multiplication, we use joint three dimensional Frobenius expansion to compute double-scalar multiplication, compared with the traditional τNAF method, and calculation speed has been obviously improved.

We arrange this paper as follows: three dimensional Frobenius expansion is introduced in section 2. We propose our new joint three dimensional Frobenius expansion in section 3. In section 4, we arrange the experiment and conclude the paper.

2 THREE DIMENSIONAL FROBENIUS EXPANSION

Frobenius map can be regarded as a complex number τ satisfying $\tau^2 + 2 = \mu\tau$, we choose $\tau = (\mu + \sqrt{-7})/2$, where $\mu = (-1)^{1-a}$, a complex number shaped like $a+b\tau (a,b \in Z)$ is called a Kleinian integer. The Paper [4] has given the definition of the $\{\tau, \tau - \mu, \tau^2 - \mu\tau - 1\}$ – Kleinian integer: a Kleinian integer ω of form $\omega = \pm\tau^x(\tau - \mu)^y (\tau^2 - \mu\tau - 1)^z$, where $x,y,z \geq 0$ is called $\{\tau, \tau - \mu, \tau^2 - \mu\tau - 1\}$ – *Kleinian* integer [4].

The three dimensional Frobenius expansion of a given integer can be represented as an equation like this:

$$k = \sum_{i=1}^{d} S_i \tau^{a_i} (\tau - \mu)^{b_i} (\tau^2 - \mu\tau - 1)^{c_i},$$

$$S_i = \pm 1, a_i, b_i, c_i \in Z \geq 0, \qquad (4)$$

where d is the length of the expansion, we readjust the equation (4) like this:

$$k = \sum_{l_2}^{\max(ci)} (\tau^2 - \mu\tau - 1)^{l_2} \sum_{l_1=1}^{\max(ci)} (\tau - \mu)^{l_2}$$
$$\times \sum_{i=1}^{\max(a_i, l_1, l_2)} \tau^{a_i, l_1, l_2}, \qquad (5)$$

where $\max(a_i, l_1, l_2)$ is the maximal power of τ that is multiplied by $(\tau - \mu)^{l_1}(\tau^2 - \mu\tau - 1)^{l_2}$ in (4).

The theorem in paper [1] shows: every Kleinian integer $\varepsilon = a + b$ can be represented as the sum of at most $O(\log N(\varepsilon)/\log\log N(\varepsilon)) \{\tau, \tau - \mu, \tau^2 - \mu\tau - 1\}$ – Kleinian integers, such that the largest power of both $\tau - \mu$ and $\tau^2 - \mu\tau - 1$ is $O(((\log N(\varepsilon))^\alpha)$ for any real constant α with $0 < \alpha < 1/2$. The detailed proof is given in Paper [1]. Applying this theorem, we describe the steps of computing kP under the condition that the $\{\tau, \tau - \mu, \tau^2 - \mu\tau - 1\}$ –expansion of k is given.

Algorithm: *Point multiplication algorithm using* $\{\tau, \tau - \mu, \tau^2 - \mu\tau - 1\}$ – *expansions.*

INPUT: Kleinian integer ε, a point P on a Koblitz curve, a real constant α with $0 < \alpha < 1/2$

OUTPUT: $Q = \varepsilon P$

1: $P_0^{(1)} = P_0^{(2)} = P$

2: **for** $i = 0, 1, ..., [((\log N(\varepsilon))^\alpha]$

$$P_i^{(1)} = (\tau - \mu)P_{i-1}^{(1)}$$

$$P_i^{(2)} = (\tau^2 - \mu\tau - 1)P_{i-1}^{(2)}.$$

3: **for** $i_1, i_2 = 0, 1, ..., [((\log N(\varepsilon))^\alpha]$

$$Qi_1, i_2 = P_{i_1}^{(1)} + P_{i_2}^{(2)}.$$

4: Compute a $\{\tau, \tau - \mu, \tau^2 - \mu\tau - 1\}$ expansion of the form (5)

5: Apply in succession τNAF based point multiplications based on (5) to compute Q.

3 JOINT THREE DIMENSIONAL FROBENIUS EXPANSION

Dimitrov et al has discussed the advantage of three dimensional Frobenius expansion in computing single scalar multiplication on Koblitz curves in paper [1]. In this paper, we applied three dimensional Frobenius expansion to the calculation of double-scalar multiplication on Koblitz curves and proposed a new algorithm to compute double-scalar multiplication. We call the new algorithm joint three dimensional Frobenius expansion. The following algorithm illustrates the procedure to compute a joint three dimensional Frobenius expansion in $Z[\tau]$ for a pair of Kleinian integers.

Algorithm: Joint three dimensional Frobenius *expansion*

INPUT: A pair of Kleinian integers $\theta_0, \theta_1 \in Z[\tau]$ window size w and precomputed table of optional joint three dimensional Frobenius expansions for all possible pairs of Kleinian integers $\sum_{i=0}^{W-1} u_{0,i}\tau^i$ and $\sum_{i=0}^{W-1} u_{1,i}\tau^i$.

OUTPUT: A pair of lists L_0, L_1 of $\{\tau, \tau - \mu, \tau^2 - \mu\tau - 1\}$ –Kleinian integer representing (θ_0, θ_1)

1: for $i = 0$ to 1 do

2: $L_i \leftarrow \varnothing$

3: compute τ –adic expansion

$$\theta_i = \sum_{j=1}^{1} \theta_{i,j}\tau^j, \quad \text{where } \theta_{i,j} \in \{0,1\}$$

4: **end for**

5: **for** $i = 0$ to $\lfloor 1/w \rfloor$ do

6: find optimal joint three dimensional Frobenius expansion of pair of

$$\sum_{j=0}^{w-1} u_{0, j+iw}\tau^j \text{ and } \sum_{j=0}^{w-1} u_{1, j+iw}\tau^j$$

7: multiply each term by τ^{iw} and add to L_0 or L_1

8: $i \leftarrow i + 1$

9: **end for**

10: return (L_0, L_1)

In the new algorithm, a window size w is fixed before running the algorithm. Before we choosing the optimal joint three dimensional Frobenius expansions, all possible pairs of $w - bit \tau$ – adic representations are precomputed and given as input.

First, we compute two τ – adic expansions $\sum_{i=0}^{1-1} u_i \tau^i$, where $u_i \in \{0,1\}$ and 1 is the length of the longer expansion. Then we arrange the two τ – adic expansions as the form in (6).

$$\begin{pmatrix} \theta_0 \\ \theta_1 \end{pmatrix} = \begin{pmatrix} \theta_{0,1-1} & \cdots & \theta_{0,1} & \theta_{0,0} \\ \theta_{1,1-1} & \cdots & \theta_{0,1} & \theta_{1,0} \end{pmatrix}, \qquad (6)$$

Table 1. Runtime (μS) comparison between τNAF and the new algorithm.

	τNAF	New algorithm	
		w = 5	w = 10
$F_{2^{163}}$	483.5	389.2	374.4
$F_{2^{233}}$	675.2	513.3	486.7

for all $0 \leq i < l$, the ith joint column in (6) has two elements $\theta_{0,i}, \theta_{1,i} \in \{0,1\}$. If one τ – adic expansion is shorter than the other, the higher bits of shorter expansion should be set to zero.

If we choose the window size w, two τ – adic expansions are separated into $\lceil l/w \rceil$ w – bit blocks. The lowest w bits of τ – adic expansions are labelled block 0 and the highest w bits have label block $\lfloor l/w \rfloor$.

The optimal joint three dimensional Frobenius expansion of ith block of θ_0 and θ_1 can be found by a look-up-table approach. Once the ith block of optimal joint three dimensional Frobenius expansion is obtained, all elements are multiplied by τ^{iw} and appended to the relevant lists. We will obtain the complete joint three dimensional Frobenius expansion after repeating this step $\lceil l/w \rceil$ times. Combined with some pre-computations such as P + Q and P – Q, we can effectively reduce the number of point addition and quickly calculate double-scalar multiplication.

4 EXPERIMENT AND CONCLUSION

In this section, we implemented our new algorithm using software. We compare the time performance of the new algorithm and the classical τNAF algorithm. The test platform is Intel Core™2 Duo 2.8GHz CPU and the operating system is Linux.

The running time of the new algorithm and the τNAF algorithm are given in Table 1. For the new algorithm, the running time of different window sizes w = 5 and w = 10 are given. The equation $y^2 + xy = x^3 + x + 1$ is considered over binary field $F_{2^{163}}$ and $F_{2^{233}}$ which are specified by NIST.

From Table 1, our new algorithm has about 20% improvement in speed compared with τNAF under the same condition. The speed of calculation is faster as the window increases, but the size of the look-up-table grows exponentially with the increase of the window size. This means that more storage is required, so the appropriate window size should be selected to balance time performance and space performance.

ACKNOWLEDGMENT

This work was supported by the Science and Technology Support Project of Tianjin under Grant No. 11ZCKFGX00500 and the Science and Technology Support Project of Tianjin under Grant No. 10ZCKFGX03700.

REFERENCES

[1] V.S. Dimitrov, K.U. Jarvinen, M.J. Jacobson, W.F. Chan, and Z. Huang, Provably sublinear point multiplication on Koblitz curves and its hardware implementation, Computers, *IEEE Transactions on, vol. 57, pp. 1469–1481, 2008.*

[2] N. Koblitz, Elliptic Curve Cryptosystems, *Math. Computation, vol. 48, pp. 203–209, 1987.*

[3] J. Solinas, Efficient Arithmetic on Koblitz Curves, Designs, *Codes and Cryptography, vol. 19, pp. 195–249, 2000.*

[4] J. Adikari, V.S. Dimitrov, and R.J. Cintra, A New Algorithm for Double Scalar Multiplication over Koblitz Curves, Proc. IEEE Int'l Symp. *Circuits and Systems (ISCAS'11), pp. 709–712, 2011.*

[5] R.M. Avanzi, V.S. Dimitrov, C. Doche, and F. Sica, Extending Scalar Multiplication Using Double Bases, Proc. 12th Int'l Conf. *Theory and Application of Cryptology and Information Security (Asiacrypt '06), vol. 4284, pp. 130–144, 2006.*

[6] D. Hankerson, Guide to Elliptic Curve Cryptography. Huanguo Zhang, translation. Beijing: Electronics industry Press, 2005.

Information Technology and Applications – Li (Ed.)
© 2015 Taylor & Francis Group, London, ISBN 978-1-138-02677-3

On minimal separation axioms in generalized topological spaces

Hui-Mei Guan & Shi-Zhong Bai

School of Mathematics and Computational Science, Wuyi University, Guangdong, P.R. China

ABSTRACT: In this paper, we introduce and study minimal μ-open sets, maximal μ-open sets, minimal separation axioms and minimal continuous maps in generalized topological spaces and obtain some of their basic properties.

Keywords: generalized topological space; minimal μ-open set; minimal separation axioms; minimal continuous map

1 INTRODUCTION & PRELIMINARIES

In [1], Benchalli S.S., Ittanagi B.M. and Walli R.S. introduced and studied minimal open sets and maximal open sets in topological spaces, which are subsets of open sets. Later, in [2], they also introduced minimal separation axioms in topological spaces. In [4], Roy B. and Sen R. introduced the concepts of maximal μ-open and minimal μ-closed sets via generalized topology. In [5], Xun GE and Ying GE introduced μ-separation axioms in generalized topological spaces.

The purpose of this paper is to study minimal μ-open sets and maximal μ-open sets in generalized topological spaces, which are the generalized notions of minimal open sets and maximal open sets. We then investigate minimal separation axioms, minimal continuous maps and minimal irresolute maps with the help of minimal μ-open sets.

Let's recall some notions needed in the following. Let X be a nonempty set and μ be a collection of subsets of X. Then μ is called a generalized topology [3] (briefly GT) on X iff $\varnothing \in \mu$ and $G_i \in \mu$ for $i \in I$ implies $G = \bigcup_{i \in I} G_i \in \mu$. A set X, with a GT μ on it is said to be a generalized topological space [3] (briefly GTS) and is denoted by (X,μ). μ is called strong if $X \in \mu$ and (X,μ) is called strong GTS. The elements of μ are called μ-open sets and the complements are called μ-closed sets. In [5], μ-T_0 space, μ-T_1 space, μ-T_2 space, μ-regular space and μ-normal space are explained in detail. A GTS (X,μ) is μ-T_1 if and only if the singletons of X are μ-closed [5]. Let μ and μ' be generalized topologies on X and Y, respectively. Then a map $f:(X,\mu) \to (Y,\mu')$ is said to be (μ,μ')-continuous if $M \in \mu'$ implies that $f^{-1}(M) \in \mu$ [3].

2 MINIMAL μ-OPEN SETS

Definition 2.1. A proper nonempty μ-open subset U of a GTS (X,μ) is said to be a minimal μ-open set (briefly m-μ) if any μ-open set which is contained in U is $\varnothing$ or U.

Definition 2.2. A proper nonempty μ-open subset V of a GTS (X,μ) is said to be a maximal μ-open set if any μ-open set which contains V is V or X.

Definition 2.3. A proper nonempty μ-closed subset F of a GTS (X,μ) is said to be a minimal μ-closed set if any μ-closed set which is contained in F is $\varnothing$ or F.

Definition 2.4. A proper nonempty μ-closed subset G of a GTS (X,μ) is said to be a maximal μ-closed set if any μ-closed set which contains G is G or X.

Example 2.5. Let $X = \{a,b,c\}$ and $\mu = \{\varnothing, \{a\}, \{a,b\}\}$. Then (X,μ) is a GTS. It is clear that $\{a\}$ is a minimal μ-open set, $\{a,b\}$ is a maximal μ-open set, $\{c\}$ is a minimal μ-closed set and $\{b,c\}$ is a maximal μ-closed set.

Remark 2.6. Let U be a subset of a GTS (X,μ).

1. U is a minimal μ-open set if and only if $X - U$ is a maximal μ-closed set.
2. U is a maximal μ-open set if and only if $X - U$ is a minimal μ-closed set.

Remark 2.7. Let (X,μ) be a non-strong GTS.

1. maximal μ-open set is the only.
2. minimal μ-closed set is the only.

Theorem 2.8. Let (X,μ) be a non-strong GTS.

1. If A is a maximal μ-open set, then $B \subseteq A$ for each μ-open set B.

2. If A is a minimal μ-closed set, then $B \supseteq A$ for each μ-closed set B.

Proof. This follows directly from the Remark 2.7 and the definitions of strong GTS.

Theorem 2.9. Let (X, μ) be a strong GTS. Then

1. The union of any two different maximal μ-open sets equals X.
2. The intersection of any two different minimal μ-closed sets equals $\varnothing$.

Proof. (1) Let A and B be two maximal μ-open sets and $A \neq B$. Then $A, B \subseteq A \cup B$. If $A \cup B \neq X$, which contradicts the fact that A and B are maximal μ-open sets. Thus $A \cup B = X$.

(2) Analogous to the proof of the (1).

3 MINIMAL SEPARATION AXIOMS

Definition 3.1. A GTS (X, μ) is said to be a minimal μ-T_0 (briefly min μ-T_0) space. If $x, y \in X$ and $x \neq y$, there exists a minimal μ-open set U such that $x \in U$, $y \notin U$ or a minimal μ-open set V such that $y \in V$, $x \notin V$.

It is easy to check that every min μ-T_0 space is a μ-T_0 space. But the converse need not be true. It is shown by the Example 3.2.

Example 3.2. Let $X = \{a, b, c\}$ be with $\mu = \{\varnothing, \{a\}, \{b, c\}, \{a, c\}, X\}$. Then (X, μ) is a μ-T_0 space but it is not a min μ-T_0 space, since $b, c \in X$ with $b \neq c$, there is no minimal μ-open set U containing b but not c neither V containing c but not b.

Definition 3.3. A GTS (X, μ) is said to be a minimal μ-T_1 (briefly min μ-T_1) space. If $x, y \in X$ and $x \neq y$, there exists a minimal μ-open set U such that $x \in U$, $y \notin U$ and a minimal μ-open set V such that $y \in V$, $x \notin V$.

It is easy to check that every min μ-T_1 space is a μ-T_1 space. But the converse need not be true. It is shown by the Example 3.4.

Example 3.4. Let $X = \{a, b, c\}$ be with $\mu = \{\varnothing, \{b\}, \{b, c\}, \{a, b\}, \{a, c\}, X\}$. Then (X, μ) is a μ-T_1 space but it is not a min μ-T_1 space, since $a, c \in X$ with $a \neq c$, there don't exist minimal μ-open sets containing a but not c.

Theorem 3.5. If a GTS (X, μ) is a min μ-T_1 space, then the singletons of X are μ-closed.

The proof is obvious.

Definition 3.6. A GTS (X, μ) is said to be a minimal μ-T_2 (briefly min μ-T_2) space. If $x, y \in X$ and $x \neq y$, there exist minimal μ-open sets U, V such that $x \in U$, $y \in V$ and $U \cap V = \varnothing$.

It is clear that every min μ-T_2 space is a μ-T_2 space. But the converse need not be true. It is shown by the Example 3.7.

Example 3.7. Let $X = \{a, b, c\}$ be with $\mu = \{\varnothing, \{a\}, \{c\}, \{a, b\}, \{b, c\}, \{a, c\}, X\}$. Then (X, μ)

is a μ-T_2 space but it is not a min μ-T_2 space, since $a, b \in X$ with $a \neq b$, there don't exist disjoint minimal μ-open sets U, V such that $a \in U$, $b \in V$.

Definition 3.8. A GTS (X, μ) is said to be a minimal μ-regular (briefly min μ-regular) space. If $x \notin F$ with F μ-closed, there exist minimal μ-open sets U, V such that $x \in U$, $F \subseteq V$ and $U \cap V = \varnothing$.

It is clear that every min μ-regular space is a μ-regular space. From Example 3.7, we obtain that (X, μ) is a μ-regular space but it is not a min μ-regular space, since $b \in X$ and $\{c\}$ is μ-closed, there don't exist disjoint minimal μ-open sets U, V such that $b \in U$, $\{c\} \subseteq V$.

Theorem 3.9. Let (X, μ) be a GTS. If X is a min μ-regular space, then for each $x \in X$ and μ-open set U containing x, there exists a minimal μ-open set V in X such that $x \in V \subseteq c_\mu(V) \subseteq U$.

Proof. Let $x \in X$ and U be any μ-open set in X such that $x \in U$. Then $X - U$ is a μ-closed set in X such that $x \notin X - U$. Since X is a min μ-regular space, there exist minimal μ-open sets V, W in X such that $x \in V$, $X - U \subseteq W$ and $V \cap W = \varnothing$. Then we have

$$V \cap W = \varnothing \Rightarrow V \subseteq X - W$$
$$\Rightarrow c_\mu(V) \subseteq c_\mu(X - W) = X - W,$$

since $X - W$ is a μ-closed set in X. Also we have

$$X - U \subseteq W \Rightarrow X - W \subseteq U.$$

Thus $c_\mu(V) \subseteq U$. Therefore, $x \in V \subseteq c_\mu(V) \subseteq U$.

Definition 3.10. A GTS (X, μ) is said to be a minimal μ-normal (briefly min μ-normal) space. If F and G are μ-closed and $F \cap G = \varnothing$, there exist minimal μ-open sets U, V such that $F \subseteq U$, $G \subseteq V$ and $U \cap V = \varnothing$.

It is obvious that every min μ-normal space is a μ-normal space. From Example 3.7, we obtain that (X, μ) is a μ-normal space but it is not a min μ-normal space, since $\{a\}$ and $\{b, c\}$ are disjoint μ-closed sets, there don't exist disjoint minimal μ-open sets U, V such that $\{a\} \subseteq U$, $\{b, c\} \subseteq V$.

Theorem 3.11. Let (X, μ) be a GTS. If X is a min μ-normal space, then for any μ-closed set F of X and a μ-open set U containing F, there exists a minimal μ-open set V in X such that $F \subseteq V \subseteq c_\mu(V) \subseteq U$.

The proof is similar to Theorem 3.9.

4 OTHER CHARACTERIZATIONS

Definition 4.1. Let μ and μ' be generalized topologies on X and Y, respectively. Then a map $f : (X, \mu) \to (Y, \mu')$ is called minimal continuous map (briefly $(\mu, \mathbf{m} - \mu')$-continuous map) if for every minimal μ'-open set M in Y, $f^{-1}(M)$ is a μ-open set in X.

Definition 4.2. Let μ and μ' be generalized topologies on X and Y, respectively. Then a map $f:(X,\mu)\to(Y,\mu')$ is called minimal irresolute map (briefly $(m-\mu,m-\mu')$-irresolute map) if for every minimal μ'-open set M in Y, $f^{-1}(M)$ is a minimal μ-open set in X.

Remark 4.3. Let (X,μ) and (Y,μ') be the generalized topological spaces. It is easy to obtain the following relationships.

1. If $f:(X,\mu)\to(Y,\mu')$ is a (μ,μ')-continuous map, then it is a $(\mu,m-\mu')$-continuous map.
2. If $f:(X,\mu)\to(Y,\mu')$ is a $(m-\mu,m-\mu')$-irresolute map, then it is a $(\mu,m-\mu')$-continuous map.

Converses of the Remark 4.3 (1) (2) need not be true which follow from the following example.

Example 4.4. Let $f:(X,\mu)\to(Y,\mu')$ be a mapping such that $f(x)=x$ for each $x\in X$. Here $X=\{a,b,c\}$, $Y=\{a,b,c,d\}$ and the corresponding generalized topology be $\mu=\{\varnothing,\{a\},\{a,b\}\}$, $\mu'=\{\varnothing,\{a,b\},\{a,b,c\}\}$.

1. Here $\{a,b\}$ is the only minimal μ'-open set and the inverse image of $\{a,b\}$ is a μ-open set, but $\{a,b,c\}$ is a μ'-open set and the inverse image of $\{a,b,c\}$ is not a μ-open set. So f is $(\mu,m-\mu')$-continuous but not (μ,μ')-continuous.
2. From (1), we obtain that f is $(\mu,m-\mu')$-continuous. Since $\{a,b\}$ is not a minimal μ-open set, f is not $(m-\mu,m-\mu')$-irresolute.

Theorem 4.5. If $f:(X,\mu)\to(Y,\mu')$ is a bijection, $(\mu,m-\mu')$-continuous map and Y is a min μ'-T_0 space, then X is a μ-T_0 space.

Proof. Let $x_1,x_2\in X$ with $x_1\neq x_2$. Since f is a bijection, there exist $y_1,y_2\in Y$ with $y_1\neq y_2$ such that $f(x_1)=y_1$ and $f(x_2)=y_2$. So $x_1=f^{-1}(y_1)$ and $x_2=f^{-1}(y_2)$. Since Y is a min μ'-T_0 space, suppose there exists a minimal μ'-open set M in Y such that $y_1\in M$, $y_2\notin M$. Since f is $(\mu,m-\mu')$-continuous, $f^{-1}(M)$ is a μ-open set in X. Now we have

$$y_1\in M\Rightarrow f^{-1}(y_1)\in f^{-1}(M)\Rightarrow x_1\in f^{-1}(M)$$

and

$$y_2\notin M\Rightarrow f^{-1}(y_2)\notin f^{-1}(M)\Rightarrow x_2\notin f^{-1}(M).$$

Hence, X is a μ-T_0 space.

Theorem 4.6. If $f:(X,\mu)\to(Y,\mu')$ is a bijection, $(m-\mu,m-\mu')$-irresolute map and Y is a min μ'-T_0 space (resp. min μ'-T_1 space, min μ'-T_2 space), then X is also a min μ-T_0 space (resp. min μ-T_1 space, min μ-T_2 space).

The proof is similar to Theorem 4.5.

Theorem 4.7. If $f:(X,\mu)\to(Y,\mu')$ is a bijection, $(\mu,m-\mu')$-continuous map and Y is a min μ'-T_1 space (resp. min μ'-T_2 space), then X is a min μ-T_1 space (resp. μ-T_2 space).

The proof is similar to Theorem 4.5.

Theorem 4.8. If $f:(X,\mu)\to(Y,\mu')$ is a homeomorphism and Y is a min μ'-regular space (resp. min μ'-normal space), then X is a μ-regular space (resp. μ-normal space).

The proof is similar to Theorem 4.5.

Theorem 4.9. If $f:(X,\mu)\to(Y,\mu')$ is a homeomorphism, $(m-\mu,m-\mu')$-irresolute map and Y is a min μ'-regular space (resp. min μ'-normal space), then X is also a min μ-regular space (resp. min μ-normal space).

The proof is similar to Theorem 4.5.

ACKNOWLEDGEMENT

This work is supported by the National Natural Science Foundation of China (No. 61070150) and the Natural Science Foundation of Guangdong Province (No. S2012010008833).

REFERENCES

[1] Benchalli, S.S. & Ittanagi, B.M. & Walli, R.S. 2011. On minimal open sets and maps in topological spaces. *J.Comp. and Math. Sci.* 2 (2): 208–220.
[2] Benchalli, S.S. & Ittanagi, B.M. & Walli, R.S. 2012. On minimal separation axioms in topological spaces. *Journal of Advanced Studies in Topology* 3 (1): 98–104.
[3] Csaszar, A. 2002. Generalized topology, generalized continuity. *Acta Math. Hungar.* 96 (4): 351–357.
[4] Roy, B. & Sen, R. 2012. On maximal μ-open and minimal μ-closed sets via generalized topology. *Acta Math. Hungar.* 136 (4): 233–239.
[5] Xun, G.E. & Ying, G.E. 2010. μ-Separations in generalized topological spaces. *Appl. Math. J. Chinese Univ.* 25 (2): 243–252.

Information Technology and Applications – Li (Ed.)
© 2015 Taylor & Francis Group, London, ISBN 978-1-138-02677-3

Analytical study on the vulnerability scanning and tools

Nafei Zhu, Yue Zhou, Li Lou & Wei Wang
National Application Software Testing Labs, Beijing, China

ABSTRACT: Vulnerability scanning can be done on the principles of information matching based on vulnerability database and simulating attacks. Practically, vulnerability scanners implement the principles in network-based and host-based scanning manners. Also, the differences between vulnerability scanning and penetration testing can be induced, as a vulnerability scanning is the process to find out the potential security issues, whereas a penetration test actually performs exploitation to prove that a security issue exists. Furthermore, penetration tools usually contain the more functions than the port scanner and vulnerability scanner, to do a further work on the basis of them. Lastly, we show some famous security testing tools with their purposes and methods in detail.

Keywords: vulnerability scanning; penetration testing; vulnerability scanner; security testing

1 DEFINITION OF VULNERABILITY SCANNING

Vulnerability is a weakness in the software or system configuration that can be exploited. Vulnerability scanning is the inspection for information system to find out the loopholes which can be used by hackers. Vulnerability scanning is actually a part of the system security assessment even for cloud security and it points out that what the possible attacks are faced by the system [1].

2 PRINCIPLES OF VULNERABILITY SCANNING

The principles of vulnerability scanning are generally referred as two types which are information matching based on vulnerability database and simulating attacks.

2.1 *Based on vulnerability database*

For the information matching based on vulnerability database scanning, the vulnerability database is the basement. We should setup the database according to the knowledge of the previous hack attacks, system configuration as well as the system loopholes. The database should contain the loopholes with the system configuration and the detecting methods, which are referred to as match pattern. The scanning is completed by the program in terms of loopholes and their match patterns. So the scanning accuracy is decided by the completeness and validity of the vulnerability database.

2.2 *Based on the simulated attacks*

Plug-in is a subprogram module written by a scripting language and can be called by the scanner to complete the detection. Adding new plug-ins means enabling the scanner to detect more vulnerability, increasing the identified types and number, and upgrading the characteristics of the vulnerabilities to make the scanning results more accurate. With the Plug-in technology, vulnerabilities scanning software upgrading becomes somewhat simple and its expansibility becomes strong.

2.3 *Summary*

At present, vulnerability database mainly includes the CGI, POP3, FTP, SSH, HTTP vulnerabilities and so on [2][3][4]. So these vulnerabilities can be detected using the scanner based on vulnerability database. While other vulnerabilities, such as the Unicode traversal directory vulnerability, FTP weak passwords, Open-Relay mail forwarding loopholes, need the simulating attacks, which called plug-ins to detect, for there are no such items in the vulnerability database. Practically, the scanner is based on the vulnerability database with the plug-in function. And we refer to such scanner in the following paper when we talk about the vulnerability scanner.

3 METHODS OF VULNERABILITY SCANNING

Aside from the two principles related in 2.1 and 2.2, there are also two methods to implement the scanning principles. Based on the underlying technologies, the two methods are network-based vulnerability scanning and host-based vulnerability scanning. That is to say there are two types of vulnerability scanners, which are different in the underlying architecture, even though they use the same scanning principle.

3.1 Network-based vulnerability scanner

Based on B/S architecture, network-based vulnerability scanner scans the vulnerabilities of the remote computer through the network. This scanner can complete network mapping and port scanning.

The scanning procedure of network-based vulnerability scanner can divided into 5 steps. Firstly, it will detect the online host, and then find out the open ports. Furthermore, it will identify what services the ports provide as well as the service editions. At the same time, the scanner identifies the kind of operating system on the host. At last, the scanner finds the possible vulnerabilities matching the information got by the four steps ahead and sends the detecting packets to the target. When it gets the replies, the scanner will analyze these responds and decide whether they belong to some known vulnerabilities.

Generally, Network-based vulnerability scanner has 5 parts and its architecture is shown in Figure 1.

First and the fundamental parts is the vulnerability database module, which includes the vulnerabilities for all kinds of operating systems as well as the detection instructions. We need to upgrade the database to contain the newest vulnerabilities.

The second part is the user configuration console module: security administrator sets up the target system and the vulnerabilities for a scanning through the configuration console.

The third part is the knowledge base module for current scanning activity. By examining the

configuration in the memory, this module monitors the scanning activities, provides relevant information to the scanning engine, and receives scanning results returned by the scanning engine.

The fourth part is scanning engine module, which is the major part of the scanner. According to the user configuration from the user configuration console module, scan engine assembles corresponding packets and sends to the target system. When it receives the reply of the target system, the scanner will compare the reply with the vulnerability features in the vulnerability database to judge whether the target contains this vulnerability.

The fifth part is results storage and report generation tool. Report generation tool generates scan report using the scanning results stored in the knowledge base. Scanning report will report the user configuration as well as the scanning results, what vulnerabilities are found, for such configuration.

3.2 Host-based vulnerability scanning

The principle of host-based vulnerability scanning is similar to that of the network-based scanning, which is information matching based on vulnerability database and simulating attacks related in part 2. But the structure of the host-based scanning is C/S which is different from B/S of network-based scanning. In host-based scanning, the target is installed an agent or service to access all the files and processes, thus to find more loopholes in the target.

Host-based vulnerability scanner is usually Client/Server three-layer architecture. The three layers are the vulnerability scanner console, a vulnerability scanner manager and a vulnerability scanner agent. Its architecture is shown in Figure 2.

The workflow of host-based vulnerability scanner is related as following.

The first step is installation. Vulnerability scanner console is installed in a computer, while

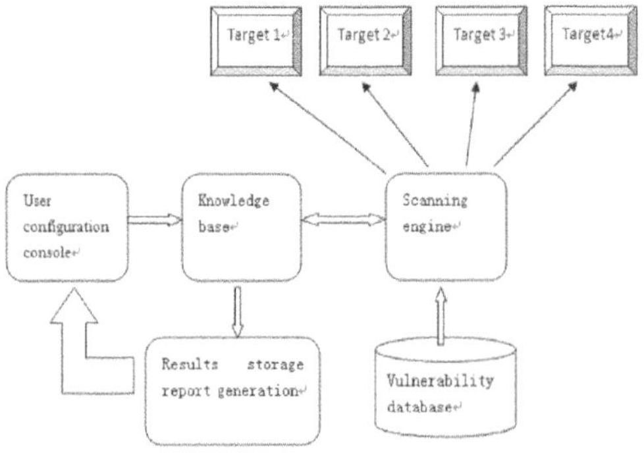

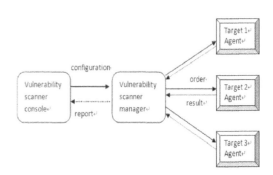

Figure 1. Architecture of the network-based vulnerability scanner.

Figure 2. Architecture of the host-based vulnerability scanner.

vulnerability scanner manager is installed in the enterprise network. All the target systems need to install the vulnerability scanner agents, which need to register to the vulnerability scanner manager.

The second step is scanning. Vulnerability scanner agent scans the target system when received the command from vulnerability scanner manager.

The third step is reporting. After the scanning is completed, vulnerability scanner agent will report the result to the vulnerability scanner manager, and then the end users can read the results through the vulnerability scanner console.

4 DIFFERENCES BETWEEN VULNERABILITY SCANNING AND PENETRATION TESTING

4.1 *Vulnerability scanning vs. penetration testing*

Penetration testing can be defined as a legal and authorized attempt to locate and successfully exploit computer systems for the purpose of making those systems more secure. The process includes probing for vulnerabilities as well as providing proof of concept attacks to demonstrate the vulnerabilities are real [5].

Depending on literatures, the process of penetration testing usually contains between four and seven steps or phases. Although the overall names or number of steps can vary between methodologies, the important thing is that the main steps are almost the same, providing a complete overview of the penetration testing process. The main steps are: reconnaissance, scanning, exploitation, and maintaining Access [6].

It is important to tell the difference between penetration testing and vulnerability assessment. A vulnerability assessment is the process of reviewing services and systems for potential security issues, whereas a penetration test actually performs exploitation and POC (Proof of Concept) attacks to prove that a security issue exists [5].

The first difference is that besides locating the vulnerability as the vulnerability assessment does, the penetration testing also exploits the vulnerability, escalates the privilege and keeps controlling the system. That is vulnerability assessment only shows there are vulnerabilities in the system, but it cannot tell what the affects of the vulnerabilities are.

The second difference is that penetration test is much more aggressive than vulnerability assessment. Penetration test simulates hacker activity and delivers live payloads to the target, while vulnerability assessment just locates and quantifies the vulnerabilities in a non-aggressive manner.

Third, the process of vulnerability assessment is one of the steps utilized to complete a penetration test. So penetration test costs more than the vulnerability assessment does.

4.2 *Port scanner vs. vulnerability scanner vs. penetration tool*

As for tools, the ports scanning, vulnerability scanning and penetration have their own tools, respectively. Ports scanner just finds out what ports are open on the target. Vulnerability scanner detects the loopholes on the target and does not provide the steps to get into the target. Tools of Penetration will exploit the loopholes to get into the target and execute remote code (payloads) on the target to control it. The differences among the three types of tools are shown in Table 1. From the table, we can get that penetration tools contains more powerful function than the vulnerability scanner and it does a further work on the basis of the work of Vulnerability scanner. It is the same to the relation between vulnerability scanner and port scanner.

5 TOOLS FOR SECURITY TESTING

For a scanning tool, availability and accuracy as well as controllability are important. The availability

Table 1. Port scanner vs. vulnerability scanner vs. penetration tool.

Tools	Port scanner	Vulnerability scanner	Penetration tool
Host scanning	√	√	√
TCP\UDP ports scanning	√	√	√
OS detection	√	√	√
Service identification	√	√	√
Vulnerability database	\	√	√
Vulnerability detection	\	√	√
Exploit code database	\	\	√
Vulnerability exploiting	\	\	√

Table 2. Tools for security testing.

Name	Purpose	Method
Nmap	Host scanning, port scanning, service identification, OS detection, Firewall detection	It uses raw IP packets in novel ways to determine what hosts are available on the network, what services those hosts are offering, what operating systems they are running, and what type of packet filters/firewalls are in use, even in secret manners.
FPort	Port scanning	It reports all open TCP/IP and UDP ports and maps them to the owning application with process ID, name and path.
SuperScan	Port scanning	It can do ping scanning, port scanning, OS detection, and Trojan detection using a Trojan ports list.
Hping	Host scanning, port scanning, OS detection, firewall detection	It sends customized ICMP/UDP/TCP packet to the destination address and show the response.
Xprobe 2	Port scanning, OS detection	With a detailed OS characteristics database, Xprobe has the reliability to OS detection.
THC-Amap	Application protocol detection, OS detection	By sending trigger packets and looking up the port response in the application fingerprint database to identify applications and services.
Friendly Pinger	Network monitor/pin	It uses friendly interface to track all the network devices availability over the network and displays all the device of the computer as small pictures on a screen.
Nessus	Vulnerability scanning	Combined with the world's largest continuously updated library of vulnerability and compliance checks, it is excellent in vulnerability scanning in depth and coverage of systems.
OpenVAS	Vulnerability scanning	The actual security scanner is accompanied with a daily updated feed of Network Vulnerability Tests (NVTs). Its core component is a server, including a set of network vulnerability testing procedures, can detect the security problems in the remote system and the application.
X-scan	Vulnerability scanning	In multithread manner, it detects vulnerabilities for CGI, IIS, RPC, SQL server, FTP server, SMTP server, POP3 server, NT server, and weak password and so on. It also supports plug-in function.
Nikto	Web server scanning	It performs generic and server type specific checks to find out dangerous files/CGIs, outdated server software and other problems, based on Whisker/libwhisker.
Metasploit	Vulnerability exploiting	It integrates the common exploit and popular shellcodes of various platforms, and constantly updates them. It provides an effective way to automatically detect and validate the vulnerabilities.

is the most important one among the three. There are a lot of such scanning tools or security testing tools for different purposes. We list some famous ones in Table 2, with their names, purposes and working methods.

6 CONCLUSION

Vulnerability scanning is actually a part of the system security assessment and it points out that what the possible attacks are faced by the system. There are usually two principles for vulnerability scanning, which are information matching based on vulnerability database and simulating attacks. In practice, the scanner is based on the vulnerability database with the plug-in function as well. At the same time, based on the underlying technologies, vulnerability scanners work in two different ways, which are network-based vulnerability scanning and host-based vulnerability scanning. The differences between vulnerability scanning and penetration testing are studied and can be induced, as a vulnerability assessment is the process to find out the potential security issues, whereas a penetration test actually performs exploitation to prove that a security issue exists. Furthermore, the comparisons among port scanner, vulnerability scanner and penetration tools are made here. Lastly, we show some famous security testing tools with their names, purposes and working methods in detail. Vulnerability scanning is the effective ways to keep the security of the information system and also an indispensible basement for further penetration testing and other security testing.

ACKNOWLEDGEMENT

This research was supported by the Research and Demonstration on the Information Technology Specifications for the Operation Dispatching Command System of the Urban Rail Transit (D141100000714003).

REFERENCES

[1] Chung, C.J., Khatkar, P., Xing, T.Y. & Lee, J. 2013. NICE: Network Intrusion Detection and Countermeasure Selection in Virtual Network Systems. *IEEE Transactions on Dependable and Secure Computing* 10(4):198–211.

[2] Qu, B. & Yang, Z.X. 2012. Design of Embedded Secure CGI Daemon. *IEEE 3rd International Conference on Software Engineering and Service Science (ICSESS)*, 22–24 June, 725–728.

[3] Shukla, R., Prakash, H.O., Bhushan, R.P. & Venkataraman, S. 2013. Sahastradhara: Biometric and EToken Integrated Secure Email System. *15th International Conference on Advanced Computing Technologies (ICACT)*. 21–22 Sept:1–4.

[4] Cinti, A. & Rizzi, A. 2013. FPGA Targeted Implementation of a Neurofuzzy System for Real Time TCP/IP Traffic Classification. *Sixth International Conference on Advanced Computational Intelligence (ICACI)*, 19–21 Oct:312–317.

[5] Engebretson, P. 2011. *The Basics of Hacking and Penetration Testing*, First ed, Elsevier. Waltham.

[6] Zhu, N.F., Zhou, Y., Wang, W. & Lou, L. 2014. Study on the Standards and Procedures of the Penetration Testing. *International Conference on Computer Science and Network Security*.

Information Technology and Applications – Li (Ed.)
© 2015 Taylor & Francis Group, London, ISBN 978-1-138-02677-3

Research of multi-user traffic identification technology on mobile Internet

Rui Li & Zhizhong Zhang
Key Laboratory on Communication Networks and Testing Technology,
Chongqing University of Posts and Telecommunications, China

ABSTRACT: This paper presents a traffic identification scheme based on the data of WCDMA network to do research of user behaviors on mobile Internet. The scheme uses an enhanced hash algorithm to improve the efficiency of data stream pre-process and uses the Deep Packet Inspection (DPI) technology to increase the accuracy of traffic identification. In this paper, we described pre-process to decode messages and generate Call Detail Records (CDR). Then, we described real time identification and elaborate identification in DPI identification process to identify the CDR and do the statistic of traffic. The statistical result of identification process is the basis for further user behaviors analyzing. The scheme has been verified in data mining system and it performs efficiently and accurately in real network data testing.

Keywords: mobile Internet, traffic identification, data mining, hash algorithm, DPI technology

1 INTRODUCTION

Currently, mobile Internet is one of the hot topics in the field of information technology. It reflects the ubiquitous network, omnipotent business; and it changes the way people live and work. Nowadays, the number of users get attached to the Internet via mobile terminals has exceeded that of those who use desktop computers. To identify and count the traffic accurately, provide new monitoring programs targetedly and improve the quality of service have become urgent problems for network operators to solve.

As the result, we select the original data from Iu-PS interface in WCDMA network to analyze the traffic of mobile Internet. This interface contains both signalling connection part and service connection part. When people surf the Internet, the two parts of this interface are created by core network. The signalling connection part contains the users' identity information such as IMSI, IMEI, MCC, MNC; and the service connection part includes the user's behavior information. Furthermore, combining DPI technology with the feature database technology, we can accurately identify the user traffic. Hence, the research of traffic identification technology on this interface is of great importance in the field of mobile Internet.

2 OVERVIEW OF IU-PS INTERFACE

Iu-PS interface is the PS domain interface connecting UTRAN and CN, which can also be seen as a reference point between SGSN and RNC. Iu-PS interface includes both signalling and service planes. Both plane consist of wireless network layer protocol and transport layer protocol.

Service plane in Iu-PS interface using GTP-U protocol implements the data transfer between UTRAN and SGSN entities. GTP-U protocol carried by UDP protocol can transmit any user data with IPv4, IPv6 and PPP protocol format. And signalling plane provides reliable transmission by using the SCTP (Stream Control Transmission Protocol). The SCTP protocol carries M3UA (MTP 3 User Adaptation) protocol, SCCP (Signal Connection Control Protocol) protocol, RANAP (Radio Access Network Application Part) protocol, GMM (GPRS Mobility Management) or SM protocol.

3 OVERALL DESIGN AND ANALYSIS OF SYSTEM STRUCTURE

According to the demand of relevant specification, traffic identification scheme of multi-user in mobile Internet ought to complete the functions as follows: data collection; pre-processing process of data; real time identification; elaborate identification. The framework design of traffic identification is showed in Figure 1.

As mentioned above, pre-process module can be divided into decoding part, composing part and associating part. Raw data captured by Ethernet card is stored in message buffer after the message

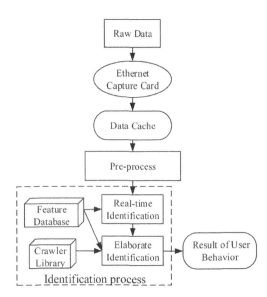

Figure 1. Overall system structure.

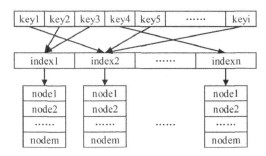

Figure 2. Bucket sort algorithm model.

filtering of Iu-PS interface. Decoding part gets message from message buffer, and then decode each protocol by level, in order to achieve call information of every message. Composing part classifies messages by signal process and service process, and links these messages together with their index. Associating part combines signal processing result and service processing result into a new CDR, in order to complete advanced process.

According to this CDR, real-time identification process will identify application type with the support of feature library. And elaborate identification process will identify the specific behavior of users by using crawler library. Meanwhile, statistic module finishes the statistics of user behaviors. These modules accomplish the relevant senior functions and save the result in the disk as files.

4 HASH BUCKET SORT ALGORITHM

In order to have a high efficiency on hash key values processing, this scheme adopts a hash bucket sort algorithm model based on the remainder of division[2]. The model to get index value from remainder of division is shown in Equation 1.

$$A_i = (H_j(x) + d_k)\%n \ (i,j,k = 1,2,...,m \,(m \leq n-1))$$
(1)

Bucket sort algorithm model is shown in Figure 2.

Each hash bucket mounts multiple nodes with the same index number. These nodes store a pointer

which points to the node structure corresponds to its address. And the head of the structure stores the key value of the node. Every time, messages come with the same index value of data, we should check whether the nodes of the index exist or not at first. If the nodes do not exist, it needs to create a new node in the bucket. And, when the nodes exist, it will add the node at the back of bucket directly. Using the above method can completely avoid the conflict of the node address. And combining with appropriate hash bucket number, it can effectively improve the efficiency.

5 DATA PRE-PROCESSING PROCESS

Pre-processing process, Figure 3, is the fundament of traffic identification scheme.

This process gets the messages from data cache, and recognizes protocol type beyond IP layer. Then it sends the messages with UDP protocol to service data decoder, while the messages with SCTP protocol are sent to signalling data decoder.

When the data entered, the decoder will identify the IE (information element) and query the protocol ID, then decode the protocol of each message. When a layer decoding finished, it will query the protocol ID of upper layer and call the function to decode. If remained only one protocol, then call the call-back function interface and update call information structure.

According to the call information structure, the same kinds of message need to have the same key characteristic value in composing process part. On signalling plane, we select OPC (Originating Point Code) and DPC (Destination Point Code) on M3UA layer; SLR (Source Local Reference) and DLR (Destination Local Reference) on SCCP layer as calculation parameters, Equation 2.

$$key = (OPC << 16 + SLR) \| (DPC << 16 + DLR) \quad (2)$$

On service plane, we select src_port (Source Port) and dst_port (Destination Port) on UDP layer;

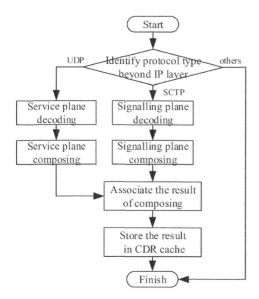

Figure 3. Pre-processing process.

src_ip (Source IP) and dst_ip (Destination IP) on IP layer as calculation parameters, Equation 3.

$$key = (src\_port \ll 32 + src\_ip)$$
$$\| (dst\_port \ll 32 + dst\_ip) \qquad (3)$$

When synthesizer gets the data from call information structure, it will identify the message type at first. Then it create the CDR structure of the message flow when the starting message comes such as CR (Connection Request) message of SCCP layer on signalling plane and HTTP request message on service plane. When a message comes next and exists the same key value, it modifies the properties of CDR structure and adds it to CDR cache. When the complete message comes such as RLC (Release Completion) message, HTTP response message comes or the message timeout, we finish the CDR composing process and save the CDR to the cache. At last, we should delete the key value in hash table and release the space of the node.

To associate CDR, we choose the IP address on third layer as the associating fields, Equation 4.

$$key = layer3\_ip\_addr \qquad (4)$$

For this part, it puts the signalling CDR and service CDR into the memory at first, then it corresponds the CDR ID with key value and stores the value in the hash table by using hash bucket sort algorithm. If the key value of signalling CDR is same to the one of service CDR, then it will generate a new CDR including both signalling and

service CDR information and delete the key values. At last, it stores the CDR in CDR cache. Through the third IP layer address, it can efficiently associate with identity information on signalling plane and behavior information of the users on service plane. And it will lay the foundation of analyzing the specific behavior of a particular user.

6 DPI IDENTIFICATION PROCESS

For further research on user traffic, accurately and efficiently identifying user behavior appears to be very important. Faced with big data, the method to deal with the data based on the feature database is adopted in this scheme. To ensure the accuracy of the data, the feature database can be updated regularly.

DPI identification process includes two parts. One is real time identification process. It can identify the application type, website name, packet number, and the traffic of data stream. The other is elaborate identification process. It can distinguish the behavior of users when they surfing the Internet, such as the novel they read, the video they watched, the key word they searched and so on.

6.1 *Real time identification process*

In this process, we obtain the user accessing characteristics by using triple IP key values, Equation 5, and quintuple IP key values, Equation 6. We select IP (Destination IP), Port (Server Port), Bear_Protocol (Bearing layer protocol) as the component of triple IP key values; and quintuple IP key values consist of Src_IP (Source IP), Dst_IP (Destination IP), Src_Port (Source Port), Dst_Port (Destination Port) and Bear_Protocol (Bearing layer protocol).

$$3\,key = ip\ \&\ port\ \&\ bear\_protocol \qquad (5)$$

$$5\,key = src\_ip\ \&\ dst\_ip\ \&\ src\_port\ \&$$
$$dst\_port\ \&\ bear\_protocol \qquad (6)$$

Real time DPI process, Figure 4, needs the support of triple key values feature database, quintuple key values feature database and flow characteristics feature database. Triple key feature database includes main application type, application subtype, server address and server port; it needs to manually configure and regularly update. Flow characteristics database includes primary classification, secondary classification, packet type, port, method type, host address, original address, resource identifier, user agent, carrying key, third layer IP and extra fields; it needs to update regularly and automatically. Quintuple key feature database as an index feature database, includes all information of triple key feature and the type of each host which stored by machine learning.

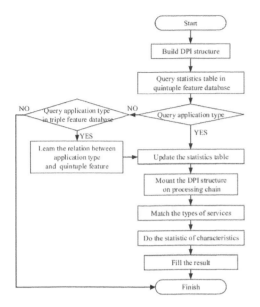

Figure 4. Real time identification process.

crawled from website by crawlers automatically and they can be updated regularly. The matching process using regular expressions which corresponds the website can match the visiting URL of website. Part of the regular expression in database in suffix feature library is showed in Figure 5.

After the real time identification process, elaborate identification process will extract the domain name at first. Then, with the domain and website type, it can search the analytical regular expression and website name in the table.

If the result is the searching website, it will use the analytical regular expression to match the URL. After that it can get feature value of URL, and then it will output the decoding result of URL. For the other kind of website, after the decoding of URL, it will query characteristic details in crawler library by searching the feature value and then it will output the result of details to the cache. In this way, it can identify the specific behavior of user accessed to the Internet.

At last, it does the statistics of matching results, and outputs the results as CSV file in the disk every

When the CDR comes, it builds the structure of key values at first. Then, it searches the application type in statistic table and query 5key values in quintuple feature database. If the key value exists and the type is available, the statistic table of packet number should be added and the amount of flow should be updated. While the key value does not exist, we should query the triple feature database and build the relationship mapping between application type and the quintuple feature of these key values. At last, we need to match the type of message, to do the statistic of the remaining characteristics and to fill the result.

When all fields match successfully, the message matching is finished; and the quintuple feature will be credited in DPI database for subsequent use.

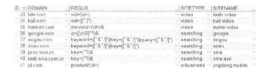

Figure 5. Regular expression table correspond the host (Part of the screenshot).

6.2 Elaborate identification process

Due to the large amount of data and limited computer processing ability, the real time identification process can only do the original identifying. Further details, such as the detailed content accessed to the Internet and the behavior of users require elaborate identification process to identify. With the support of powerful suffix feature library in feature database and crawler library, which includes the specific information corresponding to the website details such as the title and chapter on reading website, the channel and articles on portal website, and so on, we can accurately match for user behaviors. Suffix feature library is built in the way of manual analysis and crawler library is

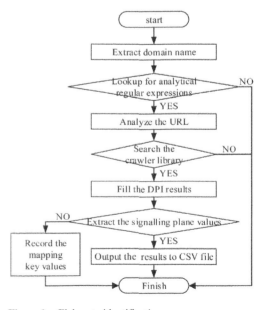

Figure 6. Elaborate identification process.

once in a while. Elaborate identification process is shown in Figure 6.

ID	UPDATET	NAME	TYPE	SUBTYPE	AUTHOR
11618	2014/4/24 1	Super Weapon	Science Fiction	Future World	Xuefei Yan

Figure 9. Detail of value in crawler library (Part of the screenshot).

7 RESULT AND ANALYSIS OF THE IDENTIFICATION

Through the above process, we can get the following matching results, Figure 7. In this figure, IMSI, IMEI, MCC and MNC fields belong to the signalling plane while the HOST, URL and some other fields belong to the service plane.

From the third line, we can see the message host is "www.google.com", the site type is "searching", and the regular expression of "www.google.com" can be found in Figure 5 as "q=([\s\S]*?)&". The valid part of the URL is "/uds/afs?q=jewellery&client=mobile-zhaopin-58&hl=zh_CN&r=m&adstyle=hm&". Through the matching of regular expression, the result of characteristic value is "jewellery". At the same time, the result of URL in browser identification also shows the searching for "jewellery". Therefore, the result of matching can be proved correctly.

From the ninth line, the host is "m.qidian.com", and the site type is "reading". The URL is "http://m.qidian.com/book/bookchapterlist.aspx?bookid=11618", and the regular expression of "m.qidian.com" can be found in regular expression table as "BookId=(\d+)|Bookid=(\d+)|bookId=(\d+)|bookid=(\d+)", Figure 8.

Through the matching of regular expression, the matching result of characteristic value is "11618", and then process will search this value in crawler library. The details of the value "11618" in crawler library is showed in Figure 9.

The result matches the part of DPI1 to DPI4 in the ninth line of CSV file. After we put this URL into mobile browser, the identification result turn out to be the same. And the result can be also proved correctly.

Figure 7. Part of result of identification in CSV file (Part of the screenshot).

ID	DOMAIN	REGUX	SITETYPE	SITENAME
118	book.qq.com	bid=(\d+)	reading	tencent reading
119	m.qidian.com	BookId=(\d+)|Bookid=(\d+)|bookId=(\d+)|bookid=(\d+)	reading	qidian reading

Figure 8. Regular expression table correspond the host (Part of the screenshot).

8 CONCLUSION

User behavior analysis is one of the most popular researches in mobile Internet. An efficient solution is designed by this paper to identify the user behavior. And it analyses the specific scheme of pre-processing process and identification process to recognize the traffic of users, which has certain reference value for the further development of user behavior analysis. At last, the identifying solution has been tested and simulated in real network. The testing results show that the program are stable and reliable, effective and feasible, which meets identifying requirements completely.

ACKNOWLEDGMENT

It is supported by the National Science and Technology Major Project (next-generation broadband wireless mobile communication network: 2012ZX03005008-004, 2012ZX03001021-004), the National High Technology Research and Development Program of China ("863" Program: 2014AA01A706), and the university innovation team in Chongqing (2013).

REFERENCES

[1] Luo, J.Z., Wu, W.J. & Yang, M. 2011. Chinese Journal of Computers. *Mobile Internet: Terminal Devices, Networks and Services* 34(11): 2029–2051.
[2] Yang, L., Huang, H. & Song, T. 2005. Journal of Software. *The sample separator based distributing scheme of the external bucket sort algorithm* 16(5): 643–651.
[3] Chen, T.S., Chou, Y.S. & Chen, T.C. 2012. IEEE Transactions on Systems, Man, and Cybernetics. *Mining user movement behavior patterns in a mobile service environment.* 42(01): 87–101.
[4] Zhao, G.F., Lai, W.J., Xu, C. & Tang, H. 2013. Chinese Journal of Computers. *Revealing Service Visit Characteristics in Mobile Internet* 36(07): 1388–1398.
[5] Cheng, F., Zhang, Z.Z. & Xie, J.F. *Mobile communication system evolution and 3G signal,* Beijing: Publishing House of Electronics Industry.
[6] Duan, H.G., Bi, M. & Luo, Y.J. 2007. *WCDMA the third generation mobile communication system protocol system and signal process.* Beijing: The Posts and Telecommunications Press.
[7] TS 25.413 V11.4.0. 2013. *3rd Generation Partnership Project; Technical Specification Group Radio Access Network; UTRAN Iu interface Radio Access Network Application Part (RANAP) signalling.* 3GPP.

Information Technology and Applications – Li (Ed.)
© 2015 Taylor & Francis Group, London, ISBN 978-1-138-02677-3

Performance analysis of two-way AF cooperative networks with relay selection over Nakagami-m fading channels

Chen-Hong Yao

State Key Laboratory of Integrated Services Networks, Xidian University, Xi'an, China
Xi'an University of Architecture and Technology, Xi'an, China

Chang-Xing Pei & Jing Guo

State Key Laboratory of Integrated Services Networks, Xidian University, Xi'an, China

ABSTRACT: In this letter, we propose a novel two-way Amplify-and-Forward (AF) relay selection scheme over independent flat Nakagami-m fading channels. Two sources select the "best" relay for each user, which brings significant gains and can be easily implemented. The performance of the proposed system is quantified by deriving the lower bound of the overall outage probability, high SNR approximation and Diversity-Multiplexing Trade-off (DMT). Simulation results are presented to verify the theoretical analysis.

Keywords: two-way relaying network; Amplify-and-Forward (AF); outage probability; Diversity-Multiplexing Tradeoff (DMT); Nakagami-m fading

1 INTRODUCTION

Two-way relaying system has attracted much research interest because of its potential in achieving higher spectral efficiency and throughput for wireless networks [1]. Based on different processing at the relay, such as decode-and-forward and amplify-and-forward, achievable rate regions for various two-way relaying systems have been studied in [2] [3]. In [4], the authors investigated the performance of two-way amplify-and-forward relaying networks over independently but not necessarily identically distributed Nakagami-m fading channels in terms of outage probability, average SER, and average sum-rate. Adaptive two-way relaying, which the relay R adaptively selects AF or DF depending on the decodability of the two bi-directional data streams received at R, has been introduced in [5]. Because of synchronization and a lot of messages exchanging among different relays, a proper relay selection scheme will play a key role in the multiple relay system performance. In [6], the authors derived a simple two-way amplify-and-forward relay selection criterion for the general case over Nakagami-m fading channels. In [7], the authors proposed the max-min sum rate selection algorithm for AF bidirectional network based on the outage probability. Two-way relay selection, which is introduced for differential modulation systems to improve system performance, has been proposed in [8].

One of the key features in the two-way relaying networks is that the two-way protocol should not only consider the downlink (broadcasting) but also the uplink (multiple-access) data flows in the same time, which can be improve the overall performance of the two directional transmission. However, most of the recent research force on the "best" single relay selection methods, which means the "best" relay for one directional transmission is probably not the "best" one for the other directional transmission.

Motivated by all of the above, this letter presents a novel two-way relay selection scheme and we investigate the performance of the proposed system over independent flat Nakagami-m fading channels. Firstly, the lower bound of the overall outage probability for the proposed system is investigated. And then, the high SNR approximations of outage probability and Diversity-Multiplexing Trade-off (DMT) are determined. Finally, we make an outage performance comparison between the proposed scheme and the "best" single selection criterion. Simulation results verify our analysis.

2 SYSTEM MODEL

Consider a two-way Amplify-and-Forward (AF) relaying system, where two sources S_1 and S_2 exchange information with the help of N cooperating AF relay nodes k, $k \in \{1,...,N\}$, over independent flat Nakagami-m fading channels. We assume that all terminals are single-antenna devices and operate in a half-duplex mode, and there is no direct path between S_1 and S_2. We also assume here the Additive White Gaussian Noise (AWGN) at all nodes is independent and identically distributed (i.i.d.) $C\mathcal{N}(0, N_0)$.

In the first time slot, S_1 and S_2 transmit their signals x_1 and x_2 to all the relays simultaneously. So the received signal at relay k can be expressed as

$$y_k = \sqrt{P_1} h_{1k} x_1 + \sqrt{P_2} h_{2k} x_2 + n_k \qquad (1)$$

where P_1 and P_2 are the transmitting power of S_1 and S_2. h_{1k} and h_{2k} stand for the channel gains of $S_1 \to k$ and $S_2 \to k$, respectively.

In the next two time slots, one or two out of N relays are selected for amplifying its received signal and forwarding it. Let i and j denote the selected relays for two source S_1 and S_2, respectively, which could be the same relay or different relays. Thus, in the second time slot, the received signal at relay i is amplified with the fixed-gain, $G = \sqrt{1/CN_0}$, whereby C is a constant, and retransmitted to S_1 while S_2 keeps silent. Then, in the last time slot, the received signal at relay j is also amplified with the fixed-gain, G, and retransmitted to S_2 while S_1 keeps silent. After the self-interference parts are subtracted [1], the received SNRs of two-links $S_2 \to i \to S_1$ and $S_1 \to j \to S_2$ are given by

$$\gamma_{1i} = \frac{P_2 \gamma_{g_{i1}} \gamma_{h_{i2}}}{\gamma_{g_{i1}} + C} \qquad (2)$$

$$\gamma_{2j} = \frac{P_1 \gamma_{g_{j2}} \gamma_{h_{j1}}}{\gamma_{g_{j2}} + C} \qquad (3)$$

respectively, where $\gamma_{g_{is}} = |g_{is}|^2 / N_0$, $\gamma_{h_{is}} = |h_{is}|^2 / N_0$, $s = 1,2$ and g_{i1} and g_{j2} stand for the channel gains of $i \to S_1$ and $j \to S_2$, respectively.

Since the overall performance of the considered system is governed by the performance of the weakest source, let $\gamma_{ij} = \min(\gamma_{1i}, \gamma_{2j})$. Thus, the relay selection criterion can be expressed as

$$(i^*, j^*) = \arg \max_{i \in \{1,...,N\}} \max_{j \in \{1,...,N\}} \{\gamma_{ij}\} \qquad (4)$$

Note, we can see that the optimal selection criterion can be only implemented in a centralised manner, in which there are too many information exchanges between terminals especially for lots of relays and that requires an exhaustive search of all possible combinations to determine the optimum subset. In order to reduce the complexity of the optimal selection criterion, we can rearranged the indexes i and j. Hence, the new selection criterion can be rewritten as

$$(i^*, j^*) = \arg\min\left\{ \max_{i \in \{1,...,N\}} \gamma_{1i}, \max_{j \in \{1,...,N\}} \gamma_{2j} \right\} \qquad (5)$$

According to (5), the selection criterion can be regarded as: the relay i is chosen to maximise γ_{1i} and the relay j is chosen to maximise γ_{2j}, respectively. In contrast with the optimal selection criterion (4), the new selection criterion can be implemented by separate terminals and can be widely used in practice.

3 PERFORMANCE ANALYSIS

In this section, performance of the above-mentioned relay selection schemes under Nakagami-m is analyzed. First of all, closed-form lower bound for the overall outage probability is derived. Then, useful insights into practical systems implementation are obtained by quantifying the diversity-multiplexing trade-off.

3.1 Outage probability

With the above optimal scheduling, the overall outage probability is given by

$$P_{out} = \Pr\left[\min\left\{ \max_{i \in \{1,...,N\}} \gamma_{1i}, \max_{j \in \{1,...,N\}} \gamma_{2j} \right\} \leq \gamma_0 \right] \qquad (6)$$

Since that is hard to obtain the exact result of (6), we derive the tight lower bound of the outage probability. First, from (2) and (3), we can obtain

$$\gamma_{1i} = \frac{G^2 |g_{i1}|^2 |h_{2i}|^2 \gamma_2}{G^2 |g_{i1}|^2 + 1} < \gamma_2 |h_{2i}|^2 \qquad (7)$$

$$\gamma_{2j} = \frac{G^2 |g_{j2}|^2 |h_{1j}|^2 \gamma_1}{G^2 |g_{j2}|^2 + 1} < \gamma_1 |h_{1j}|^2 \qquad (8)$$

Then, the overall outage probability is given by

$$P_{out} > P_{out}^{lb} = \Pr\left[\min\{\gamma_{1i^*}, \gamma_{2j^*}\} \leq \gamma_0 \right]$$
$$= F_{\gamma_{1i^*}}(\gamma_0) + F_{\gamma_{2j^*}}(\gamma_0) - F_{\gamma_{1i^*}}(\gamma_0) F_{\gamma_{2j^*}}(\gamma_0) \qquad (9)$$

where $\gamma_{1i^*} = \max_{i \in \{1,...,N\}} \gamma_2 |h_{2i}|^2$, $\gamma_{2j^*} = \max_{j \in \{1,...,N\}} \gamma_1 |h_{1j}|^2$. $F_{\gamma_{1i^*}}(\gamma_0)$ and $F_{\gamma_{2j^*}}(\gamma_0)$ are the CDF of γ_{1i^*} and γ_{2j^*}, respectively.

Under Nakagami-m fading, $|h_{1k}|^2$ and $|h_{2k}|^2$ for any $k \in \{1,...,N\}$ are independent but not necessarily identically distributed with shape parameter m_{1k}, m_{2k} and scale parameter $\Omega_{1k} = E[|h_{1k}|^2]$, $\Omega_{2k} = E[|h_{2k}|^2]$, respectively, where $E(\cdot)$ denotes the expectation operator. Hence, the Cumulative Distribution Function (CDF) of the fading amplitude $|h_{sk}|^2$, $s = 1, 2$, can be given by

$$F_{h_{sk}}(x_s) = 1 - \frac{\Gamma\left(m_{sk}, \frac{m_{sk}}{\Omega_{sk}}x_s\right)}{\Gamma(m_{sk})} \qquad (10)$$

where $\Gamma(\cdot)$ is the gamma function, $\Gamma(\cdot, \cdot)$ is the incomplete gamma function [11].

The CDF for γ_{1i*} can be expressed by

$$F_{\gamma_{1i*}}(\gamma_0) = \Pr\left[\max_{i\in\{1,...,N\}} \gamma_2 |h_{2i}|^2 \leq \gamma_0\right]$$

$$= \prod_{i=1}^{N} \Pr\left[|h_{2i}|^2 \leq \frac{\gamma_0}{\gamma_2}\right] = \prod_{i=1}^{N}\left[1 - \frac{\Gamma\left(m_{2i}, \frac{m_{2i}\gamma_0}{\Omega_{2i}\gamma_2}\right)}{\Gamma(m_{2i})}\right] \qquad (11)$$

By following the similar derivation procedure of the CDF for γ_{1i*}, the CDF for γ_{2j*} can be expressed by

$$F_{\gamma_{2j*}}(\gamma_0) = \Pr\left[\max_{j\in\{1,..,N\}} \gamma_1 |h_{1j}|^2 \leq \gamma_0\right]$$

$$= \prod_{j=1}^{N} \Pr\left[|h_{1j}|^2 \leq \frac{\gamma_0}{\gamma_1}\right] = \prod_{j=1}^{N}\left[1 - \frac{\Gamma\left(m_{1j}, \frac{m_{1j}\gamma_0}{\Omega_{1j}\gamma_1}\right)}{\Gamma(m_{1j})}\right] \qquad (12)$$

By substituting (11) and (12) into (9), the desired result of the lower bound of the overall outage probability P_{out}^{lb} can be derived as in (13).

$$P_{out}^{lb} = \prod_{i=1}^{N}\left[1 - \frac{\Gamma\left(m_{2i}, \frac{m_{2i}\gamma_0}{\Omega_{2i}\gamma_2}\right)}{\Gamma(m_{2i})}\right] + \prod_{j=1}^{N}\left[1 - \frac{\Gamma\left(m_{1j}, \frac{m_{1j}\gamma_0}{\Omega_{1j}\gamma_1}\right)}{\Gamma(m_{1j})}\right]$$

$$- \prod_{i=1}^{N}\prod_{j=1}^{N}\left[1 - \frac{\Gamma\left(m_{2i}, \frac{m_{2i}\gamma_0}{\Omega_{2i}\gamma_2}\right)}{\Gamma(m_{2i})}\right]\left[1 - \frac{\Gamma\left(m_{1j}, \frac{m_{1j}\gamma_0}{\Omega_{1j}\gamma_1}\right)}{\Gamma(m_{1j})}\right] \qquad (13)$$

To tackle the product term in (13), we invoke useful formula given by

$$\prod_{i=1}^{N}[1 - \lambda_i] = 1 + \sum_{i=1}^{N}\sum_{n_1=1}^{N-i+1}\sum_{n_2=n_1+1}^{N-i+2}\cdots\sum_{n_i=n_{i-1}+1}^{N}(-1)^i\prod_{p=1}^{i}\lambda_{n_p} \qquad (14)$$

which is a summation over the subset of all combinations of elements λ_i.

Finally, the overall outage probability P_{out}^{lb} is given in closed-form by applying (14) into (13), which results in

$$P_{out}^{lb} = 1 - \sum_{i=1}^{N}\sum_{n_1=1}^{N-i+1}\sum_{n_2=n_1+1}^{N-i+2}\cdots\sum_{n_i=n_{i-1}+1}^{N}(-1)^i\prod_{p=1}^{i}\Lambda_{n_p}$$

$$\times \sum_{j=1}^{N}\sum_{n_1=1}^{N-j+1}\sum_{n_2=n_1+1}^{N-j+2}\cdots\sum_{n_j=n_{j-1}+1}^{N}(-1)^j\prod_{q=1}^{j}\Lambda_{n_q} \qquad (15)$$

where $\Lambda_i = \dfrac{\Gamma\left(m_{2i}, \frac{m_{2i}\gamma_0}{\Omega_{2i}\gamma_2}\right)}{\Gamma(m_{2i})}$ and $\Lambda_j = \dfrac{\Gamma\left(m_{1j}, \frac{m_{1j}\gamma_0}{\Omega_{1j}\gamma_1}\right)}{\Gamma(m_{1j})}$.

3.2 Diversity and coding gains

At high SNR, the outage probability P_{out} can be characterized by two parameters: diversity G_d and coding gain G_c [9], denoted by

$$P_{out} \approx (G_c \times \mathbf{SNR})^{-G_d} \qquad (16)$$

We assume $P_1 = \alpha_1 P$, $P_2 = \alpha_2 P$, and $\mathrm{SNR} = P/N_0$ [11], we can obtain that $\Gamma(s,x)/\Gamma(s) = 1 - \gamma(s,x)/\Gamma(s)$. For sufficiently high SNR, by using the fact that $\gamma(s,x) \to x^s/s$ when $x \to 0$, the overall outage probability can be approximated by

$$P_{out} \approx \Theta \times \left(\frac{\gamma_0}{\mathrm{SNR}}\right)^{G_d} \qquad (17)$$

where the diversity order is given by

$$G_d = \min\left\{\sum_{i=1}^{N}m_{2i}, \sum_{j=1}^{N}m_{1j}\right\} \qquad (18)$$

and Θ is given by

$$\Theta = \begin{cases} \prod_{i=1}^{N}\dfrac{m_{2i}^{m_{2i}}}{\Gamma(m_{2i}+1)\Omega_{2i}^{m_{2i}}\alpha_2^{m_{2i}}}, & \sum_{i=1}^{N}m_{2i} < \sum_{j=1}^{N}m_{1j} \\[2em] \prod_{i=1}^{N}\dfrac{m_{2i}^{m_{2i}}}{\Gamma(m_{2i}+1)\Omega_{2i}^{m_{2i}}\alpha_2^{m_{2i}}} \\ \quad + \prod_{j=1}^{N}\dfrac{m_{1j}^{m_{1j}}}{\Gamma(m_{1j}+1)\Omega_{1j}^{m_{1j}}\alpha_1^{m_{1j}}}, & \sum_{i=1}^{N}m_{2i} = \sum_{j=1}^{N}m_{1j} \\[2em] \prod_{j=1}^{N}\dfrac{m_{1j}^{m_{1j}}}{\Gamma(m_{1j}+1)\Omega_{1j}^{m_{1j}}\alpha_1^{m_{1j}}}, & \sum_{i=1}^{N}m_{2i} > \sum_{j=1}^{N}m_{1j} \end{cases} \qquad (19)$$

So the overall coding gain is given by

$$G_c = \Theta^{\left(-1/\min\left\{\sum_{i=1}^{N} m_{2i}, \sum_{j=1}^{N} m_{1j}\right\}\right)} \tag{20}$$

3.3 Diversity-Multiplexing Trade-off

The Diversity-Multiplexing Tradeoff (DMT) proposed by Zheng and Tsein [10] is a measure which governs the fundamental trade-off between achievable data rate and reliability. Consider a system that can achieve an outage probability of P_{out} and an averaged transmission rate of R bit/s/Hz, the diversity gain d and multiplexing gain r are denoted by

$$d = -\lim_{SNR \to \infty} \frac{\ln P_{out}}{\ln SNR} \text{ and } d = \lim_{SNR \to \infty} \frac{R}{\ln SNR},$$

respectively. The balance between d and r is called the DMT.

For large SNR, from the above-considered system, we get $\gamma_0 = 2^{3r \ln SNR} - 1 \approx SNR^{3r}$. By substituting $\gamma_0 \approx SNR^{3r}$ into (13) the diversity-multiplexing trade-off cant be quantified by using its definition as

$$G_d = \min\left\{\sum_{i=1}^{N} m_{2i}, \sum_{j=1}^{N} m_{1j}\right\}(1-3r) \tag{21}$$

As a result, the maximum multiplexing gain of the considered protocol is 1/3, which is smaller than that of the Single Relay Selection (SRS) protocol [6]. This is because, although the considered protocol supports two traffic flows concurrently as the SRS protocol, to avoid interference, each traffic flows takes three times lots to complete the transmission while the SRS protocol takes two time slots. However, the maximum multiplexing gain of the considered protocol is still larger than that of the SRS protocol.

Since $\min\left\{\sum_{i=1}^{N} m_{2i}, \sum_{j=1}^{N} m_{1j}\right\} \geq \sum_{k=1}^{N} \min\{m_{1k}, m_{2k}\}$. quality holds if and only if $m_{1k} \leq m_{2k}$ or $m_{1k} \geq m_{2k}$ for all k.

4 NUMERICAL RESULTS

In this section, we present Monte-Carlo simulations to confirm the derived analytical results. We assume that the average SNR in links from two sources to relay is equal, i.e. $\alpha_1 = 1$, $\alpha_2 = 1$. The distance between two sources S_1 and S_2 is normalised to 1, the normalised distance from S_1 to relay k ($k = 1, ..., N$) is d_{1k}, and the path loss exponent is 3. Thus, $\Omega_{1k} = d_{1k}^{-3}$, $\Omega_{2k} = (1-d_{1k})^{-3}$. Without loss of generality, we set $d_{1k} = 0.3$ for all the relays. Furthermore, we assume $m_{1k} \in \{0.5, 1.0, 1.5, 2.0\}$, $m_{2k} \in \{0.5, 1.0, 1.5, 2.0\}$, $k = 1, ..., N$, and the targeted transmission rate $R = 2$ bit/s/Hz. Figure 1

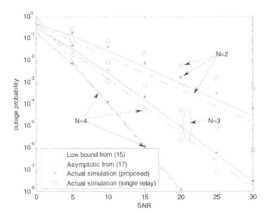

Figure 1. Outage probability against SNR for $N = 2, 3, 4$.

shows the theoretical results for the overall outage probability, asymptotically high SNR approximations and actual simulation of the proposed scheme. In order to compare the outage performance, we also show the actual simulation of the "best" single relay selection protocol with $N = 2, 3, 4$. From figure, it is observed that analytical results exactly coincide with Monte Carlo simulations and therefore verify the accuracy of our analysis. Asymptotic results verify the diversity order. As expected, we can see that the number of relay N also affects the difference of the overall outage probability between the proposed scheme and the "best" single relay selection protocol, that is, as the number of relay N gets bigger, the difference becomes more obvious. The performance of the proposed scheme is much better than that of the "best" single relay selection protocol.

5 CONCLUSIONS

In this letter, we propose a novel two-way relay selection scheme and investigate the performance of the proposed system over independent flat Nakagami-m fading channels by deriving the lower bound of the overall outage probability, asymptotically high SNR approximations and Diversity-Multiplexing Trade-off (DMT). And then, numerical results verify that the proposed relay selection scheme has better outage performance than that of the single relay selection protocol.

REFERENCES

[1] B. Rankov & A. Wittneben. 2007. Spectral efficient protocols for half-duplex fading relay channels. *IEEEJ. Select. Areas Commun.* 25(2):379–389.

[2] S.J. Kim, P. Mitran & V. Tarokh. 2008. Performance bounds for bidirectional coded cooperation protocols. *IEEE Trans. Inform. Theory*. 54:5235–5241.

[3] B. Rankov & A. Wittneben. 2006. Achievable rate regions for the two-way relay channel. *Proc. IEEE ISIT*. Seattle, WA, USA. 1668–1672.

[4] Jing Yang, Pingzhi Fan, Trung Q. Duong & Xianfu Lei. 2011. Exact Performance of Two-Way AF Relaying in Nakagami-m Fading Environment. *IEEE Trans. Wireless Commun*. 10(3):980–987.

[5] Q. Li, S.H. Ting, A. Pandharipande & Y. Han. 2009. Adaptive two-way relaying and outage analysis. *IEEE Trans. Wireless Commun*. 8(6):3288–3299.

[6] E.Y. Li & S.Z. Yang. 2012. Simple relay selection criterion for general two-way opportunistic relaying networks. *Electron. Lett*. 48(14):881–882.

[7] X. Zhang & Y. Gong. 2009. Adaptive poweral location in two-way amplif-and-forward relay networks. *Proc. IEEE ICC*. Dresden, Germany. 1–5.

[8] L. Song, Y. Li, H. Guo & B. Jiao. 2010. Differential bidirectional relay selection using analog network coding. *Proc. IEEE WCNC*. Sydney, Australia. 1–5.

[9] Z. Wang & G.B. Giannakis. 2003. A simple and general parameterization quantifying performance in fading channels. *IEEE Trans. Commun*. 51(8): 1389–1398.

[10] L. Zheng & D.N.C. Tse. 2003. Diversity and multiplexing: A fundamental tradeoff in multiple antenna channels. *IEEE Trans. Inform. Theory*. 49:1073–1096.

[11] I.S. Gradshteyn & I.M. Ryzhik. 2000. Table of Integrals(6). *Series and Products*. Academic Press.

Information Technology and Applications – Li (Ed.)
© 2015 Taylor & Francis Group, London, ISBN 978-1-138-02677-3

Changes in cortical network topology with brain aging

L. Lin, C. Jin, X. Xu & S. Wu
Biomedical Research Center, College of Life Science and Bioengineering, Beijing University of Technology, Beijing, China

ABSTRACT: The last decade has seen an increase in neuroimaging studies examining brain connectivity. The current study examined differences of structural Connectome for 57 normal aging subjects. 3T MRI acquisitions included high resolution 3D T1, EPI sequences for Diffusion Tensor Imaging (DTI). The main processing procedure included three main steps: (1) preprocessing; (2) brain parcellation and structural connection matrix construction; and (3) topological properties extraction. Our findings suggest that the topology of the local network architecture in frontal lobe and temporal lobe may be more vulnerable to aging effects than other parts of brain.

Keywords: aging; connectivity; connectome; DTI

1 INTRODUCTION

Increase in the life expectancy has caused ageing population in the world. China is the country that faces a situation where the largest population cohort will be those over 70 years old and average age approach 50 years old in next few decades. Cognitive health has consistently been cited as a key factor for quality of life [10] and as an important contributor to late life functioning [2, 5, 11]. Thus, a key element in dealing with the aging is to discover ways to optimize cognitive performance when people are aging.

The brain is a neural system of interlinked structural and functional regions. The set of connections in this neural complex network, now called the Connectome [13], has been the focus in recent years due to advances in neuroimaging technique and the availability of new network analysis methods that can examine human brain connectivity systematically. The NIH Blueprint for Neuroscience Research, Human Connectome Project (HCP) [12, 19, 20] was launched in 2009 as an ambitious 5-year effort to characterize brain connectivity and function in healthy adults. Aging is not only a major risk factor for human cognition, but also a predictor of conversion to other neurodegenerative diseases. While much research in Connectome has focused on diseases of aging, such as Mild Cognitive Impairment (MCI) [16, 17] Alzheimer's Disease (AD) [15] etc, there are few Connectome studies on the brain aging. Diffusion Tensor Imaging (DTI) is a non-invasive imaging technique that can be used to investigate White Matter (WM) microstructures.

In this study, we employed DTI to investigate the topological WM changes in the normal aging subjects. This study hypothesized that: When healthy people are aging, they will show a topological pattern revealed by characteristics in the WM networks.

2 SUBJECTS AND METHOD

2.1 Subjects

57 right-handed healthy elderly subject were included in this study. Participants will be eligible to participate in the this study if they do not meet research NINCDS-ADRDA criteria of a diagnosis of probable AD or DSM-IV criteria for any type of dementia or research criteria for mild cognitive impairment; do not have a significant neurological, psychiatric, or medical disorder or injury that would affect cognitive function; and have a score ≥ 28 on the Folstein MMSE and a score ≤ 10 on the Hamilton Depression Rating Scale. Subject characteristics are shown in Table 1. MRI scans for the healthy elderly subject were performed with

Table 1. Subject characteristics.

N	57
Age (years)	59.7 ± 4.7
Age range (years)	50.4–69.9
Gender, M/F	30/27
Education (years)	16.6 ± 2.1
MMSE, mean ± SD	29.2 ± 0.5

a 3T GE Signa II scanner with an eight-channel phased array coil (HD Signa Excite, General Electric, Milwaukee, WI). The T1-weighted data were acquired using 3D Spoiled Gradient-echo (3DSPGR) sequence (TR = 5.3 ms, TE = 2.0 ms, TI = 500; flip angle = 15°; matrix = 256 × 256; FOV = 256 × 256 mm²) with 204 contiguous 1.0 mm thick coronal slices. The DTI data were acquired using a single-shot spin echo diffusion sensitized Echo-Planar Imaging (EPI) sequence, TR = 12500 ms, TE = 71 ms. Fifty-nine volumes were collected including 8 without diffusion weighting ($b = 0$ s/mm²) and 51 with diffusion weighting ($b = 1000$ s/mm²). The slice thickness was 2.6 mm with 58 slices covering the whole-brain, FOV = 250 × 250 mm², and an image matrix of 128 × 128 after k-space reconstruction.

2.2 *Image preprocessing*

T1-weighted MRI images are processed using Statistical Parametric Mapping (SPM8, Welcome Department of Imaging Neuroscience, London; http://www.fil.ion.ucl.ac.uk/spm) running on Matlab 7.9.0 (Math-Works, Natick, MA, USA). T1-weighted images are segmented using new segmentation algorithm in SPM8, an updated version of the unified segmentation algorithm [1] into Gray Matter (GM), WM and Cerebrospinal Fluid (CSF).

DTI images are processed by using a pipeline toolbox for analyzing brain diffusion images [4] (PANDA, State Key Laboratory of Cognitive Neuroscience and Learning at Beijing Normal University, http://www.nitrc.org/projects/panda/) following these steps: converting DICOM files into NIfTI images; manually inspecting for large artifacts; applying brain mask from new segmentation; correcting eddy-current induced image distortion and simple head motion using an affine registration; calculating voxel-wise diffusion tensor matrix by weighted linear least squares fitting to the log-transformed diffusion data and registering images to a standardized template in the MNI space by FSL (FMRIB Software Library, FMRIB, Oxford) [9]. The nonlinear deformation was initialized with an affine registration generated with the FSL's registration tool FLIRT8).

2.3 *Brain parcellation and structural connection matrix*

An important step to define brain networks is to determine the nodes in the network. To do so, it is necessary to divide brain at the desired scale according to specific strategies for anatomical partitioning. The Automated Anatomical Labeling (AAL) atlas is used in this study to parcellate the cerebral cortex into 90 standardized cortical regions (45 in each

hemisphere) [14]. The inverse of the estimated deformation field is applied to the AAL labels, producing masks for the 90 cortical and subcortical regions in DTI native space. Deterministic tractography assumes a deterministic fiber orientation at every nodes during tracking, and tracking ends up with 3D trajectories for WM tracts. Based on the linking fibers, three kinds of basic weighted matrices: Fractional Anisotropy (FA) matrix, Fiber Number (FN) matrix and Fiber Length (FL) matrix are extracted. The weighted values of FA matrix indicate the averaged FA between two nodes. In FN, weights indicate the number of joint fibers connected between anatomical nodes. And in FL, the weighted values indicate averaged length of linking fibers that reached from one node to another.

2.4 *Topological properties of structural brain networks*

In brief, we first determined the most consistent connections, which are the backbone of the network. For the FL and FN matrixes, if at least three fibers are located in a region, the weight of edge would be kept; otherwise, the weight would be set to zero. For the FA matrix, if at least three fibers are located in a region and FA is greater than 0.2, the weight of edge would be kept; otherwise, the weight would be set to zero. Moreover, the three weighted matrices can be also binarized based on whether their values are greater than zero.

Next, the topological properties of anatomical networks are analyzed quantitatively using graph theory. Six nodal measures (degree, strength, local efficiency, local clustering coefficient, shortest path length, and node between centrality) are examined to describe the regional properties of the anatomical networks. Further preprocessing of topological properties are processed by the Brain Connectivity Toolbox (BCT, http://www.brain-connectivity-toolbox.net).

Statistical analyses are performed using the statistic toolbox in Matlab and SPSS 20 (IBM SPSS Inc., Armonk, NY). Shapiro–Wilks tests shows that the ages of subjects are normally distributed. Possible effects of age on the topological properties derived from the binary and the weighted anatomical brain networks are evaluated by the Pearson's correlation coefficients. The relationship between the brain network measurements and age are examined with partial correlation analyses controlled for gender and education.

3 RESULTS

We extracted 14 features based on the criteria of correlation coefficient in both bivariate correlation

Table 2. Network characteristics with correlation coefficient greater than 0.3 and p < 0.01, with co-variables of gender and education.

AAL region name	Connection matrix	Network measure
Left superior frontal gyrus, dorsolateral	Weighted FN	Strength
Right superior frontal gyrus, dorsolateral	Weighted FN	Strength
Left superior frontal gyrus, orbital part	Weighted FL	Strength
Left middle frontal gyrus, lateral part	Weighted FN	Strength
Left middle frontal gyrus, orbital part	Weighted FL	Local clustering coefficient
Left opercular part of inferior frontal gyrus	Weighted FN	Strength
Left supplementary motor area	Weighted FL	Local clustering coefficient
Left superior frontal gyrus, medial part	Weighted FN	Strength
Right superior frontal gyrus, medial orbital part	Weighted FL	Local clustering coefficient
Right angular gyrus	Weighted FA	Local efficiency
Left transverse temporal gyri	Binary FA	Degree
Right transverse temporal gyri	Weighted FA	Local clustering coefficient
Right transverse temporal gyri	Weighted FA	Local efficiency
Right transverse temporal gyri	Weighted FA	Strength

and partial correlation, the negative correlation coefficients are less than 0.3 and p value are less than 0.01. Age is associated with changes in weighted structural connection metrics, with decreases from middle age people to old people. The result reveals that prominent age-related network changes happen on frontal lobe and temporal lobe (Table 2).

4 DISCUSSIONS

Growing evidence suggests that cognition of human suffers continuous decline during the lifecycle and brain degeneration is not uniform (frontal and temporal cortices are the brain regions most sensitive to the negative effects of aging [6, 7, 18] Age related decreases in cognitive performance are not uniform, attention, memory and executive function are the most affected. Our finding in network analysis is consistent with previous studies of normal age-related brain anatomical and functional changes [3]. For example, the network measure of Dorsolateral Prefrontal Cortex (DLPFC) is highly corrected with age in this study. DLPFC's main functions are the executive functions, such as working memory, cognitive flexibility, planning, inhibition, and abstract reasoning which decrease with aging. All complex mental activity requires the additional cortical and subcortical circuits with which the DLPFC is connected. Our findings highlight that altered structural connectivity between nodes of the network in frontal lobe and temporal lobe regions.

In this study, we successfully constructed binary and weighted brain anatomical networks from 57 normal elderly subjects using DTI. Network

topological properties were analyzed based on graph theory, and fourteen features were extracted with high association according to the correlation analysis with age. The findings highlight the advantage of conceptualizing brain aging as a result of disturbances in an interconnected network, rather than isolated sub-sets of brain regions.

ACKNOWLEDGEMENTS

This work was supported by grants from Natural Science Foundation of Beijing (7143171).

REFERENCES

[1] Ashburner J. & Friston K.J., Unified segmentation. *Neuroimage*, 26 (3):839–85.1, 2005.
[2] Baltes M.M., Wahl H.W., Schmid-Furstoss U., The daily life of elderly Germans: Activity patterns, personal control, and functional health. *J Gerontol*, 45: 173–179, 1990.
[3] Convit A., Wolf O.T., de Leon M.J., et al., Volumetric analysis of the pre-frontal regions: findings in aging and schizophrenia. *Psychiatry Res.*, 107 (2): 61–73, 2001.
[4] Cui Z., Zhong S., Xu P., et al., PANDA: a pipeline toolbox for analyzing brain diffusion images. *Front Hum Neurosci.* 21;7:42. doi: 10.3389/fnhum.2013.00042, 2013.
[5] Depp C.A., Jeste D.V., Definitions and predictors of successful aging: A comprehensive review of larger quantitative studies. *Am J Geriatr Psychiatry*, 14: 6–20, 2006.
[6] Dirnberger G., Lalouschek W., Lindinger G., et al., Reduced activation of midline frontal areas in human elderly subjects: a contingent negative variation study *Neurosci. Lett.*, 280: 61–64, 2000.

[7] Fjell A., Walhovd K., Structural brain changes in aging: courses, causes and cognitive consequences. *Rev Neurosci.* 21:187–221, 2010.

[8] Jenkinson M., Smith S., A global optimisation method for robust affine registration of brain images. *Medical image Analysis*, 5: 143–156, 2001.

[9] Jenkinson M., Beckmann C.F., Behrens T.E.J., et al., FSL. *Neuroimage*, 62 (2): 782–790, 2012.

[10] Reichstadt J., Depp C.A., L.A. Palinkas, D.P., et al., Building blocks of successful aging: A focus group study of older adults' perceived contributors to successful aging. *Am J Geriatr Psychiatry,* 15: 194–201, 2007.

[11] Rowe J.W., Kahn R.L., Successful aging. *Gerontologist,* 37: 433–440, 1997.

[12] Sotiropoulos S.N., Jbabdi S., Xu J., et al., WU-Minn HCP Consortium Advances in diffusion MRI acquisition and processing in the Human Connectome Project. *NeuroImage*, 80: 125–143, 2013.

[13] Sporns O., Tononi G., Kötter R., The human connectome: a structural description of the human brain. *PLoS Comput. Biol.,* 1: e42, 2005.

[14] Tzourio-Mazoyer N., Landeau B., Papathanassiou D., et al., Automated Anatomical Labeling of activations in SPM using a Macroscopic Anatomical Parcellation of the MNI MRI single-subject brain. *NeuroImage,* 15 (1): 273–289, 2002.

[15] Wang J., Zuo X., Dai Z., et al., Disrupted Functional Brain Connectome in Individuals at Risk for Alzheimer's Disease *Biological Psychiatry*, 73 (5): 472–481, 2013.

[16] Wang Y., West J., Shen Li, et al., Altered connectome mapping in people with mild cognitive impairment and older adults with cognitive complaints. *Alzheimer's & Dementia*, 9(4 supply): 276, 2013.

[17] Wee C.Y., Yap P.T., Li W., et al., Enriched white matter connectivity networks for accurate identification of MCI patients *NeuroImage*, 54 (3): 1812–182, 2011.

[18] West R., An application of prefrontal cortex function theory to cognitive aging. *Psychol. Bull.,* 120: 72–292, 1996.

[19] Van Essen D.C., Ugurbil K., Auerbach E., et al., WU–Minn HCP Consortium The human connectome project: a data acquisition perspective. *NeuroImage,* 62: 2222–2243, 2012.

[20] Van Essen D.C., Smith S.M., Barch D.M., et al., The WU-Minn Human Connectome Project: An overview. *NeuroImage*, 80: 62–79, 2013.

Information Technology and Applications – Li (Ed.)
© 2015 Taylor & Francis Group, London, ISBN 978-1-138-02677-3

Research of the measure method of identity trustworthiness based on software code homology detection

Xianghui Zhao, Hui Liu, Mingguo Zou, Lin Liu & Lei Zhang
China Information Technology Security Evaluation Center, Beijing, China

Yafang Huang
Tsinghua University, Beijing, China

ABSTRACT: Along the occurrence and intensifying of software crisis, the program of software trustworthiness has been attracted by many researchers and organizations at home and abroad. This paper briefly analyzes the current technology status of the software code homology detection. In this paper, a measurement of identity trustworthiness approach based on software code homology detection was proposed. Based on the definition of identity trustworthiness, available software source code detection tools and principles are used to figure out the total loss caused by identity problems in the measurement of identity trustworthiness. The proposed approach was evaluated by statistical experiment using some samples. The result shows that the approach is accurate and effective and also some identity trustworthiness problems were explored.

Keywords: software trustworthiness; identity trustworthiness; loss; source code detection

1 INTRODUCTION

Nowdays, plenty of huge data loss were brought by software failures, so that many countries, organizations and researchers have begun to focus on how to guarantee the correct and efficient operation of softwares, especially software trustworthiness problems. Many researchers have carried out extensive and in-depth research in the field of software trustworthiness [1–3]. Because of the more open, dynamic and ever-changing internet environment, software development process is facing a series of challenges, such as complexity, openness and evolution which makes software trustworthiness research more complicated compare to other software related researchers [4–6]. As a new concept, software trustworthiness is a comprehensive of software correctness, reliability and security and many other attributes. Relevant researchers attempt to define software trustworthiness containning attributes and scope which committed to establishing a standard definition and attribute partition of the software trustworthiness. Therefore, plenty of definitions of software trustworthiness have been proposed. However, the lack of a unified view makes relation research work of the software trustworthiness field difficult to be carried out smoothly unity [7–11].

Meanwhile, with the development of the software industry, software reusage is constantly raised, and the number and quality of open source software continues to improve, making the similarities between different software code rates increased. On one hand, these accelerates the speed of software development and reduce the cost of development at meanwhile, but on the other hand, software reusage or directly copying involves the issue of intellectual property, and reused code security problems which prompted the development of software plagiarism detection tools or software homology detection. Individual and research organizations have proposed different methods and improvement approaches for the software code homology detection. The research of source code homology detection early started in a foreign country, the first scholar who started to do the research of program code similarity detection technology and software begun in the early 1970s [12–17]. Current homology detection methods are mainly divided into three types: text-based similarity detection, token-based similarity detection and syntax structure-based similarity detection. For the above three ways, a lot of tools and software are also available, including UltraCompare [18], WinMerge [19], CP-Miner [20], CCFinder [21], Jplag [22] and CloneDr [23], and so on.

This paper briefly analyzes the current technology status of the software code homology detection. The measure method of identity trustworthiness based on software code homology detection are proposed. Based on the definition of identity trustworthiness, available software source code detection tools and principles are used to figure out the total loss caused by identity problems in the measurement of identity trustworthiness. The proposed approach was evaluated by statistical experiment using some samples. The result shows that the approach is accurate and effective and also some identity trustworthiness problems were explored.

2 OVERVIEW OF THE MEASURE METHOD OF SOFTWARE IDENTITY TRUSTWORTHINESS

References [24] sociologically speaking, in order to understand concept of the word "trustworthiness", it is given the definition of the software trustworthiness, at the same time, pointed out that the software identity trustworthiness is clear, truthful and reliable that of the itself and source code of the software T_a. Software trustworthiness measurement is usually start from the measurement of the trustworthiness attributes. However, different attributes such as reliability, security, availability, and so on which are inconsistent measure scale, which includes the probability scale, the number of holes, and so on, the status of this inconsistent are to bring the software cannot be unified measure of credibility problems. The loss ω is due to the presence of untrustworthy software elements, while making software users need to pay additional cost (money, maintenance workload, the number of lines of code, and so on). The composition of these costs includes the following three parts: because the software does not meet the specified criteria due to the problem so that the system cannot be trusted to run properly. The additional cost to restore the software to run is needed to pay. The loss is not normal operation of the software to work to bring business users, the additional cost is required to reach the required payment cost. Specifically, the loss of identity trustworthiness is the source of the source code in order to solve the problem with unreliable factors such as the unreliable and unclear software source code itself, and sometimes the need to pay an additional cost and operational work brings loss cost to users.

In the view of practical application, the loss that caused by the software identity trustworthiness is the attributes of each class software. This paper attempts to use the loss as unified metric scale. Taking the additional costs for the loss of

credibility into account due to a software problem, to bring the software to run and maintain, measure the amount of actual costs may be used, or the software development and maintenance, such as the amount of code required. In this paper, based on the loss in the scale of software running market realities to consider, this is a unified measure important properties of various types of trustworthiness scale, the credibility of key importance for the evaluation of software. Therefore, this paper will lose as the measure index of identity trustworthiness, expected by the measure method of this paper, to get the amount of loss due to software trustworthiness.

The measure method of identity trustworthiness in this paper isn't trustworthiness for the software to meet the level of the property, but rather due to the properties of the resulting amount of untrusted software to carry out the measure and predict losses. The resulting overall loss of credibility as a measure standard of the software trustworthiness. The larger the amount of the loss by the model calculated, then the lower the trustworthiness of the software. We give a definition of identity trustworthiness and mathematical description as the following.

Definition 1: Identity trustworthiness, its credibility $WS(T_a(t))$ with said means at the moment t, the software T_a for itself and the source code is not trusted, contrary to clear, truthful and reliable principles. To compensate for these problems, we need to pay additional fees or costs. And the amount of the loss to the loss of part of the cost of non-standard work to bring business users, that is caused by the loss. The higher the value of the $WS(T_a(t))$, indicating that the greater the loss due to problems caused by identity trustworthiness, the lower the credibility of identity.

Among them, which $T_a(t)$ refers to the state of the software at the time t. Different moments, the state in which different software, the software code itself will change. Surrounding environment in which even the policies and guidelines in field is changing. Therefore, adding credibility parameter t indicates the software trustworthiness is a time-related function.

3 THE MEASURE METHOD OF IDENTITY TRUSTWORTHINESS BASED ON SOFTWARE CODE HOMOLOGY DETECTION

The software code homology detection tools in this paper is self-developed product (other software code homology detection tools that method mentioned above can also be used, the measure method is the same). According to this tool, history software and similar percentage can be detected which is similar to software code to be detected. If T_i is i software, ω_i shows the loss total amount of the corresponding software which is in the running life cycle. (million units, abbreviated as W). The Table 1 gives the historical data sample of software loss.

According to the historical data and the results obtained based on tools, to unify the trustworthiness measurement value of the software T_a. Here are the underlying assumptions of the measure and the measure method of the identity trustworthiness.

3.1 Specific analysis examples

Before giving the model method of the measure method of the identity trustworthiness, the following assumption is put forward.

Assumption 1: The actual software and the to be tested software has the same set of source code, which is similar to a collection of software using code homology detection tool, that there is no code homology detection tools that cannot detect homologous software.

Assumption 2: If calculating identity is trustworthy, the loss of every line code of the software is same.

Assumption 3: If calculating identity is trustworthy, assumed that the software is running, the probability of running into each line of code is the same.

To analyze assumption 3, as usual, the probability that running to each line of code is not the same, the code homology detection tool in this paper gets the same code, which will occupy a certain threshold above of the to be tested software. For simplifying and unifying its calculations, that the probability of the similar part code to be run is similar to the proportion of its code is considered. Based on the above assumption, to give a sample: software A, using the code homology detection tool, detects the software B, C and D, which has the similar code as software A. The similar value are 50%, 30% and 70%. The similar part code, which has the proportion of A's share of the software are 10%, 20% and 15%. Its schematic shown in Figure 1.

Among them, the grey area represents the software is similar to software A which is detected by

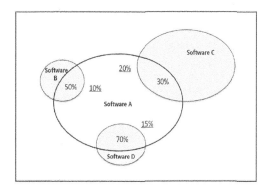

Figure 1. The result sample of software code homology detection.

the code homology detection tool. The percentage of the grey area is proportion of the similar code for similar software, and the percentage of the grey area beside with underlined is representation the proportion of the similar code accounts for software A. If the loss of the software B, C and D produced in the life cycle of operation are: 50 W, 300 W and 100 W.

Figure 1 is an example of the analysis, as the software trustworthiness is to investigate the identity of the trusted software source code itself and clear source of truth, reliability. Its loss is the amount of loss caused by the breach of authenticity. Because of the source code of the software A from the software B, C and D, that itself is not entirely credible. These three software code there are some flaws. Therefore, it produces a certain amount of loss in their running life cycle. It is because of the source code of software B, C and D cannot be trusted, and so make software A also exists credible identity problem. According to the assumption 2 that the amount of loss by each row of the software is the same, the amount of loss is 100 W for the software, the loss of 50% software code is 50 W. Thus, in Figure 1, the amount of loss 50% software code caused by software B is 50 W*50% = 25 W, similar, the amount of loss caused by software C and D which are similar to the software A is 90 W and 70 W. According to the assumption 3, when the software A runs, the probability of running into each line of code is the same, there is a 10% probability to run the same software code portions with software B. Therefore, the untrusted code of software A is appeared in the software B, and the amount of loss bring to software A is 25 W*10% = 2.5 W, similar, the amount of loss that software C and D code for software A is 18 W and 10.5 W. Accordingly, as the amount of loss that the source code of software A is from untrusted software B, C and D, is 2.5 W + 18 W + 10.5 W = 31 W.

Table 1. The historical data sample of software loss.

T_i	ω_i
Software 1	100 W
Software 2	2 W
Software 3	30 W
...	...
Software n	83 W

3.2 The idea of the measure method of identity trustworthiness

Based on the analysis of the above example, in order to the measure process of mathematical that the identity trustworthiness, the following definitions are given special instructions:

1. T_a indicates that the current to be evaluation software.
2. T_i *representative* of software that is similar to the use of code homology detection tool detected in software T_a. $i = 1, 2, ..., N$, N is detected with software T_a similar software number, correspondingly, ω_i is history loss, the measure method is the sum of the amount of loss in the run lifecycle of software T_i.
3. α_i is the *similarity* part codes of accounts for the percentage of all the codes of T_i, that is similarity rate.
4. β_i is the similarity part codes of accounts for the percentage of all the codes of T_a.

The *value* of the identity trustworthiness $WS(T_a(t))$ can be expressed as follows:

$$WS(T_a(t)) = \sum_{i=1}^{N} \omega_i \alpha_i \beta_i \qquad (1)$$

The above equation (1) indicates that the identity trustworthiness value of the software T_a is the amount of its losses caused, all software codes which is similar to this software code. The amount of loss caused by the code of software T_i for the code of software T_a. That has relations with similarity codes of accounts for the percentage of T_i and T_a.

This paper presents the measure method of identity trustworthiness, which measure steps are summarized as follows:

First, testing the similar software T_i with software T_a and calculating a similar percentage α_i, which is using the code testing tools.

Second, calculating β_i which the similarity part codes of accounts for the percentage of all the codes of T_a.

Final, according to the equation (1), calculating $WS(T_a(t))$ which the value of the identity trustworthiness base on the above two steps and the historical data ω_i of the loss of the software identity trustworthiness.

4 EXPERIMENTAL ANALYSIS AND VERIFICATION

According to the measure method of identity trustworthiness based on software code homology detection, need the help of software code homology detection tools, testing the similar software with

software T_i and calculating the similar results. Calculating $WS(T_a(t))$ based on the historical data ω_i of the software T_i, when measure the results of identity trustworthiness for the software T_i (the units of the loss is million). For a detailed description of the measure method of identity trustworthiness, the following experimental examples are given to illustrate. The detecting result is obtained in Table 2.

According to the equation (1), calculating the value of the identity trustworthiness for the software T_a.

$$WS(T_a(t)) = \sum_{i=1}^{N} \omega_i \alpha_i \beta_i = 100 * 20\% * 60\%$$
$$+ 2 * 50\% * 35\% + 30 * 30\% * 10\% = 3.05$$

By calculation, the amount of loss of software identity trustworthiness is 3.05 million.

Starting from the definition of the software identity trustworthiness, using code homology detection tools, testing similar code with the detection software, investigating the amount of the loss of these codes which cannot be trusted because of their origin, the measure method of identity trustworthiness based on software code homology detection is proposed, its meaning reflected in the following three main areas:

1. Objectivity. The code itself is real and trustworthiness based on subjective judgments way, and this paper uses tools to standard measure and test, the evaluation results can be traced. Then relying on the fixed historical data can get the objective test results, this method has overcome the subjectivity of the original measure.
2. Mature theory. Because the source code detection tools rely on mature homology detection theory, it has by a third party evaluation agencies conducted in-depth research and standards verification, therefore, this method has a more mature theory and practice.
3. Uniformity. The loss of the software identity trustworthiness is dependent on the software code that are similar to itself code. The loss of the software identity trustworthiness is near which the software has similar code.

Table 2. The result sample of software code homology detection.

T_i	β_i	α_i	ω_i
Software 1	20%	60%	100 W
Software 2	50%	35%	2 W
Software 3	30%	10%	30 W

Because measure standard uniform, making measurement results with unity. Assumptions software A and B, the value of the software identity trustworthiness is 100 W and 200 W. Because these two values depend on the same historical set data of software loss, do not appear the problem which the value of the software identity trustworthiness by the method of this paper cannot comparable.

5 CONCLUSION AND PROSPECT

This paper briefly analyzed the current technology status of the software code homology detection. The identity trustworthiness measurement based on software code homology detection was proposed. Based on the definition of identity trustworthiness, available software source code detection tools and principles are used to figure out the total loss caused by identity problems in the measurement of identity trustworthiness. The proposed approach was evaluated by statistical experiment using some samples. The result shows that the approach is accurate and effective and also some identity trustworthiness problems were explored. Although the identity trustworthiness measurement based on software code homology detection are presented, considering the situation in accordance with different software systems and the environment of the software systems. Further refinement of methods and detailed rules of the loss of data collection and other problems need to be further studied.

ACKNOWLEDGMENT

This work is supported by the project of the State Key Program of National Natural Science Foundation of China (No. 90818021). Meantime, it is supported by a grant from the National High Technology Research and Development Program of China (863 Program) (No. 2012AA012903). The authors are also thankful to the anonymous reviews for their constructive comments.

REFERENCES

[1] Nami M., Suryn W. Software Trustworthiness: Past, Present and Future [M]//Trustworthy Computing and Services. Springer Berlin Heidelberg, 2013: 1–12.

[2] Keivanloo I., Rilling J. Software trustworthiness 2.0—A semantic web enabled global source code analysis approach [J]. Journal of Systems and Software, 2013.

[3] Del Bianco V., Lavazza L., Morasca S., et al. A survey on open source software trustworthiness [J]. Software, IEEE, 2011, 28(5): 67–75.

[4] Yamada S., Ohba M., Osaki S. S-Shaped Reliability Growth Modeling for Software Error Detection [J]. IEEE Transactions on Reliability, 1983, 32(5): 475–478.

[5] Huang Yafang, Liu Yanzhao, Luo Ping. SSRGM: Software strong reliability growth model based on failure loss [C] // Proceedings of the 2012 Fifth International Symposium on Parallel Architectures, Algorithms and Programming (PAAP 2012), Taibei, China: Parallel Architectures, Algorithms and Programming (PAAP), 2012: 225–261.

[6] Koshy L.M., Conrad M., Shukla M., et al. Identity Implies Trust in Distributed Systems—A Novel Approach [M]//Trust and Trustworthy Computing. Springer Berlin Heidelberg, 2013: 269–270.

[7] Kraxberger S., Toegl R., Pirker M., et al. Trusted Identity Management for Overlay Networks [M]// Information Security Practice and Experience. Springer Berlin Heidelberg, 2013: 16–30.

[8] Trusted Computing Group: TCG infrastructure specifications, https://www.trustedcomputinggroup.org/specs/IWG/

[9] Voltz R.M. Cyberspace Trusted Identity (CTI) Module: U.S. Patent 20, 130, 219, 481[P]. 2013-8-22.

[10] Trusted Computing Group. Specification Architecture Overview Specification, Revision 1.4 [R]. Beaverton, USA: Trusted Computing Group, Incorporated, 2007.

[11] International Organization for Standardization (ISO)/International Electrotechnical Commission (IEC). Information Technology-Security Techniques—Evaluation Criteria for IT Security [R]. Geneva, Switzerland: International Organization for Standardization (ISO)/International Electrotechnical Commission (IEC), 2005: 1–72.

[12] Tang Yongxin, Liu Zengliang. Progress in software trustworthiness metrics models [J]. Computer Engineering and Applications, 2010, 46(27): 12–15.

[13] Wang Huaimin, Tang Yangbin, Yin Gang, et al. Trustworthiness mechanism of Internet software [J]. Science in China:Ser. E Information Sciences, 2006, 36(10): 1156–1169.

[14] Lingxiao Jiang, Ghassan Misherghi, Zhengdong Su, et al. DECKARD: Scalable and Accurate Tree-based Detection of Code Clones. ICSE'07 Proceedings of the 29th international conference on Software Engineering, 2007: 96–105.

[15] Wang F., Cui B. Similarity Detection and Used in Static Source Code Analysis [J]. Software, 2012, 12: 071.

[16] Keivanloo I. Source Code Similarity and Clone Search [D]. Concordia University, 2013.

[17] Jiang L., Misherghi G., Su Z., et al. Deckard: Scalable and accurate tree-based detection of code clones [C]//Proceedings of the 29th international conference on Software Engineering. IEEE Computer Society, 2007: 96–105.

[18] UltraCompare [DB/OL]. http://www.ultraedit.com/products/ultracompare.html

[19] Christian List, Dean Grimm, et al. Winmerge [DB/OL]. http://winmerge.org/

[20] Li Z., Lu S., Myagmar S., et al. CP-Miner: Finding copy-paste and related bugs in large-scale software code [J]. Software Engineering, IEEE Transactions on, 2006, 32(3): 176–192.

[21] Kamiya T., Kusumoto S., Inoue K. CCFinder: a multilinguistic token-based code clone detection system for large scale source code [J]. Software Engineering, IEEE Transactions on, 2002, 28(7): 654–670.

[22] Malpohl G. JPlag: detecting software plagiarism [J]. URL http://www.ipd.uka.de, 2006, 2222.

[23] Baxter I.D., Quigley A., Bier L., et al. CloneDR: clone detection and removal [C]//Proceedings of the 1st International Workshop on Soft Computing Applied to Software Engineering. 1999: 111–117.

[24] Anderson J.P. Computer Security Technology Planning Study. Technical Report ESD-TR-73-51, Air Force Electronic Systems Division, Hanscom AFB, Bedford, MA, October 1972.

[25] China Information Technology Security Evaluation Cente. "The platform of software trustworthiness measurement and evaluation based on network environment", 2014.

Information Technology and Applications – Li (Ed.)
© 2015 Taylor & Francis Group, London, ISBN 978-1-138-02677-3

Learning hierarchical feature representation for RGB-D object recognition

S.-Q. Tu, Y.-G. Xue, X. Zhang, X.-L. Huang & H.-K. Lin
College of Information, South China Agricultural University, Guangzhou, P.R. China

ABSTRACT: RGB-D that is recently introduced, to record high quality RGB and depth information can improve image recognition. By analyzing deep learning framework in image signals, we can learn image features with convolution and subsample for classification instead of using hand-crafted descriptors like SIFT or Hog. On the other hand, group sparse representation is a promising approach for image classification. In this paper we have proposed a new method to learn hierarchical feature representation from RGB-D data. First, we learn the first layer features with Convolutional and Recursive Neural Networks (CNN-RNN) from raw RGB-D images directly by unsupervised algorithm. Then, we extract meaning feature representation for classification by Block Group Sparse Coding (BGSC) in the following feature extraction processes. Experimental results show that the proposed method can achieve better accuracy on RGB-D object dataset in comparison to convolutional and recursive neural networks, Group Sparse Coding (GSC), label-consistent K-SVD (LC-KSVD) and Sparse Representation base Classification (SRC).

Keywords: RGB-D; block group sparse coding; convolutional and recursive neural networks; classification recognition; feature representation

1 INTRODUCTION

Object recognition is an important capability in computer science appealing to researchers from different fields such as computer vision, machine learning and robotics. In the past decades, a variety of features and classification algorithms have been proposed and applied to this problem, resulting in significant progress in object recognition capabilities, as can be observed from the steady improvements on standard benchmarks such as Caltech101 [1] and CIFAR [2]. New sensing technologies, the rapidly maturing technologies of RGB-D (Kinect-style) and depth cameras [3] provide synchronized videos of color and depth with high quality, presenting a great opportunity for combining color and depth based on recognition. Most recent methods for object recognition with RGB-D images used hand-designed features which include SIFT for 2d images [5], Spin Images [6] for 3D point clouds, specific color, shape and geometry features [7], and a Learned Feature Descriptor [8][9].

Sparse representation has been a powerful image representation model for variety of problems, including image tag, image annotation [4] and image classification [13, 18]. The Sparse Representation Classifier (SRC) [18] employs the entire set of training samples as the dictionary for classifying image, and gets impressive performances on face recognition. But the sparse representation does not

take advantage of any structural information from train data. These features representation may have a group clustering in which the nonzero elements exist in the union of subspaces. The group sparse coding in which a block structure is imposed on the dictionary [12] is widely used in image classification [13] [14]. Therefore, we introduce Convolutional and Recursive Neural Networks (CNN-RNN) [10] for extracting the first layer features from raw RGB-D images and structured group sparse representation for learning higher representation.

In this paper, we proposed an efficient model to learning hierarchical feature representation from RGB-D images. The CNN-RNN can extract low-level invariant features from raw RGB-D images, and the features are then given as inputs to Block Group Sparse Representation (BGSC) to compose higher features. The objective of this study is to learn hierarchical feature representation by applying the CNN-RNN and BGSC method for RGB-D object recognition.

2 LEARNING HIERARCHICAL FEATURE REPRESENTATION WITH CNN-RNN AND BGSC

In this section, we will first describe a model based on CNN-RNN for learning lower features. Then, we will adopt block group sparse representation for high-level feature and linear SVM for RGB-D object recognition. The overview of our method is shown in Figure 1. We will discuss each part in detail in the following content.

2.1 CNN-RNN model for features of the first layer

A model based on CNN-RNN [10] for learning the first layer features is established. Figure 2 illustrates the structure of CNN-RNN. We convolve each image of size (height and width) d_I with K square filters of size d_p, which results in K filter responses and each of dimensionality $d_I - d_P + 1$. We average pool them with square regions of size d_ℓ and a stride size of s, to obtain a pooled response with width and height equal to $r = (d_I - d_l)/S + 1$. So the output X of the CNN layer applied to one image is a $K \times r \times r$ dimensional 3D matrix. After getting a 3D matrix $X \in R^{K \times r \times r}$ for each image, a block will be defined as a parent vector $p \in R^K$ by merging a list of adjacent column vectors in our RNN architecture. The same procedure is applied to both color and depth images separately. The CNN-RNN layer learns the first-level features which are input into block group sparse coding to compose features of the second layer.

We follow the procedure described in [15] to learn filters for the CNN-RNN. First, random patches are extracted into two sets (RGB and Depth). Each set of patches are normalized and whitened. The pre-processed patches are clustered by running k-means. Figure 3 shows the resulting filters which have captured standard edges and color features. CNN-RNN is used to convolve filters over the input image to extract features for the inputs of block group sparse coding layer.

2.2 Learning group sparse representation

Block group sparse representation is obtained via incorporating both group structure in data and block structure in dictionary, and similar to BGSC [14]. Figure 4 illustrates the BGSC algorithm in which a group of data $X(g)$ on the left is sparsely coded with respect to the dictionary D with block structure $D[1] \dots D[b]$.

Given a collection of low-level feature data $X = \{x1, \dots, xl\}$, the $X(g)$ is vector/matrix generated from the data X; $C(g)$ is sparse coefficient vector; and D is the learned dictionary which has a block structure. The sparse representations for a group of data samples X are computed by using the minimum number of blocks from D to solve the following minimization problem:

$$p_{I_{o,p}} : \min_c \sum_i I(\| c_{[i]} \|) \quad s.t. X = DC \tag{1}$$

where $I(\cdot)$ is an indicator function, $p = 1, 2$ and $C[i]$ is the i-th block (sub-matrix) of C corresponding to the i-th block of D as shown in Figure 4. This

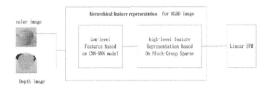

Figure 1. An overview of hierarchical feature representation method.

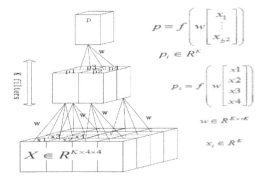

Figure 2. The pooled CNN and RNN structure with blocks of 4 children.

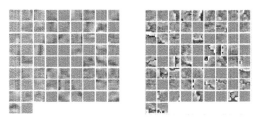

Figure 3. Visualization of the K-means filters after pre-training. **Left**: Standard RGB filters capture edges and colors. **Right**: The depth images filters using k-means method.

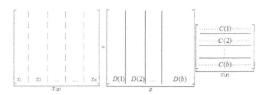

Figure 4. The block group sparse representation.

324

combinatorial problem is NP-hard. The L1-relaxed version of the above program is:

$$p_{l_{o,p}}: \min_c \sum_i (\| c_{[i]} \|) \quad s.t. X = DC \qquad (2)$$

The program $p_{l_{o,p}}$ in equation (2) can be cast as an optimization problem which minimizes the objective function:

$$e_c(C; X, D) = \sum_g e_c(c^{(g)}; X^{(g)}, D)$$
$$= \sum_g \left(\frac{1}{2} \| X^g - DC^{(g)} \|_F^2 + \lambda \sum_i \| C_{[i]}^{(g)} \|_p \right) \qquad (3)$$

The optimization steps are presented only for one specific group of data X and its corresponding sparse coefficients C. Equation (3) can be written as follows:

$$\frac{1}{2} \| X - DC \|_F^2 + \lambda \sum_i \| C_{[i]} \|_2$$
$$= \frac{1}{2} \left\| X - \sum_{i \neq r} D_{[i]} C_{[i]} - D_{[r]} C_{[r]} \right\|_F^2 + \lambda \| C_{[r]} \|_2 + c \qquad (4)$$

Here c includes the terms not depending on $C[r]$. When p is equals to 1, the objective function is separable. Iterates of elements in $C[r]$ can be solved using a method similar to [12]. When p is equals to 2, it is only block-wise separable. After computing the gradient of Equation (4) with respect to $C[r]$, the optimal solution is obtained by adopting sub-gradient algorithms similar to [14].

2.3 Updating dictionary D

To initialize the dictionary, we use the K-SVD method. The initial sub-dictionary D_i for class i is obtained by several iterations within each training feature class. The input dictionary $D^0 (D^0 = [D_1, D_2, ..., D_N])$ is initialized by combining all the individual class dictionaries.

Intra-Block Coherence Suppression Dictionary Learning (ICS-DL) algorithm is used for dictionary learning. The intra-block coherence can be defined as follows:

$$\mu_s(D) = \max_i \left(\max_{p,q \in I(i), p \neq q} \frac{|d_p^T \cdot d_q|}{\|d_p\| \|d_p\|} \right)$$

where $I(i)$ is the index of the atoms in block i. Inter-block coherence is defined as:

$$\mu_B(D) = \max_{i \neq j} \left(\left(\frac{1}{n_a} \sigma_1 \left(D_{[i]}^T D_{[j]} \right) \right) \right)$$

where σ_1 is the largest singular value and n_a is the size of block. It is necessary to have a dictionary updating algorithm that minimizes the intra-block coherence. The dictionary updating algorithm has the following objective function:

$$e_d(D, X, C) = \sum_g \frac{1}{2} \| X^{(g)} - DC^{(g)} \|_F^2$$
$$+ \gamma \sum_{k=1}^{|D|} \| d_k \|_2 + \beta \sum_b \left(\sum_{p,q \in t(b), p \neq q} \| d_p^T d_q \|^2 \right)$$
$$+ \lambda \Omega(C) \qquad (5)$$

The third term minimizes the intra-block coherence. The updated formula is derived to be required in optimizing the objective function above for one group of features. The dictionary construction process is summarized in Algorithm 1.

Algorithm 1: Dictionary Learning via block-gradient descent

Input: Data X and Parameters λ, Υ, β
Output: D, C
Initialize: Initial Dictionary D^0, $\varepsilon_d = 10^{-5}$
While not converged, $i \leq$ maxIterD do
find C with respect to D^i using equation (3)
fix C and update D^{i+1} by equation (5)
check the convergence conditions:
$\quad \| D^{i+1} - D^i \|_\infty < \varepsilon_d$
End while

3 EXPERIMENTS ON RGB-D DATASET

This research focuses on object recognition using RGB-D data which is a newly established field mainly popularized by the availability of new and cheap sensors such as the Microsoft Kinect. The dataset, called RGBD dataset, contains 41,877 RGB-D images of 300 physically distinct everyday objects recorded from three viewpoints [5]. The objects are classified into 51 categories. The objects in the dataset are segmented and cropped from the background by combining color and depth segmentation cues. The dataset used contains 10 object instances. The objects picked are shown in Figure 5. The dataset is subsampled to get 5880 RGB-D images by taking every fifth frame, resulting in 5880 RGB-D images. Average accuracies and standard deviation for each experiment were retrieved across 10 trials.

In this work, we focus on the problem of instance recognition and use the same setup 10 random splits as [10]. All development is carried out on a separate split and model ablations are run on each of the 10 splits. We follow the testing procedure described: randomly leaving half objects of each instance for testing and training the classifiers on the half objects.

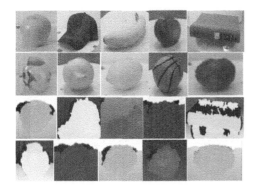

Figure 5. Object categories from RGB-D dataset: apple, cap, banana, peach, binder, pitcher, onion, lemon, ball, and lime.

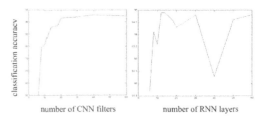

Figure 6. Classification accuracy on different values of N1 and N2. **Left:** Increasing the number of N1 improves performance. **Right:** Impact of N2(RNN layers) when N1 equal 40.

3.1 *Learning hierarchical feature representation for RGBD images*

Before the images are given to the CNN they are resized to $d_I = 148$. The $9 \times 9 \times 3$ patches for RGB and 9×9 patches for depth are individually normalized by subtracting the mean and divided by the standard deviation of its elements. Unsupervised pre-training for CNN filters is performed for all experiments by using k-means. In addition, ZCA whitening is performed to de-correlate pixels and gets rid of redundant features in raw images [15]. A valid convolution is performed implemented with the size of K = 40 and 9×9 filter. Average pooling is then performed with the size of pooling regions being of size $d_I = 10$ and stride size being $s = 5$ to produce a 3D matrix of with $40 \times 27 \times 27$ for each image. The final features with 4736 dimensions are obtained by concatenating those of RGB and depth. N1 and N2 are the numbers of CNN filters and RNN layers respectively. When N2 equals 2, the impacts of N1 for classification accuracy is shown in Figure 6 (left). Impact of N2 is demonstrated in Figure 6 (right) when N1 equals 40. When the numbers of N1 and N2 equal 40 and 12 respectively, our method obtains the best classification results. Therefore, we select 40 of CNN filters and 12 of RNN layers in our experiments.

After obtaining the features of first layer, we learn high-level feature by completing the steps as follows:

Step 1. Project the samples to m (m = 500) dimensions by performing PCA on features CNN-RNN model.

Step 2. By K-SVD method, generate initial dictionary D with nb blocks (nb = 50) and each block contains 20 columns atoms (na = 20), and dictionary D contains 1000 (nb × na) columns.

Step 3. Iteratively compute coefficients C applying BGSC method and update the dictionary D using dictionary learning algorithm 1.

Step 4. Compute sparse coefficients of the testing samples using BGSC, GSC and SC, respectively.

Step 5. Compute the class label of each column of X individually using line SVM classifier:

3.2 *Comparison with other methods*

We evaluate our method and compare with a number of approaches: CNN-RNN [8], label-consistent K-SVD (LC-KSVD) [13], the same produce in 3.1 except using different sparse coding methods including Sparse Representation based Classification (SRC) [18] and Group Sparse Coding without block structure in dictionary (GSC) [12].

The classification accuracies of the four methods are shown in Table 1. Table 2 shows classification accuracy on each kind of object by our method. The results indicate that group regularized methods are generally better than SRC. Our method outperforms others by a significant margin of 6%. CNN-RNN exploit the strong spatially local correlation present in natural images by obtaining low-level features from raw images, and the size of feature dimensions is still large and the features represent local aspects of the images. Sparse representation can however reflect whole characters of RGB-D images and represent these structures of features in a compact and sparse representation compared with RNN.

Compared with SRC, group sparse coding during dictionary learning of per class can promotes the use of the same dictionary words for all images of the same class, which yields better performance on time (a smaller dictionary yields faster image encoding) and space (one can store more images when they take less space). From Table 1, we find that GSC and our method are better than SC and LC-KSVD2.

We conducted experiment to test whether the additional depth information can promote

Table 1. Classification accuracy on the RGB-D dataset.

Classifier	RGB (%)	RGB+Depth (%)
CNN-RNN [8]	86.3 ± 1.7	93.4 ± 2.6
SRC [18]	89.5 ± 2.1	95.2 ± 2.8
GSC [12]	90.6 ± 1.5	96.4 ± 1.9
LC-KSVD2 [13]	90.1 ± 1.6	94.1 ± 2.3
Our method	**91.8 ± 2.6**	**97.6 ± 2.1**

Table 2. Classification accuracy on each kind of object by our method.

Name	Classification accuracy	
	RGB (only)	RGB-D
Apple	73.0 ± 1.0	73.7 ± 1.7
Ball	97.7 ± 0.5	98.9 ± 0.3
Banana	35.9 ± 2.1	38.3 ± 1.5
Lemon	35.0 ± 1.6	35.6 ± 1.9
Lime	52.1 ± 0.3	52.5 ± 0.6
Binder	81.7 ± 1.3	84.6 ± 1.5
Peach	80.0 ± 1.8	80.0 ± 1.8
Pitcher	74.5 ± 1.0	74.8 ± 1.0
Cap	62.5 ± 0.3	63.8 ± 1.2
Onion	70.0 ± 0.9	71.1 ± 1.1

recognition accuracy. From Table 1 and Table 2, There are two important observations: One, Using all available channels, accuracy was significantly higher (97.6%) by our method, ignoring depth information (color only), the recognition accuracy was 91.8%; Two, every instance object using RGB and depth information achieved better result of classification compared with our method used only RGB information. This shows that the additional depth channel does carry additional information which can improves recognition accuracy.

4 CONCLUSION

In this paper, hierarchical features have been learned from RGB-D using CNN-RNN model combined with block group sparse coding method. By incorporating both the group structure for the input data and the block structure for the dictionary in the learning processes, discriminative feature representations have been obtained by raw RGB-D images. Experimental results show that our method can get significantly higher accuracy than CNN-RNN method, and our method using both depth and color information outperforms that only using color information (RGB) from the images by more than 8%.

ACKNOWLEDGEMENTS

This work was supported by National Science and Technology Support Program of china (2013BAJ13B05).

REFERENCES

[1] F.F. Li, R. Fergus, and P. Perona. One-shot learning of object categories. IEEE Transactions on Pattern Analysis and Machine Intelligence, 28(4):594–611, 2006.

[2] A. Krizhevsky. Learning Multiple Layers of Features from Tiny Images. Master thesis, Department of Computer Science, University of Toronto, 2009.

[3] Microsoft Kinect. http://www.xbox.com/en-us/kinect.

[4] Y. Yuan, F. Wu, J. Shao, et al. Image annotation by semi-supervised cross-domain learning with group sparsity[J], Journal of Visual Communication and Image Representation. 2013, 24(2):95–102.

[5] K. Lai, L. Bo, X. Ren, and D. Fox. A Large-Scale Hierarchical Multi-View RGB-D Object Dataset, In ICRA, 2011. pp.1817–1824.

[6] A. Johnson. Spin-Images: A Representation for 3-D Surface Matching. PhD thesis, Robotics Institute, Carnegie Mellon University, 1997.

[7] H.S. Koppula, A. Anand, T. Joachims, and A. Saxena. Semantic labeling of 3d point clouds for indoor scenes, In NIPS, 2011.

[8] L. Bo, X. Ren, and D. Fox. Depth kernel descriptors for object recognition, In IROS, 2011.

[9] M. Blum, J.T. Springenberg, J. Wlfing, and M. Riedmiller. A Learned Feature Descriptor for Object Recognition in RGB-D Data, In ICRA, 2012.

[10] R. Socher, B. Huval, B. Bhat, et al. Convolutional-Recursive deep learning for 3D object classification, In NIPS 2012.

[11] J. Yang, K. Yu, Y. Gong, et al. Linear spatial pyramid matching using sparse coding for image classification. 2009.

[12] S. Bengio, F. Pereira, Y. Singer, and D. Strelow. Group sparse coding, Advances in NIPS, 22:82–89, 2009.

[13] Z. Jiang, Z. Lin, and L.S. Davis. Label consistent K-SVD: learning a discriminative dictionary for recognition, IEEE Transactions on Pattern Analysis and Machine Intelligence, vol. 35, no. 11, pp. 2651–2664, 2013.

[14] Y.T. Chi, M. Ali, A. Rajwade, et al. Block and Group Regularized Sparse Modeling for Dictionary Learning, In CVPR, 2013.

[15] A. Coates, A.Y. Ng, H. Lee. An Analysis of Single-Layer Networks in Unsupervised Feature Learning, Journal of Machine Learning Research Proceedings Track: AISTATS, 2011.

[16] E. Elhamifar, R. Vidal. Robust classification using structured sparse representation, In CVPR, 2011, pages 1873–1879.

[17] G.E. Hinton. Learning multiple layers of representation. Trends in cognitive sciences, 11: 428–34, October 2007.

[18] J. Wright, A. Yang, A. Ganesh, S. Sastry, and Y. Ma. Robust face recognition via sparse representation. Pattern Anal. Mach. Intell, IEEE Trans. on, 31(2):210–227, 2009.

Information Technology and Applications – Li (Ed.)
© 2015 Taylor & Francis Group, London, ISBN 978-1-138-02677-3

An image registration algorithm based on FAST, RANSAC and SURF

Weisheng An, Rangming Yu & Yuling Wu
School of Mechanical Engineering, Southwest Jiaotong University, Chengdu, Sichuan, China

ABSTRACT: In view of image registration an algorithm was proposed to realize fast registration in this paper, including detecting interest points using FAST detector, calculating feature vectors using SURF description and matching the interest points through fast approximate nearest-neighbor search. First a Gaussian scale pyramid of the reference image and the image to be matched is established. FAST corners are detected from each level in the pyramid. Then every interest point is assigned an orientation based on information from a circular region around the interest point. SURF description vector is extracted by a constructed square region aligned to the selected orientation. Finally the original matching points are determined through fast approximate nearest neighbor search, and RANSAC algorithm is adopted to exclude false matching points. The experiments show that the algorithm performs better than SIFT and SURF in feature detection speed and matching accuracy.

Keywords: image registration; FAST; SURF; Approximate Nearest Neighbor (ANN); RANSAC

1 INTRODUCTION

Image registration is the process of finding correspondence between all points in two or more images of the same scene from different time, different views or different sensors and taking space transformation processing to make each image corresponding to each other on the geometry. Its application scope is quite widespread, including: computer vision and pattern recognition (such as image segmentation, 3D reconstruction, object recognition, motion estimation and tracking, etc.), medical image analysis, remote sensing image processing, image fusion, etc. Image registration methods can be divided into two broad categories: image registration based on gray level information and image registration based on feature. With good robustness and adaptability to image transformation, image registration based on feature is more and more widely applied. It is the focus and hotspot in the research of the image registration nowadays.

The Scale Invariant Feature Transform (SIFT) algorithm proposed by Lowe, which combines a scale-invariant region detector and a descriptor based on the gradient distribution in the detected regions, has been proved to be invariant to image translation, rotation and scaling, and partially invariant to illumination changes, noise and affine or 3D projection. But the high dimensionality of the descriptor is a drawback of SIFT at the matching step. The interest point detection and the descriptor extraction are computationally inefficient. The SIFT algorithm is not suitable for real-time applications. A replacement called Speeded Up Robust Feature (SURF) with lower computation cost was proposed by Bay et al. SURF approximates SIFT with respect to repeatability, distinctiveness and robustness, yet can be computed and compared much faster by relying on the integral image for image convolutions, by using a Fast Hessian matrix for the detector and by using Haar wavelet responses for the descriptor. But the matching speed is slow through greedy nearest neighbor search. Rosten et al. proposed a feature detection algorithm called Features from Accelerated Segment Test (FAST), which is very fast in terms of feature detection, yet not invariant to image scale, susceptible to noise and illumination changes.

Feature point detection and matching are the two basic steps of image registration. This paper put forward an algorithm based on the FAST corner detection, SURF description vector and fast approximate nearest neighbor search algorithm on the basis of previous work.

2 FEATURE DETECTION

The purpose of Feature point detection is to select "interest points" at distinctive location in the image, such as corners, blobs and T-junctions. FAST does not produce multi-scale features. A Gaussian scale pyramid of the image to be detected was established in this paper. FAST was applied to locate features at each level in the pyramid. To improve the invariance to illumination changes the image was preprocessed with histogram equalization before detection.

2.1 Build a Gaussian image pyramid

Detecting feature points that are invariant to scale change of an image can be accomplished by searching for stable features across all possible scales, using a continuous function of scale known as scale space. It has been shown by Lindeberg[6] that under a variety of reasonable assumptions the Gaussian function is the only possible linear scale-space kernel. The scale space of an image is defined as a function, $L(x, y, \sigma)$, that is produced from the convolution of a variable scale Gaussian, $G(x, y, \sigma)$, with an input image, $I(x, y)$:

$$L(x, y, \sigma) = G(x, y, \sigma) \otimes I(x, y), \tag{1}$$

where $\otimes$ is the convolution operation in x and y. The Gaussian convolution kernel function is defined as:

$$G(x, y, \sigma) = \frac{1}{2\pi\sigma^2} e^{-(x^2 + y^2)/2\sigma^2}, \tag{2}$$

where σ is the scale factor which indicates the degree of image is smoothed. The bigger the value, the more fuzzy the image.

In order to improve the efficiency of algorithm, a Gaussian pyramid was set up with three groups and three layers in each group, totally 9 levels. Through experimental comparison, the first level image of the pyramid has a smooth scale $\sigma_0 = 1.4$. Let k is a scale ratio between adjacent layers, and $k = 2^{1/S}$, where S is the total number of each group. Any level's scale can be computed through $\sigma_s = \sigma_0 k^s$, where s is the ordinal level number in the pyramid, $s = \{0,1,2,...,9\}$. The image of the first layer in each group is from the sample of the image of the last layer in previous group. So each group needs one more layer, the last layer is not for feature point detection. In order to get more feature points, the original image can be amplified 2 times as the first group images.

2.2 FAST feature detection

FAST determines whether a candidate pixel p is a corner or not by testing pixel p is brighter or darker than the pixels around it. It is very fast because of its simple calculation.

The segment test is performed for a feature at a pixel p by examining a circle of 16 pixels (a Bresenham circle of radius 3) surrounding p. FAST detector classifies p as a corner if there exists a set of n contiguous pixels in the circle which are all brighter than the intensity of the candidate pixel I_p plus a threshold t, or all darker than $I_p - t$, as illustrated in Figure 1.

$$I_p = \begin{cases} d & I_x \le I_p - t & \text{(darker)} \\ s & I_p - t < I_x < I_p + t & \text{(similiar)}, \\ b & I_p + t \le I_x & \text{(brighter)} \end{cases} \tag{3}$$

The highlighted squares are the pixels used in the feature detection. The pixel at p is the centre of a detected corner. The arc is indicated by the dashed line passes through 9 contiguous pixels which are brighter than p by more than the threshold. Where I_p is the grey value of the center point p, I_x is the grey value of the point on the circle, and t is intensity threshold.

It has been shown that the best repeatability is achieved while n takes 9 in literature. FAST does not produce a measure of corner, and it has large responses along edges. A Harris corner measure was employed to order the FAST key points. For a target number N of key points, first a threshold low enough was set to get more than N key points, then the key points were ordered according to the Harris responses. The top N key points were picked at last. The points across the edge can be discarded through using a 2×2 Hessian matrix,

$$H = \begin{bmatrix} D_{xx} & D_{xy} \\ D_{xy} & D_{yy} \end{bmatrix}, \tag{4}$$

where D_{xx}, D_{xy} and D_{yy} are the second derivatives of candidate corner. The derivatives can be estimated by taking difference of neighboring sample point. The principal curvature of D is proportional to the eigenvalues of H. In order to avoid calculating the eigenvalues directly, the ratio between the eigenvalues is computed. Let α be the eigenvalue with the largest magnitude and β be the smaller one. Then, the sum of the eigenvalues can be computed from the trace of H, and the product from the determinant:

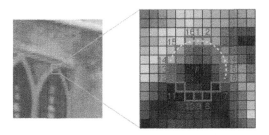

Figure 1.   FAST feature detection in an image patch.

$$Tr(\boldsymbol{H}) = D_{xx} + D_{yy} = \alpha + \beta,$$
$$Det(\boldsymbol{H}) = D_{xx}D_{yy} - (D_{xy})^2 = \alpha\beta. \quad (5)$$

Let r be the ratio of the eigenvalues, so $\alpha = r\beta$. Then,

$$\frac{Tr(\boldsymbol{H})^2}{Det(\boldsymbol{H})} = \frac{(\alpha + \beta)^2}{\alpha\beta} = \frac{(r+1)^2}{r}. \quad (6)$$

The quantity $(r + 1)^2/r$ is at a minimum when the two eigenvalues are equal and it increases with r. In this paper the experiments use a value of $r = 10$. When

$$\frac{Tr(\boldsymbol{H})^2}{Det(\boldsymbol{H})} < \frac{(r+1)^2}{r}, \quad (7)$$

The feature point is retained, or given up.

3 SURF DESCRIPTOR

3.1 *Orientation assignment*

In order to be invariant to image rotation, every feature point has to be assigned a reproducible orientation. For that purpose, first the Haar wavelet responses in x and y direction within a circular neighborhood of radius $6s$ around the interest point (corner) are calculated, with s the scale at which the interest point was detected. The sampling step is scale dependent and chosen to be s. The size of the wavelets is also scale dependent and set to a side length of $4s$. Then the wavelet responses are weighted with a Gaussian ($\sigma = 2.5$) centered at the interest point, which makes the responses near the interest point contribute much, the responses away from the interest point little.

Following the responses are represented as points in a space with the horizontal response strength along the abscissa and the vertical response strength along the ordinate. The dominant orientation is estimated by calculating the sum of all responses within a sliding orientation window of size 60 degrees (a step of 5 degrees). The horizontal and vertical responses within the window are summed. The two summed responses then yield a local orientation vector. The longest such vector over all windows defines the orientation of the interest point.

3.2 *Descriptor extraction*

For the extraction of the descriptor, the first step consists of constructing a square region centered on the interest point and oriented along the orientation selected in previous section. The side length

of the square is $20s$. The region is regularly divided into smaller 4×4 square sub-regions. For each sub-region, the Haar wavelet responses at 5×5 regularly spaced sample points are computed. The Haar wavelet response in horizontal direction (direction parallel to the interest point orientation) is simply called d_x and the Haar wavelet response in vertical direction (direction perpendicular to the interest point orientation) d_y. The size of the wavelets is $2s$. To increase the robustness of geometric transformation and localization errors, the responses d_x and d_y are weighted with a Gaussian ($\sigma = 3.3s$) centered at the interest point. Then, the wavelet responses d_x and d_y, and the absolute values of the responses $|d_x|$ and $|d_y|$ are summed up over each sub-region and form a first set of entries in the feature vector. Hence, each sub-region has a 4D descriptor vector V for its underlying intensity structure $V = [\sum d_x, \sum d_y, \sum |d_x|, \sum |d_y|]$. Concatenating this for all 4×4 sub-regions, this results in a descriptor vector of length 64. In order to achieve invariance to contrast the descriptor vector has to be turned into a unit vector.

4 FEATURE MATCHING

The purpose of the feature matching is to estimate the optimal transformation model parameters according to the feature information extracted from the reference image and stay registration image and the images' properties. The parameters reflect the geometric relationship between each other. Generally, the Euclidean distance between feature vectors is used to measure the similarity of feature points. SURF feature vector is a high dimensional vector. The nearest neighbor matching method of greedy search results accurately, but matching speed is quite slow. In literature approximate nearest neighbor matching algorithm in computer vision for high-dimensional spaces has been evaluated. On the basis of the comparison, Muja and Lowe proposed that for the problem of approximate nearest neighbor search in high dimensional space hierarchical k-means tree and multiple randomized k-d tree has good performance, and implemented the algorithm to automatically choose the best algorithm and parameter values according to the user input data and expected accuracy. This paper uses the characteristics of Fast Approximate Nearest neighbor search algorithm for feature matching. It is based on the latest OpenCV implementation of FLANN (Fast Library for Approximate Nearest Neighbors).

Given a key point P in the reference image, the Euclidean distances from it to the first nearest neighbor P1 and the second nearest neighbor P2 in the stay registration image are d1 and d2, respectively. FLANN search is used to find out d1 and d2. If it holds d1/d2 < threshold, T, the point P is matched to its first nearest neighbor P1. The threshold is usually 0.6~0.8, the greater the threshold value is, the more the matching points.

There are always some false matches in the coarse matching result because the assignment only takes the local content of the images and ignores the spatial relationship of the key points. This paper uses the Random Sample Consensus (RANSAC) algorithm to remove the outliers.

5 EXPERIMENTAL RESULTS

The experiment was performed on a 2.10 GHz Intel Core i3-2310 with 2 GB dynamic memory. The test images are all real scene images from a set of publicly available dataset[1]. FAST corner detection threshold, t, was set to 20, and d1/d2 threshold, T, 0.65.

The algorithm was compared to SIFT and SURF, whose implementation was based on the latest OpenCV. The number of interest points (corners) selected is on average similar for all detectors (see Table 1 for an example). The thresholds were adapted according to the number of interest points found with the multi-scale FAST detector. The matching results of two boat images were shown in Table 2 and Figure 2 (c). The experiment time shown in the table is mean time for five experiments.

As can be seen from Tables 1 and 2, the multi-scale FAST detector is faster than SURF and SIFT, and the matching accuracy is higher also.

The images with viewpoint change, scaling, rotation and illumination change were used to test the

Table 1.   Corner detection time comparison.

	Number of points		Time (ms)	
Algorithm	Image 1	Image 2	Image 1	Image 2
This paper	1200	1200	173	152
SURF	1201	1207	231	228
SIFT	1200	1200	822	481

*Image 1 and image 2 resolution 850×680.

Table 2.   Feature matching performance comparison.

Algorithm	Number	Time (ms)	Accurancy (%)
This paper	48	128	93.7
SURF	102	157	87.2
SIFT	124	236	85.4

(a) illumination

(b) illumination (histogram equalization)

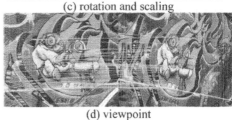

(c) rotation and scaling

(d) viewpoint

Figure 2.   Matching results in variant changes.

performance of the algorithm in the experiment. Some parts of the experimental results are shown in Figure 2.

From Figure 2 (a) and (b) we can see that the matching points are more evenly distributed after histogram equalization processing in the case of illumination change. Figure 2 (c) is in the case of scaling and rotation, and Figure 2 (d) in the case of viewpoint change. The experimental results in Figure 2 demonstrate that the algorithm is robust to scaling, rotation and illumination change, and performs well on multi-view images.

6 CONCLUSIONS

Based on the characteristics of FAST detection, SURF feature description vector and fast approximate nearest neighbor search, an algorithm for image registration is put forward. The experimental results show that it is invariant to image translation, rotation, and partially invariant to image scale, illumination change and view change. It performs well on feature point detection speed and matching accuracy. It can be used to the application with the images existing scaling, rotation, illumination change and little viewpoint change.

ACKNOWLEDGEMENT

This research was supported by the Fundamental Research Funds for the Central Universities (No.2682013CX024).

REFERENCES

[1] Chen Xiuxin, Xing suxia. *Image/video retrieval and image fusion* [M]. Mechanical industry publishing house, 2012: 141–154.

[2] Lowe, D.G. *Object Recognition from Local Scale-Invariant Features*[C]. Proceeding of ICCV'99, Keykyra, Greece: IEEE Computer Society, 1999: 1150–1157.

[3] Lowe, D.G. *Distinctive Image Features from Scale-Invariant Keypoints*[J]. International Journal of Computer Vision, 2004, 60(2): 91–110.

[4] Bay, H., Ess, A., et al. *SURF: Speeded-Up Robust Features*[J]. Computer Vision And Image Understanding, 2008, 110(3): 346–359.

[5] Rosten, E. *Machine learning for high-speed corner detection*[C]. European Conference on Computer Vision, 2006: 430–443.

[6] Lindeberg, T. *Scale-space theory: A basic tool for analyzing structures at different scales*[J]. Journal of Applied Statistics, 1994. 21(2): 224–270.

[7] Muja, M., Lowe, D.G. *Fast Approximate Nearest Neighbors With Automatic Algorithm Configuration*[C]. Proc of the VISAPP International Conference on Computer Vision Theory and Application. 2009: 331–340.

Information Technology and Applications – Li (Ed.)
© 2015 Taylor & Francis Group, London, ISBN 978-1-138-02677-3

The light source system for star simulator with multi-color temperature and multi-magnitude

Xiao-ni Li
Xi'an Institute of Optics and Precision Mechanics, Chinese Academy of Sciences, Xi'an, China
University of Chinese Academy of Sciences, Beijing, China

Cui-gang Wu, Zhen-hua Lu & Lai-yun Xie
Xi'an Institute of Optics and Precision Mechanics, Chinese Academy of Sciences, Xi'an, China

ABSTRACT: In order to obtain the star simulator light source with the energy distribution of more color temperatures and magnitudes in the broad band, this paper introduces a new simulator light source system based on least-square method mixed light source through learning from the amplitude and colorimetry. The system uses a variety of different narrowband LEDs and two different sizes of integrating spheres to achieve the different color temperatures and magnitudes spectral distribution by adopting the method of digital dimming to control the light intensity output of each LED light source. Make experiments to test and verify the method in the 400 nm–900 nm wavelength range of color temperature of 3900k, 4800k and 6500k as well as the magnitudes of −1 mv to 6 mv. Experimental results show that the error of spectral distribution is less than 10% and the error of magnitude is less than 0.08 mv. Compared with the traditional star simulator, the system is simpler with stable performance, high measurement accuracy, strong practicability and other characteristics.

Keywords: star simulator; spectral distribution; color temperature; magnitude; least square algorithm

1 INTRODUCTION

With the increasing development of the aerospace industry, the star sensor which used to make aircraft spatial position and posture positioning has been large scale use in the aerospace field. The star sensor mainly measures the light emitted at different locations of the stellars by the optical system, produces the observed star map, and make feature matching with the star map stored in the navigation system to determine the space craft attitude information. As we all know, one key technology of the star sensor is receiving the light emitted from the nearby stars. So the correction and calibration of the equipment receiving an light signal is very important. However, due to difficulties in the implementation of the air calibration exit, the reliability is not high and the price is very expensive so that the star sensor must be calibrated on the ground [1]. With the increasing high requirements on position accuracy of star sensor, put forward higher requirements on the star simulator which is used for laboratory testing and calibration of the star sensor. Star simulator is that providing a bundle of parallel light for star sensor measurements to simulate parallel light emitted by the star at infinity. Due to the different stellar brightness level and temperature of the star's age, mass, pressure and chemical composition of different stellars, the spectrum of light emitted is not in the same. Therefore, the star will have a different magnitude and spectral type. The main problems existing in the current star simulator is narrow spectral bands, single color temperature, small magnitude ranges. Zhang Jie et al have developed a simulator systems with multi-color temperature and multi-magnitude output [2]. The source of the system makes use of xenon lamp and LED as a compensation. To achieve more accurate color temperature star, spectral distribution fitting through the band intensity controller between the light source and the six prism integral rod in the wide band ranges. Therefore the structure of system is complex and difficult to achieve, moreover the spectral energy distribution of the error is large, difficult to control and, difficult to debug and calibration [3–4].

From the basic knowledge of the amplitude and chroma [5–6], this paper presents the light source system of a simple, stable performance and high accuracy. Using the ideal blackbody spectral radiation distribution as the spectral radiance of the stellar surface, use ideal monochromatic light spectral distribution and the data of the narrow band monochromatic LED collected actually. Respectively, using Matlab software simulation of the source system based on the least squares method. Finally, validate by experimental analysis, the experimental results agree well with the theoretical analysis and to verify the correctness and applicability of this method. The system has the characteristics of wide spectral, multicolor temperature, multi-magnitude and ideal spectral energy distribution and so on, to meet the needs of star simulator for research and production.

2 PRINCIPLES

2.1 *Definition of stellar magnitude*

The magnitude of stars is expressed the bright star level numerically. In 1830, the British scholar Herschel discovered the illuminance ratio of two adjacent magnitude is 2.512, further more set the illumination of zero star as 2.65, and the magnitude of the star brighter than zero star as the negative, the order of −1MV, −2MV, −3MV......then the stars darker than the zero-magnitude star are positive, followed by +1MV, +2MV, +3MV......[6]. As the receiving energy of the band is different, so there is the apparent magnitude, the magnitude of radiation, thermal magnitude, absolute magnitude and UBV magnitude system depending on the measurement frequency band magnitude different categories. As used herein are the apparent magnitude, that is using the human eye as a detector, at the center wavelength of 550 nm, establishing the star standards in the visible band.

2.2 *Blackbody radiation and color temperature*

The so-called blackbody, all means the object that can absorb any energy of any wavelength radiation at any temperature. The star is not really blackbody, but in the visible band and near infrared band, the radiation can be simulated approximately by a blackbody radiation at a temperature. The maximum radiation wavelength λ_p and surface temperature T of the star conform to the laws of displacement Wayne:

$$\lambda_p \times T = 2898 \ \mu m \cdot K \tag{1}$$

Generally based on Planck's formula of blackbody radiation, use the radiant existence to express the blackbody radiation:

$$M_{\lambda B} = \frac{2\pi hc^1}{\lambda^2} \cdot \frac{1}{e^{hc\lambda kBT} - 1}$$
$$= c_1 \lambda^{-5} \left[\exp(c_2/\lambda T) - 1 \right]^{-1}. \tag{2}$$

Among them, λ—radiation wavelength (unit: μm);

T—The thermodynamic temperature of blackbody (unit: K);

h—Planck's constant, $h = 6.625 \times 10^{-34} \ W \cdot s^2$;

k—The boltzmann constant, $k = 1.38054 \times 10^{-23} \ W \cdot s/K$;

c—The propagation velocity of light in a vacuum, $c = 2.99793 \times 10^{10} \ cm/s$;

c_1—The first radiation constant, $c_1 = 2\pi hc^2 = 3.7411 \times 10^4 \ W \cdot \mu m^4/cm^2$;

c_2—The second radiation constant, $c_2 = ch/k = 1.43867 \times 10^4 \ \mu m \cdot K$.

3 THEORETICAL ANALYSIS

The most critical star light simulator system mentioned in this paper is the realization of different spectral distribution under different color temperature and different magnitude. Therefore, this method is first analyzed theoretically and simulated basing on the above basic principles.

3.1 *The spectral distribution of different color temperature*

According o the theory of blackbody radiation, uses the Planck's formula to describe the blackbody spectrum radiation, that is

$$M_{\lambda B}(\lambda, T) = \frac{c_1}{\lambda^2} \cdot \frac{1}{\exp\left(\frac{c_1}{\lambda, T}\right) - 1} \tag{3}$$

Among them, $c_1 = 2\pi hc^2 = 3.741832 \times 10^{-6} (W \cdot m^2)$, $c_2 = hc/kB = 1.438786 \times 10^{-2} \ (W \cdot K)$ and c is the propagation velocity of light in a vacuum.

According to (2), it is concluded that the blackbody spectrum distribution under different color temperature curve using the Matlab simulation, as shown in Figure 1.

From Figure 1 it can be observed that, using a single light source can not reach the star simulator under different color temperature and spectral distribution. According to the spectral energy distribution of 4800k as shown in Figure 1. This distribution can be observed as the superposition of an infinite number of enough narrow spectral energy. Therefore, from the angle of the mathematical model, it may be used an infinite number of monochromatic light sources to obtain the desired spectral energy distribution of the superimposition [7]. But it cannot get enough of monochromatic light source for a wide spectral distribution of high accuracy simulation in practice. To solve the above problems, this paper

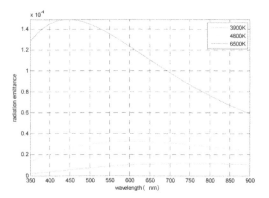

Figure 1. The radiation emission curve of the black-body under different color temperatures.

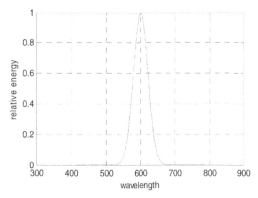

Figure 2. The spectral energy distribution curve of the LED light source.

uses multiple narrower spectral distribution of the LED light source to get the spectral distribution of the inside certain limits, and the integrating spheres are used as the equipment of mixing and equalizing light so that the intensity of the output light has high uniformity. Typical monochrome LED light output is a narrow band spectrum, its bandwidth (half peak width) is in 20 nm–50 nm commonly. According to the physical characteristics of LED light source, the single LED spectrum model of the radiation power distribution can be approximated by Gaussian distribution function in optical axis direction unit solid angle, as shown in the following type:

$$L(\lambda) = \alpha \cdot \exp\left[\frac{-(\lambda - \lambda_w)^1}{w^2}\right] \quad (4)$$

$$W_{FWHM} = 2w\sqrt{ln4} \quad (5)$$

Different monochromatic ideal LED light sources (see Fig. 2) of peak wavelength work together to form the spectral power distribution which is shown in Figure 1, and according to the superposition principle of the spectra, the basic mathematical model of LED synthesis spectrum is as shown in the following:

$$\hat{L}(\lambda) = \sum K_i L_i(\lambda) \quad (6)$$

Among them, $L_i(\lambda)$ are on behalf of a single LED spectral distribution and K_i represents a scale factor [9].

Although the actual spectrum is continuous, the experimental measured data are discrete data for particular wavelengths. Assume that discrete distribution of the simulate spectrum is $\{\lambda_i, M^m(\lambda_i, T)\}(i=1,2,...,m)$, among them,

$M^m(\lambda_i, T)$ is the spectral energy distribution curve in optical system pupil with LED spectral radial brightness as the basis fitting function $\{L_j(\lambda)\}(j=1,2,...,n)$. When fitted with the function $f(\lambda) = k_1 L_1(\lambda) + k_2 L_2(\lambda) + ... + k_n L_n(\lambda)$. To properly select the fitting parameters $k_1 \cdot k_2 ... k_n$ can improve the fitting precision. Remember the vector $L_1 = [L_1(\lambda_1), L_1(\lambda_2), ..., L_1(\lambda_n)]^T$, matrix $A = [L_1, L_2, ..., L_n]$ $b = M^m(\lambda_1, T), M^m(\lambda_2, T), ..., M^m(\lambda_m, T), X = [K_1, K_2, ..., K_n]^T$. When $m > n$, overdetermined equation is obtained as follows:

$$AX = b \quad (7)$$

Calculate the approximate solution of overdetermined equations. In practical applications, the proportion coefficient can only take the non-negative, so get the linear nonnegative least squares solution of overdetermined equation.

This paper uses the monochrome LEDs of half peak width w = 25 nm and different peak wavelength interval to make Matlab simulation of the spectral energy distribution. Figures 3–5 are spectra for peak wavelength interval d = 25, color temperatures T = 3900k, 4800k and 6500k; the spectrum of peak wavelength interval d = 20, color temperature T = 4800k is in Figure 6 and all the error bounds are 10%. It can be observed from Figures 3–5, the errors of synthesized spectral distribution using the 23 kinds monochromatic LED light sources with peak wavelength interval d = 25 nm and the half peak width w = 25 nm are bigger.

It can be observed in the Figures 3–5, the synthesis spectral distribution has obvious jagged fluctuations, moreover the error of low band range (350 nm to 450 nm) is more obvious at T = 3900k. And color temperature of the synthesis

337

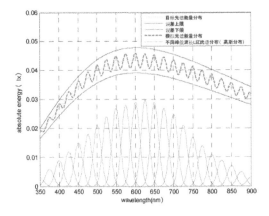

Figure 3. The spectral energy distribution simulated by using the ideal Gaussian distribution under T = 4800k, W_{FWMM} = 25 nm, d = 25 nm.

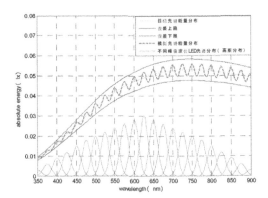

Figure 4. The spectral energy distribution simulated by using the ideal Gaussian distribution under T = 3900k, W_{FWMM} = 25 nm, d = 25 nm.

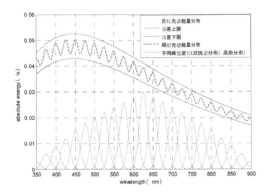

Figure 5. The spectral energy distribution simulated by using the ideal Gaussian distribution under T = 6500k, W_{FWMM} = 25 nm, d = 25 nm.

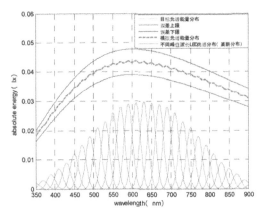

Figure 6. The spectral energy distribution simulated by using the ideal Gaussian distribution under T = 4800k, W_{FWMM} = 25 nm, d = 20 nm.

4800k 25 nm the error of the spectral distribution which is synthesized by using 29 kinds monochromatic LED light sources that the half width and peak wavelength interval are 25 nm and 20 nm is small in Figure 6. The maximum errors of the three spectral fitting are 3.10%, 3.69% and 4.44% when the color temperatures are 3900k, 4800k and 6500k.

Therefore, select the LED light sources, use the LEDs whose error of the measured peak wavelength and the nominal value is smaller to complete the synthesis of target spectrum in this paper, and reduce the sawtooth wave of the spectral distribution, so that to make the spectral distribution relatively smooth, and further reduce the simulation error of the spectral energy distribution curve. By increasing the types of LED light source, choosing the LED light source with small peak interval and wider half peak width. We found market sales peak interval of less than 25 nm and half peak width of 25 nm LED light emitting diode through research. As the impact of the manufacturing process and the driving current, there will be some deviation between the nominal values and peak wavelength and width at half maximum [8]. As shown in Figure 7, when using the 29 single-color LED light source, if the peak wavelength of 610 nm LED changes to 614 nm, the biggest error of the synthetic spectral distribution which changes from 4.44% to 10.52% is beyond the error limit of 10%.

3.2 The implementation of different magnitudes

The control of magnitudes is actually the control of the location of light intensity. Its implementation can generally adopt the methods, including the use of neutral attenuation slice and polaroid,

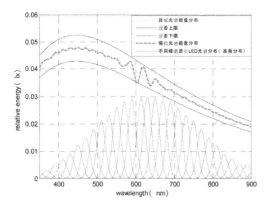

Figure 7. The spectral distribution curve when the peak wavelength is changed under the condition of 6500k.

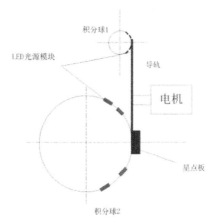

Figure 8. The schematic diagram of the light source system.

the change of the diaphragm aperture and light output power and so on. Due to the large scope of magnitudes control and need to simulate stellar emission spectrum in this paper, it cannot regulate a wide range of LED output power. If using a neutral attenuator or polarizer to complete magnitude control, it is difficult to achieve satisfactory neutral requirements in the high rate attenuation, and requires more neutral attenuation or polarizer [9]. Therefore, this paper uses many monochromatic LED light sources to simulate the light spectra by adding white light LED and two different sizes of integral sphere.

4 EXPERIMENTS AND ERROR ANALYSIS

As shown in Figure 8, different sizes of integrating sphere and LED light source are composed of the light part of star simulator by the orbit. Different integral sphere controls different magnitude range, in order to realize the large range of magnitudes. Finally, in the 400~900 nm range, such experiments with the color temperature of 3900k, 4800k and 6500k and the 8 magnitudes of −1 to +6 mv were carried out respectively.

First, select the monochrome LEDs to collect the spectral distributions, and use these LED spectral data to simulate to get A matrix of the formula (7) by Matlab, that is the number of each color LED used actually. As shown in Figure 9(a) and (b) and (c), they are the fitting results of the 2MV spectral distribution for 400 nm~900 nm in three different color temperature under the condition of using 30 kinds of monochromatic LED light sources in 390 nm~910 nm range. The green line represents the target spectral energy distribution, the black line represents the target 10% of the upper and lower error, the blue line shows the measured

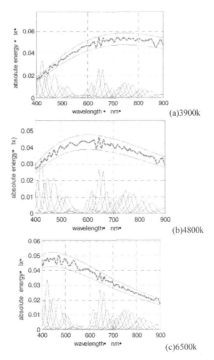

Figure 9. The +2MV spectral distribution experimental figures under three different color temperatures.

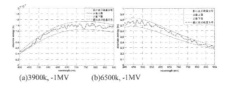

(a)3900k, -1MV   (b)6500k, -1MV

Figure 10. The spectral distributions measured in the sphere outlet.

Table 1. The comparison of theoretical and experimental values in the different color temperature.

M	−1MV	0MV	+1MV	+2MV	+3MV	+4MV	+5MV	+6MV
S_{3900}	−0.9341MV	−0.0641MV	1.0659MV	2.0658MV	3.0657MV	4.0657MV	4.9334MV	5.9344MV
W_{3900}	0.0659MV	−0.0641MV	0.0659MV	0.0658MV	0.0657MV	0.0657MV	−0.0666MV	−0.0656MV
S_{4800}	−0.9323MV	0.0656MV	1.0624MV	2.0648MV	2.9342MV	4.0654MV	4.9289MV	6.0554MV
W_{4800}	0.0677MV	0.0656MV	0.0624MV	0.0648MV	−0.0658MV	0.0654MV	−0.0711MV	0.0554MV
S_{6500}	−0.9315MV	0.0653MV	0.9361MV	1.9341MV	3.0652MV	4.0658MV	4.9340MV	5.9333MV
W_{6500}	0.0685MV	0.0653MV	−0.0639MV	−0.0659MV	0.0652MV	0.0658MV	−0.0660MV	−0.0667MV

spectral energy distribution of each monochrome LEDs, and the red line shows the spectral energy distribution are obtained by monochromatic light fitting. Figure10(a) and (b) are the spectral energy distribution measured at the outlet of integrating sphere for different color temperatures and different magnitudes when making experiments (red line indicates). We compared the experimental value and the theoretical value under the condition of different color temperature of different magnitude, as shown in Table 1, where M is the target value, S represents the experimental value and W is the error value.

By the experimental data, the method to get the fitting error of light source spectrum less than 10%, and the magnitude error is less than 0.08 MV.

5 CONCLUSION

This paper used MATLAB to fit the spectral curve of the star in different color temperature and different magnitude by basing on the least squares method, at the same time, used two different sizes of spheres to achieve a wide range of magnitude, and the results of the fitting results were analyzed and verified. Experimental results show that the method used in this paper formed a good fit for the light system of the star simulator, and the error of the spectral distribution and the error of magnitude is respectively ±10% and ±0.08 MV, be able to meet the demand for use, and are conducive to the realization of the star simulator. At the same time, using this method can simulate the light source spectrum with any color temperature, and can further facilitate the use of star sensor calibration.

The key in this experiment is to control the luminous intensity of the LED light source, as well as the switching between the two integrating

spheres for different magnitudes. Meanwhile, due to the large number of LED light sources are used, choosing correctly the size of the integrating spheres is very important, especially the size of the small integrating sphere. We must ensure that the integrating sphere can contain a sufficient number of LED light source module under the condition of no affecting its properties.

REFERENCES

[1] Lu Xin, Fang Rongchu. The star seusor technology [J]. *Control Engineering*, 2001, 5(6): 69–78. (in Chinese)
[2] Zhang Jie, Zhang Tao, Xiao Peng. Opto-mechanical system of single-star simulator with multi-color temperature and multi-magnitude output [J]. *Applied optics*, 2012, 33(5): 949–953. (in Chinese)
[3] Liu Yangping, Li Juan, Zhang Hong. Design and calibration of star simulator [J]. *Infrared and Laser Engineering*, 2006, 35(Sup): 331–334. (in Chinese)
[4] Sun Gaofei, Zhang Guoyu, Jiang Huilin, et al. Design of very high accuracy star simulator [J]. *Optics and Precision Engineering*, 2011, 19(8): 1730–1735. (in Chinese)
[5] Zhou Taiming, Zhou Xiang, Cai Weixin. Principle and Design of Light Sources [M]. Shanghai: Fudan University Housing Press, 2006: 1–13, 423–452. (in Chinese)
[6] Jin Weiqi, Hu Weijie. Radiometric Photometric Colorimetric and Measurement [M]. Beijing: Beijing Institute of Technology Press, 2007: 34–50.
[7] Sperier Ingo, Salsbury Marc. Color Temperature Tunable White Light LED System [J]. *Proc. of SPIE*, 2006, 6337: 63371F.
[8] Li Qingyang, Wang Nengchao, Yi Dayi. Numerical Analysis [M]. Wuhan: Huazhong science and technology press, 2008: 64–70.
[9] Chen Yuan, Zhang Wenming. Tunable Color Temperature Light Source for Star Simulator [J]. *Opto-Electronic Engineering*, 2010, 37(8): 24–28.

Information Technology and Applications – Li (Ed.)
© *2015 Taylor & Francis Group, London, ISBN 978-1-138-02677-3*

Real-time moving detection algorithm based on YUV color space

Qi Ke
School of Information and Statistics, Guangxi University of Finance and Economics, Nanning, China

Fu-yuan Zhang
Guangxi Zhuang Autonomous Region Public Security Department, Nanning, China

Fang-xiong Xiao
School of Information and Statistics, Guangxi University of Finance and Economics, Nanning, China

ABSTRACT: Illumination dramatically changes in outdoor scene is one of the difficult problems in background modeling. This paper proposed a real-time moving objects detection algorithm which is based on intensity and chroma of YUV color space. The YUV background model is initialized by average background method, and then detected moving objects via the intensity and chroma difference of current and background image. Finally, the model is dynamically updated by mean shift method. Experiment results show the proposed algorithm can effectively suppress dramatic illumination changes in outdoor scene and meet the requirements of real-time detection.

Keywords: background modeling; YUV; mean shift; illumination

1 INTRODUCTION

Moving object detection is an important step in intelligent video surveillance system. However, how to deal with the dramatic changes in outdoor light is still one of the hot topics in moving object detection. The main approach to solve the illumination change includes the following: average background method (Zheng & Fan 2010) can detect moving object in real-time, but it will cause pseudo foreground point problem when the illumination dramatically change. Gaussian mixture model (Bing & Vemuri 2011) can handle illumination changes, background change and other problems, but it needs a lot of computation and cannot automatically select the refresh rate of background. Non-parametric technique (Alireza et al. 2009) can model rapidly changing scenario, but it requires large memory space to store L frame image sequences. Hidden Markov model (Aeschliman et al. 2010) can solve illumination changes in outdoor environment, but its need large calculation, and cannot meet real-time detection. Codebook model (Munir et al. 2011) can built several codewords which contain background history information for each pixel, but it cannot adapt to dynamic background very well.

This paper presents a moving detection algorithm based on YUV color space to solve illumination change problem in outdoor scene. The algorithm initializes average deviation value and average value of background through the study of the history video sequences. Then by using the characteristic of Y channel's value will dramatically change when moving objects appear, this method can effectively suppress illumination change and extracts the foreground pixels.

2 BACKGROUND MODELING BASED ON YUV COLOR SPACE

2.1 Color space select discussion

Currently, RGB and HSV are the most popular color spaces used in motion detection. But compared with these two spaces, YUV color space has the following advantages:

(1) In YUV color space, Y component and U, V components are separated, which means that the luminance signal independence from chrominance signal. (2) Compared with the nonlinear transformation of RGB to HSV color space, RGB to YUV color space is a linear transformation. We can observe that from the equations 1 to 3. (3) YUV color space has a better performance in foreground and shadow detection (Kumar et al. 2002). Therefore, the algorithm chooses YUV color space as we discuss.

Color space conversion from RGB to YUV described below:

$$Y = 0.257R + 0.504G + 0.098B + 16 \qquad (1)$$

$$U = 0.439R - 0.368G - 0.071B + 128 \qquad (2)$$

$$V = -0.148R - 0.291G + 0.439B + 128 \qquad (3)$$

2.2 Comparative experiment on pixel's change over a period in YUV color space

Through the observation of history video sequences, we found that pixel value dramatically change according to different pixel and scenes. The comparative experiment video comes from ftp://ftp.pets.rdg.ac.uk/pub/PETS2001/. Figure 1 is the original frame. Figure 2 is the values change of different pixel in the original frame over 50 frames. From the experiments we can draw the following conclusions:

1. Pixel Y channel's value will significantly change from 0 to 255 along with the moving object's color from black to white.
2. Pixel Y channel's value will instantly descend when shadow passing by which is very similar to the real moving object passing by condition.
3. Pixel's U channel and V channel value distance will be reduced when moving object or shadow passing by, and U channel has less sensitive to the color's change which caused by moving object, while V channel is very sensitivity.

2.3 Model description

Our algorithm can detect moving object by using the characteristic when moving object passing

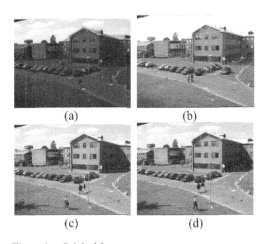

(a)　　　　　　(b)

(c)　　　　　　(d)

Figure 1.　Original frame.

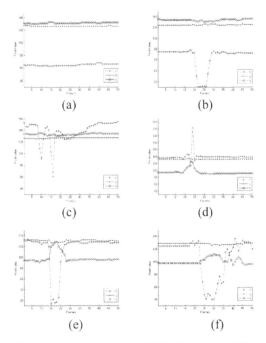

(a)　　　　　　(b)

(c)　　　　　　(d)

(e)　　　　　　(f)

Figure 2. Comparison of YUV changes of different pixels in a period of time. (a) YUV three channels' change of non-human. (b) YUV three channels' change of human areas when illumination becomes darker. (c) YUV three channels' change of human areas when illumination keeps stable. (d) YUV three channels' change of white clothes on the grass. (e) YUV three channels' change of black trousers on the grass. (f) YUV three channels' change of girl's shadow on the grass.

by the pixel's Y channel and V channel will dramatically change under YUV color space. Assumed that $P = \{p_1, p_2, ..., p_N\}$ as a pixel sample training sequences, and $p_t (t = 1, 2, ..., N)$ as a YUV vector. YUV model for each pixel is described as a 5-tube $M = \langle Y_{curr}, Y_a, Y_{pa}, Y_{ad}, V_a \rangle$. Y_a, Y channel average value; Y_{pa}, Y channel's short term average value; Y_{ad}, Y channel average standard deviation and V_a, V channel average value, respectively. YUV model generation process as follows:

Step 1: Initialize each variable to 0.

Step 2: Through the study of video sequences, average background and average deviation background image can be generated. They respectively stand for the average value and deviation from the average value for each pixel during the study period. The average background AVG(x,y) and average deviation background image AbsDiff(x,y) can be calculated as follows:

$$\mathrm{Avg}(x,y) = \frac{1}{n}\sum_{k=1}^{n}\left|f_k(x,y) + f_{k-1}(x,y)\right| \qquad (4)$$

$$\text{AbsDiff}(x,y) = \frac{1}{n}\sum_{k=1}^{n}|f_k(x,y) - f_{k-1}(x,y)| \qquad (5)$$

where $f_k(x,y)$ and $f_{k-1}(x,y)$ are consecutive frames, and n is the number of learning frames.

Step 3: Assuming Y_c is the current pixel value of Y channel. Then each variable of YUV model can be assigned as follows:

$$Y_{curr} = Y_c, \quad Y_a = Y_{pa} = Y_{avg} \qquad (6)$$

$$Y_{ad} = Y_{AD}, \quad V_a = V_{av} \qquad (7)$$

3 FOREGROUND DETECTION

The subtraction operation FD(x) of incoming pixel value x = (Y,U,V) is defined as:

3.1 Calculating the difference between the x and corresponding YUV model M

$$YLD = Y - Y_a; \quad YMD = Y - Y_{pa} \qquad (8)$$

$$YSD = Y - Y_{curr}; \quad VMD = V - V_a \qquad (9)$$

where YLD, YMD, YSD respectively stands for the long-term, mid-term and short-term difference of channel, and VMD is the mid-term difference of channel.

Update the variables by mean shift method:

$$Y_{curr} = Y; \quad Y_a = \lambda_1 Y_a + (1-\lambda_1)Y \qquad (10)$$

$$Y_{pa} = \lambda_2 Y_{pa} + (1-\lambda_2)Y \qquad (11)$$

$$Y_{ad} = \lambda_3 Y_{ad} + (1-\lambda_3)YSD \qquad (12)$$

$$V_a = \lambda_4 V_a + (1-\lambda_4)V \qquad (13)$$

where we take $\lambda_1 = 0.99$, $\lambda_2 = \lambda_4 = 0.8$, $\lambda_3 = 0.9$.

3.2 Background subtraction

Background subtraction operation is defined as:

a. When YLD > 0:

$$BS(\mathbf{x}) = \begin{cases} \text{foreground, } YLD > \alpha Y_{ad}, YMD > \beta, \\ \quad |VMD| > \lambda \\ \text{background, other} \end{cases}$$

$$\qquad (14)$$

b. When YLD < 0:

$$BS(\mathbf{x}) = \begin{cases} \text{foreground, } |YLD| > \delta Y_{ad}, |YMD| > \varepsilon, \\ \quad |VMD| > \lambda \\ \text{background, other} \end{cases}$$

$$\qquad (15)$$

where $\alpha = \delta = 4$, $\beta = \varepsilon = 15$, $\lambda = 5$. YLD > 0 means that the illumination becomes brighter, on the other hand YLD < 0 means that the illumination becomes darker. VMD is absolute value in equation 14 to 15, because it is irrelevant to the illumination but sensitive to the moving area.

4 EXPERIMENT RESULTS AND ANALYSIS

The experiment focused on comparison between the traditional motion detection method and the proposed YUV model. Hardware environment: 512M RAM, Pentium 4 CPU clock speed: 2.66 GHz.

Table 1 is the comparison of the speed of moving detection and sensitivity under illumination change between Average model, Gaussian model, Gaussian mixture model, Codebook model and proposed model. The proposed method needs small amount of computation both in training and detecting stage. Meanwhile by introducing V channel's property which makes the model has a better robustness to the light change. It can be observed that the proposed motion detection method not only has good performance in shadow depression, but also has very good speed in training and detecting which meets the real-time detection requirement.

Figure 3 is the foreground detection comparison of multiple moving targets when the light becomes brighter. From Figure 3 it can be observed that after updating a period of time the Average model, Gaussian mixture model and Codebook model Pseudo foreground point phenomenon has been improved, but still has a certain amount of pseudo foreground point. Gaussian model in Figure 3(c) can suppress the illumination change and has no Pseudo foreground point, but it only detects single moving object. The proposed method in Figure 3(f) first using the characteristic of long-term average of Y component to suppress the light influence, and then use the short-term average of Y component to detect foreground points, therefore compared with other algorithms, this algorithm has higher detection rate and lower false positive rate.

Table 1. Comparative table of the speed of moving detection and sensitivity under illumination change.

Algorithm	Training time	Detecting time
Average model	–	4.83 ms
Gaussian model	–	21.70 ms
Gaussian mixture model	–	30.55 ms
Codebook model	66.22 ms	38.02 ms
Proposed	9.36 ms	17.54 ms

343

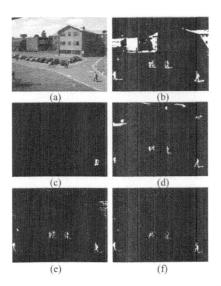

Figure 3. Foreground detection of multi-objects when illumination becomes stronger. (a) Original frame. (b) Average background method. (c) Gaussian model. (d) Gaussian mixture model. (e) Codebook model. (f) Proposed method.

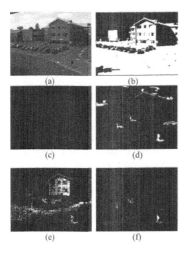

Figure 4. Foreground detection of multi-objects when illumination becomes darker. (a) Original frame. (b) Average background method. (c) Gaussian model. (d) Gaussian mixture model. (e) Codebook model. (f) Proposed method.

Figure 4 is the foreground detection comparison of multiple moving targets when the light becomes darker. It can be observed that Average model, Gaussian mixture model and Codebook model has the same pseudo foreground point phenomenon because of updating rate problem. Gaussian model in Figure 4(c) can suppress the illumination

change, but causes miss detect detects condition which consider the real moving targets as background. Gaussian mixture model in Figure 4(d) has very good detecting rate for moving objects, but it has more shadow points when compared with the Codebook and proposed method. The proposed method in Figure 4(f) not only can suppress the illumination change, but also maintain very good performance on correct detecting rate.

5 CONCLUSION

In this paper, we presented a novel method for background modeling by analyzing the properties of luminance and chrominance of moving object under YUV color space. The proposed method needs less memory and has good performance on suppressing illumination change in outdoor scenes and extracting moving objects. The further research will be focus on suppressing shadow problem of moving objects in outdoor environment.

ACKNOWLEDGMENT

The financial support of the National Natural Science Foundation of China (Grant number 61262002) and the Guangxi scientific research and technology development project (Grant number 2013ZD060) are gratefully acknowledged.

REFERENCES

Aeschliman C., J. Park, A.C. Kak, 2010, A probabilistic framework for joint segmentation and tracking, *IEEE Conference on Computer Vision and Pattern Recognition*, 1371–1378.

Alireza Tavakkoli, Mircea Nicolescu, George Bebis, Monica N. Nicolescu, 2009, Non-parametric statistical background modeling for efficient foreground region detection, *Machine Vision and Applications*, 395–409.

Bing Jian, Baba C. Vemuri, 2011, Robust Point Set Registration Using Gaussian Mixture Models, *IEEE Transactions on Pattern Analysis and Machine Intelligence*, 33(8) 1633–1645.

Kumar P., K. Sengupta, A. Lee, 2002, A comparative study of different color spaces for foreground and shadow detection for traffic monitoring system, Proceeding IEEE 5th Conference on Intelligent Transportation Systems, 100–105.

Munir Shah, Jeremiah Deng, Brendon Woodford, 2011, Enhanced Codebook Model for Real-Time Background Subtraction, *In proceeding of: Neural Information Processing 18th International Conference*, 449–458.

Zheng yi, Fan Liangzhong, 2010, Moving object detection based on running average background and temporal difference, International Conference on Intelligent Systems and Knowledge Engineering, 270–272.

Information Technology and Applications – Li (Ed.)
© 2015 Taylor & Francis Group, London, ISBN 978-1-138-02677-3

Research on frequent itemsets mining algorithms based on MapReduce in cloud computing environment

Ying Zhang
The Business School of Hohai University, Nanjing, Jiangsu, China

Zhi Gang Zhang
The Computing School of Nanjing Normal University, Nanjing, Jiangsu, China

ABSTRACT: Frequent itemsets mining is an important task of data mining. In this paper, frequent itemsets mining algorithms based on MapReduce in cloud computing environment are introduced, including algorithms based on Apriori and FP-Growth. Improvement strategies of the algorithms are presented in this paper.

Keywords: cloud computing; MapReduce; frequent itemsets; Apriori; FP-Growth

1 INTRODUCTION

With the rapid development of computer technology, cloud computing has become the development direction of distributed parallel computing. As a new commercial calculation model, cloud computing sends the task of calculation to the resource pool composed of large numbers of computers, so all kinds of application systems can get strong computing power and plenty of storage space. The novelty of cloud computing lies in the fact that it can supply cheap storage service but with strong computing capacity. So it has unique advantages for massive data storage and processing. MapReduce [1] proposed by Google is a typical programming model based on mapreduce. It is suitable for distributed processing datasets with high efficiency of computing.

Frequent itemsets mining is an important task of data mining. The main mining method are Apriori algorithm [2] and FP-Growth [3] algorithm and their derivation algorithm. How to use the computing model of MapReduce to mine frequent itemsets in cloud computing environment is a hot research topic. At present, the scholars in domestic and abroad put forward several algorithms (4–12). This paper gives detailed review of Research on Frequent Itemsets Mining Algorithms Based on MapReduce in Cloud Computing Environment, and its advantages and disadvantages, and puts forward the corresponding improved algorithm.

2 RELEVANT KNOWLEDGE

2.1 *MapReduce computing model*

Briefly, MapReduce is the "decomposition of the tasks and summarizing of the results.". The processing above can be highly abstracted as two functions: Map and Reduce.

The Map is responsible for decomposing the task into multiple ones. In the Map function, we usually get a series of intermediate results like [<K2,V2>] by dealing with <K1,V1> we put in, then the intermediate results of the key value will be sorted, finally the key value with the same key will be delivered to the task of Reduce in the form of <K3, list <V3>>.

Reduce is responsible for summarizing the results decomposed, Reduce function merges all values in < K3, list <V3>>, creating new key-value pairs.

2.2 *Frequent intemsets*

Let us assume that $I = \{i_1, i_2, \ldots, i_n\}$ is an itemset, transactional databases $D = \{t_1, t_2, \ldots, t_m\}$ is composed of a series of transactions with unique identification TID, in addition, each transaction is corresponding to a subset of I.

Definition 1: Suppose $I_1 \subseteq I, I_1 \subseteq I$, support $(I_1) = \| \{t \in D \mid I_1 \subseteq t\} \| / \| D \|$.

Definition 2: Itemsets with minimum sustain degree users required or the non-void subsets greater than or equal to the minimum support degree are called frequent itemsets.

3 THE FREQUENT ITEMSETS MINING

Nowadays, there are 3 kinds of frequent itemsets mining algorithms based on MapReduce. They are (1) direct solution [4] (2) parallel frequent itemsets based on Apriori algorithm [5–9]. (3) parallel frequent itemsets based on FP-Growth algorithm [10–12].

3.1 Directi solution

Algorithm description:
Map (key, value = itemset in transaction t):
Input: a database partition D_i
1. for each transaction $t_j \in D_i$ do
2. num = the numbers of items in t_j
3. for (k = 1; k < num; k++)
4. Calculating all k-itemsets t_j can produce, and putting the results in itemsets collection set
5. for each items in set do
6. Output < items, 1>; //putting out each k-itemset in set
7. end
8. end
9. end

Reduce (key = itemset, value = count):
1. for each key y do
2. for each value v in y's value list do
3. y.count + = v; // mergering the key value pairs of the same key
4. end
5. if y.count ≥ minimum support count do
6. output <y, y.count>; // putting out the itemsets with greater or equal to the minimum supporting degree
7. end
8. end

The advantage of the algorithm is that we can fulfill the task just by scanning the files once, while the disadvantage is obvious that it not only greatly increases the burden of network transmission, but also brings side effects on the efficiency of algorithm.

3.2 The parallel strategy based on Apriori algorithm

The literature [7] puts forward the parallel algorithm SPC firstly.
The description of SPC algorithm

//Phase-1
Map (key, value = itemset in a transaction t):
Input: a database partition D_i

1. for each transaction $t_j \in t_j$ do
2. for each item $i \in t_j$ do

3. output<i,1>; // putting out each item
4. end
5. end

Reduce (key = itemset, value = count):
1. for each key y do
2. for each value v in y's value list do
3. y.count + = v; // calculating the frequency of item
4. end
5. if y.count > = minimum support count do
6. output <y, y.count>; // putting out the itemsets with the supporting degree greater or equal to the minimum one
7. end
8. end

// Phase-2
Map (key, value = itemset in a transaction t):
Input: a database partition D_i and $L_{k-1}(k≥2)$
1. read L_{k-1} from Distributed Cache;
2. C_k= apriori − gen(L_{k-1}); // generating C_k according to
3. for each transaction $t_j \in D_i$ do // scanning transactions on the partition
4. $C_t = subset(C_k, t_j)$; // getting the subsets of t_j
5. for each candidate $c \in C_t$ do
6. output< c,1>;
7. end
8. end

Reduce (key = itemset, value = count): // the same as the Reduce in phase 1
1. for each key y do
2. for each value v in y's value list do
3. y.count + = v;
4. end
5. if y.count > = minimum support count do
6. output <y, y.count>;
7. end
8. end

Algorithm SPC
1. Phase-1: find L_1; // calculating itemset 1
2. Phase-2: find L_2; //calculating itemset 2
3. for (k = 3; $L_{k-1} \neq \varnothing$; k++)
4. Map function // the Map in Phase-2
5. Reduce function // the Reduce in Phase-2
6. end

The literature [7] points out 2 kinds of improved algorithms FPC and DPC, decreasing iterations. FPC uses fixed step size 3, which is improved in the map of phrase 2 in SPC, generating C_k, C_{k+1}, C_{k+2} according to L_{k-1} every time.
The description of FPC algorithm:
Map (key, value = itemset in transaction t): //the map process in iteration
Input: a database partition D_i and $L_{k-1} (k \geq 2)$

1. read L_{k-1} from Distributed Cache;
2. C_k= apriori − gen(L_{k-1}); // generating C_k according to L_{k-1}
3. C_{k+1} = apriori − gen(C_k); //generating C_k according to C_{k+1}
4. C_{k+2}= apriori − gen(C_{k+1}); // generating C_{k+2} according to C_{k+1}
5. for each transaction $t_j \in D_i$ do
6. C_t = subset($C_k \cup C_{k+1} \cup C_{k+2}, t_j$) // getting the candidating subsets generating by t_j
7. for each candidate $c \in C_t$ do
8. output <c,1>
9. end
10. End

Algorithm FPC
1. Phase-1: find L_1;
2. Phase-2: find L_2; // the first two phases are the same as SPC
3. for(k = 3; $L_{k-1} \neq \varnothing$; k += 3)
4. Map function //the improved map
5. Reduce function // the same as SPC
6. end

The description of the algorithm this paper puts forward:
// Phase-1
Map (key = transaction id, value = itemset in a transaction t):
Input: a database partition D_i
1. for each transaction $t_j \in D_i$ do
2. for each item $i \in t_j$ do
3. output <i:1,key>; // serving item and the frequency as key, the original key serves as the new value
4. end
5. end

Reduce (key, value = transaction id)
1. for each key y do
2. for each value v in y's value list do
3. y.count + = 1; // the number of adding is as to the number of values
4. end
5. if y.count > = minimum support count do
6. Get itemname i from key // extracting frequent itemsets from keys, serving them as new keys
7. output <i, y.count>;
8. for each value v in y's value list do set. add(v);
9. end
10. end
11. end
12. writing all values in set to the new distributed file f, if the f has existed, deleting f firstly, and then creating a new file.// ensuring that the saved records can cover last ones//Phase-2

Map (key = transaction id, value = itemset in a transaction t):
Input: a database partition D_i, L_{k-1} (k≥2), distribute file f
1. reading-out all values from the distributed file f, and putting the values in a global collection variables.
2. if (s.contain(key) = = false) do // leaving the transaction unhandled if the transaction can not produce L_{k-1}
3. return;
4. end
5. read L_{k-1} from distributed cache; // reading-in L_{k-1} from distributed buffers
6. C_k = apriori − gen(L_{k-1}); // generating C_k according to L_{k-1}
7. for each transaction $t_j \in D_i$ d
8. construct all k-items set from t_j; // generating all k-candidating sets according to the transaction t
9. for each k-item i do
10. if Ck.contain(i) do
11. Output < i:1, key>;
12. end
13. end
14. end

//the algorithm of this paper

1. Phase-1:find L_1;
2. for(k = 2; $L_{k-1} \neq \varnothing$; k++)
3. Map function // Phase-2
4. Reduce function // the same as the Reduce in Phase-1
5. end

The algorithm used this paper decreases the accessing to unnecessary transactions, cutting down the intermediate results, and reducing the burden of network communication and shortening the operation time.

3.3 *The parallelization strategies based on algorithm FP-Growth*

The PFP algorithm can be divided into 5 steps, the following are the descriptions of each step:
The first step: database partition.
The second step: parallel statistics.
The third step: grouping all items.
The fourth step: parallel FP-Growth is the most critical step in PFP. The description of MapReduce is as following:
Map(key, value = itemset in transaction t): //the map process in iteration
Input: a database partition D_i
1. Load G-list; //reading in grouped information

2. Generate Hash Table H from G-list; // the hash value representing the group number
3. $a[] = Split(t)$; // getting all items from t, and being ranked according to the specified order
4. for $j = |t| - 1$ to 0 do // $|t|$ represents the number of items in t
5. $HashNum = getHashNum(H, a[j])$; // getting the value of $a[j]$
6. if($HashNum \neq Null$) then do
7. Delete all pairs which hash value is $HashNum$ in H;
8. output<$HashNum$, a[0]+a[1]+...+a[j]>;
9. end
10. end

Reduce (key = gid, value = DB_{gid}):
7. Load G-list;
8. $nowGroup = G - list_{gid}$;
9. for each t in DB_{gid} do
10. insert-build-fptree($LocalFPtree$, t);
11. end
12. for each a_i in $nowGroup$ do
13. Define and clear a size K max heap:HP;
14. $TopKFPGrowth(LocalFPtree, HP)$;
15. for each v_i in HP do
16. output(null, v_i +supp(v_i));
17. end
18. end

The fifth step: results aggregation. The description of the algorithm is as following:
Map (key, value = v+ supp(v)):
1. for each item a_i in v do
2. output< a_i, v+ supp(v)>

Reduce (key = a_i, value = $S(v + supp(v))$):
1. Define and clear a size K max heap:HP;
2. for each pattern v in v+ supp(v) do
3. if $|HP|<K$ then
4. insert v+ supp(v) into HP;
5. else
6. if supp($HP[0].v$) <supp(v) then do
7. delete top elements in HP;
8. insert v+ supp(v) into HP;
9. end
10. end
11. end

4 CONCLUSION

This paper clarifies frequent itemsets mining algorithms based on MapReduce in cloud computing environment. Firstly, it introduces the direct method, then it introduces the parallel strategy based on Apriori and FP-Growth. Meanwhile, the paper puts forward improved strategy, and introduces some most outstanding algorithms.

REFERENCES

[1] J. Dean, S. Ghemawat. MapReduce: Simplified data processing on large clusters[J]. Communication of ACM, 2008, 51(1): 107–113.
[2] R. Agrawal, R. Srikant. Fast Algorithms for Mining Association Rules[C]. Proceedings of the 20th International Conference on Very Large Data Base. San Francisco: Morgan Kaufmann, 1994:287–299.
[3] J. Han, J. Pei, Y. Yin. Mining Frequent Patterns Without Candidate Generation[C]. Proceedings of the ACM SIGMOD International Conference on Management of Data, 2000:1–12.
[4] Jongwook Woo, Yuhang Xu. Market Basket Analysis Algorithm with Map/Reduce of Cloud Computing[C]. Proceedings of the 2011 International Conference on Parallel and Distributed Processing Techniques and Applications, Las Vegas:Springer, 2011:211–228.
[5] XinYue Yang, Zhen Liu, Yan Fu. MapReduce as a Programming Model for Association Rules Algorithm on Hadoop[C]. Proceedings of 2010 3rd International Conference on Information Sciences and Interaction Sciences, Chengdu: IEEE, 2010:99–102.
[6] Lingjuan Li, Ming Zhang. The Strategy of Mining Association Rule Based on Cloud Computing[C]. Proceedings of 2011 International Conference on Business Computing and Global Information. Shanghai: IEEE, 2011:475–478.
[7] Mingyen Lin, Peiyu Lee, Suechen Hsueh. Apriori-based Frequent Itemset Mining Algorithms on MapReduce[C]. Proceedings of the 6th International Conference on Ubiquitous Information Management and Communication. 2012: No. 76.

Information Technology and Applications – Li (Ed.)
© 2015 Taylor & Francis Group, London, ISBN 978-1-138-02677-3

Energy-saving optimization in data center based on computing resource scheduling

Shunling Ruan, Caiwu Lu & Yujuan Guan
School of Management, Xi'an University of Architecture and Technology, Xi'an, China

ABSTRACT: Since the server resources utilization is low, resources load is not balanced, and the energy waste is serious, a power-aware scheduling method of server based on virtualization migration is proposed in this paper. The method in guarantee of the computing resources is available, to dispatch the resources of server according to virtualization migration strategies. Thus, to improve the server resource utilization and to make the resources load is balancing, and minimize the energy consumption of data center. Experimental results show that the proposed method can effectively reduce the energy consumption of data center.

Keywords: data center; energy-efficiency optimization; energy-saving scheduling; virtual migration

1 INTRODUCTION

With the rapid development of enterprise information and cloud computing [1] application, the size and complexity of data center is growing, and the energy cost of data center growing at an average of 20% per year, and the energy consumption of server accounted for more than 40% [2]. However, according to IDC, most of energy is consumed by idle servers, the average utilization rate of servers in most data center only about 30% [3], Virtual migration.

Technology [4] can be used on server migration and integration [5], reduce the energy cost of the server. Reference [6] provides an energy-saving method based on virtual machine migration, which can optimize the energy consumption when servers load is imbalance. Reference [7] used the mechanism of limit violation to control virtual machine migration, and saving the energy by close the idle server. Reference [8] proposed an algorithm to find the best allocation between cost and energy consumption.

However, the above research needs to be improved in the following aspects: the virtual machine migration time, the target server selection, and the migration copy method. When the re-source utilization is not stable, perhaps lead to frequent migration of virtual machine. In view of the above questions, the method of data center energy saving scheduling based on virtual migration was proposed.

2 VIRTUAL SCHEDULING MODEL

The virtual scheduling model is shown in Figure 1. A Virtual Device Group (VDG) including servers, Air Conditioning (AC) and UPS. The migration scheduling includes the following 4 steps.

Step 1, Choose the migration time. Assume that a collection of servers as $S = \{s_1, s_2, ..., s_n\}$, on the basis of migration selection conditions, select a proper server S_i to migrate from the collection.

Step 2, Select the virtual machine to be migrated. Choose a proper virtual machine vm_j form the server S_i, and transfer the virtual machine vm_j to another server.

Step 3, Select the target server. Select a server from collection $D = \{d_1, d_2, ..., d_{n-1}\}$, $D \in S$ and $S_i \notin D$.

Step 4, Migration of virtual machine scheduling. Copy the virtual machine vm_j form the original server to the target server based on the copy algorithm.

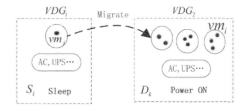

Figure 1. Virtual scheduling model.

3 VIRTUAL MIGRATION SELECTION METHOD

3.1 *Migration trigger strategy*

If the migration timing was not right, it will cause the virtual machine migration frequently, reduce the performance of the server, and increased the energy consumption of server migration. Therefore, in the choice of migration time, it should obey the following rules.

Condition 1, Resource Requirement. Requirement of the upper and lower resources load rate, here mainly refers to the CPU (C_{utili}) and memory (M_{utili}) load rate. The upper and lower load rate requirement of CPU is C_{RMin} and C_{RMax}. The upper and lower load rate requirement of memory M_{RMin} and M_{RMax}.

Condition 2, Threshold Hysteresis. The acceptable load rate which above or below the threshold. The CPU threshold hysteresis is C_{Offset}. The memory threshold hysteresis is M_{Offset}.

Condition 3, Time Threshold. To forecast the server load changes according to the current and historical load rate. T_1 is the first time threshold, which is the duration to reach threshold, T_2 is the second time threshold, which is the duration after the first time to reach threshold. $T_{Max2} > T_{Max1}$, and $T_{Offset} = T_{Max2} - T_{Max1}$ is the time difference between two time threshold.

The migration time can be chosen by three reasons, the definition of it as following three contexts.

Scenarios 1, The migration reason is: server resources insufficient, or server failure warning, or environment abnormal, and the duration more than T_{Max1}. For example: $C_{utili} > C_{RMax} + 2C_{Offset}$;

Scenarios 2, The migration reason is: server load is too high, or environment abnormal, and duration more than T_{Max2}. For example:

$$C_{utili} > C_{RMax} + C_{Offset}$$

Scenarios 3, The migration reason is: server load is more than threshold hysteresis, and the duration more than T_{Max2}. For example:

$$C_{utili} > C_{RMax} \ or \ C_{utili} < C_{RMin};$$

3.2 *Virtual machine selection*

Assume the virtual machine priority value range is $P \in [1,10]$, when the priority is the highest $P = 10$, it means that the virtual machine must be migrating. The virtual machine priority evaluation as following steps:

Step 1, Get the status information of servers and virtual machines, and processing the state information according to the membership functions of fuzzy. The state information includes CPU and memory utilization. The membership function of fuzzy can be expressed as formula (1):

$$P_{1,i} = F_i(x_i) \tag{1}$$

x_i is the input information of virtual machine state.

Step 2, Get the correlation between virtual machine state and dispatching rules according to the membership function of fuzzy, the calculation results is proportion to the correlation degree. The function can be expressed as formula (2):

$$P_{2,i} = R_i = F_{A_i}(x_i)F(y_j) \tag{2}$$

$F(y_j)$ is the member function of scheduling rules. The correlation degree evaluation principles are:

1. Resource matching, the demand and release resource more close, the correlation degree is bigger.
2. Migration cost minimization, when there are multiple virtual machines meet the migration conditions, as $S_1 - M_1 \geq 0$, the lower memory usage, and the bigger correlation degree.
3. Last-In First-Out. For the virtual machine that often migration will be scheduling firstly.

Step 3, Get the correlation-weighted output of each scheduling rules according to the migration conditions. The calculation function can be expressed as formula (3):

$$P_{3,i} = R_i \omega_i \tag{3}$$

For example: If migration reason is memory factor, the weight of memory will be more than other factors.

Step 4, Get the value of correlation degree through precise calculation, thus get the scheduling priority P_x. The calculation function can be expressed as formula (4):

$$P_{4,i} = P_{x_i} = \sum R_i \omega_i \tag{4}$$

Step 5, Get the virtual machine arrangement by calculated the priority, and to scheduling the virtual machine by the priority.

3.3 *Target server selection*

3.3.1 *Selection rules*

Rule 1, Resource Requirement Matching. The available resource of target server must meet the resource requirement of virtual machine. For example, the CPU resource requirement as formula (5, 6, 7, 8), if there have other requirements, should be put into it.

$$C_{\text{Requir}} \leq Min(C_{\text{Usable}}, C_{\text{Usable}+i}) \tag{5}$$

$$C_{\text{RMin}} \geq (C_{\text{Usable}} - C_{\text{Requir}} - C_{MaxOffset}) \tag{6}$$

$$M_{\text{Requir}} \leq Min(M_{\text{Usable}}, M_{\text{Usable}+i}) \tag{7}$$

$$M_{\text{RMin}} \geq (M_{\text{Usable}} - M_{\text{Requir}} - M_{MaxOffset}) \tag{8}$$

C_{Usable} is the target server available CPU resources.

$C_{\text{Usable}+i}$ is prediction available CPU resources of target server in the time T_i.

C_{Requir} The CPU resources requirements of virtual machine that be migrate.

Rule 2, Anti-resource competition. When virtual machines have the same special resource requirement or have some strategy conflict, those machines cannot be deployed to the same server.

Rule 3, Migration in one time. The maximum number of virtual machine that allows migrates into the same target server at the same time. It can be described as: $\sum VM_i \leq N_{SMax}$, V_i is the virtual machine which suitable for the target server.

S_{Max} is the maximum number limitation.

Rule 4, Priority within group. When choosing the target server, the server within the same VDG have the priority that get selected. If cannot meet the requirements, to lookup in other VDG.

Rule 5, when there is multiple target servers: $S_i = (S_1, S_2, ..., S_n)$ meets the requirement, in order to improve server utilization, Select the target server with minimum S_{C_i}. If the same size, and select the target server with smaller S_M.

$$S_{C_i} = S_{CPU} - (C_{\text{Usable}} + C_{\text{Requir}} + 3C_{\text{Max}Offset}) \tag{9}$$

$$S_{M_i} = S_{Memory} - (M_{\text{Usable}} + M_{\text{Requir}} + 3M_{\text{Max}Offset}) \tag{10}$$

3.3.2 Target server selection algorithm

Main steps of the selection algorithm:

1. Begin Format (VM, Scheduling rules…)
2. Do {
3. If (Match the Requirements)
4. Compute $C_{\text{Usable}+i}$ and $M_{\text{Usable}+i}$
5. If (($C_{\text{Requir}} \leq C_{\text{Usable}+i}$) && ($M_{\text{Requir}} \leq M_{\text{Usable}+i}$))
6. Push S_i into the Server Candidate List;
7. End if
8. End if
9. } until (Iterates through all the servers of S_i)
10. For each (Server Candidate List) Do
11. Get the resources information of servers and server group
12. For each (Resource Require List) Do
13. If (resources conflict OR infrastructure less resources OR immigration limit)
14. Remove S_i form Server Candidate List;
15. Else
16. Update the list of servers that to be migrate;
17. End if
18. End for
19. End for
20. If (Length (Server Candidate List) == 1) Return D_k, $D_k \in S$
21. Else If (Length (Server Candidate List) > 1) Return S_i which selected by Rule 5
22. Else
23. If (Have dormant server)
24. Select a server of D_k, $D_k \in S$
25. Else
26. Sleep (60 Seconds), Repeat (1) – (3) step
27. End if
28. End if

4 VIRTUAL MACHINE MIGRATION METHOD

The migration of virtual machines can be divided into two stages. The first phase completed the preparatory work before migration; the second stage is the key stage in the whole migration process. And the second stage of Virtual machine migration can be divided into the following four stages:

1. Pre-copy phase: In this phase, all the state data will be copy from the original server to the target server.
2. Iterative copy phase: After completed pre-copy, in current iteration only copies the pages that are updated in the last iteration. Means, the pages are copied in the N wheel are the pages, which are updated in the $N - 1$ wheel. When meet the iteration termination condition, frozen the original server and enter the stop-copy phase. The paper marks the dirty pages by bitmap in each iterative. The mark method is shown in Figure 2.

Data_{last} is the dirty pages which generated in the previous iteration.

$\text{Data}_{current}$ is the dirty pages which generated in the current iteration.

Data_{fix} is the dirty pages which needed to migrate in the stop-copy stage.

In the migration of copy algorithm, the only condition for dirty page migration is: $\text{Data}_{last} = 1$ and $\text{Data}_{current} = 0$, in other cases, consider that the dirty page has not been modified or have not completed change, not to migrate the pages. Refer Table 1.

3. Stop-Copy phase: In this phase, all the dirty pages and running state of the final round will

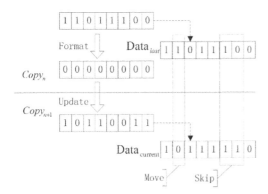

Figure 2. Dirty pages marking method.

Table 1. Migration principle of dirty pages.

Last data	Current data	Migration
1	1	No
1	0	Yes
0	1	No
0	0	No

be copied to the target server, and complete the migration of virtual machine.

4. Service transition phase: When complete the copy of dirty pages in original host, to activate the virtual machine in the target server, and complete the whole service migration.

5 EXPERIMENTAL ANALYSIS

The experimental is in a Modular Data Center (MDC), the environment is as follows, UPS: 2, Air-conditioning: 2, Temperature and humidity sensor: 38. Rack: 16, Server: 122. The configuration of virtual management as following:

$$RC_{Min} = 2\% \qquad RC_{Max} = 98\%$$
$$RM_{Min} = 40\% \qquad RM_{Max} = 85\%$$
$$T_{Max1} = 2 \text{ minute} \qquad T_{Max2} = 5 \text{ minute}$$
$$C_{MaxOffset} = 3\% \qquad C_{MinOffset} = 2\%$$
$$C_{Offset} = 10\% \qquad M_{Offset} = 10\%$$

5.1 Server utilization statistics

Assume that the server resource utilization of CPU and memory achieve to the conditions as $AVG.C_{utili} \leq 2\%$ and $AVG.M_{utili} \leq 5\%$, and the duration more than T_{Max2}, the server will be marked as Idle. Otherwise, marked as Primary.

Based on the analysis of server running statistics, the server utilization comparison chart can be obtained which is before and after virtual scheduling. Refer Figure 3 and Figure 4.

These results suggest that using static management will leads to about 30% of the server is idle. After using virtual migration and integration, server idle rate dropped to about 3% and about 25% servers is down or sleep. However, the number of primary server did not change basically.

From the statistical results of server resources utilization, using virtual scheduling can improve the server CPU and memory utilization, and the tasks number in a single server has increased significantly, thus the whole server utilization was improved. The average utilization of server shows as Table 2.

5.2 Energy efficiency statistics

The data center is usually using Power Usage Effectiveness [9,10] (PUE) as the energy utilization evaluation method, and the computational methods as given below:

$$PUE = \frac{\text{Data Center Energy Consumption}}{\text{IT Equipment Energy Consumption}} \quad (11)$$

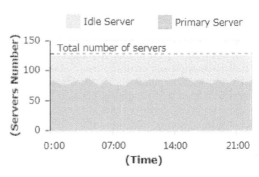

Figure 3. Server utilization before virtual migration.

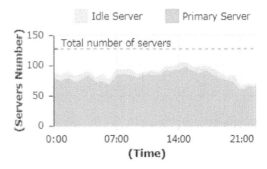

Figure 4. Server utilization after virtual migration.

352

Table 2. Details of average server utilization.

Virtualization	Avg. tasks number	Avg. CPU utilization	Avg. memory utilization
Before	6	8.85%	12.63%
After	11	39.51%	53.72%

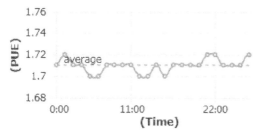

Figure 5. PUE statistics before virtual migration.

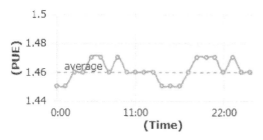

Figure 6. PUE statistics after virtual migration.

Based on the statistics of energy efficiency, the server power usage effectiveness comparison chart can be obtained which before and after virtual scheduling. Refer Figure 5 and Figure 6.

These results suggest that using static management will leads to about 30% of the server is idle, and almost all the server are in a running state. It leads to the PUE as high as 1.71, more than the average value of about 1.5. After using virtual migration and integration, because the number of running server was reduced by virtual integration and the server utilization was improved, the average PUE is 1.46, the energy saving nearly to 35%.

The experimental show that the scheduling method can effectively reduce the server energy consumption through server migration and integration, and saving MDC energy cost.

6 CONCLUSION

Since the server resources utilization is low and the energy waste is serious, a virtual migration and integration scheduling method of server is proposed. This method thoroughly analyzed the server resources utilization, to choose the migration time, select the virtual machine to be migrated, select the target server and schedule the virtual machine. Through the experimental application, the method can reduce the number of running server, and reduce the energy consumption of data center.

REFERENCES

[1] Miller M. Cloud computing [M]. Beijing: Machinery Industry Press, 2009.
[2] KeYong Li, Li Wang, JingWen Xu. Study on energy saving technology of data center air conditioner [J]. Energy research and utilization, 2012,(2):29–31.
[3] Lamia Lafdil, Marcus Torchia, Jill Feblowitz. Worldwide Utility Industry IT Spending Guide [R], July, 2012, 2010–2015, Version 2.
[4] Hermenier F., Lorca X., Menaud J.M. A consolidation manager for clusters [C]//Proc. of the 2009 ACM SIGPLAN/SIGOPS International Conference on Virtual Execution Environments (VEE 09). Washington, USA, 2009:41–50.
[5] Orgerie A.C., Assuncao M.D., Lefevre L. Energy Aware Clouds [J]. Grids, Clouds and Virtualization, 2011,143–166.
[6] ZhiGang Hu, Cheng OuYang, ChaoKun Yan, Resource Load Balancing Method for Energy-consumption Reducing in Cloud Environment [J]. Computer Engineering, 2012,38(5):53–55.
[7] Ding Yan, Liu Jiang-chuan, Wang Dan. Peer to peer video on demand with scalable video coding [J]. Computer Communication, 2010,33(14):1589–1597.
[8] F. Yao, A. Demers, S. Shenker. A Scheduling Model for Reduced CPU Energy [J]. Annual Symposium on Foundations of Computer Science, 1995,374–381.
[9] Dan Azevedo, Mark Blackburn, Jud Cooley, Data Center Efficiency Metrics [R]. Green Grid, Tech Rep: 2011.
[10] Jon Haas, Mark Monroe, John Pflueger, Andy Rawson, Freeman Rawson, Proxy proposals for measuring data center productivity [R]. Green Grid, Tech Rep: 2009.

Information Technology and Applications – Li (Ed.)
© 2015 Taylor & Francis Group, London, ISBN 978-1-138-02677-3

Joint sink mobility and node deployment for prolonging lifetime in wireless sensor networks

Jingguo Dai & Huiyong Yuan

Department of Computer Science, Shaoguan University, Shaoguan, Guangdong, China

ABSTRACT: When cluster heads transmit their data to the sink via multi-hop mode, the cluster heads closer to the sink are burdened with heavy relay traffic and tend to die early. In this paper, taking both sink mobility and node deployment into account, we investigate the energy-hole problem of data gathering in wireless sensor networks, and propose a new node deployment strategy for multi-hop wireless sensor networks with mobile sink. Simulation experiments show that the proposed strategy can effectively balance the sensors energy consumption and prolong the network lifetime.

Keywords: wireless sensor network; node deployment; mobile sink; network lifetime

1 INTRODUCTION

With the development of the MEMS (Micro-Electro-Mechanical-Systems), memory technologies and recent advances in microprocessor and wireless communication technologies, it's possible to produce micro sensors [1]. For some applications (such as forest monitoring), a large-scale sensor network can be deployed, with thousands or more sensors densely distributed in the interested area and working cooperatively for data collection and transmission.

Due to the limited computing power, sensing range, and transmission range of individual sensors, the sensor network is formed to detect the indicated phenomenon and to deliver the collected data to the sink via multiple hops. In order to prolong network lifetime, the sensors can be organized hierarchically by grouping them into clusters. In clustered wireless sensor networks, the sensors do not transmit their collected data to the sink, but to designated cluster heads which aggregate the data packets and send them to the sink via multiple hops.

For multi-hop mode, the cluster heads closest to the sink are burdened with a heavy relay traffic load and die first [2]. In case of sensors failure or malfunctioning around the sink, the sensor network connectivity and coverage may not be guaranteed. No matter how many remaining sensors are still active, none of them can communicate with the sink. As a result, the system lifetime becomes short.

There are a lot of interests of introducing sink mobility into wireless sensor networks for lifetime improvement. With a mobile sink, the cluster heads around the sink always changes, thus balancing the energy consumption in the entire network and improving the network lifetime [3].

Deployment of sensors in a wireless sensor network is a critical task as deployment should be optimum to increase network lifetime [4]. In most current designs, random and uniform deployment is popular proposed schemes due to their simplicity. However, the traditional random and uniform deployments are not suitable because of the energy-hole problem.

In this paper, taking both sink mobility and node deployment into account, we investigate the energy-hole problem of data gathering in wireless sensor networks, and proposed a new node deployment strategy for multi-hop wireless sensor networks. Simulation experiments show that the proposed strategy can effectively balance the sensors energy consumption and prolong the network lifetime.

In Section II, we discuss the most related work. We introduce the system model and problem statement in Section III. In Section IV, we give the solutions for the problem. Finally, in Section V we present the simulation results compared with other known algorithm. We conclude with Section VI.

2 RELATED WORK

Much work has been done during recent years to increase the lifetime of sensor networks.

Node deployment has received considerable attention recently. Yuh-Ren Tsai [5] focuses on the node deployment problem for large-scale randomly distributed wireless sensor networks, and propose a non-uniform sensor deployment strategy for the multi-hop routing. In [6], the authors exhibit the weakness of the uniform distribution by disclosing the fatal sink routing-hole problem, and propose power-aware sensor deployment scheme based on a general sensor application model. The authors of [7] conclude that in a circular multi-hop sensor network with non-uniform sensor distribution, the unbalanced energy consumption among all the sensors in the network is unavoidable, and propose a non-uniform sensor distribution strategy to achieve nearly balanced energy consumption in the network. However, all these proposals assume that sensor networks are modeled with only static sink, i.e., sink do not have mobility.

In order to tack the unbalanced energy depletion and extend the network lifetime, other introduce mobility into the WSNs. The authors of [8] study the maximum lifetime problem of wireless sensor networks where a mobile sink can visit only small number of locations. The sensors near the sink would change overtime with a sink moving in the network, thus mitigating the energy imbalance around the sink. In [3], the authors show that the network lifetime can be extended significantly if the mobile sink moves around the periphery of sensor networks. They assume that, if the mobile sink can balance the traffic load of the sensors, the lifetime of the network can increase. Therefore, they propose an optimization problem for choosing a mobility strategy that minimizes the maximum traffic load of the sensors.

The authors of [9] make use of a mobile relay to prolong the network lifetime. They state that the mobile relay only needs to stay with in two hops away from the sink to enhance the network lifetime by a factor of nearly four. They also propose two joint mobility and routing algorithms capable of attaining the claimed results.

Our work differs from the above works since we take both sink mobility and node deployment into account. We first derive the optimal cluster radius and then propose a new node deployment strategy for wireless sensor network with mobile sink.

3 SYSTEM MODEL AND FORMULATION

3.1 Energy consumption model

We assume there is an energy-efficient MAC protocol in the underlying MAC layer, energy will be consumed only when performing sensing task, processing raw data, and transmitting and receiving data for itself and other sensors. The radio model discussed in [10] can be used to evaluate

energy consumption of data transmission. In this model, a radio dissipates E_{elec}, defined for the transmitter or receiver circuitry, and E_{amp}, defined for the transmitter amplifier.

We assume all sensors have transmit power control and can use just the minimum required energy to send information to the recipients. The equations used to model energy consumption of a sensor for communication are given below.

The energy consumption for transmitting sensor:

$$E_{TX}(len,d) = E_{elec} \times len + E_{amp} \times len \times d^2 \qquad (1)$$

The energy consumption for receiving sensor:

$$E_{RX}(len,d) = E_{elec} \times len \qquad (2)$$

Here d is the distance between two sensors, len is the number of bits of information sent, and E_{elec} and E_{amp} are the constants as previously defined.

The energy dissipation is a second order function of distance. So the data routing with multiple shorter nearby hops will be more efficient than directly transmitting between two far sensors. The energy consumption is also a linear function of len which is bits of information transmitted through the sensor network.

3.2 System assumptions

In this paper, we assume that N sensors are deployed in a rectangular $2W \times W$ observation region. The movement trajectory of the mobile sink is on concentric of the rectangular, because it is the only symmetric strategy that we can have within the network region. The mobile sink collects data from cluster heads. We also assume that all sensors have the same initial energy E_{init} and the energy of the mobile sink is unlimited. Each sensor generates a raw message packet with the same size len-bits. Sensor network is organized into clusters whose radius is equal to r. Cluster heads can aggregate their members data into single-sized data packet and forward it to the mobile sink. Figure 1 depicts an application where sensors periodically transmit their data to the sink. The figure illustrates that cluster heads transmit the aggregated data in multi-hop mode.

To facilitate our discussion, we divide sensors to different sets according to their distance to the sink. The set P_i contains all sensors which can reach the mobile sink with minimal hop count i. The sensor S_j will be in the set P_i if $(i-1) \times \sqrt{2}r < dist(S_j,\text{sink}) \le i \times \sqrt{2}r$, where $dist(S_j,\text{sink})$ is the Euclidean distance between sensor S_j and the mobile sink. Thus, the sensors in

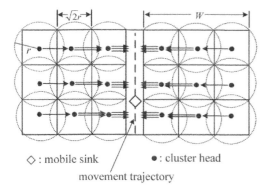

◇ : mobile sink   ● : cluster head

movement trajectory

Figure 1. Mobile sink network model.

set P_i will be in the i-th cell to the movement trajectory of the mobile sink.

3.3 Problem statement

In our network model, the sensors at the same cell will die at almost the same time. Maximizing the network lifetime is equivalent to minimizing the average energy consumption of the sensors. We also know that an optimal deployment should have an equal lifetime for all sensors in the network.

The lifetime T_i of sensors in set P_i is defined as the expected time for the battery energy E_{init} to be exhausted, that is, $T_i = E_{init}/E_i$, where E_i is the average energy consumption per round. We define network lifetime $T_{network}$ of a sensor network as the minimum lifetime of all sensors in the network, that is

$$T_{network} = min\{T_i\} \tag{3}$$

Based on equations (3), the maximum network lifetime problem can be written as follows

$$
\begin{aligned}
Maximize \quad & T_{network} \equiv min\{T_i\} \\
Subject\ to \quad & T_i = E_{init}/E_{i\_average}, \quad 1 \le i \le k \\
& k = \left\lceil \frac{W}{\sqrt{2}r} \right\rceil
\end{aligned}
\tag{4}
$$

Our objective is to find the optimal cluster radius r and the number of sensors in set P_i.

4 SOLUTIONS FOR THE PROBLEM

4.1 The optimal cluster radius

Mhatre and Rosenberg in [11] define characteristic distance as that distance which when used as the inter-sensor distance, minimizes the energy consumption for sending a data packet from a source sensor to a destination sensor. This characteristic distance, d_{char} is

$$d_{char} = \sqrt{2E_{elec}/E_{amp}} \tag{5}$$

Only width d_{char} can the energy consumption rate be minimized. A good clustering algorithm may average the workload on each sensor and the generated clusters should have the same cluster sizes. Thus, we set the optimal cluster radius $r=d_{char}$.

4.2 Solutions for the problem

In this section, we first study the energy consumption. Let us consider a single cluster with radius r. We assume that N_k denotes the number of sensors in set P_k. The average number of sensors in a cluster is

$$N_{k\_chu} = N_k/W \times \sqrt{2}r \tag{6}$$

To keep the total energy dissipation within the cluster as small as possible, the cluster head should be positioned at the centroid of the cluster area A. In this case, the square of distance between cluster members and the cluster head is given as:

$$d_{to\_cluster}^2 = \iint_A (x^2 + y^2)dxdy = r^2/2 \tag{7}$$

The transmitter energy consumption of cluster members in set P_k is given by

$$
\begin{aligned}
& E_{k\_member} \\
& = len(N_{k\_chu} - 1)\left(E_{elec} + E_{amp}d_{to\_cluster}^2\right)
\end{aligned}
\tag{8}
$$

where we denote len as the length of the data packet.

Each cluster head receives the data from all member nodes in the cluster, aggregates and transmits to the sink. The energy required is

$$E_{k\_cluster} = len(N_{k\_chu} - 1)E_{elec} + lenE_{amp}\left(\sqrt{2}r\right)^2 \tag{9}$$

Thus, the average energy consumption of set P_k in a round is

$$E_{k\_average} = \left(E_{k\_member} + E_{k\_cluster}\right)/N_{k\_chu} \tag{10}$$

We assume that N_i denotes the number of sensors in set P_i. Let us consider the average energy consumption of set P_i in a round. The average number in a cluster of set P_i is

$$N_{i\_clu} = N_i / W \times \sqrt{2}r \qquad (11)$$

The transmitter energy consumption of cluster members in a cluster of set P_i is

$$E_{i\_member} = len(N_{i\_clu} - 1)\left(E_{elec} + E_{amp}d_{to\_cluster}^2\right) \qquad (12)$$

The energy depletion of cluster head for receiving data is given by

$$E_{i\_cluster\_rec} = len(N_{i\_clu} - 1)E_{elec} + len(k - i)E_{elec} \qquad (13)$$

The energy consumption for transmitting data to set P_{i-1} is given by

$$E_{i\_cluster\_trans} = len(k - i + 1)\left(E_{elec} + E_{amp} \times \left(\sqrt{2}r\right)^2\right) \qquad (14)$$

Thus, the average energy consumption of set P_i in a round is

$$E_{i\_average} = \left(E_{i\_member} + E_{i\_cluster\_rec} + E_{i\_cluster\_trans}\right) / N_{i\_clu} \qquad (15)$$

To fully utilize total energy allocated to all sensors in the deployment region and avoid energy waste, we want to ensure that all sensors have the same lifetime, which is equal to the network lifetime, T_{net}. Consequently, the number of sensors that resides in P_i, denoted by N_i, can calculated by solving $E_{i\_average} = E_{k\_average}$.

The total number of nodes deployed in the network is given by

$$N = \sum_{i=1}^{k} N_i \qquad (16)$$

4.3 Clustering algorithm

Our clustering algorithm is a distributed cluster heads competitive algorithm, here cluster head selection is primarily based on the residual energy and position of each sensor. Every sensor become a tentative cluster head with the same probability p which is a predefined threshold. The algorithm pseudocode for an arbitrary sensor i is given in Figure 2.

```
state ← candidate;
broadcast Residual_Energy_Msg
receive Residual_Msg;
update neighborhood table NT[];
t ← the broadcast delay time for competing a cluster head;
while (the timer1 (T) for cluster head election is not expired)
{   if (CurrentTime<t)
    {      if (a Head_Msg is overheard from a neighbor NT[i])
        {   state ← plain;   NT[i].state=head;   }
        else   continue;
    }
    else   if (state=candidate)
    {      state ← head;      broadcast Head_Msg;   }
}
while (the timer2 for cluster join is not expired)
{    if (state=plain && have not sended Join_Msg)
        Send (Join_Msg to the nearest cluster head);
    else   receive (Join_Msg from its neighbor plain nodes);
}
```

Figure 2. Cluster head selection pseudocode.

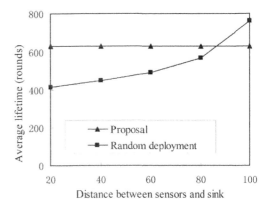

Figure 3. Sensors lifetime comparison. We assume $W = 100$ m, $N = 100$.

5 SIMULATION AND RESULTS

In this section, we conduct extensive simulations to evaluate the performances of our proposed node deployment strategy and compared it with random deployment. We use the same energy consumption model as [10]. Every result shown below is the average of 100 independent experiments.

We first compare the sensors lifetime of our proposed strategy and random deployment. As shown in Figure 3, in random deployment, the lifetime of a sensor decreases with the decreasing distance between the mobile sink and sensor. However, in our proposed node deployment strategy, different sensor densities are assigned to different cells to balance the energy depletion of each cell. Thus, our proposed strategy can ensure that all sensors have almost same lifetime, which is equal to the network lifetime.

We then observe the effect of network region size on the network lifetime. In this experiment, we compare the network lifetime with varied W from 50 to 200 meters. To deploy sensors with relatively high density, there are 25 to 400 sensors.

Figure 4 shows the effect of different network region sizes on the network lifetime of two strategies. It can be seen that network region size does not have great impact on network lifetime of our proposed strategy, but the network lifetime of uniform distribution drops sharply as the network region size expends. Therefore, our proposed strategy leads to a better performance in term of scalability.

Finally, we study the impact of sensor density on the network lifetime. Figure 5 shows the comparison of network lifetime as we set the number of sensors at 100 and 500, using 50 sensors spacing. The network lifetime is the number of rounds until the first sensor dies. As expected, network lifetime using our proposed strategy is efficiently prolonged compared with random deployment.

6 CONCLUSION

In this paper, we investigate the energy-hole problem of data gathering in wireless sensor networks, and proposed a node deployment strategy for wireless sensor networks with mobile sink. Simulation experiments show that the proposed strategy can effectively balance the energy consumption and prolong the network lifetime.

ACKNOWLEDGEMENTS

This research has been supported by the science and technology project of Shaoguan (Grant No. 2012 CX/K91) and Science Foundation of Hunan Province (Grant No. 11JJ3074).

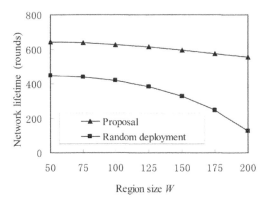

Figure 4. Network lifetime comparison with different region size W.

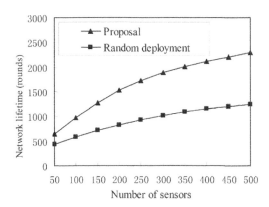

Figure 5. Network lifetime comparison with different sensor density. We assume $W = 100$ m.

REFERENCES

[1] I.F. Akyildiz, W. Su, Y. Sankarasubramaniam, E. Cayirci, Wireless sensor networks: A survey, Comput. Netw, 2002, pp. 393–422

[2] S. Olariu and I. Stojmenovic, Design guidelines for maximizing lifetime and avoiding energy holes in sensor networks with uniform distribution and uniform reporting, In proceedings of INFOCOM 2006, pp. 1–12, Apr. 2006.

[3] J. Luo and J. Hubaux, "Joint mobility and routing for lifetime elongation in wireless sensor networks," Proc. Of the 24th Annual Conference of the IEEE Computer and Communication Societies (INFOCOM'05), vol. 3, pp. 1735–1746, Miami, Fla, USA, 2005.

[4] Sha Chao and Wang Ru-chuan, "Energy-efficient node deployment strategy for wireless sensor networks." The Journal of China Universities of Posts and Telecommunications, vol. 19, no. 10 pp. 110–120, 2013.

[5] Yuh-Ren Tsai, Kai-Jie Yang and Sz-Yi Yeh, "Nonuniform node deployment for lifetime extension in large-scale randomly distributed wireless sensor Networks," In proceedings of AINA. 2008, pp. 517–524, 2012.

[6] Y. Liu, H. Ngan and L.M. Ni, "Power-aware node deployment in wireless sensor networks." In proceedings of IJDSN. 2007, pp. 225–241.

[7] G. Chen, S.K. Das and X. Wu. "Avoiding energy holes in wireless sensor networks with nonuniform node distribution." IEEE Trans. On Parallel and Distributed System, vol. 19, no. 5 pp. 710–720, 2008.

[8] I. Papadimitriou and L. Georgiadis, "Maximum lifetime routing to mobile sink in wireless sensor networks," in the 13th IEEE SoftCom, 2005.

[9] W. Wang, V. Srinivasan, and K.C. Chua. Using mobile relays to prolong the lifetime of wireless sensor networks. In Proc. Of the 11th ACM Mobi-Com, 2005.

[10] W.R. Heinzelman, A.. Chandrakasan, "An application specific protocol architecture for wireless micro-sensor networks," IEEE Transactions on Wireless Communications, 1(4):660–669, 2002.

[11] V. Mhatre and C. Rosenberg, "Design guidelines for wireless sensor networks: communication, cluster-ing and aggregation," Ad Hoc Networks Journal, vol. 2, no. 1, pp. 45–63, 2004.

Information Technology and Applications – Li (Ed.)
© 2015 Taylor & Francis Group, London, ISBN 978-1-138-02677-3

A remote sensing image fusion method based on wavelet transform

Ling Han
School of Geology Engineering and Geomatics, Chang'an University, Xi'an, China

Dawei Liu
School of Geology Engineering and Geomatics, Chang'an University, Xi'an, China
Department of Information Engineering, Engineering University of CAPF, Xi'an, China

ABSTRACT: This paper presents a new remote sensing image fusion algorithm based on wavelet transform for the fusion of multispectral image and panchromatic image. The algorithm fully uses the advantage of wavelet transform on maintaining spectral information. Firstly, the Intensity -Hue-Saturation (IHS) transform is applied to the multispectral image. Then, the monolayer wavelet transform is applied respectively to the panchromatic image and the transformed I component, and the wavelet coefficients are obtained. Different strategies are adopted to fuse the high frequency coefficients and the low frequency ones after that. The fusion principle of low frequency coefficients is based on maximizing the regional variance; meanwhile the principle of high frequency coefficients is based on getting the larger absolute value. Finally, the inverse wavelet transform and inverse IHS transform are used to obtain the fusion image. The experimental results show that the algorithm retains the maximum spectral information, while preserving the detail and edge information.

Keywords: image fusion; IHS transform; wavelet transform; variance significance

1 INTRODUCTION

It is one of the research hotspots of multi-source remote sensing image fusion technology that how to effectively fuse the high resolution panchromat-ic images and the low resolution multispectral images and balance the spatial and spectral information of the results [1–5]. So far, people have proposed many kinds of remote sensing image fusion technologies, such as IHS method, Principal Component Analysis (PCA) method, high-pass filter method and the fusion method based on wavelet transform [6]. Among these methods, IHS and PCA have larger spectral distortion, and high-pass filter method loses a lot of texture information in the process of image filtering [7].

But in image fusion process of wavelet transform, because the wavelet transform has the property of multi-scale, abilities of showing local characteristics of signals in time domain and frequency domain and multi resolution analysis [8], therefore, image fusion based on wavelet transform can handle the image edges and details in different scales, different sizes and the directions. It can decompose the image into high frequency information in the horizontal, vertical and diagonal directions and low frequency profile information with lower resolution. It can preserve well the spectral information of multispectral images. So that image fusion method based on wavelet transform can

preserve more original spectral information while providing high spatial resolution.

2 IMAGE DECOMPOSITION AND RECONSTRUCTION BASED ON WAVELET TRANSFORM

Wavelet transform is an orthogonal transform. In addition to have the advantages of the traditional Fourier transform, it also solves well the problems of Fourier transform in time domain and frequency domain. The original image is decomposed into low frequency image and high frequency image by wavelet transform, the low frequency image can continually be decompose into sub images which contain spatial structure information of original image. We use $A0\,(m, n)$ to represent the image, then the following formulas are used to make wavelet decomposition with scale and wavelet functions:

$$A_k(i, j) = \sum_m \sum_n A_{k-1}(m, n)h(2m - i)h(2n - j) \quad (1)$$

$$D_k^H(i, j) = \sum_m \sum_n A_{k-1}(m, n)h(2m - i)g(2n - j) \quad (2)$$

$$D_k^V(i, j) = \sum_m \sum_n A_{k-1}(m, n)g(2m - i)h(2n - j) \quad (3)$$

$$D_k^D(i, j) = \sum_m \sum_n A_{k-1}(m, n)g(2m - i)g(2n - j) \quad (4)$$

The reconstruction process is:

$$A_{k-1}(m,n) = A_k(m,n) + D_k^H(m,n) \\ \qquad + D_k^V(m,n) + D_k^D(m,n).$$

Image obtained by formula (1) reflects the general situation of the source image, and we called it low frequency image. Images obtained by formula (2)–(4) reflect the detailed information of the source image, and they are respectively high-frequency sub images in directions of horizontal, vertical and diagonal. The low frequency image reflects the original approximation and the average property, which has most information of the original image. The high frequency subs reflect the edge, region boundary information of the original image.

3 FUSION ALGORITHM AND RULES BASED ON WAVELET TRANSFORM

3.1 Fusion algorithm

The algorithm is used to fuse multispectral image and panchromatic image after their geometric corrections and registrations. The process of the algorithm is shown in Figure 1. Firstly we apply the IHS transform to the multispectral image, then we make the transformed component I and panchromatic image to have histogram matching, after that step the matched component I and panchromatic images are applied monolayer wavelet transform, then different fusion strategies are used respectively to high frequency part and the low frequency part of wavelet coefficients for fusion. Finally inverse wavelet transform and inverse HIS transform are applied to the fused coefficients, and the fusion image is obtained.

3.2 Fusion rules

In this paper the panchromatic image and the I component of the multispectral image are supposed

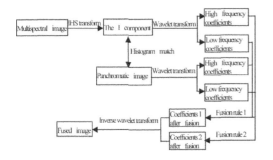

Figure 1. The flow diagram of the algorithm.

to represented by image A and image B respectively, and the fusion image is represented by image F. $M(X)$ represents the wavelet coefficients matrix of image X (X {A, B}), and $p = (m, n)$ indicates the spatial location of the wavelet coefficients, so that $M(X, p)$ represents the value of the element in the wavelet coefficients matrix with index (m, n). For the high frequency coefficients, we use the rule of getting the larger absolute value to fuse image:

$$M(F,p) = \begin{cases} M(A,p), |M(A,p)| \ge |M(B,p)| \\ M(B,p), |M(A,p)| < |M(B,p)| \end{cases}$$

For the low frequency coefficients, the fusion rule is based on the principle of maximum regional variance. $S(X, p)$ is supposed to represent the variance significance of $3*3$ area Q with p as the center in the low frequency wavelet coefficient matrix of image X, and $u(X, p)$ is supposed to represent the mean value of area Q with p as the center in the low frequency coefficient matrix of image X, so that:

$$S(X,p) = \sum_{q \in Q} w(q)(M(X,q) - u(X,p))^2,$$

where $w(q)$ represents the weight, if point q is closer to the point p, the weight is greater. In this paper $w(q) = e^{d(p,q)}$, where $d(p, q)$ is the Euclidean distance between point p and point q.

We use $N(P)$ to represent the regional variance matching degree in point p between the low frequency coefficients matrixes of image A and B, and

$$N(p) \\ = \frac{2 \sum_{q \in Q} w(q) |S(A,q) - u(A,p)||S(B,q) - u(B,p)|}{S(A,p) + G(B,p)}$$

where $S(A, p)$ and $S(B, p)$ respectively represent the regional variance significance of the low frequency coefficients matrixes of image A and image B. The value of $N(p)$ is between 0 and 1, and it indicates that the coefficients correlation of the two images is lower when the value is smaller. We use T to represent the matching threshold, which is generally between 0.5 and 1.

If $N(p)<T$, we use optional fusion strategy:,

$$M(F,p) = \begin{cases} M(A,p), S(A,p) \ge S(B,p) \\ M(B,p), S(A,p) < S(B,p) \end{cases},$$

If $N(p) \ge T$, we use optional fusion strategy:

$$M(F,p) \\ = \begin{cases} w_{\max} M(A,p) + w_{\min} M(B,p), S(A,p) \ge S(B,p) \\ w_{\min} M(A,p) + w_{\max} M(B,p), S(A,p) < S(B,p) \end{cases}$$

where $w_{max} = 0.5 - 0.5\dfrac{1-N(p)}{1-T}$ and $w_{min} = 1 - w_{max}$.

4 EXPERIMENTAL RESULTS AND ANALYSIS

In order to validate the correctness and effectiveness of the algorithm, we do some image fusion experiments using TM and Spot images. Figure 2(c)–(e) are respectively the fusion results of IHS method, PCA method and the method of this paper.

In this paper, we use the features of Information Entropy (IE), Image Definition (ID) and Correlation Coefficient (CC) to make objective evaluations to the fusion images obtained by various methods. The result is shown in Table 1. The value of each feature is the mean value of the three R, G, and B bands. The IE reflects the rich degree of the information captained in the image and the ID reflects the clear degree of the image. The CC reflects the

Table 1. Evaluation results of the fused image.

Images	IE	ID	CC
Multispectral image	5.392	2.485	–
IHS fusion result	6.171	3.832	0.728
PCA fusion result	6.712	3.536	0.701
Result of the method in this paper	6.925	3.941	0.737

similar degree between fusion image and original multispectral image. From Table 1, we can find that all the three features have been promoted by the method of this paper.

5 CONCLUSIONS

This paper proposes a new algorithm for remote sensing image fusion based on the multi-resolution analysis of wavelet transform and IHS transform. Through experiments, we find that the algorithm has achieved good results, which not only preserves the spectral information of multispectral images, but also effectively improves the spatial information amount of fusion image. This method can provide strong support for remote sensing image enhancement and target extraction.

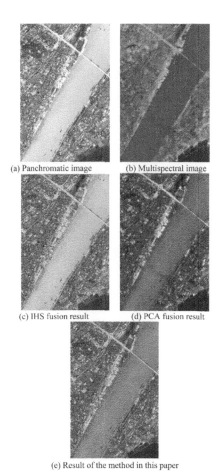

(a) Panchromatic image (b) Multispectral image

(c) IHS fusion result (d) PCA fusion result

(e) Result of the method in this paper

Figure 2. Fusion results of various methods.

REFERENCES

[1] Wald L., Ranchin T., Mangolini M. Fusion of Satellite Images of Different Spatial Resolutions: Assessing the Quality of Resulting Images [J]. Photogram, Engine, Remote Sensing, 1997, 63 (6):691–699.
[2] Haydn R., Dalke G.W. Application of IHS Color Transform to the Processing of Multisensors Data and Image Enhancement [A]. Proceedings of the International Symposium on RS of A—rid and Semi-Arid Lands [C]. 1982.
[3] Te-Ming, Shun-Chi, Hsuen-Chyun Shyu, et al. A New Look at IHS-like Image Fusion Methods [J]. Information Fusion, 2001, 2: 177–186.
[4] Jorge N., Xavie O., et al. Multi-resolution-based Imaged Fusion with Additive Wavelet Decomposition [J]. IEEE Tran on Geoscience and Remote Sensing, 1999, 37(3): 1204–1211.
[5] Argenti F., Alparone L. Filer banks Design for Multisensor Data Fusion [J]. IEEE Signal Processing Letters, 2000, 2(5): 100–103.
[6] Jing Z.L., Xiao G., Li Z.H. Image Fusion-Theory and Applications [M]. Beijing: High Education Press, 2007: 1–262.
[7] Ting L., Jian C. Remote sensing image fusion with wavelet transform and sparse representation [J]. Journal of Image and Graphics, 2013, 18(8): 1045–1053.
[8] Mallat S.G. A theory for multiresolution Signal decomposition: the wavelet representation [J]. IEEE Transactions on Pattern Analysis and Machine Intelligence, 1989, 11(7): 674–693.

Information Technology and Applications – Li (Ed.)
© 2015 Taylor & Francis Group, London, ISBN 978-1-138-02677-3

A Network-on-Chip based homogeneous many-core Digital Signal Processor framework with distributed shared memory

Qinhong Zhang
Institute of Microelectronics, Tsinghua University, Beijing, China

Fang Wang & Zhaolin Li
Research Institute of Information Technology, Tsinghua University, Beijing, China

ABSTRACT: With the rapid development of microprocessors, hundreds of cores will be integrated on a single chip. Network-on-Chip (NoC) is the substitute for bus interconnection on microprocessor framework. In the paper, we propose a NoC based many-core Digital Signal Processor (DSP) framework with distributed shared memory. By designing the Network Interface (NI) between NoC and DSP/Random Access Memory (RAM), we integrate the DSP simulator and RAM model into the NoC, and the whole system platform are written in SystemC language. The experiment result of multi-thread Fast Fourier Transform (FFT) program shows that the framework is working efficiently with high performance.

Keywords: NoC; many-core; DSP; distributed shared memory

1 INTRODUCTION

With the rapid development of computing demands, more and more processors are integrated on a single chip to achieve higher computing performance [1–3]. Besides, with the development of communication technology, the chip communicates more efficiently [4]. However, when more processors are integrated, the conventional bus interconnection becomes the bottleneck, which has many problems: with constant bus bandwidth, each core will have less bandwidth; the communication latency and power consumption will be much higher [5]. NoC has much higher bandwidth compared with bus, as routers are used for communication. As a result, NoC is widely used in many-core system.

DSPs are microprocessors specifically designed to handle digital signal processing tasks, which are carried out by mathematical operations [6]. Distributed shared memory architecture uses a logically centralized while physically distributed memory [7]. With distributed shared memory, the network load will be more balance and different memory block will be parallelly accessed.

In the paper, we propose a NoC based many-core DSP framework with distributed shared memory. By designing the Network Interface (NI) between NoC and DSP/RAM, we integrate the DSP simulator and RAM model into the NoC, and the whole system platform is written in SystemC language.

The rest of the paper is organized as follows: the section 2 describes the implementation of the proposed framework. The operating mechanism of the many-core DSP framework is described in section 3. The experiment results are shown in section 4 and section 5 give the conclusion of the work.

2 DESIGN AND IMPLEMENTATION OF THE PROPOSED FRAMEWORK

2.1 *The proposed many-core framework*

The proposed many-core DSP framework is built on a 5×5 mesh topology NoC which is consist of twenty-one DSPs and four RAMs, as shown in Figure 1. Every node in the network is made up of two major parts: the router and the Processing Element (PE), which can be either DSP or RAM. The network is connected by the routers, contain five different channels: local, north, south, east, and west. In addition, the routers are implemented by fixed five-stage pipeline at clock accurate level.

As exhibited in Figure 1, the dark color blocks represent the RAM cores whereas the light color blocks represent the DSP cores. In the distributed shared memory architecture, the four memory blocks share one global address space and every RAM takes one part of the whole address space. Thus, the RAMs are distributed in the NoC uniformly, which is a valid way to solve the network

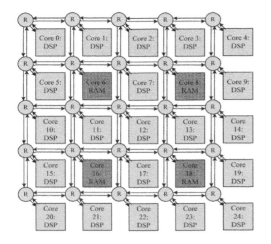

Figure 1. The proposed many-core DSP framework.

congestion problem when a large amount of DSP cores are accessing the same external memory. In addition, in this way, the shared data in the external memory will be un-cached.

To implement the proposed system, we build a platform by integrating a NoC simulator called Nirgam [8], a DSP simulator model and a memory model. As these three components are written in SystemC language, the platform is clock accuracy. Hence, it can simulate the real system accurately. The proposed NoC simulator provides various NoC implementations with variable parameters, such as topology, network size, routing algorithm, buffer depth, virtual channels, and so on. At the meantime, the PE can be extended in the NoC simulation platform. Based on the TMS320C66x DSP, we design a DSP simulator, which implements the basic communication between DSP core and the external memory by using load/store instruction. Correspondingly, a RAM model is built in behavior level to implement the memory accessing.

2.2 Design of the network interface

The NoC transfers data in the form of package, which is made up of flits. The routing information and data payload are contained in the flit. The routing information indicates the destination node of the package and the type of the flit. There are four kinds of flits, head flit, body flit, tail flit and head-tail flit. The head flit, which contains the routing information, is the first flit in the package and indicates the start of the package. On the contrary, the tail flit indicates the end of the package. The head-tail flit is a distinctive flit that contains all the information when there is only one flit in the package. The data payload are contained in the

body flit and part of head/tail flit. In the paper, the DSP simulator supports single data load/store in one instruction, so the package contains one flit, which is the head-tail flit.

As a consequence, in order to connect the DSPs and RAMs to the network, the NI is designed in SystemC language. Besides, two kinds of NIs are presented, they are NI in DSP node and NI in RAM node.

2.2.1 NI in DSP node

The NI in DSP node transforms the communication data between DSP and network, as the two components have different communication manner. When executing the load/store instruction to access the external memory, the load/store signals are given by the DSP, which are shown in Figure 2.

The "load" and "store" signals indicate the instruction type and the moment instruction are executed. The "data" signal indicates the data value. The "address" signal indicates the address of the data. The "load respond" signal will be availed while the required "data" are feeding back to the DSP from the external memory. Similarly, the "store respond" signal will be availed while the store respond is feed back to the DSP. The flits are travelled through the "flit_out" channel and the "flit_in" channel, and the "credit_in" signal shows whether the input channel buffer of the router is free.

The transmitting section of the NI follows the state transition exhibited at the top of Figure 3.

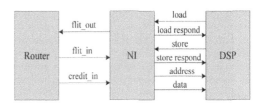

Figure 2. The NI architecture in DSP node.

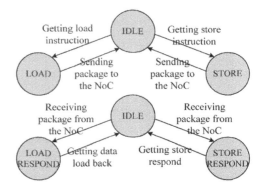

Figure 3. State transition of NI in DSP node.

When the NI gets a load instruction, it turns to "LOAD" state from "IDLE" state and embeds the load information in the flits. After sending the load package to NoC, the NI turns back to "IDLE" state. It does the same way when the NI translating the store instruction to the store package. Accordingly, as to the receiving section of the NI, the moment NI receiving a load respond package, it turns to "LOAD ERSPOND" state from "IDLE" state. While the NI getting the feedback data, it turns back to "IDLE" state as exhibited at the bottom of Figure 3. Besides, the NI deals with the store respond package the same way as the load respond package.

In the NI, the routing information is determined by the data address in the load/store instruction. In the proposed framework, the four memory blocks share the 1G bit address space from 0x00000000 to 0x3FFFFFFF. The address spaces of the RAMs, which are shown in Table 1, determine the routing destination node.

2.2.2 NI in RAM node

The NI in the RAM node transforms the load/store package to the information accessing the RAM and sends the respond package back to the source DSP node as shown in Figure 4. The "read" and "write" signal differentiate the different accessing of the RAM, and the "chip select" signal is used to indicate the accessing time. The "address" and "data" channels are connected to the NI. In addition, the signals between the NI and the NoC follow the same rule as in the DSP node.

Figure 5 exhibits the way that NI works in RAM node, which is similar with the NI in DSP node. While the NI receiving a load package from

Table 1. Routing destination depending on RAM address.

Routing destination	Address space
Core 6	0x00000000–0x0FFFFFFF
Core 8	0x10000000–0x1FFFFFFF
Core 16	0x20000000–0x2FFFFFFF
Core 18	0x30000000–0x3FFFFFFF

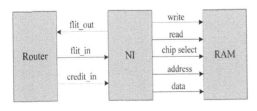

Figure 4. The NI architecture in RAM node.

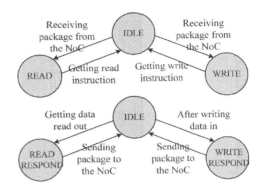

Figure 5. State transition of NI in RAM node.

the NoC, the receiving part of the NI turns to "READ" state from "IDLE" state, and after read accessing the RAM, the NI turns back to "IDLE" state, as shown at the top of Figure 5. As a result, the NI translates the load package to the read accessing in RAM. Besides, it does the same way while the NI translating a store package to a write accessing in RAM. As to the transmitting section of the NI, while the NI getting the data read out, it turns to "READ ERSPOND" state from "IDLE" state. After sending out the read respond package, the NI turns back to "IDLE" state, as shown at the bottom of Figure 5. As a result, the NI translates the data read out to the load respond package. Similarly, the NI uses the same way to generate the write respond package.

3 OPERATING MECHANISM OF THE MANY-CORE DSP FRAMEWORK

3.1 Execution of DSP's load/store instruction

The DSP has two kinds of major instructions, the regular instructions and the load/store instruction. The regular instructions such as arithmetic instructions and logic instructions, which only cause several cycles delay and do not affect the network performance. However, the load/store instruction will incur transmission delay while the DSP accessing the external memory on the NoC. As a result, the execution of DSP's load/store instruction on the NoC can be described like the way demonstrated below.

For instance, the DSP in node 13 loads the data in address 0x00000001 at first, then executes several arithmetic instructions, and stores the result to address 0x30000001 at last. As exhibited in Figure 6, the execution follows the sequence like this: 1) The node 13 sends load package to the RAM in node 6 after the DSP generated a load instruction. 2) After reading access the RAM,

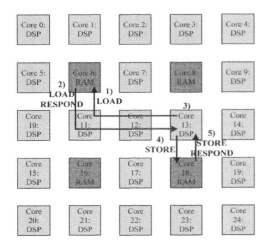

Figure 6. Execution of DSP's load/store instruction.

the node 6 sends load respond package, which contains the feedback data, back to the node 13. 3) The DSP in node 13 executes several arithmetic instructions. 4) The node 13 sends store package to the RAM in node 18 after the DSP generated a store instruction. 5) After writing access the RAM, the node 18 sends store respond package, which contains the writing success or failure information, back to the node 13.

3.2 Execution of multi-thread program example on the proposed framework

In a many-core framework, programs can be divided into multi-thread and mapped on the cores. Such as, the JPEG encoding program, the H.264 encoding program, the MPEG-4 encoding program and the FFT program, which are highly parallel. As a result, the program will be executed on many cores concurrently. To demonstrate the execution of multi-thread program on the proposed framework, we present an example multi-thread program. For example, the FFT program, which can be divided into multi-thread and executed on many cores. As a matter of convenience, we divide the FFT program into 16-thread and map it on 16 cores. For N-point FFT, the number of points to be calculated on each DSP core will be N/16.

In the paper, we take the 64-point FFT as an example to demonstrate the operating mechanism of the proposed many-core system. In the experiment, the data volume is not very large so that it can be stored in one RAM. However, while processing a huge number of data, the memory accessing will be distributed to make the network more efficient.

4 EXPERIMENT RESULT

On the proposed platform we built, we run the 64-point FFT and get the performance of the framework. Some parameters of the NoC are shown in Table 2.

The 16 cores that the program is mapped on, are the 16 DSP cores among node 0–node 19. The average network latency (clock cycles) per flit is shown in Figure 7, and the average network throughput (Gbps) is shown in Figure 8. In Figures 7 and 8, the left figures indicate the east and west channels (X-direction) whereas the right figures indicate the east and west channels (Y-direction) of the network. The result shows that each node has low average network latency, whereas the average latency around node 6 is larger than others, which indicates the traffic congestion near the RAM caused by memory accessing. The average network throughput of the network is uniformly distributed, which indicates the uniform thread dispatch of the program on the DSP cores. Besides, each node's average network throughput is on the level

Table 2. Some parameters of the NoC.

Topology	5 × 5 mesh
System frequency	1 GHz
Routing algorithm	Odd even
Virtual channels number	4
Buffer depth	16
Flit size	8 bytes

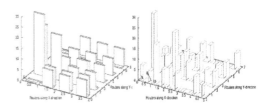

Figure 7. The average latency (clock cycles) per package of the network.

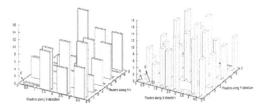

Figure 8. The average throughput (Gbps) of the network.

of Gbps, which demonstrates the high performance of the framework.

5 CONCLUSION

In the work, we proposed a NoC based many-core
DSP framework with distributed shared memory.
By designing the NI between NoC and DSP/
RAM, we integrated the DSP simulator and RAM
model into the NoC, and the whole system platform are written in SystemC language. The experiment result of multi-thread FFT program shows
that the framework is working efficiently with high
performance.

REFERENCES

[1] Cao, S., Energy-efficient Stream Task Scheduling
Scheme for Embedded Multimedia Applications
on Multi-Issued Stream Architectures. *Journal of
Systems Architecture* 59(4–5): 187–201, 2013.

[2] Jiang, G., A High-Utilization Scheduling Scheme
of Stream Programs on Clustered VLIW Stream
Architecture. *IEEE Transactions on Parallel and
Distributed Systems* 25(4): 840–850, 2014.

[3] Cao, S., Compiler-Assisted Leakage- and
Temperature-Aware Instruction-Level VLIW
Scheduling. *IEEE Transactions on Very Large Scale
Integration Systems* 22(6): 1416–1428, 2014.

[4] Zhang, Q., A high performance inter-chip fiber
communication scheme with short frame protocol.
*Proceedings of the 2013 International Conference on
Information Science and Technology*: 850–853, 2013.

[5] Bjerregaard, T. & Mahadevan, S., A Survey of
Research and Practices of Network-on-Chip. *ACM
Computing Surveys* 38(1): 1-es, 2006.

[6] Smith, S.W. *The scientist and engineer's guide to digital
signal processing*. San Diego: California Technical
Publishing, 1997.

[7] Hennessy, J.L. & Patterson, D.A., *Computer
Architecture: A Quantitative Approach*. San Francisco:
Morgan Kaufmann Publishers Inc., 2003.

[8] NIRGAM, http://www.nirgam.ecs.soton.ac.uk.

Information Technology and Applications – Li (Ed.)
© 2015 Taylor & Francis Group, London, ISBN 978-1-138-02677-3

Research and design of Network Intrusion Detection System based on Multi-Core Tilera64

Liang Zhang, Peiyi Shen, Luobing Dong, Juan Song, Yong Feng & Qiangqiang Liu
Xidian University, Xi'an, Shanxi, China

Kang Yi
Xi'an Communication Institute, Xi'an, Shanxi, China

Lukui Zhi & Wentao Zhao
The Public Security Bureau of Shaanxi Province, China

ABSTRACT: In high-speed network environment, influence factors for the general Network Intrusion Detection System's performance mainly include the capability of network packets capturing and the speed of data processing. According to the issue, this paper designs a parallel Network Intrusion Detection System based on Multi-Core Tilera64. First, using pipeline decomposition and protocol flow distribution, we deal with the process of high-speed data capture. Second, based on different configuration requirements for the intrusion detection, we come up with the distribution strategy for data assignment. At last, a palatalization and verification system for Snort is realized on Multi-Core Tilera64 platform. From the evaluation, we find that the system has strong scalability, and its overall performances can be rising with the number of distribution tiles increasing.

Keywords: Snort; network intrusion detection system; Multi-Core Tilera64; high-speed network

1 INTRODUCTION

Snort is the Network Intrusion Detection System for rule matching on network data. When detecting network data, comparing data in the repository is needed for each match. Therefore, the realization of Network Intrusion Detection System requires many comparisons on string operations. If system has weak computing capabilities, packets' loss will occur. In the meantime, the under-reporting network intrusion will happen. Especially in heavy traffic, high-load network environment, the package's false negative rate is quite higher. Resulting for this phenomenon mainly comes from as follows:

a. In high throughput network environment, due to limitations of the hardware and the NIC driver, improving speed for receiving data is difficult after reaching a certain peak. Once data stream skyrockets on the network, the module for receiving data has to discard many packets. Therefore, incomplete data and omission of intrusion comes up.

b. The time of detecting one network packet's attribution is a little bit long, which is decided by the pattern matching algorithms. On the one hand, if one packet is detected by a long time, subsequent packets will be lately received. Hence, packages' discard and under-reporting intrusion can happen. On the other hand, in reality, the intrusions cannot be disguised with one packet, but many packets correlate. Discarding any one of them can cause the false negative.

Therefore, many implementations for Network Intrusion Detection System utilize multi-core processor. Teams such as Snort, Haagdorens, Intel and Giorgos, have developed and tested for Snort on multi-core platform.

Recently, the multi-core processing platform not only provides a good environment for the Network Intrusion Detection System, but also puts forward to high demands on the parallel Network Intrusion Detection System.

This paper implements the parallelization for Network Intrusion Detection System on high-performance multi-core processing Tile64 platform. First, the efficient capability for network data reception guarantees the integrity of data input in Network Intrusion Detection Systems. Second, according to rules' requirements for the Network Intrusion Detection System, the rules are statically planned. Thus, system distributes the same protocol data stream on one core as possible and shares many cores on large amount of detection data or the rules for complexity detection process. Meanwhile, the balanced strategy for network data dynamically adjusts network data to flow on different cores. The parallelization method in which Snort designs and implements on Tilera64 is described in the paper.

The remainder of this paper is organized as follows: section II introduces related approaches, section III gives the analysis and design of our approach, and section IV is the evaluation. Section V is the conclusion on this research of Network Instruction Detection system.

2 TECHNICAL METHOD

2.1 Snort's system framework

In order to realize the parallel Network Intrusion Detection System, we must have in-depth analysis for the system framework. Using plug-in mechanism makes a better effect on system's coupling and scalability, and the concise design makes it easier to improve the performance for system. Without modifying any core code, the system's detection function are supported and improved by the interface dynamic loaded modules using plug-in technology. This way protects not only the relation between plug-in program and system's core modules, but also the good scalability for detection system. In Snort framework, plug-in mode manages relation for modules, which are the capture/decode module, the preprocessing module, the detection module and the alarm output module. Based on Snort-2.9.1.2 and multi-core architecture, this paper implements a parallel Network Intrusion Detection System.

2.2 Snort system basic process

The intrusion detection process for Snort is as follows:

First, the network packets are excavated and used to find the characteristics for offensive network packets' rules.

Next, the network packets conformed to these rules, are taken as knowledge library for detection engine in the Network Intrusion Detection System.

At Last, network data packages are matched by these rules.

The process for Snort can be divided in two phases. In the first phase, the system mainly finishes global data initialization, rules' file parsing, plug-in registration, rule tree's construction, and capturing the network data. In the second phase,

the system mainly and recursively complete to receive network packets, analyze data, do preprocessing, detect intrusion and output alarm, and so on.

Based on defined rules, the second phase is the key to detect abnormal behavior on Network Intrusion Detection System. For the purpose of beneficial analysis in this part, there is need to understand two mechanisms that are the plug-in and events; they are concerned as focus objects to achieve system's parallelism. The different network data require different detection methods, and the plug-in mechanism can meet changeable demands. The results for detection are inserted into the unified event queue. Finally, event management module calls the relational output plug-in to alarm.

3 IMPLEMENT

3.1 The implement of parallel intrusion detection system, Snort

In the high-speed network environment, the performance bottlenecks for Network Intrusion Detection System, Snort, mainly reflect in two aspects: one is the capture and distribution for network data; the other is the detection rules for network data.

According to above analysis, the architecture for parallel Network Intrusion Detection System uses techniques of the pipe decomposition and the protocol stream distribution. Using parallel processing method, the system's performance can be rapidly improved. In this situation, in order to support multi-core, the system uses multi-thread parallel model and combines with parallel networks for data reception library, Libpcap. Snort must be dispatched on multi-core, which simultaneously detects network data rules. As shown in Figure 1.

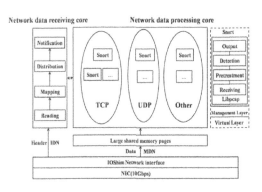

Figure 1. Parallel Network Intrusion Detection System for Snort.

3.2 Snort of data-dependent changes

If running on multi-core by simply copying, the program for Snort does not definitely improve the computing performance. Taking these into consideration, programs should have special optimization:

1. Eliminate data dependency on global. Each processing core executes Snort as a process; the data privacy feature in each process leads to easily solve data dependencies on global.
2. Eliminate file resource dependence. The Snort's running process records in log file. When many cores operate the same file, serialization problem happens. However, due to unique identifier characteristics, system binds the log file with the identification for the processing core. Therefore, it makes each Snort having its own log files. Thus, the problem for file resource dependency is solved.

After the revise above, Snort can run on multi-core platform in parallel. Therefore, this not only can improve the performance for detection system, but also makes the high scalability. However, if supported by distribution strategy, the performance for Network Intrusion Detection System can be increased linearly with the number of handling core rising.

3.2.1 Distribution policy design for Snort

Network Intrusion Detection System, Snort, uses Libpcap as a tool to receive network data. Meanwhile, the system obeys its respective rules for Libpcap. Rules for the library are as follows:

Rule 1: application should bind single core. Non-binding or rebinding will lead to invoke error for library API.

Rule 2: application invokes API for parallel library through common procedure. Dispatch strategy can be registered all the time and take immediate action.

Rule 3: using the design mode for data distribution strategy dispatch. According to the network situation, user can set distribution policy by special interface.

In order to satisfy Rule 1, we make a unique Snort process run on a processing core through interfaces for processing core. Based on the distribution mechanisms for network data-driven and intrusion detection application's demand, system configures for Rule 2 and Rule 3. And then, we implement the distribution policy.

Handling process is as follows: first, processing cores are registered; second, the distribution strategy should be configured. Doing distribution policy, it requires configuring the queue label, the group hash and the hash slot.

3.2.2 Distribution strategy I

Network data receiving core equally sends network data to each processing core. Therefore, the design for distribution describes as follows:

1. Each core is assigned a queue tag.
2. The group number for group hash is unified to zero.
3. According to the number of registered processing cores, hash slots were equally divided. Based on static equilibrium strategies, the system can equally process network packets. As shown in Figure 2, core 2 and core 3 are registered by queue labels, but they are not established the corresponding relation with slot hashes. In this case, core 2 and core 3 cannot receive any network data. Therefore, it can impact the load balancing.

According to the strategy I, all network data are distributed to a unique queue tag by the matching algorithm.

We assume that a mark, TCP1, is a TCP packet on network. First, the result maps mark 0 in the hash group; and then, according to distribution mapping, the result is mark 2 after the slot hash; finally, as we can observe in Figure 2, due to mark 2 establishes mapping with the queue tag 0, therefore, the TCP1 is dealt with the Snort on the core 0.

Again, we assume that a mark, TCP2, is another TCP packet (not the same stream with TCP1) on the network. First, the result maps mark 0 after the group hash; and then, according to distribution mapping, the result is mark 3 after the slot hash; finally, as we can observe in Figure 2, due to mark 3 establishes mapping with the queue tag 1, therefore, the TCP2 is dealt with the Snort on the core 1.

Similarly, other types for data packets follow the same procedure above; finally data are dealt with the Snort on a single core. Thus, the mapping distribution algorithm for calculating hash slots is the key to determine direction for data flow.

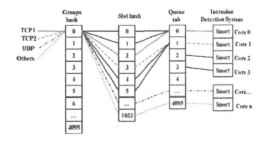

Figure 2. Data distribution on strategy I.

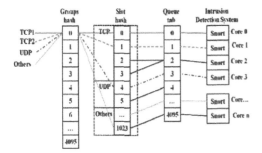

Figure 3. Data Distribution for Strategy II.

3.2.3 *Distribution strategy II*

First, the system divides the network data by protocols. Second, according to the distribution strategy, network data are sent to relative process cores.

The strategy I just take data stream integrity on single core into consideration, but am regardless the specific data type. Due to distribution algorithm used by hash map, the range for source data types is much larger than the scale for hash results. The drawer principle proves that different types of network data will be mapped into the same hash result. That is to say, different types of network data are distributed to the same core. Therefore, we can use strategy II to perfectly avoid this phenomenon.

Specific designs are described as follows:

1. Every core for processing network data is assigned to an independent queue label.
2. According to the protocol type, the group hash table is configured using the design of Base-Bound technology.
3. Hash slot is distributed by protocols, as shown in Figure 3. In the meantime, the strategy creates relational mappings between queue labels and hash slots.

Compared with the strategy I, the strategy II shows that all packets should be selected the appropriate hash slot by protocol type in group hash.

For example, as Figure 3 shows, first, TCP1 is selected group hash number 0 in group hash. Second, using distribution algorithm selects the relational hash slot. Finally, according to the mapping relations between hash slots and queue labels, data are sent to the appropriate processing core.

4 EXPERIMENT

4.1 *Functional testing based on protocol flow distribution strategy*

Based on the size of data customized by user, this paper designed two distribution mechanisms to meet the needs.

Scenario 1: when the scale of data is small and application can be processed on single core, system adds direct mapping configuration table. Therefore, the data type specified by user is sent to the user layer early;

Scenario 2: when the scale of data is large and application needs to be processed on many cores, hash slot is segmented by treating protocol as group number. Then, using hash algorithm in the segmentation, specified types are distributed to the relational processing core.

The system uses xcap, to construct packet structure, which can test effectiveness for distribution on these two strategies.

1. Test designed by direct distribution map.

Three processing cores (A, B, and C) run program to receive data. The system configures the distribution strategy using A. Meanwhile, each processing cores is assigned different label (0, 1, and 2). Observing results for receiving data can be used to verify the validity of the distribution strategy. Specific distribution strategy is configured in Table 1.

The tool, xcap, has constructed three types in data TCP, UDP, and ICMP. Sending 10,000 packages in turn for testing, the results for receiving data are shown in Table 2.

From the above table, when the distribution configuration number is 1, B and C separately receive data packets in different type. But A does not receive data. It verifies that the data distribution is correct according to the protocol configuration rules. However, results for distribution configuration number 1, 2, 3, and 4 shows that the core configured by protocol can receive the same protocol type data. However, the core without configuration does not receive any data. It also verifies that the data distribution is correct according to the protocol configuration rules.

2. Test based on group hash for the distribution protocol design.

While protocols are associated with group number, the default mapping relationship is shown in Table 3. The group number is mapped protocol type for TCP, UDP, and ICMP.

Three processing cores (A, B, and C) run program to receive data. The system configures the distribution strategy using A. Meanwhile, each processing core is assigned different label (0, 1, and 2). Observing results for receiving data can verify the validity for distribution strategy.

The tool, xcap, has constructed three types in data, TCP, UDP, and ICMP. Sending 10,000 packages in turn for testing, the results for receiving data are shown in Table 4.

The queue tags for core B and C are number 1 and 2. The two labels are respectively mapped with

Table 1. Direct mapping distribution configuration table.

No.	Distribution strategy	Description
1	1. Group hash as default settings uniformly treats as zero. 2. In slot hash, the hash slot 1 binds with the queue tag 1; the hash slot 2 binds with the queue tag 2; the rest of all bind with the queues label 0. 3. In direct mapping table, the first hash TCP binds the hash slot 1; the second UDP binds the hash slot 2.	The settings for B and C have the specified protocol, but A is not specified. The queue label for A is 0, and apart from B and C occupies three queue labels, the hash slot binds with the queue label 0. Therefore, all other data distributes on A.
2	1. Group hash as default settings uniformly treats as zero. 2. In slot hash, the hash slot 1 binds with the queue tag 1; the hash slot 2 binds with the queue tag 2; the rest of all bind with the queues label 0. 3. In direct mapping table, the first hash TCP binds the hash slot 1.	The TCP type is specified in direct mapping table.
3	1. Group hash as default settings uniformly treats as zero. 2. In slot hash, the hash slot 1 binds with the queue tag 1; the hash slot 2 binds with the queue tag 2; the rest of all bind with the queues label 0. 3. In direct mapping table, the first hash TCP binds the hash slot 1.	The UDP type is specified in direct mapping table.
4	1. Group hash as default settings uniformly treats as zero. 2. In slot hash, the hash slot 1 binds with the queue tag 1; the hash slot 2 binds with the queue tag 2; the rest of all bind with the queues label 0. 3. The direct mapping table is empty.	The direct mapping table does not enable.

Table 2. Test results for small-scale data in single core.

No.	A packages' number	B packages' number	C packages' number
1	10000 (ICMP)	10000 (TCP)	10000 (UDP)
2	10000 (UDP/ICMP)	10000 (TCP)	0
3	10000 (TCP/ICMP)	0	10000 (UDP)
4	20000 (TCP/UDP/ICMP)	0	0

Table 3. Mapping protocols with the group number.

Protocol type	Group number
TCP	1
UDP	2
ICMP	3

Table 4. Test results for group hash distribution protocols.

A receiving packages' number	B receiving packages' number	C receiving packages' number
10000 (UDP)	10000 (TCP)	0

4.2 Performance testing for parallel network intrusion detection system

Using techniques for pipeline decomposition and protocol flow distribution, the Snort Network Intrusion Detection program can run on Tilera64 in parallel. In order to analyze performance for the system, this paper uses the performance enhancing factor. The concrete method for calculation is as follows:

$$perfup(x) = \frac{1 - Lostp(x)}{1 - Lostp(1)} \tag{1}$$

hash slot 1 and 2; meanwhile the hash slot 1 and 2 maps with group hash 1. Therefore, when data for TCP come, group number is set with 1. According to distribution rules, the TCP data are distributed to hash slot 1 or 2, which means that core B or C can receive TCP data types only. In Table 4 shows that the hash for protocol distribution is verified right by core B receiving 10000 TCP data packets.

In equation (1), $perfup(x)$ represents the performance factor when using the core number of x. $Lostp(x)$ is a data loss rate when using the core number of x. For example, $Lostp(1)$ represents a data loss rate when using one core.

Through testing, with increasing the data processing core number, lost count for data is shown in Table 5.

Table 5. Parallel program for Snort.

Core's number	1	2	4
Lost count for data	245456	381646	801162
Core's number	8	12	20
Lost count for data	1014654	841445	649154

Table 6. Revised parallelization Snort.

Core's number	1	2	4
Lost count for data	243927	392988	802087
Core's number	8	12	20
Lost count for data	1432347	1856443	2858964

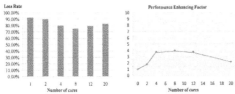

Figure 4. Loss rate for parallelization Snort.

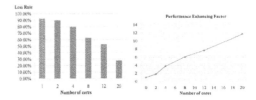

Figure 5. Modified parallelization Snort.

Table 7. Parallelization snort system's loss under different throughput.

| Working core | Throughput | | | | | |
	100 Mbps	200 Mbps	300 Mbps	400 Mbps	500 Mbps	600 Mbps
1	156735	187321	195587	199469	201776	204805
2	104487	157366	176763	185289	189192	190961
8	110	43273	96345	117889	128575	137950
20	0	2989	37861	76483	88955	101706

Figure 4 shows the running situation for Snort program in parallel. At the beginning, with the number of data processing core increasing, the data loss rate decreases.

However, once processing cores are increasing to the peak, the data loss rate will rise. It describes that data dependency has little effect on system performance.

In order to make system's performance rise when the number of processing core increases, it is necessary to eliminate data dependent for Snort on each core.

This paper solves this problem for data dependent as follows: First, each Snort is designed as a separate process. Second, its resources bind with the processing core. The test results are shown in Table 6.

As Figure 5 shows, with the increasing of processing core, data loss rate decreases after the parallel Snort system modified. The performance factor curve graph shows that with the number of processing core increasing, the system performance almost linearly improves.

The tool, tcp replay, is playback for the IDS evaluation data set. Meanwhile, it tests the performance for parallelization Snort system. The results for data loss are shown in Table 7.

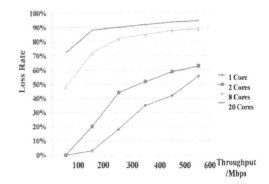

Figure 6. Loss rate with core's number and throughput.

The file size for IDS evaluation data is 226M, which contains a total of 217,452 packets. The data loss rate on different core data and different throughput are shown in Figure 6.

As Figure 6 shows, under certain data throughput, the data loss is significantly lower with increasing of working core. This phenomenon validates the effectiveness of multi-core parallelism Snort.

5 CONCLUSION

First, this paper analyzes the framework design for Network Intrusion Detection System. Second, this paper implements the Snort on Multi-Core Tilera64 and verifies them by experiments. Third, in order to improve the performance, system runs each Snort as a separate process on each core. Finally, using the resource binds with running cores to solve data-dependent problems.

The experimental results show that the system's performance for Snort will be linearly increasing with the number of processing cores rising.

ACKNOWLEDGMENT

This project is supported by NSFC Grant (number 61305109 and number 61072105), by 863 Program (2013AA014601), and by ShanXi Scientific research plan (2013K06-09).

REFERENCES

[1] P. Kermani and L. Kleinrock, 1979, "Virtual cut-through: a new computer communication switching technique".
[2] W. Bolosky, R. Fitzgerald, and M. Scott, 1989, "Simple but effective techniques for NUMA memory management".
[3] Y. Tamir and G.L. Frazier, 1992, "Dynamically-allocated multi queue buffers for VLSI communication switches". IEEE Transactions on Computers.
[4] M.E. Wolf, D.E. Maydan and D.-K. Chen, 1996, "Combining loop transformations considering caches and scheduling", in IEEE Computer Society, Washington, DC, USA.
[5] L.N. Bhuyan, H. Wang, R. Iyer, and A. Kumar, 1998, "Impact of switch design on the application performance of cache coherent multi processors", in Orlando.
[6] C. Brunschen and M. Brorsson, 2000, "OdinMP/CCp—a portable implementation of OpenMP for C, Concurrency—Practice and Experience".
[7] Jeff Chase, Andrew Gallatin, and Ken Yocum, 2000, "End-system optimizations for high-speed tcp", IEEE Communications Magazine.
[8] Handley M., Paxson V., Kreibich C., 2001, "Network intrusion detection: Evasion, traffic normalization, and end-to-end protocol semantics", in Washington, DC.
[9] Kruegel C., Valeur F., Vigna G., Kemmerer R.A., 2002, "Stateful intrusion detection for high-speed networks" Oakland, CA.
[10] Schuehler D.V., Lockwood J.W., 2003, "Tcp-splitter: A TCP/IP flow monitor in reconfigurable hardware", Hot Interconnects, Stanford, CA.
[11] Attig M.E., Lockwood J., 2005, "SIFT: Snort intrusion filter for TCP. Symposium on High Performance Interconnects (HotI)", Stanford, CA.
[12] R. Jin and G. Agrawal, 2005, "A methodology for detailed performance modeling of reduction computations on SMP machines".
[13] Kruegel C., Mutz D., Robertson W., Vigna G., Kemmerer R., 2005, "Reverse engineering of network signatures", Gold Coast, Australia.
[14] White Paper, 2006, "Super-linear Packet Processing Performance with Intel Multi-core Processors".
[15] Nathan L. Binkert, Lisa R. Hsu, Ali G. Saidi, Ronald G. Dreslinski, Andrew L. Schultz, and Steven K. Reinhardt, 2005, "Performance analysis of system overheads in tcp/ip workloads", in Parallel Architectures and Compilation Techniques.
[16] Michael Attig and John Lockwood, 2005, "Snort intrusion filter for tcp", in IEEE Symposium on High Performance Interconnnects.
[17] B. Chapman, L. Huang, H. Jin, G. Jost and B.R. de Supinski, 2006, "Toward enhancing OpenMP's work-sharing directives".
[18] E.M. Nahum, D.J. Yates, J.F. Kurose, and D. Towsley, 2007, "Performance issues in parallelized net-work protocols".
[19] C. Liao and B. Chapman, 2007, " A compile-time cost model for OpenMP".
[20] Z. Jia, Z. Liang, and Y. Dai, 2008, "Scalability evaluation and optimization of multi-core SIP proxy server".
[21] D. Patterson, 2008, "The parallel revolution has started: are you part of the solution or the problem", In USENIX ATEC.
[22] CHEN Yuan zhi, 2009, "Milestone of multi-core processor-Tile64".
[23] A. Pesterev, N. Zeldovich, and R.T. Morris, 2010, "Locating cache performance bottlenecks using data profiling", in Paris, France.

Information Technology and Applications – Li (Ed.)
© 2015 Taylor & Francis Group, London, ISBN 978-1-138-02677-3

The research and realization of the shortest path planning algorithms in map navigation area

Shuhao Wen, Guangzhao Feng, Jinyu Zhan & Ming Sun
University of Electronic Science and Technology of China, Chengdu, China

ABSTRACT: The purpose of this paper is to optimize original A* algorithm. New algorithm can improve the efficiency of finding the shortest path in map navigation. The paper introduced two new methods: first, setting indirect target, Figure 1; second, angle optimization, Figure 2. Using these new methods can highly reduce the searching-area of nodes. Moreover, we compared binary heap and merging algorithm, and choosed binary heap as the data structure of OPEN table. Experiment results, Table 1 and Table 2, show that new algorithm cost less time than A* with same conditions and obviously more efficient.

Keywords: the shortest path; map navigation; A* algorithms improvement

1 INTRODUCTION

Nowadays, electronic map has been widely used in many ways in human's life. For example: the traffic management system, urban-land planning, agricultural irrigation system, and so on. People always need the navigation function of their E-map to find the best way home or the shortest path of the pipeline. Therefore the shortest path problem is the core issue of the map navigation system, and that is the reason it is the topic of this paper. In this text, the scope is limited to general problem of the shortest path between two nodes in transportation network. Although some relevant problems, such as the Chinese postman problem, are in demanding of practical applications, this will not be discussed.

There exists so many researches [1] about the shortest path problem. The Dijkstra's algorithm [2, 3, 4], which proposed by Dijkstra at 1959, is a graph search algorithm that solves the single-source shortest path problem for a graph with non-negative edge path costs, producing a shortest path tree. The latecomers optimize it in many ways to fit different situations. Some relevant algorithm will be discussed in this paper with explanations to tell features and disadvantages. After that, the heuristic A* algorithm will be discussed. We will optimize the A* in two ways: on the one hand, the angle optimization technique is added to the heuristic function, to improve search efficiency; sorting algorithm is added to the OPEN table, to improve the access speed of the table. The improvement and related experiment results are provided in the end.

2 RELATED WORKS

Path planning plays an important role in the map navigation system. Its major function is finding an optimal path by E-maps and related data. According to people different needs on their transportation, the criteria selected by the path planning algorithm is varied. Some common criteria are the shortest distance, the lowest time cost or the least fees. No matter what criteria are selected, the path planning problem can turns into the shortest path problem of graph theory.

The Dijkstra's algorithm, proposed by Netherland computer scientist Dijkstra, is very useful to solve the problem of the shortest path between two nodes in vector graph. Because of the high complexity ($O(n^2)$), another algorithm[6] has been proposed. This algorithm uses redistribution heap and Fibonacci heap as the data structure and reduces the complexity to ($O(m+n\sqrt{logC})$). In addition, Bellman [7] proposed a successive approximation algorithm to solve the problem, which combined in functional equation technique, used in dynamic [8] planning, and the similarity concept of policy space. Pape and Pallottino [9, 10] raised the Graph-Growth algorithm that aims to relative efficiency. The Graph-Growth algorithm proved that it is useful to use different data structure and techniques in algorithm. Moreover, Hart et al. came up with the A* algorithm [11], which is a heuristic algorithm that used the heuristic factors to accelerate searching process, in order to reduce the time cost in searching graph.

3 NEW ALGORITHM BASED ON A* ALGORITHM

3.1 A* Algorithm

In order for a heuristic algorithm with the strategy of $f(n) = g(n) + h(n)$ to become the A* algorithm, these following sufficient conditions must be met:

1. There is at least on optimal path, from the starting point to the finish point, exists on the searching tree.
2. Problem domain is limited.
3. Son nodes' searching value of all nodes is greater than zero.
4. $h(n) \leq$ the real distance from node n to its aim point.

Only all the four conditions are meet can a heuristic algorithm with the strategy of $f(n) = g(n) + h(n)$ to become the A-star algorithm and find the optimal solution for sure.

The A* algorithm is one of the most effective algorithms to solve the problem searching paths on maps of video games. Designing a cost function $f(n) = g(n) + h(n)$, $f(n)$ is the heuristic function of current node n, which is the cost from START to n, $g(n)$ is the real cost from START to n, $h(n)$ is the estimated cost from START to n. With this equation we can calculate the cost to each nodes, and evaluate the price to find the node with lowest cost.

3.2 Basic steps of A* algorithm

1. Set the start node S, and objective node D. Create OPEN and CLOSE, and initialize CLOSE empty.
2. Put S into OPEN.
3. Check OPEN for empty. If yes, stop run. If not, find nodes n with the minimum $f(n)$, and put it into OPEN.
4. Move n to CLOSE.
5. Check n for D. If yes, success. If not run 6.
6. Expand n, find its son nodes m with the minimum f(m). If m is not in both OPEN and CLOSE, put it into OPEN and give it a pointer aim to n, the parent node. If m is in OPEN, compare the new and the old f(m). If the new one is smaller, replace the old one with the new one. If m is in CLOSED, drop m and go back to 6.
7. Go back to 3 until find solutions or all nodes are in CLOSE.

3.3 Improvement to A* algorithm

Two kinds of optimizations are implemented in new A* algorithm.

The $h(n)$ is optimized to reduce numbers of the searching nodes in map. And proper sorting algorithm is added to OPEN to accelerate access speed to OPEN.

. $h(n)$ is always one of Euclidean distance, Manhattan Distance and diagonal distance. In common navigation process, the distance between S and D is not very long, so Euclidean distance with angle optimization technique and indirect target selection technique.

3.4 Indirect target selection technique

As it is shown in analog map, A-M is nodes of the graph, curve segments between nodes is paths, and particularly K-L is bridge ($\alpha < \beta < 90°$). Even using angle optimization technique, extra nodes would appear, such as inserting A, B, C, and D to OPEN and compare their $f(n)$. To avoid this situation, we first should check path between A and M for existing barrier, mountains or rivers. If yes, finding the nearest path through the barrier from A and set it as the indirect target, K in the figure. Then A-K direction becomes new searching direction. That can reduce some redundant nodes searching and improve the efficiency.

3.5 Angle optimization technique

Purely using Euclidean distance to calculate $f(n)$ and choose nodes still bring a large searching area.

As it is shown in angle diagram, S1 is the real distance between A and B, S2 is the real distance between A and C, S3 is the Euclidean distance between B and D, S4 is the Euclidean distance between C and D,
$\alpha = \angle BAD$ and $\beta = \angle CAD$ ($\alpha < \beta < 90$, S1+S3 < S2+S4), A is the start and D is the end.
According to $f(n) = g(n) + h(n)$,
$f(B) = S1+S3$; $f(C) = S2+S4$;

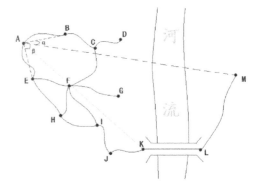

Figure 1. Analog map.

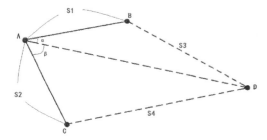

Figure 2. Angle diagram.

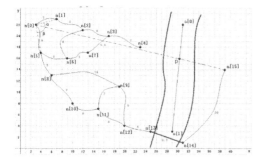

Figure 3. Experiment map.

After angle optimization:
f(B) = S1+S3+k × sinα; f(C) = S2+S4+k × sinβ;
(k is a constant needing calculation to determine).

3.6 *Optimization in OPEN*

Every parent node moving from OPEN to CLOSE will bring in some son nodes, which need to be sorted and store in OPEN. And these son nodes need to be sorted with other nodes stored in OPEN. So merge sort can be a good choice, for sorting two arrays. And min heap is also a good way to solve the problem.

3.7 *Merge sort*

Conceptually, a merge sort works as follows:

1. Divide the unsorted list into n sub lists, each containing 1 element (a list of 1 element is considered sorted).
2. Repeatedly merge sub lists to produce new sorted sub lists until there is only 1 sub list remaining. This will be the sorted list.

For nodes inserted into OPEN are uncertain, merge sort is just fit this situation, which means the time complexity is stable. And quick sort is add in, the combination of quick sort and merge sort can have a high efficiency.

3.8 *Binary heap*

A binary heap is a heap data structure created using a binary tree. It can be observed as a binary tree with two additional constraints: the tree is a complete binary tree; that is, all levels of the tree, except possibly the last one (deepest) are fully filled, and, if the last level of the tree is not complete, the nodes of that level are filled from left to right. Nodes are either [greater than or equal to] or [less than or equal to] each of its children, according to a max heap or min heap.

In A*' OPEN, we need to find the smallest f(n), so the min heap [12] is better.

4 EXPERIMENT RESULTS

4.1 *Experiment environment*

Hard ware:
  64 bit 2.30GHz CPU; 6GB RAM.
  Operating System: Windows 7
  IDE: Microsoft Visual Studio
  Node:
  struct node{
    int number;
    double x,y;
    double fn; // f(n) of a node
    double gn; //real cost
    int fnumber; //parent node
    int location; //node's location in OPEN
    bool open,close;
  };
  Barrier node:
  struct obstacle{
    double k,b;
    int T[];
  };
Experiment map is shown above, including 16 nodes, n[0]-n[15], barrier nodes, o[1] and o[2], 18 paths and a path through barrier, n[13]-n[14], cost is noted by each path.

Using two-dimensional array to store path cost:
double distance[16][16] = {
{0,5,-1,-1,6,-1,-1,-1,-1,-1,-1,-1,-1,-1,-1,-1},
{5,0,6,-1,-1,-1,-1,-1,-1,-1,-1,-1,-1,-1,-1,-1},
{-1,6,0,6,-1,-1,8,-1,-1,-1,-1,-1,-1,-1,-1,-1},
{-1,-1,6,0,7,-1,-1,5.6,-1,-1,-1,-1,-1,-1,-1,-1},
{-1,-1,-1,7,0,-1,-1,-1,-1,-1,-1,-1,-1,-1,-1,-1},
{6,-1,-1,-1,-1,0,6,-1,5,-1,-1,-1,-1,-1,-1,-1},
{-1,-1,8,-1,-1,6,0,5,-1,-1,-1,-1,-1,-1,-1,-1},
{-1,-1,-1,5.6,-1,-1,5,0,-1,-1,-1,-1,-1,-1,-1,-1},
{-1,-1,-1,-1,-1,5,-1,-1,0,16,7,-1,-1,-1,-1,-1},
{-1,-1,-1,-1,-1,-1,-1,-1,16,0,-1,7,8,-1,-1,-1},
{-1,-1,-1,-1,-1,-1,-1,-1,7,-1,0,6,-1,-1,-1,-1},
{-1,-1,-1,-1,-1,-1,-1,-1,-1,7,6,0,6,-1,-1,-1},
{-1,-1,-1,-1,-1,-1,-1,-1,-1,8,-1,6,0,6,-1,-1},
{-1,-1,-1,-1,-1,-1,-1,-1,-1,-1,-1,-1,6,0,6.3,-1},

{−1,−1,−1,−1,−1,−1,−1,−1,−1,−1,−1,−1,−1,6.3,0,20},
{−1,−1,−1,−1,−1,−1,−1,−1,−1,−1,−1,−1,−1,−1,20,0}

The cost of path between node and itself is 0, and the cost of impasses is −1.

4.2 Implementation of indirect target selection technique

We add a decision condition to original A* algorithm:

If there are barriers (mountains, rivers, and so on), new algorithm would search the passing path of the barrier and choose the nearest path from the start node as the indirect target, the searching direction would turn to the new node quickly.

And when the algorithm run on the target node, the direction will turn again. The direction will turn to the start node. If finding other barrier between target node and start node, choose function will be called again to find proper indirect target.

The function is used to judge if there is an intersection point between the segment, between start node and target, and the barrier:

```
bool Is_intersect(double k1,double b1,double
k2,double b2,node*c,node*d){//k1,k2,b1,b2 are
the equation of two segments' line node*c, node*d
are segment's end points
    if(k1 = = k2)
        return false;
    else{
        if(MIN(c->x,d->x)<(b2-b1)/(k1-k2)&&(b2-
b1)/(k1-k2)<MAX(c->x,d->x))
            return true;
        else
            return false;
    }
}
```

Choose function:

S is the source node, T[] stores barrier nodes, a[] stores nodes

```
//The function of find_T is to find the nearest
path from source node.
    int find_T(int s,int T[],node a[]){
        double min = 0;int j;
        for(int i = 0;i < sizeof(T)/2;i++)
        if((a[s].x-a[T[i]].x)*(a[s].x-a[T[i]].x)+(a[s].y-
a[T[i]].y)*(a[s].y-a[T[i]].y)< = min){
            min = (a[s].x-a[T[i]].x)*(a[s].x-
a[T[i]].x)+(a[s].y-a[T[i]].y)*(a[s].y-a[T[i]].y);
            j = i;
        }
        return j;}
```

4.3 Implementation of angle optimization technique

Formula ($\tan\alpha = |k1-k2|/|1+k1 \times k2|$) has been used to calculate the angle between the source-target line

and the source-other line. Then put $\tan\alpha$ into heuristic function ($f(n) = g(n)+h(n)+k \times \tan\alpha$), which replace the way in original A* algorithm only using $h(n)$ to get the value of $f(n)$. And that makes the new algorithm even smarter.

```
//get slopes of two lines
    double get_slope(node*c,node*d){
        return (c->y-d->y)/(c->x-d->x);}
//get tanα
    double get_angle(double k1,double k2){
        return abs(k1-k2)/abs(1+k1*k2);}
//get f(n) using f(n) = g(n) + h(n) + k × tanα
    n[i].fn = n[i].gn+dis(&n[i],&n[end])+get_angle(k1,
k2);
```

4.4 Optimization of the open table

Two sort algorithms have been tested. Compared with the merge one, min heap has approximately the same time complexity and obviously smaller space complexity.

```
// min heap
```

Use add_node function (void add_node(node* a[],node* b,int length)) to add node in the heap, length is the length of the heap. Then declare variable.

node* n;bool flag = true;int temp = length;//The new insert node position at the moment

If node is not in OPEN and CLOSE, insert the node into heap.

```
    if(!b->close&&!b->open)//If insert node in the
table has at the close directly out of the way
    {
        if(*length+1 = = 1){//The first node processing
            a[0] = b;flag = false;
            b->open = true;b->location = 1;
        }
        while(temp! = 1&&flag){
            if(b->fn < a[int(temp/2)-1]->fn){
                a[int(temp/2)-1]->location = temp;
                b->location = int(temp/2);
                n = a[int(temp/2)-1];
                a[int(temp/2)-1] = b;
                a[temp-1] = n;
                temp = temp/2;//Insert the node position
after exchange
                b->open = true;
            }
            else{
                a[temp-1] = b;flag = false;
            }
        }
        (*length)++;}
```

If the inserting node has been already in OPEN, the node need to be compared not only with original $f(n)$, but also with the $f(n)$ of parent nodes and son nodes.

```
        else if(b->open&&!b->close){//whether to up-
date the value of f(n) new node
```

382

```
if(b->fn < a[int(b->location/2)-1]->fn){
    n = a[int(b->location/2)];
    a[int(b->location/2)] = b;b = n;}
if(int(b->location*2+1)< = length-1){
    temp  =  min_node(a,b->location*2,b-
>location*2+1);
    if(b->fn > a[temp]->fn){
        n = a[temp];a[temp] = b;b = n;}
}
else if(int(b->location*2)< = length-1){
    if(b->fn > a[int(b->location*2)]->fn){
        n = a[int(b->location*2)];
        a[int(b->location*2)] = b;b = n;}
    }
}
```

4.5 *Explanation of results*

In time cost table of optimized A* algorithm, optimized A* algorithm's time cost is obviously less than the A* algorithm's, the ratio nearly equal to 1:2.

Table 1. Time cost table of optimized A* algorithm.

	Time cost with new algorithm	Time cost with A*	Ratio
1	0.012000	0.022000	1:1.833
2	0.014000	0.023000	1:1.643
3	0.011000	0.023000	1:2.091
4	0.013000	0.025000	1:1.923
5	0.013000	0.027000	1:2.077
6	0.015000	0.020000	1:1.333
7	0.014000	0.030000	1:2.143
8	0.011000	0.026000	1:2.364
9	0.012000	0.030000	1:2.500
10	0.010000	0.021000	1:2.100
Avg	0.012500	0.024700	1:1.976

*All units of time are in seconds.

Table 2. Time cost table of min heap and merge sort.

	Time cost with new algorithm and min heap	Time cost with new algorithm and merge sort	Ratio
1	0.010000	0.012000	1:1.200
2	0.012000	0.013000	1:1.083
3	0.011000	0.013000	1:1.182
4	0.015000	0.014000	1:0.933
5	0.012000	0.012000	1:1.000
6	0.011000	0.011000	1:1.000
7	0.013000	0.013000	1:1.000
8	0.011000	0.012000	1:1.091
9	0.009000	0.013000	1:1.444
10	0.012000	0.011000	1:0.917
Avg	0.011600	0.012400	1:1.068

*All units of time are in seconds.

As it is shown above, the time complexities of merge sort and min heap are equal to each other. And we can observe the ratio approximately equals to 1, although we add quick sort to merge sort, which slightly improve the time complexity. On the other hand, the space complexity of merge sort is bigger than min heap, therefore we choose min heap to store nodes of OPEN table.

5 CONCLUSION

This paper studied the shortest path problem in map navigation. Optimized the A* in two ways: adding angle optimization technique to the heuristic function and adding sorting algorithm to the OPEN table. The numbers of nodes that in A*algorithm needs to be compared are decrease, for using indirect target selection technique and angle optimization technique. Moreover, the nodes of OPEN table store in min heap, which improve the input–output efficiency remarkably. The original method is optimized on algorithm, improving the search speed and the optimization level of solution path. The new algorithm can complete the purpose that finding the shortest path efficiently through the experiment results.

ACKNOWLEDGEMENTS

This work is supported by the Fundamental Research Funds for the Central Universities of China under Grant number ZYGX2012 J076.

REFERENCES

[1] Cherkassky B.V., Goldberg A.V., Radzikt. Shortest paths algorithms Theory and experimental evaluation [J]. Mathematical Programming, 1996, 73:129~174.

[2] Weimin Yan, Weimin Wu. Data Structure (Second Edition) [M]. (Chinese) Beijing: Tsing Hua University Press, 1997:159~281.

[3] Shichang Fang. Discrete Mathematics [M] (Chinese). Xi'an: Xidian University Press, 1995:254~256

[4] Peyer S., Rautenbach D., Vygen J. A generalization of Dijkstra's shortest path algorithm with application to VLSI routing[J]. Journal of Discrete Algorithms, 2009, 7(4):377–390.

[5] Kumar Pawan. Efficient Path Finding for 2D Games. Proceedings of CGAIDE"2004[C]. Game Simulation and Artificial Intelligence Centre (GSAI), 2004,:265–266.

[6] Ahuja R.K., Mehlh0rn K., 0rlin J.B. Faster algorithms for the shortest path problem[J]. Journal of the ACM, 1990, 37(2):213–223.

[7] Artmeier A., Haselmayr J., Leucker M., Sachenbacher M. Optimal routing for Electric Vehicles [J]. KI 2010: Advances in Artificial Intelligence Lecture Notes in computer Science, 2010, 6935:309–316.

[8] Vliet D.V. Improved shortest path algorithms for transport networks [J]. Transport Research, 1978, 12(1):7–20.

[9] Pape U. Implementation and efficiency of moore algorithms for the shortest route problem [J]. Mathematical Programming, 1974, 7(1):212–222.

[10] Pallottino S. Shortest path methods: complexity, interrelations and new propositions [J]. Networks, 1984, 14(2):257–267.

[11] Hart P., Nilsson N., Raphael B. Correction to a formal basis for the heuristic determination of minimum cost paths[J]. SIGART Newsletters, 1972, 37:38–39.

[12] Lester P. Using Binary Heaps in A * Path finding [N]. 2003.

Information Technology and Applications – Li (Ed.)
© 2015 Taylor & Francis Group, London, ISBN 978-1-138-02677-3

Research of message queues stream processing method based on cloud services

Xianghui Zhao, Lin Liu, Yuangang Yao, Lei Zhang & Juan Li
China Information Technology Security Evaluation Center, Beijing, China

ABSTRACT: This paper proposes a message queues stream processing method based on cloud services, which establish the message queue framework as the system core and construct eight subsystem structure, namely the portal subsystem, terminal access subsystem, internal service subsystem, professional service subsystem, the public version of the service subsystem, service management subsystem, system management subsystem and database subsystem comprehensive service. Finally, by implementation of this method on actual information systems in information security field, this approach is accurate and effective verified through some experiments and samples, and the cloud push based on message queue can become a platform or service for end user to realize the convenient cross-device message push.

Keywords: cloud services; message queues; stream processing; intelligent terminal

1 INTRODUCTION

The development and application of next-generation internet, big data, mobile internet, cloud computing, internet of things and intelligent control technology is driving profound changes in information technology [1]. Cloud services bring a significant change from equipment-centered times to information-centered times. Equipment also includes applications, and people will keep the investment of the information for a long time [2]. The idea of using cloud services for message queues processing and high performance computing has been around for several years [3] [4]. There have been many research works about utilizing cloud environment on scientific computing and high performance computing and message queues processing [5]. Most of these works show that cloud services was not able to perform well running scientific applications [6] [7]. Most of the existing research works have taken the approach of exploiting the cloud using as a similar resource to traditional clusters and super computers and message queues processing. Using cloud such as Amazon as a job execution resource could be complex for end-users if it only provided raw Infrastructure as a Service (IaaS) [8]. It would be very useful if users could only login to their system and submit jobs without worrying about the resource management. Reference [9] via the message service provided by the cloudqueues platform, which has store and forward, reliable transmission across the internet and other characteristics, the cloudqueuqe can provide high performance, reliable, high-capacity internet-oriented messaging services in the era of cloud computing. Reference [10] explores the feasibility of transport streaming data (or file) through the advanced message queuing protocol messaging.

More and more people get the information by pushing method at work and spare time [11] [12]. Information is transmitted to the user's person computer or smart terminal in real-time. Therefore, here comes a problem that different information are pushed to different devices, or the same information is pushed to different devices. Whether we could make all these pushed information gathered at one device has become an obvious problem. At the same time, the development of information pushing application gets more complex. Due to the type of mobile devices is expanding, developers have to code and maintain different versions of the same application, and the devices including Android, iPhone, iPad, Windows phone, other smart phones and tablet devices. To solving these problems, we did research on the information system in information security field. Finally, we made some changes to the existing message queue processing method, and put forward a kind of message queue stream processing method based on cloud services in this paper.

2 MESSAGE QUEUE STREAM PROCESSING METHOD BASED ON CLOUD SERVICES

This paper proposed a message queue stream processing method based on cloud services. The method can help end users, information security professionals and security vendors to deal with their challenges, by making it more easily to access, organize and store notifications and cutting down the cost and complexity of the development. Developers can create a unified and convenient notification service to transmit notifications to all types of devices, such as mobile phone, tablet, PC, laptop and smart TV. The main implementation steps of message queue stream processing method are based on cloud services as follows.

2.1 *Taking a system in the information security field as an example, we designed the system framework of the message queue stream processing method based on cloud service*

The framework has eight subsystems, including portal subsystem, terminal access subsystem, internal version service subsystem, professional version services subsystem, public version service subsystem, service management subsystem, system management subsystem and integrated services database subsystem. The portal subsystem acts as unified the web portal entrance of internal edition, professional edition and public version. The terminal access subsystem includes the tablet or smart phone terminals of internal edition, professional edition and public version. The internal version service subsystem works as a comprehensive service platform for company leaders, staff and administrator. The professional version of service subsystem works as a comprehensive service platform for typical industries, government agencies and corporate users. The public version of the service subsystem works as a comprehensive service platform for the public, information security personnel and institutions. The service management subsystem is responsible for the management of the

subsystems of three classes user (mainly including service subscription, service delivery, information publish and mail pushing, and so on). The system management subsystem is responsible for the comprehensive service platform for the whole system management (mainly including permissions, user management, access logs, audit, and so on). The comprehensive services database subsystem stores huge amounts of data such as pushing data, user data, and service data of comprehensive services platform. The service management subsystem is connected with the management subsystem and system comprehensive service database subsystem, and the internal service subsystem, professional service subsystem and the edition of the public version of the service subsystem are connected with the terminal subsystem separately. The data resources platform provides the data source for information pushing.

2.2 *Taking the message queue framework as the core of the information system*

The components include message producer, message consumer and message queuing service. Message producer includes information security website. The professionals and vendors can act as a producer by calling the APIs, and the common users can also become a producer by publishing security information via web and intelligent terminal. The message consumers have a variety of message consumer ways, including the portal website, SNS, tablet computer, intelligent mobile phone, and so on. Consumers have registration, subscription, automatically receiving push message rights. The message queuing service provides base message queue service, including the message pushing, subscription, and filtering function. Especially, the steps to establish the message queuing service are as follows:

1. The client sends a message sending request.
2. Redis Master stores the message.
3. The Redis database records the transaction log.
4. System parses the transaction log, and provides the RabbitMQ Master as a message producer.
5. RabbitMQ Master message queue is copied to multiple Rabbit Slave.
6. RabbitMQ message queue is consumed by MySQL, memory cached and Redis.
7. MySQL provides a comprehensive query.
8. Memory cached provides various terminal cache.
9. Redis provides real-time push function.

2.3 *To conveniently push messages*

Message queue adopts the push model. That is to say, when the client sends a message, the message

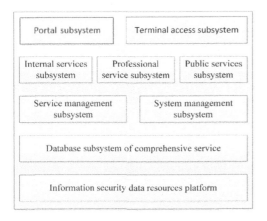

Figure 1. The system platform framework of the message queues stream processing method based on cloud services.

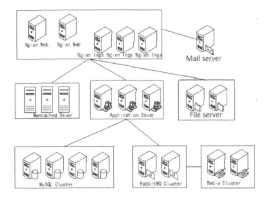

Figure 2. The implement logic diagram of the message queues stream processing method based on cloud services.

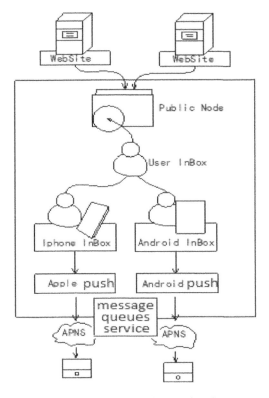

Figure 3. The flow chart of release and push messages based on cloud services.

queuing server first stores the message into the queue to be released, and the message queue back-stage service process the ready-issued message queue. Each ready-issued message is taken from the queue, and the message is stored. The message queuing service registers a pointer of the message on the queue of all the subscription queues.

All subscription clients can obtain the push message in real-time. The process of push realized multiple devices sharing management, which allowed users in all their devices to share and manage their notice.

Open API of message queue stream processing method of message queue based on the cloud service mainly realizes the following functions:

1. Establishment of a MQS connection
   SQS sqs = new SQSClient(new BasicMQSCredentials (MQS_KEY, MQS_SECRET));
   In MQS API, the call to create queue does not create a new queue every time. If the queue already exists, it returns the handle of the queue. In SQS, a queue only contains some URLs. Therefore, processing the queue is just processing a URL. Note that queue URL is a String type, not Java URL type in MQS SDK API.
2. Creation of a message queue
   String url = sqs.createQueue(new CreateQueueRequest ("a_queue")).getQueueUrl();
   With the queue, you can write a message to it.
3. Sending a message by SQS
   sqs.sendMessage(new SendMessageRequest (url, "BrowserCRM multiple SQL injection vulnerability"));
4. Retrieving a message by SQS
   while (true) {
     List<Message> msgs = sqs.receiveMessage(
     new ReceiveMessageRequest(url).withMaxNumberOfMessages (1)).getMessages();
     if (msgs.size() > 0) {
     Message message = msgs.get(0);
     System.out.println("The message is" + message.getBody());
     sqs.deleteMessage(new DeleteMessageRequest (url, message.getReceiptHandle()));
     } else {
     System.out.println("nothing found, trying again in 30 seconds");
     Thread.sleep(3000);
     }

3 ADVANTAGE ANALYSIS OF MESSAGE QUEUES STREAM PROCESSING METHOD BASED ON CLOUD SERVICES

After the implementation of some specific project, we notice that the message queues stream processing method based on cloud services has the following advantages:

1. The cloud services platform of message queue can make the message queues push as a cloud based service, for end user development platform or service;

2. It has an open protocol, defining the user and content model, which allows third parties to prepare both interoperable federal server implementation and client;

3. It can realize the real-time push notifications passed directly to the user, regardless of whether it is the mobile, desktop or media device, and also the push across devices whenever and wherever possible. This solution can reduce the cost and complexity for developers, while providing greater convenience and value-added functions for end users.

4 CONCLUSION AND PROSPECT

This paper proposed a message queues stream processing method based on cloud services, which establish the message queue framework as the system core and construct eight subsystem structure, namely the portal subsystem, terminal access subsystem, internal service subsystem, professional service subsystem, the public version of the service subsystem, service management subsystem, system management subsystem and database subsystem comprehensive service. By implementation of this method on actual information systems in information security field, the cloud push based on message queue can become a platform or service for end user to realize the convenient cross-device message push. This method also provides a reference for other industries to imply the message queuing stream processing based on cloud services.

ACKNOWLEDGMENT

This work is supported by a grant from the National High Technology Research and Development Program of China (863 Program) (No. 2012AA012903). The authors are also thankful to the anonymous reviews for their constructive comments.

REFERENCES

[1] Yang W, Liu X, Zhang L, et al. Big Data Real-Time Processing Based on Storm [C]//Trust, Security and Privacy in Computing and Communications (TrustCom), 2013 12th IEEE International Conference on. IEEE, 2013: 1784–1787.

[2] Noor T.H, Sheng Q.Z, Ngu A H H, et al. Analysis of Web-Scale Cloud Services [J]. Internet Computing, IEEE, 2014, 18(4): 55–61.

[3] Sadooghi I, Palur S, Anthony A, et al. Achieving Efficient Distributed Scheduling with Message Queues in the Cloud for Many-Task Computing and High-Performance Computing [C]//14th IEEE/ACM International Symposium on Cluster, Cloud and Grid Computing (CCGrid). 2014.

[4] Dixon S, Huband S.T, McKenna L.R. Message queue transaction tracking using application activity trace data: U.S. Patent 8,683,489 [P]. 2014–3–25.

[5] Davies P.J, Zaheer N, Crozier S.L. Distributed processing of binary objects via message queues including a failover safeguard: U.S. Patent 8,484,659 [P]. 2013–7–9.

[6] Ramakrishnan L. et al. Evaluating Interconnect and virtualization performance for high performance computing [J]. ACM Performance Evaluation Review, 40(2), 2012.

[7] G. Wang and T.S. Eugene. The Impact of Virtualization on Network Performance of Amazon EC2 Data Center [R]. In IEEE INFOCOM, 2010.

[8] Amazon Dynamo DB (beta), Amazon Web Services, [online] 2014 [DB/OL], http://aws.amazon.com/dynamodb.

[9] Shi Dongdong. Cloudqueue: an internet-scale messaging infrastructure based on hadoop [D]. Dong hua university master dissertation, 2012.

[10] Zhang yifang, Yu Zhi-an. Exploration on advanced message queue protocol [J]. Computer knowledge and technology, 2014, 10(1):47–50.

[11] Zhao Xianghui, Liu Hui, et al. cloud computing service security and access: from the providers and customers' perspective [C]. International Conference on Information Technology and Applications, 2013, 379–383.

[12] Wang C, Jiao R L. Quick Task Queue Management Software Design [J]. Applied Mechanics and Materials, 2014, 519: 357–360.

Author index

`*9781138026773*`

An environmentally friendly book printed and bound in England by www.printondemand-worldwide.com

PEFC Certified

This product is
from sustainably
managed forests
and controlled
sources

www.pefc.org

PEFC/16-33-415

This book is made of chain-of-custody materials; FSC materials for the cover and PEFC materials for the text pages.

#0106 - 020216 - C0 - 246/174/22 [24] - CB - 9781138026773